Statistics for Industry and Technology

Advances in Ranking and Selection, Multiple Comparisons, and Reliability

Methodology and Applications

N. Balakrishnan
N. Kannan
H.N. Nagaraja
Editors

Birkhäuser
Boston • Basel • Berlin

N. Balakrishnan
Department of Mathematics and Statistics
McMaster University
Hamilton, Ontario L8S 4K1
Canada

N. Kannan
Department of Management
Science and Statistics
University of Texas at San Antonio
6900 N. Loop 1604 W
San Antonio, TX 78249-0632
U.S.A.

H.N. Nagaraja
Department of Statistics
Ohio State University
1958 Neil Avenue
Cockins Hall, Room 404
Columbus, OH 43210-1247
U.S.A.

AMS Subject Classifications: 62F03, 62F07, 62F10, 62J15, 62N01, 62N05

Library of Congress Cataloging-in-Publication Data

Advances in ranking and selection, multiple comparisons and reliability
/ N. Balakrishnan, N. Kannan, H.N. Nagarjuna, editors.
p. cm. – (Statistics for industry and technology)
Includes bibliographical references and index.
ISBN 0-8176-3232-8 (alk. paper)
1. Ranking and selection (Statistics) 2. Statistical hypothesis testing. 3. Statistical decision. I. Balakrishnan, N., 1956- II. Kannan, N. III. Nagaraja, H. N. (Haikady Navada), 1954- IV. Series.

QA278.75.A38 2004
519.5–dc22 2004062299

ISBN 0-8176-3232-8 Printed on acid-free paper.

Birkhäuser

Printed in the United States of America.

9 8 7 6 5 4 3 2 1 SPIN 10925294

www.birkhauser.com

In Honor of S. Panchapakesan

S. Panchapakesan

Contents

Preface

Prof. S. Panchapakesan has made significant contributions to the area of ranking and selection. Besides ranking and selection, he has also published in many other areas of statistics including order statistics, reliability theory, stochastic inequalities, and inference.

In order to reflect his diverse interests and also to recognize his important contributions to different areas, we invited a number of authors to write articles for this volume. These authors form a representative group from coauthors, friends, colleagues and other close professional associates of S. Panchapakesan, in addition to being experts working in one or more of the above-mentioned areas. All the articles present here have been peer reviewed and carefully organized into 20 chapters. For the convenience of the readers, this volume has been divided into the following parts:

- INFERENCE
- RANKING AND SELECTION
- MULTIPLE COMPARISONS AND TESTS
- AGREEMENT ASSESSMENT ANALYSIS
- RELIABILITY
- BIOSTATISTICS

The above list has taken into account various types of inferential problems of interest. This volume is *not* a proceedings, although many of the authors were present at an International Conference held in honor of S. Panchapakesan during December 2002 in Chennai, Tamilnadu, India.

Our sincere thanks go to all the authors who have contributed to this volume. They all share our admiration and appreciation of S. Panchapakesan for all his contributions and sincere work during the past 35 years, and have given us their full cooperation and support in bringing this volume out. We are also indebted to the referees for helping us in the evaluation of the manuscripts and in improving the quality of this publication. In particular, we thank Professors Dipak Dey, Wen-Tao Huang and Nitis Mukhopadhyay for their assistance in the editorial process.

Special thanks are due to Mrs. Debbie Iscoe for the excellent typesetting of the entire volume. Finally, we thank Mr. Thomas Grasso (Editor, Birkhäuser, Boston) for the invitation and encouragement to undertake this project.

With great pleasure, we dedicate this volume to our beloved friend and colleague, S. Panchapakesan.

N. Balakrishnan
McMaster University, Hamilton, Canada
N. Kannan
University of Texas, San Antonio, U.S.A.
H. N. Nagaraja
The Ohio State University, Columbus, U.S.A.

October 2004

S. Panchapakesan
Career and Accomplishments

Subramanian Panchapakesan was born on August 27, 1933, to Subramanian and Visalakshi, in Mylapore, Madras (currently known as Chennai). In 1938, his whole family moved to Pudukkottai in Tamilnadu where he completed the first thirteen years of his studies – first to eleventh standards of his school in an institution run by the Church of Swedish Mission, and then two years of Intermediate in Arts and Sciences at the Rajah's College. He then moved to Madras for further studies at Vivekananda College, obtaining a B.A. (Honours) in Mathematics from the University of Madras in 1954. This B.A. (Honours) degree was equivalent to the M.A. degree, but he had to wait for a year to get the M.A. degree in Mathematics because of some technical reasons.

During 1955–1960, he served as a Lecturer in Mathematics at Islamiah College in Vaniyambadi, Tamilnadu. He left this position in 1960 to join the Indian Statistical Institute (ISI), Calcutta, where he obtained a M.Stat. degree in 1962. For the next two years, he held a Research Assistantship in the Research and Training School of the ISI. During this period, he also spent six months in Hyderabad, Andhra Pradesh, as Officer-in-Charge of Evening Centre run by the ISI and the rest of the period as an instructor at the International Statistical Education Centre (ISEC). This training centre was run by the ISI under a United Nations program. In February 1965, he joined the technical staff in the Data Processing Unit of the Research and Training School of the ISI. After six months, he resigned to go to Purdue University, West Lafayette, Indiana, to do his graduate study in statistics. He received his Ph.D. in Mathematical Statistics in 1969 for his thesis entitled *Some Contributions to Multiple Decision (Selection and Ranking) Procedures* written under the guidance of Professor Shanti Swarup Gupta. With this, he started his illustrious career and made pioneering contributions to the area of Ranking and Selection Methodology and many other areas of Statistics.

Subsequent to his Ph.D., he took a one-year visiting Assistant Professorship in the Department of Statistics at Purdue University. In 1970, he joined the Department of Mathematics at Southern Illinois University, Carbondale, Illinois, as an Assistant Professor. He was promoted to the rank of Associate Professor in 1974 and then to Full Professor in 1980. After a 28-year service there, he retired on June 1, 1998, and currently holds the title of Professor Emeritus in that department.

He had visiting appointments in the Department of Statistics at Purdue University during the fall term of 1975, and the spring terms of 1984 and 1986. He also went as a Visiting Expert to the Institute of Mathematics at Academia Sinica, Taipei, Taiwan, during the spring term of 1980.

He has provided valuable service to many research journals in various capacities. Included in this list are: Member of the International Editorial Board of *Communications in Statistics* during 1985–1994, Associate Editor of *Journal of Statistical Planning and Inference* during 1984–2000, Member of the Editorial Board of *American Journal of Mathematical and Management Sciences* since 1993, and Associate Editor of *Communications in Statistics* since 2001.

S. Panchapakesan, through his pioneering research in the area of ranking and selection over the last 35 years, has made a significant impact in this area. He has not only inspired and encouraged, but also mentored and helped numerous young researchers.

Now that he is retired and is free from teaching and administrative duties, we are confident that he will continue to make fine contributions to the field with renewed interest, enthusiasm and energy. In addition, he will also have more time to enjoy the Indian classical music, his lifelong interest.

Publications

Books

1. *Multiple Decision Procedures: Methodology of Selecting and Ranking Populations* (co-authored with S. S. Gupta), New York: John Wiley & Sons, 1979. Republished as: Classics in Applied Mathematics, **44**, *SIAM*, Philadelphia, 2002.

2. *Advances in Statistical Decision Theory and Applications* (co-edited with N. Balakrishnan), Birkhäuser, Boston, 1997.

Articles in Journals/Books and Technical Reports

1967

1. Order statistics arising from independent binomial populations (with S. S. Gupta), *Mimeograph Series No. 120*, Department of Statistics, Purdue University, West Lafayette, Indiana.

1969

2. Some selection and ranking procedures for multivariate normal populations (with S. S. Gupta), In *Multivariate Analysis II: Proceedings of the Second International Symposium on Multivariate Analysis* (Ed., P. R. Krishnaiah), pp. 475–505, Academic Press, New York.

3. On the distribution of the maximum and minimum of ratios of order statistics (with R. E. Barlow and S. S. Gupta), *Annals of Mathematical Statistics*, **40**, 918–934.

4. Selection and ranking procedures (with S. S. Gupta), In *The Design of Computer Simulation Experiments* (Ed., T. H. Naylor), pp. 132–160, Duke University Press, North Carolina.

5. Some contributions to multiple decision (selection and ranking) procedures, *Mimeograph Series No. 192*, Department of Statistics, Purdue University, West Lafayette, Indiana.

1971

6. On a subset selection procedure for the most probable event in a multinomial distribution, In *Statistical Decision Theory and Related Topics* (Eds., S. S. Gupta and J. Yackel), pp. 275–298, Academic Press, New York.

7. Contributions to multiple decision (subset selection) rules, multivariate distribution theory and order statistics (with S. S. Gupta), *Technical Report ARL 71-0218*, Aerospace Research Laboratories, Dayton, Ohio.

1972

8. On a class of subset selection procedures (with S. S. Gupta), *Annals of Mathematical Statistics*, **43**, 814–822.

9. On multiple decision (subset selection) procedures (with S. S. Gupta), *Journal of Physical and Mathematical Sciences*, **6**, 1–72.

1973

10. On order statistics and some applications of combinatorial methods in statistics (with S. S. Gupta), In *A Survey of Combinatorial Theory* (Ed., J. N. Srivastava), pp. 217–250, North-Holland Publishing Company, Amsterdam.

11. On order statistics from equally correlated normal random variables (with S. S. Gupta and K. Nagel), *Biometrika*, **60**, 403–413.

12. Inference for restricted families: (A) multiple decision procedures; (B) order statistics inequalities (with S. S. Gupta), In *Reliability and Biometry: Statistical Analysis of Life Length* (Eds., F. Proschan and R. J. Serfling), pp. 503–596, SIAM, Philadelphia.

1974

13. On moments of order statistics from independent binomial populations (with S. S. Gupta), *Annals of the Institute of Statistical Mathematics Supplement*, **8**, 95–113.

1975

14. On a quantile selection procedure and associated distribution of ratios of order statistics from a restricted family of probability distributions (with S. S. Gupta), In *Reliability and Fault Tree Analysis* (Eds., R. E. Barlow, J. B. Fusell and N. D. Singpurwalla), pp. 557–576, SIAM, Philadelphia.

1976

15. A modified subset selection formulation with special reference to one-way and two-way layout experiments (with D.-Y. Huang), *Communications in Statistics—Theory and Methods*, **5**, 621–633.

1977

16. Subset selection procedures for Δ_p-superior populations (with T. J. Santner), *Communications in Statistics—Theory and Methods*, **6**, 1081–1090.

1978

17. A subset selection formulation of the complete ranking problem (with D.-Y. Huang), *Journal of the Chinese Statistical Association*, **16**, 5801–5810.

18. On a monotonicity property relating to gamma distributions, *Journal of the Chinese Statistical Association*, **16**, 6003–6005.

1980

19. Some statistical techniques in climatological data (with S. S. Gupta), In *Statistical Climatology: Developments in Atmospheric Science*, **13** (Eds., S. Ikeda et al.), pp. 35–48, Elsevier Scientific Publishing Company, Amsterdam.

1981

20. ASA and statistical education (with S. S. Gupta), In *Proceedings of the American Statistical Association - Section on Statistical Education*, pp. 27–29, Alexandria, Virginia.

1982

21. On eliminating inferior regression models (with D.-Y. Huang), *Communications in Statistics—Theory and Methods*, **11**, 751–759.

22. Some locally optimal subset selection rules based on ranks (with D.-Y. Huang), In *Statistical Decision Theory and Related Topics-III* (Eds., S. S. Gupta and J. O. Berger), Vol. 2, pp. 1–14, Academic Press, New York.

1984

23. Some locally optimal subset selection rules for comparison with a control (with D.-Y. Huang), *Journal of Statistical Planning and Inference*, **9**, 63–72.

24. On some inequalities and monotonicity results in selection and ranking theory (with S. S. Gupta and D.-Y. Huang), In *Inequalities in Statistics and Probability* (Ed., Y. L. Tong), IMS Lecture Notes - Monograph Series, Vol. 5, pp. 211–227, Institute of Mathematical Statistics, Hayward, California.

25. An estimation problem relating to subset selection from normal populations (with S. Jeyaratnam), In *Design of Experiments: Ranking and Selection* (Eds., T. J. Santner and A. C. Tamhane), pp. 287–302, Marcel Dekker, New York.

26. Edgeworth expansions in statistics: some recent developments (with S. S. Gupta), In *Colloquia Mathematica Societatis János Bolyai: 34. Limit Theorems in Probability and Statistics, Vol. I* (Ed., P. Révész), pp. 519–565, North-Holland Publishing Company, Amsterdam.

1985

27. On the distribution of the studentized maximum of equally correlated normal random variables (with S. S. Gupta and J. K. Sohn), *Communications in Statistics—Simulation and Computation*, **14**, 103–135.

28. Subset selection procedures: review and assessment (with S. S. Gupta), *American Journal of Mathematical and Management Sciences*, **5**, 235–311.

29. Shanti S. Gupta: an appreciation (with T. J. Santner), *American Journal of Mathematical and Management Sciences*, **5**, 347–369.

30. Estimation after subset selection from exponential populations (with S. Jeyaratnam), *Communications in Statistics—Theory and Methods*, **15**, 3459–3473.

1986

31. Estimation of Pr(X > Y) for gamma distributions (with R. Ismail and S. Jeyaratnam), *Journal of Statistical Computation and Simulation*, **26** 253–267.

1987

32. Statistical selection procedures in multivariate models (with S. S. Gupta), In *Advances in Multivariate Statistical Analysis* (Ed., A. K. Gupta), pp. 141–160, D. Reidel Publishing Company, Dordrecht, Holland.

1988

33. Inference about the change-point in a sequence of random variables: a selection approach (with T. Liang), In *Statistical Decision Theory and Related Topics-IV* (Eds., S. S. Gupta and J.O. Berger), Vol. 2, pp. 79–87, Springer-Verlag, New York.

34. Selection and ranking procedures in reliability models (with S. S. Gupta), In *Handbook of Statistics 7: Quality Control and Reliability* (Eds., P. R. Krishnaiah and C. R. Rao), pp. 131–156, North-Holland Publishing Company, Amsterdam.

35. Prediction intervals for balanced one-factor random models (with S. Jeyaratnam), In *Probability and Statistics: Essays in Honor of F. A. Graybill* (Ed., J. N. Srivastava), pp. 161–170, North-Holland Publishing Company, Amsterdam.

36. Selection from uniform populations based on sample midranges and an associated estimation after selection (with S. Jeyaratnam), *Communications in Statistics—Theory and Methods*, **17**, 2303–2314.

37. Prediction intervals for the random intercept linear model (with S. Jeyaratnam), *Communications in Statistics—Theory and Methods*, **17**, 3067-3073.

1989

38. Entropy based subset selection from Bernoulli populations (with S. Jeyaratnam), In *Computing and Information* (Eds., R. Janicki and W. W. Koczkodaj), Vol. II, pp. 202–204, Canadian Scholars' Press Inc., Toronto.

1990

39. Preliminary test based sometimes-pool estimator of the Bernoulli entropy function (with S. Jeyaratnam), In *Advances in Computing and Information* (Eds., S. G. Akl, F. Fiala and W. W. Koczkodaj), pp. 16–18, Canadian Scholars' Press Inc., Toronto.

1991

40. On sequential ranking and selection procedures (with S. S. Gupta), In *Handbook of Sequential Analysis* (Eds., B. K. Ghosh and P. K. Sen), pp. 363–380, Marcel Dekker, New York.

41. An empirical Bayes procedure for selecting the most homogeneous multinomial population according to the Gini-Simpson index (with T. Liang), In *The Proceedings of the 1990 Taipei Symposium in Statistics* (Eds., M. T. Chao and P. E. Cheng), pp. 447–460, Institute of Statistical Science, Academia Sinica, Taipei, Taiwan.

1992

42. Ranking and selection procedures, In *Handbook of the Logistic Distribution* (Ed., N. Balakrishnan), pp. 145–167, Marcel Dekker, New York.

43. On a monotone empirical Bayes test procedure in geometric model (with T. Liang), *Annals of the Institute of Statistical Mathematics*, **44**, 133–140.

44. Isotonic selection with respect to a control: A Bayesian approach (with T. Liang), In *The Frontiers of Modern Statistical Inference Procedures, II* (Eds., E. Bofinger et al.), pp. 273–285, American Sciences Press, Syracuse, New York.

1993

45. A two-stage procedure for selecting δ^*-optimal guaranteed lifetimes in the two-parameter exponential model (with T. Liang), In *Multiple Comparisons, Selection, and Applications in Biometry: A Festschrift in Honor of Charles W. Dunnett* (Ed., F. M. Hoppe), Chapter 20, pp. 353–365, Marcel Dekker, New York.

46. Selection and screening procedures in multivariate analysis (with S. S. Gupta), In *Multivariate Analysis: Future Directions* (Ed., C. R. Rao), North-Holland Series in Statistics and Probability, Vol. 5, Chapter 12, pp. 223–262, Elsevier Science Publishers, Amsterdam.

47. Robustness of selection procedures: an overview, In *Selection Procedures I: Proceedings of the 3rd Schwerin Conference on Mathematical Statistics* (Eds., G. Herrendorfer and K. J. Miescke), pp. 134–146, Agricultural University, Dummerstorf-Rostock, Germany.

48. Multistage subset selection procedures for normal populations and associated second-order asymptotics (with N. Mukhopadhyay), *Metron*, **21**, 25–42.

1994

49. An integrated formulation for selecting the best normal population and eliminating bad ones (with P. Chen), In *Compstat 1994* (Eds., R. Dutter and W. Grossman), pp. 18–19, University of Technology and University of Vienna, Vienna, Austria.

50. Selecting among the multinomial losers (with P. Chen and M. Sobel), *Sequential Analysis*, **13**, 177–200.

1995

51. Estimation of the location and scale parameters of the extreme value distribution based on multiply type-II censored samples (with N. Balakrishnan and S. S. Gupta), *Communications in Statistics—Theory and Methods*, **24**, 2105–2125.

52. Estimation of the mean and standard deviation of the logistic distribution based on multiply type-II censored samples (with N. Balakrishnan and S. S. Gupta), *Statistics*, **27**, 127–142.

53. Selection of the normal population with the largest absolute mean, In *Bulletin of the International Statistical Institute, Contributed Papers of the 50th Session*, Book 2, pp. 955–956.

54. Selection and ranking procedures, In *The Exponential Distribution: Theory, Methods and Applications* (Eds., N. Balakrishnan and A. P. Basu), Chapter 16, pp. 259–278, Gordon and Breach Science Publishers, Newark, New Jersey.

55. Exponential classification and applications (with N. Balakrishnan and Q. Zhang), In *The Exponential Distribution: Theory, Methods and Applications* (Eds., N. Balakrishnan and A. P. Basu), Chapter 32, pp. 525–546, Gordon and Breach Science Publishers, Newark, New Jersey.

56. Multiple decision procedures in analysis of variance and regression analysis (with S. S. Gupta and D.-Y. Huang), *Technical Report No. 95-44c*, Department of Statistics, Purdue University, West Lafayette, Indiana.

1996

57. A review of robustness of selection procedures, *Journal of Statistical Planning and Inference*, **54**, 279–290.

58. Design of experiments with selection and ranking goals (with S. S. Gupta), In *Handbook of Statistics 13: Design and Analysis of Experiments* (Eds., S. Ghosh and C. R. Rao), Chapter 17, pp. 555–585, Elsevier Science Publishers, Amsterdam.

59. δ-exceedance records (with N. Balakrishnan and K. Balasubramanian), *Journal of Applied Statistical Sciences*, **4**, 123–132.

1997

60. An integrated formulation for selecting the best from several normal populations in terms of the absolute values of their means: common known variance case (with S. Jeyaratnam), In *Advances in Statistical Decision Theory and Applications* (Eds., S. Panchapakesan and N. Balakrishnan), Chapter 19, pp. 277–289, Birkhäuser, Boston.

1998

61. Inverse sampling procedures to test for homogeneity in a multinomial distribution (with A. Childs, B. H. Humphrey and N. Balakrishnan), In *Handbook of Statistics 17: Order Statistics and Their Applications* (Eds., N. Balakrishnan and C.R. Rao), Chapter 14, pp. 259–265, Elsevier Science Publishers, Amsterdam.

62. A two-stage procedure for selecting from normal populations the one with the largest absolute mean: common unknown variance case (with S. Jeyaratnam), In *Proceedings of the 3rd St. Petersburg Workshop on Simulation* (Eds., S. M. Ermakov, Y. N. Kashtanov and V. B. Melas), pp. 259–265, Saint Petersburg University Press, St. Petersburg, Russia.

2000

63. Selecting from normal populations the one with the largest absolute mean: common unknown variance case (with S. Jeyaratnam), In *Advances in Stochastic Simulation Methods* (Eds., N. Balakrishnan, S. M. Ermakov and V. B. Melas), Chapter 16, pp. 283–292, Birkhäuser, Boston.

2001

64. Simultaneous selection of extreme populations from a set of two-parameter exponential populations (with K. Hussein), In *Handbook of Statistics 20: Advances in Reliability* (Eds., N. Balakrishnan and C. R. Rao), Chapter 33, pp. 813–830, Elsevier Science Publishers, Amsterdam.

2002

65. On selection from normal populations in terms of the absolute values of their means (with K. Hussein), In *Advances on Theoretical and Methodological Aspects of Probability and Statistics* (Ed. N. Balakrishnan), Chapter 25, pp. 371–390, Taylor and Francis Publishers, New York.

66. Selecting the normal population with the largest mean: a restricted subset selection rule (with L. Hsu), In *Statistical Methods and Practice: Recent Advances* (Eds., N. Balakrishnan, N. Kannan and M.R. Srinivasan), pp. 145–161, Narosa Publishing House, New Delhi, India.

67. Estimation of the mean and standard deviation of the normal distribution based on multiply type-II censored samples (with N. Balakrishnan and S. S. Gupta), *Journal of Statistical Studies, Special Volume in honor of Mir Masoom Ali* (Ed., M. F. Hossain), pp. 307–320.

68. Professor Shanti Swarup Gupta (with T. J. Santner), *American Journal of Mathematical and Management Sciences*, **22**, 173–198.

2003

69. Sequential procedures for selecting the most probable multinomial cell when a nuisance cell is present (with M. Aoshima and P. Chen), *Communications in Statistics—Theory and Methods*, **32**, 893–906.

2004

70. Detecting signals simultaneously at k sites (with P. Chen), *Communications in Statistics—Theory and Methods*, **33**, 1667–1688.

71. Selection procedures for type I extreme value populations in terms of location parameters and a related homogeneity test (with S. Jeyaratnam), In *Extreme Value Distributions: Theory, Methods and Applications* (Ed., N. Balakrishnan), Taylor and Francis Publishers, New York (to appear).

72. Gupta, Shanti Swarup, In *Encyclopedia of Statistical Sciences, Second Edition* (Eds., N. Balakrishnan, C. B. Read and B. Vidakovic), John Wiley & Sons, New York (to appear).

73. Signal Processing, Selection approaches in (with P. Chen), In *Encyclopedia of Statistical Sciences, Second Edition* (Eds., N. Balakrishnan, C. B. Read and B. Vidakovic), John Wiley & Sons, New York (to appear).

74. Ranking and Selection procedures, In *Encyclopedia of Statistical Sciences, Second Edition* (Eds., N. Balakrishnan, C. B. Read and B. Vidakovic), John Wiley & Sons, New York (to appear).

75. On some variable selection procedures based on data for regression models (with D.-Y. Huang and R.-F. Lee), *Journal of Statistical Planning and Inference* (to appear).

76. A monotonicity property for gamma distributions (with G.C. McDonald), *Journal of Statistical Planning and Inference* (to appear).

77. A nonparametric procedure based on early failures for selecting the best population using a test for equality (with H. K.T. Ng and N. Balakrishnan), *Journal of Statistical Planning and Inference* (to appear).

78. A restricted subset selection procedure for selecting in terms of means at least one of the t best from k normal populations with common known variance (with L. Hsu), *In this volume.*

79. Restricted subset selection procedures for normal means: A brief review with a fresh look at the classical formulation of Bechhofer and Gupta, *Communications in Statistics—Theory and Methods* (to appear).

80. A restricted subset selection procedure for selecting in terms of means at least one of the t best from k normal populations with common known variance, II (with L. Hsu), *Under preparation.*

Book Reviews

1. *Selecting and Ordering Populations: A New Statistical Methodology*, by J. D. Gibbons, I. Olkin and M. Sobel (John Wiley & Sons), *Journal of Qualtiy Technology*, **12** (1980), 239–240.

2. *The Complete Categorized Guide to Statistical Selection and Ranking Procedures*, by E. J. Dudewicz and J. O. Koo (American Sciences Press), *Mathematical Reviews*, **84g** (1984).

3. *Selected Tables in Mathematical Statistics, Volume 11*, Eds., R. F. Odeh, J. M. Davenport and N. S. Pearson (American Mathematical Society), *Mathematical Computation*, **51** (1989), 252–253.

4. *Modern Mathematical Statistics*, by E. J. Dudewicz and S. N. Mishra (John Wiley & Sons), *Mathematical Reviews*, **89g:62002** (1989).

5. *Fundamentals of Mathematical Statistics, Vol. I: Probability for Statistics, Vol. II: Statistical Inference*, by H.T. Nguyen and C. S. Rogers (Springer-Verlag), *Mathematical Reviews*, **90i:62003a,b** (1990).

6. *A Statistical Model: Frederic Mosteller's Contributions to Statistics, Science, and Public Policy*, Eds., S. E. Fienberg, D. C. Hoaglin, W. H. Kruskal and J. M. Tanur (Springer-Verlag), *Mathematical Reviews*, **92a:01079** (1992).

7. *A First Course in Order Statistics*, by B. C. Arnold, N. Balakrishnan and H. N. Nagaraja (John Wiley & Sons), *Mathematical Reviews*, **94a:62076** (1994).

8. *Multistage Selection and Ranking Procedures*, by N. Mukhopadhyay and T. K. S. Solanky (Marcel Dekker), *Mathematical Reviews*, **95h:62032** (1995).

9. *Measurement, Regression, and Calibration*, by P.J. Brown (Oxford University Press), *SIAM Review*, **37** (1995), 635–636.

10. *Design and Analysis of Experiments for Statistical Selection, Screening, and Multiple Comparisons*, by R. E. Bechhofer, T. J. Santner and D. M. Goldsman (John Wiley & Sons), *Mathematical Reviews*, **96k:62001** (1996).

11. *Theory of Statistics*, by M. J. Schervish (Springer-Verlag), *Mathematical Reviews*, **96m:62001** (1996).

12. *Tables for the Use of Order Statistics in Estimation*, by H. L. Harter and N. Balakrishnan (CRC Press), *Mathematical Reviews*, **97d:62097** (1997).

13. *Robust Diagnostic Regression Analysis*, by A. Atkinson and M. Riani (Springer-Verlag), *Mathematical Reviews*, **2003a:62001** (2003).

Contributors

Aoshima, Makoto Institute of Mathematics, University of Tsukuba, Ibaraki 305-8571, Japan
aoshima@math.tsukuba.ac.jp

Balakrishnan, N. Department of Mathematics and Statistics, McMaster University, Hamilton, Ontario, Canada L8S 4K1
bala@univmail.cis.mcmaster.ca

Chen, Pinyuen Department of Mathematics, Syracuse University, Syracuse, NY 13244-1150 U.S.A.
Pinchen@syr.edu

Choudhary, Pankaj K. Department of Mathematical Sciences, University of Texas at Dallas, Richardson, TX 75083-0588 U.S.A.
pankaj@utdallas.edu

Cicconetti, Greg Department of Mathematical Sciences, Muhlenberg College, Allentown, PA 18104 U.S.A.
cicconet@muhlenberg.edu

Dey, Dipak K. Department of Statistics, University of Connecticut, Storrs, CT 06269 U.S.A.
dey@stat.uconn.edu

Dhariyal, Ishwari D. Department of Mathematics, Indian Institute of Technology Kanpur, Kanpur 208 016, India
idd@iitk.ac.in

Dohi, T. Department of Information Engineering, Hiroshima University, 4-1 Kagamiyama 1 Chome, Higashi-Hiroshima 739-8527, Japan
dohi@rel.hiroshima-u.ac.jp

Ghosh, Samiran Department of Statistics, University of Connecticut, Storrs, CT 06269 U.S.A.
samiran-g@msn.com

Hayter, Anthony J. School of Industrial and Systems Engineering, Georgia Institute of Technology, Atlanta, GA 30332 U.S.A.
ajh@isye.gatech.edu

Hsu, Lifang Department of Mathematics, LeMoyne College, Syracuse, NY 13214 U.S.A.
hsu@lemoyne.edu

Huang, Deng-Yuan Department of Business Administration, Fu-Jen Catholic University, Taipei, Taiwan, R.O.C.
xz028554@ispeed.com.tw

Huang, Wen-Tao Department of Management Sciences and Decision Making, Tamkang University, Tamsui, Taipei, Taiwan, R.O.C.
wentao@mail.tku.edu.tw

Indurkhya, Alka Department of Society, Human Development and Health, Harvard School of Public Health, Harvard University, Boston, MA 02115 U.S.A.
aindurkh@hsph.harvard.edu

Kaio, N. Department of Economic Informatics, Hiroshima Shudo University, 1-1-1 Ozukahigashi, Asaminami-ku, Hiroshima 731-3195, Japan

Kim, J.-A. Department of Mathematics and Statistics, McMaster University, Hamilton, Ontario, Canada L8S 4K1
kimja3@univmail.cis.mcmaster.ca

Kundu, Debasis Department of Mathematics, Indian Institute of Technology Kanpur, Kanpur 208 016, India
kundu@iitk.ac.in

Kushida, Takuya Graduate School of Pure and Applied Sciences, University of Tsukuba, Ibaraki 305-8571, Japan
kushida19512@nissay.co.jp

Lai, Yao-Tsung Department of Mathematical Statistics and Actuarial Science, Aletheia University, Tamsui, Taipei, Taiwan, R.O.C.

Lee, Ren-Fen National Kaohsiung University of Applied Sciences, Kaohsiung, Taiwan, R.O.C.

Liang, Tachen Department of Mathematics, Wayne State University, Detroit, MI 48202 U.S.A.
liang@math.wayne.edu

Mallick, Madhuja Department of Statistics, University of Connecticut, Storrs, CT 06269 U.S.A.
madhujamallick@hotmail.com

Misra, Neeraj Department of Mathematics, Indian Institute of Technology Kanpur, Kanpur 208 016, India
neeraj@iitk.ac.in

Mukhopadhyay, Nitis Department of Statistics, University of Connecticut, Storrs, CT 06269-4120 U.S.A.
mukhop@uconnvm.uconn.edu

Nagaraja, H. N. Department of Statistics, The Ohio State University, Columbus, OH 43210-1247 U.S.A.
hnn@stat.ohio-state.edu

Osaki, S. Department of Information and Telecommunication Engineering, Nanzan University, Nagoya, Japan
osaki@nanzan-u.ac.jp

Panchapakesan, S. Department of Mathematics, Southern Illinois University, Carbondale, IL 62901-4408 U.S.A.
spkesan@yahoo.com

Patra, Kaushik Department of Statistics, University of Connecticut, Storrs, CT 06269-4120 U.S.A.

Raghavachari, M. Decision Sciences and Engineering Systems, Rensselaer Polytechnic Institute, Troy, NY 12180 U.S.A.
ragavm@rpi.edu

Rao, C.R. Department of Statistics, Pennsylvania State University, State College, PA 16802 U.S.A.
crr1@psu.edu

Ravishanker, Nalini Department of Statistics, University of Connecticut, Storrs, CT 06269-4120 U.S.A.
nalini@stat.uconn.edu

Rollin, Linda Richard L. Gelb Center for Pharmaceutical Research and Development, Bristol-Myers Squibb Company, Wallingford, CT 06492-7660 U.S.A.
Linda.rollin@bms.com

Roy, Supratik Department of Statistics, University College of Cork, Cork, Republic of Ireland
supratik@stat.ucc.ie

Sampson, Allan R. Department of Statistics, University of Pittsburgh, Pittsburgh, PA 15260 U.S.A.
asampson@pitt.edu

SenGupta, Ashis Applied Statistics Division, Indian Statistical Institute, Kolkata 700 035, India
ashis@isical.ac.in

Tamhane, Ajit C. Department of IE/MS and Department of Statistics, Northwestern University, Evanston, IL 60208 U.S.A.
ajit@iems.northwestern.edu

Yothers, Greg National Surgical Adjuvant Breast and Bowel Project (NSABP), Pittsburgh, PA 51213 U.S.A.
Yothers@nsabp.pitt.edu

List of Tables

List of Figures

Advances in Ranking and Selection, Multiple Comparisons, and Reliability

PART I

INFERENCE

1

Score Test: Historical Review and Recent Developments

C.R. Rao

Department of Statistics, Pennsylvania State University, State College, Pennsylvania, U.S.A.

Abstract: The three asymptotic tests, Neyman and Pearson Likelihood Ratio (LR), Wald's statistic (W) and Rao's score (RS)are referred to in statistical literature on testing of hypotheses as the *Holy Trinity*. All these tests are equivalent to the first-order of asymptotics, but differ to some extent in the second-order properties. Some of the merits and defects of these tests are presented.

Some applications of the score test, recent developments on refining the score test and problems for further investigation are presented.

Keywords and phrases: Composite hypothesis, Lagrangian multiplier (LM) test, Likelihood ratio (LR), Neyman's C(α), Neyman-Rao test, Rao's score (RS), Wald's statistic (W)

1.1 Introduction

The Score test was introduced in Rao (1948) as an alternative to the likelihood ratio test of Neyman and Pearson (1928) and Wald (1943) test. A few years later Aitchison and Silvey (1958) and Silvey (1959) gave an interpretation of the score statistic in terms of a Lagrangian Multiplier used in optimizing a function subject to restrictions, and called it the Lagrangian Multiplier (LM) test.

The score (RS) test went unnoticed for a number of years after it was introduced. The first application of the score test, apart from the examples given in Rao (1948, 1950, 1961) appeared in econometric literature [Byron (1968)]. During the late 1970s and 1980s, the RS test was applied to a variety of problems in econometrics. Reference may be made to survey papers by Breusch and Pagan (1980), Engle (1984), Kramer and Sonnberger (1986), and Godfrey (1988). Most of the recent textbooks on econometrics also discuss the RS test. Some of them are by White (1984, pp. 72–74), Amemiya (1985, pp. 141–146),

Judge et al. (1985, pp. 182–187), Kmenta (1986, pp. 493–495), Spanos (1986, pp. 326–336), Maddala (1988, pp. 137–139), Green (1990, pp. 357–359), and Harvey (1990, pp. 169–177).

The distributional aspects of the RS statistic are covered in books by Rao (1973, pp. 418–419), Serfling (1980, pp. 156–160), Godfrey (1988, pp. 13–15), Lehmann (1999, pp. 451, 529, 532, 534, 539, 541, 570), and Bickel and Doksum (2001, pp. 335–336, 399–402).

The study of the power properties of the RS test started with a paper by Chandra and Joshi (1983) and continued by Chandra and Mukherjee (1984, 1985), Chandra and Samanta (1988), Ghosh (1991) and others. Reference may be made to Peers (1971) for a comment on a conjecture I made about the local properties of the LR test, which motivated the work of others on power properties.

In this chapter, a brief review is given of the RS statistic and its merits and demerits in terms of power properties compared to LR and W are discussed. Some of the recent developments and refinements and modifications of the RS statistic are presented and some problems for future research are indicated.

1.2 Asymptotic Tests of a Simple Hypothesis

1.2.1 Notation

Let $X = (x_1, \ldots, x_n)$ be an iid sample of size n from the density function $p(x, \theta)$ where θ is a p-vector parameter, and denote the joint density by $P(X, \theta) = p(x_1, \theta) \ldots p(x_n, \theta)$ and the log likelihood by $L(\theta|X) = \log P(X, \theta)$. The score vector of p components, as defined by Fisher, is

$$s(\theta) = \frac{1}{P}\frac{\partial P}{\partial \theta} = (s_1(\theta), \ldots, s_p(\theta))', \tag{1.1}$$

$$s_i(\theta) = \frac{1}{P}\frac{\partial P}{\partial \theta_i}, \; i = 1, \ldots, p.$$

The Fisher information matrix of order $p \times p$ is defined by

$$ni(\theta) = I(\theta) = E[s(\theta)s'(\theta)] = (i_{rs}(\theta)) \tag{1.2}$$

where $i_{rs}(\theta) = E\left[s_r(\theta)s_s(\theta)\right]$. The maximum likelihood estimate of θ is obtained as a solution of the p equations

$$s_i(\theta) = 0, \; i = 1, \ldots, p \tag{1.3}$$

which we represent by $\hat{\theta}$. Under suitable regularity conditions [Lehmann (1999, pp. 499–501)], using the multivariate cental limit theorem

$$n^{-1/2}s(\theta_0) \sim N_p\left(0, i(\theta_0)\right) \tag{1.4}$$

where θ_0 is the true value, and

$$n^{1/2}\left(\hat{\theta}-\theta_0\right) \sim N_p\left(0,[i(\theta_0)]^{-1}\right) \tag{1.5}$$

where $N_p(0, A)$ is a p variate normal distribution with mean zero and covariance matrix A.

1.2.2 Three possible tests of a simple hypothesis: The Holy Trinity

Let $H_0 : \theta = \theta_0$ (a specified p-vector) be the null hypothesis to be tested. Three tests which are in current use are as follows.

1. Likelihood ratio test [Neyman and Pearson (1928)]

$$LR = 2\left[L(\hat{\theta}|X) - L(\theta_0|X)\right] \tag{1.6}$$

   where $L(\theta|X) = \log P(X, \theta)$.

2. Wald test [Wald (1943)]

$$W = (\hat{\theta}-\theta_0)' I(\hat{\theta})(\hat{\theta}-\theta_0). \tag{1.7}$$

3. Rao Score test [Rao (1948)]

$$RS = [s(\theta_0)]' \left[I(\theta_0)\right]^{-1} [s(\theta_0)]\,. \tag{1.8}$$

All the three statistics known as the *Holy Trinity* have an asymptotic chi-square distribution on p degrees of freedom.

1.2.3 Motivation for the score test of a simple hypothesis

Consider the case of a single parameter θ and $H_0 : \theta = \theta_0$. If $w \subset R^n$ is the critical region of size α in the sample space, then the power of the test is

$$\pi(\theta) = \int_w P(X, \theta) dv \text{ with } \pi(\theta_o) = \int_w P(X, \theta_0) dv = \alpha.$$

To find a locally most powerful one-sided test $(\theta > \theta_0)$ we maximize

$$\pi'(\theta_0) = \int_w P'(X, \theta_0) dv$$

subject to $\pi(\theta_0) = \alpha$. Using the Neyman–Pearson Lemma, the optimal region is defined by

$$\frac{P'(x, \theta_0)}{P(x, \theta_0)} \geq \lambda \text{ or } s(\theta_0) \geq \lambda$$

where λ is chosen such that the size of the region is α, as shown in Rao and Poti (1946). The test can be written in the form

$$\frac{s(\theta_0)}{\sqrt{I(\theta_0)}} \geq \lambda \text{ or } \frac{[s(\theta_0)]^2}{I(\theta_0)} \geq \lambda^2. \tag{1.9}$$

In the multiparameter case, the slope of the power function in the direction $a = (a_1, \ldots, a_p)'$, at θ_0 is

$$a_1 s_1(\theta_0) + \cdots + a_p s_p(\theta_0) = a' s(\theta_0) \tag{1.10}$$

and the statistic (1.9) takes the form

$$\frac{[a' s(\theta_0)]^2}{a' I(\theta_0) a}. \tag{1.11}$$

Maximizing with respect to a yields the statistic

$$[s(\theta_0)]'[I(\theta_0)]^{-1}[s(\theta_0)] \tag{1.12}$$

which is the same as (1.8).

1.2.4 Test of a composite hypothesis

Under the same setup as in Section 1.2.1, let the hypothesis to be tested be $H_0 : h(\theta) = c$, where h is an $r \times 1$ vector function of the p-vector θ with $p \geq r$ and c is a given r-vector of constants. The corresponding *Holy Trinity* is as follows:

1. Likelihood ratio test [Neyman and Pearson (1928)]

$$LR = 2\left[l(\hat{\theta}|X) - l(\tilde{\theta}|X)\right] \tag{1.13}$$

   where $\tilde{\theta}$ is the ml of θ under the restriction $h(\theta) = c$.

2. Wald test [Wald (1943)]

$$W = \left[h(\hat{\theta}) - c\right]' \left[A(\hat{\theta})\right]^{-1} \left[h(\hat{\theta}) - c\right] \tag{1.14}$$

   where

$$A(\theta) = [H(\theta)][I(\theta)]^{-1}[H(\theta)]',$$
$$H(\theta) = (\partial h_i(\theta)/\partial\theta_j), h(\theta) = (h_1(\theta), \ldots, h_r(\theta))',$$

   and $I(\theta)$ is as defined in (1.2).

3. Rao Score test [Rao (1948)]

$$RS = [s(\tilde{\theta})]'[I(\tilde{\theta})]^{-1}[s(\tilde{\theta})]. \tag{1.15}$$

All the three statistics have an asymptotic chi-square distribution on r degrees of freedom.

An alternative way of expressing the RS statistic is as follows. Note that $\tilde{\theta}$, the restricted ml of θ, is a solution of the equation

$$s(\theta) + [H(\theta)]'\lambda = 0, \; h(\theta) = c$$

where λ is an r-vector of the Lagrangian Multiplier so that $[s(\tilde{\theta})]' = -\lambda' H(\tilde{\theta})$. Substituting in (1.15) we have

$$\text{RS} = \lambda' H(\tilde{\theta})[I(\tilde{\theta})]^{-1}[H(\tilde{\theta})]'\lambda = \lambda'[A(\tilde{\theta})]\lambda \tag{1.16}$$

where $A(\theta)$ is as defined in (1.14). Silvey (1959) expressed the RS statistic (1.15) in the form (1.16) and called it the Lagrangian Multiplier (LM) test. (In econometric literature, the RS test is generally referred to as the LM test.)

1.2.5 Special form of composite hypothesis

In many problems, the p-vector parameter θ consists of two parts, θ_1 an r vector and θ_2 a $(p-r)$ vector and the null hypothesis is of the form $H_0 : \theta_1 = \theta_{10}$ (a specified vector) and θ_2 (known as a nuisance parameter) is arbitrary. This becomes a special case of the composite hypothesis considered in Subsection 1.2.4 if we take $h(\theta) = \theta_1$. Denote the unrestricted ml of (θ_1, θ_2) by $(\hat{\theta}_1, \hat{\theta}_2)$ and its asymptotic covariance matrix by

$$\begin{aligned} cov(\hat{\theta}, \hat{\theta}) &= [I(\theta)]^{-1} \\ &= \begin{pmatrix} I_{11}(\theta) & I_{12}(\theta) \\ I_{21}(\theta) & I_{22}(\theta) \end{pmatrix}^{-1} = \begin{pmatrix} A & B \\ B' & C \end{pmatrix} \end{aligned}$$

where the partitions of the information matrix, I_{11}, I_{12}, and I_{22} are matrices of orders $r \times r$, $r \times (p-r)$ and $(p-r) \times (p-r)$, respectively, The Wald statistic can be written as

$$\begin{aligned} W &= (\hat{\theta}_1 - \theta_{10})'\hat{A}^{-1}(\hat{\theta}_1 - \theta_{10}), \; \hat{A} = A(\hat{\theta}) \\ &= (\hat{\theta}_1 - \theta_{10})' I_{1.2}(\hat{\theta})(\hat{\theta}_1 - \hat{\theta}_{10}) \end{aligned} \tag{1.17}$$

where

$$I_{1.2} = I_{11} - I_{12}I_{22}^{-1}I_{21}$$

the Schur complement of I_{22}.

To compute LR and RS statistics, we need to find the restricted ml estimates of θ_1, θ_2 under the restriction $\theta_1 = \theta_{10}$. Using the Lagrangian multiplier we have to maximize

$$L(\theta|x) - \lambda(\theta_1 - \theta_{10})$$

with respect to θ. The estimating equations are

$$s_1(\tilde{\theta}) = \lambda,\ s_2(\tilde{\theta}) = 0,\ \tilde{\theta}_1 = \theta_{10}.$$

The Rao score statistic is

$$\begin{aligned} \text{RS} &= [s_1(\tilde{\theta})', 0']'[I(\tilde{\theta})]^{-1}[s_1(\tilde{\theta})', 0'] \\ &= [s_1(\tilde{\theta})]'[I_{1.2}(\tilde{\theta})]^{-1}[s_1(\tilde{\theta})] \\ &= \lambda'[I_{1.2}(\tilde{\theta})]^{-1}\lambda. \end{aligned} \tag{1.18}$$

The LR statistic is

$$\text{LR} = 2\left[L(\hat{\theta}) - L(\tilde{\theta})\right] \tag{1.19}$$

All the three statistics have asymptotically chi-square distribution on r d.f.

1.3 Neyman's $C(\alpha)$ Test and Neyman-Rao Test

Neyman (1959, 1979) considered the problem of testing the hypothesis $H_0 : \theta_1 = \theta_{10}$ (given) and $\theta_2, \ldots, \theta_p$ are arbitrary (nuisance) parameters. Hall and Mathiason (1990) considered the more general problem of testing the composite hypothesis

$$H_0 : \theta_1 = \theta_{10}, \ldots, \theta_q = \theta_{q0} \text{ and } \theta_{q+1}, \ldots, \theta_p$$

are arbitrary by generalizing Neyman's results using the type of the argument used in Rao (1948) as in Section 1.2.3. Consider the slope of the power curve in the direction $(a_1, \ldots, a_q, 0, \ldots, 0)$

$$a_1 s_1 + \cdots + a_q s_q$$

where s_i is the derivative of the log likelihood with respect to θ_i, and define the Neyman statistic N as

$$N = \max_a \frac{(a_1 s_1 + \cdots + a_q s_q)^2}{V(a_1 s_1 + \cdots + a_q s_q)} \tag{1.20}$$

subject to

$$cov(s_i, a_1 s_1 + \cdots + a_q s_q) = 0,\ i = q + 1, \ldots, p. \tag{1.21}$$

Using notation

$$\begin{aligned} S_1 &= (s_1, \ldots, s_q)', \, S_2 = (s_{q+1}, \ldots, s_p)', \\ a &= (a_1, \ldots, a_q)', \\ E(S_1 S_1') &= I_{11}, \; E(S_1 S_2') = I_{12}, E(S_2 S_2') = I_{22}, \end{aligned}$$

the problem (1.20), (1.21) can be written as

$$N(\Theta_{10}, \Theta_2) = \max_a \frac{(a' s_1)^2}{a' I_{11} a} \tag{1.22}$$

subject to $I_{21}a = 0$, where $\Theta_{10} = (\theta_{10}, \ldots, \theta_{q0})', \Theta_2 = (\theta_{q+1}, \ldots, \theta_p)'$. Using standard algebra, the optimum N is obtained as

$$N(\Theta_{10}, \Theta_2) = (S_1 - I_{12} I_{22}^{-1} S_2)'(I_{1.2})^{-1}(S_1 - I_{12} I_{22}^{-1} S_2) \tag{1.23}$$

where $I_{1.2} = I_{11} - I_{12} I_{22}^{-1} I_{21}$.

Neyman chose $\sqrt{n}$ as the consistent estimate of Θ_2 to obtain his statistic

$$N = N\left(\Theta_{10}, \tilde{\Theta}_2\right). \tag{1.24}$$

This form of the N statistic, obtained as a generalization of Neyman's single parameter test, is called the Neyman–Rao test by Hall and Mathiason (1990). The asymptotic distribution of N as in (1.24) is chi-square on q degrees of freedom. If Θ_2 is estimated by the constrained ml method, the test reduces to the RS test (1.19).

1.4 Some Examples of the RS Test

Godfrey (1988) gives a comprehensive account of the applications of the RS test in econometrics. A few examples mentioned in the paper by Bera and Ullah (1991) are as follows.

Chi-square goodness-of-fit: Given a parametric specification of the cell probabilities in a multinomial distribution, Pearson developed the chi-square goodness-of-fit test based on observed frequencies. This test can be seen to be the RS test of a composite hypothesis [Rao (1948)].

Linear model: The analysis of the linear model $y_i = x_i'\beta + \epsilon_i$, $i = 1, \ldots, n$, is based on four basic assumptions: correct linear functional form, normality of the distribution of the error term, homoscedasticity and serial independence. The RS test for normality has been derived by Bera and Jarque (1981), for homoscedasticity by Breusch and Pagan (1979). for serial independence by Breusch (1978) and Godfrey (1978a,b) and for linearity by Byron (1968).

For further examples and interpretation of several well-known tests in terms of the score functions, reference may be made to Bera and Ullah (1991) and the papers in the special issue on Rao's score test, Vol. **97**, pp. 1–200 of *Journal of Statistical Planning and Inference* (2001).

1.5 Some Advantages of the RS Test

1. In general, it is simple to compute the RS statistic as it depends only on estimates of parameters under H_0.

2. The test is invariant under transformation of the parameters, unlike the Wald test (see Section 1.6 for examples). Transformation of parameters may simplify the estimation of parameters without effecting the value of the statistic.

3. The RS test has the same local efficiency as the Wald and LR tests.

4. The distribution of RS is not affected by parameters being on the boundary of the parameter space under H_0. In such a case the LR test, and in some cases the W test, is not applicable.

5. There are situations where nuisance parameters are not identifiable under H_0 leading to singular information matrix. In such cases the RS test can be suitably modified as illustrated in Davies (1977, 1987).

1.6 Some Anomalies

1.6.1 Behavior of the power function

The LR, W and RS tests are consistent in the sense that for a fixed alternative to the null hypothesis the power tends to unity as the sample size $n \to \infty$. However, for a fixed sample size, the power function may not be monotonically increasing with increase in the distance (defined in some sense) of the alternative hypothesis from the null.

Example 1.6.1 Let $x_1, \ldots, x_n$ be an iid sample from the Cauchy distribution with density $\pi^{-1}[1 + (x - \theta)^2]^{-1}$.

The RS test for $H_0 : \theta = \theta_0$ against the alternatives $\theta > \theta_0$ rejects when

$$2\sqrt{\frac{2}{n}}\sum_{i=1}^{n}\frac{(x_i - \theta_0)}{1 + (x_i - \theta_0)^2} \geq u_\alpha. \tag{1.25}$$

As the alternative $\theta \to \infty$, $\min(x_i - \theta_0) \to \infty$ in probability, so that for fixed n, the left-hand side of (1.25) tends to zero. Since $u_\alpha > 0$ (for $\alpha < 1/2$), the power of the test as $\theta \to \infty$ for fixed n tends to zero. [See Lehmann (1999, p. 532) for further details].

Example 1.6.2 Let $x_1, \dots, x_n$ be independent binary response variables such that

$$P(x_i = 1) = \left[1 + \exp\left(-\sum_{i=1}^{q}\beta_i z_{ij}\right)\right]^{-1}, \quad i = 1, \dots, n, \tag{1.26}$$

where $z_{i1} = 1$ and $z_{i1}, \dots, z_{iq}$ are observations on q covariables. To test the hypothesis $H_0 : \beta_q = 0$ against the alternative $H : \beta_q \neq 0$, the Wald statistic is

$$W = \hat{\beta}_q^2 / \hat{i}_{qq} \tag{1.27}$$

where $\hat{\beta}_q$ is the ml estimate of β_q and $\hat{i}_{qq}$ is the estimated variance of $\hat{\beta}_q$. Hauck and Donner (1977) show that for fixed n, $W \to 0$ as $\beta_q \to \infty$ for fixed $\beta_1, \dots, \beta_{q-1}$, so that the power of the test decreases as β_q increases. For further examples of such anamolies associated with Wald's statistic, reference may be made to Vaeth (1985) and Le Cam (1990).

The above examples do not contradict the claims made about RS and W about the local power of the tests. Nonetheless, they suggest a caution in the use of these tests [see Mantel (1987)]. It would be of interest to construct an example of the type of anomaly noted above for the RS and W tests in the case of the LR test.

1.6.2 Examples of non-invariance of the Wald test

The Wald test is not invariant for transformations of the parameter while the LR and RS statistics are. Different choices of parameters using the Wald statistic may lead to different inferences.

Example 1.6.3 Consider the likelihood $P(X, \theta)$ based on observed data X and a single unknown parameter θ. Let $\hat{\theta}$ be the ml estimate of θ and $I(\hat{\theta})$, the estimated information.

The Wald statistic for testing the hypothesis $H_0 : \theta = 0$ is

$$\hat{\theta}\sqrt{I(\hat{\theta})} \tag{1.28}$$

which is asymptotically distributed as $N(0,1)$. An equivalent hypothesis is $H_0 : \theta^3 = 0$ and the Wald test based on the parameter θ^3 (using the δ-method to compute the variance of θ^3, Rao (1973, p. 388) is

$$\left(\hat{\theta}^3/3\hat{\theta}^2\right)\sqrt{I(\hat{\theta})} = \frac{\hat{\theta}}{3}\sqrt{I(\hat{\theta})} \tag{1.29}$$

which is asymptotically normal as $N(0,1)$. The p-values based on (1.28) and (1.29) can be quite different.

Example 1.6.4 [Gregory and Veal (1985)]. Consider the linear model

$$y = \beta x + \gamma z + u, \quad u \sim N(0, \sigma^2) \tag{1.30}$$

and tests based on n observations. Let $\hat{\beta}$ and $\hat{\gamma}$ be the maximum likelihood estimates (MLEs) of β and γ with the estimated variance-covariance matrix

$$\hat{\sigma}^2 \begin{pmatrix} w_{11} & w_{12} \\ w_{21} & w_{22} \end{pmatrix}$$

where $\hat{\sigma}$ is the least squares estimate of σ. To test the hypothesis $H_0 : \beta\gamma = 1$, the Wald statistic is

$$\frac{(\hat{\beta}\hat{\gamma} - 1)^2}{\hat{\sigma}^2(\hat{\gamma}^2 w_{11} + 2\hat{\beta}\hat{\gamma}w_{12} + \hat{\beta}^2 w_{22})} \tag{1.31}$$

which is asymptotically chi-square on 1 d.f., while the test for the equivalent hypothesis $\beta = \gamma^{-1}$ is

$$\frac{(\hat{\beta} - \hat{\gamma}^{-1})^2}{\hat{\sigma}^2(w_{11} + 2\hat{\gamma}^{-2}w_{12} + \hat{\gamma}^{-4}w_{22})} = \frac{(\hat{\beta}\hat{\gamma} - 1)^2}{\sigma^2(\hat{\gamma}^2 w_{11} + 2w_{12} + \hat{\gamma}^{-2}w_{22})} \tag{1.32}$$

which is different from (1.31) and is also asymptotically chi-square on 1 d.f.

For another example of non-invariance of Wald's test, reference may be made to Fears, Benichow and Gail (1996) and Pawitan (2000).

1.6.3 Weak dependence of the RS statistic on alternatives to the null hypothesis

In general, when the null hypothesis is rejected, one looks for alternative stochastic models for the observed data. The score test depends on the slope of the likelihood function at the null hypothesis. There may be different likelihoods all giving the same score statistic. If the score test is significant, there is no way of knowing what the alternative is.

Test for normality: Suppose we start with the Pearson family or Gram-Charlier type of distributions and construct a test for normality. The same RS statistic is obtained for both alternatives [Bera and Bilias (2001)].

Test for homoscedasticity: The RS statistic for testing homoscedasticity is the same for alternatives such as multiplicative and additive homoscedasticity [Breusch and Pagan (1979) and Godfrey and Wickens (1981)].

Testing for serial independence: The RS statistic for testing serial independence is the same whether we consider as alternatives the pth order autoregressive or pth order moving average model [Breusch (1978), Godfrey (1978a)].

Such difficulties may exist with other test criteria and it would be of interest to construct some examples.

1.7 Power Comparisons

The following is a summary of numerous papers devoted to power comparisons of LR, W and RS tests.

Taniguchi (1988; 1991): The first-order local powers are the same for all the tests. The second-order local powers are different but no one dominates the other.

Taniguchi (2001): In terms of Bahadur efficiency, they are the same up to the second order.

Bing Li (2001): They are all sensitive to changes in the values of the nuisance parameters.

Chandra and Joshi (1983): Rao's test is more powerful to the order $(1/n)$ than LR and W, when one modifies the critical regions to have the same size up to order $(1/n)$.

Ghosh and Mukherjee (2001): RS is more (or equally) efficient than LR and W under the criteria of maximinity and average local power. See also Mukherjee (1990, 1993) for results on asymptotic efficiency of Rao's Score.

Further investigation of power properties of LR, W and RS tests would be of interest.

1.8 Some Recent Developments

In this Section, we consider some modifications and refinements made on the RS statistic and indicate the need for further research in some cases.

1. In testing a composite hypothesis the estimated score vector $s(\tilde{\theta})$, where $\tilde{\theta}$ is the restricted ml estimate of θ under the hypothesis, is used in computing the RS statistic. It was argued that $s(\tilde{\theta})$ is close to zero if the hypothesis is true. But $E[s(\tilde{\theta})]$ may not be zero unless the null hypothesis is a simple one. In such a case Conniffe (1990) suggested the use of the quadratic form

$$\left[s(\tilde{\theta}) - Es(\tilde{\theta})]'J(\tilde{\theta})[s(\tilde{\theta}) - Es(\tilde{\theta})\right] \tag{1.33}$$

   where J is the inverse of the covariance matrix of $s(\tilde{\theta}) - E[s(\tilde{\theta})]$. The computation of (1.33) and its improvement over the RS statistic needs further study.

2. White (1982) developed score type of statistics based on estimating equations and the quasi-likelihood functions. This introduces some robustification in inference procedures. See also Godfrey and Orme (2001).

3. Several authors tried to adjust the RS statistic similar to a Bartlett (1937) type of adjustment to the LR statistic. Harris (1985) suggested an adjustment based on Edgeworth-type expansion. Dean and Lawless (1989) suggested a different type of adjustment in certain models. Ghosh and Mukherjee (2001) developed a method of adjustment when the RS statistic is based on quasilikelihood. This is an area where further research is needed. Reference may also be made to a recent contribution by Tu, Chen and Shi (2004) on Bartlett type correction to the Score test in the Cox regression model.

4. The RS statistic (1.8) for testing a simple hypothesis $H_0 : \theta = \theta_0$ is

$$[s(\theta_0)]' \, [I(\theta_0)]^{-1} \, [s(\theta_0)]$$

   which involves the computation of the information matrix

$$I(\theta_0) = E\left[s(\theta_0)s(\theta_0)'\right].$$

   Instead of $I(\theta_0)$, one could use the $p \times p$ matrix of second derivatives of the log likelihood with a minus sign

$$A(\theta) = -\left(\frac{\partial^2 L}{\partial\theta_r \partial\theta_s}\right) \tag{1.34}$$

leading to the statistic

$$[s(\theta_0)]'\,(A(\theta_0))^{-1}\,[s(\theta_0)]\,. \tag{1.35}$$

Terril (2001) suggests further simplification by using what he calls the gradient statistic

$$F^2 = [s(\theta_0)]'(\hat{\theta} - \theta_0) \tag{1.36}$$

where $\hat{\theta}$ is the ml estimate of θ. The suggestion by Terril is attractive as it is simple to compute. It would be of interest to investigate the performance of the statistic (1.36).

5. In considering the score statistic, Rao (1948) used the ml estimates of parameters. A similar theory can be developed using BAN estimators.

6. Rao (1951) suggested the use of score tests in sequential analysis for testing a simple null versus a simple alternative hypothesis. Bradley (1953) considered a nice application of Rao's sequential test in clinical trials. An application in quality control is given by Box and Ramirez (1992). For some comments on sequential score test and possible applications reference may be made to Sen (1997).

Acknowledgments: I am gratified to see the large number of papers contributed by econometricians on the application of the score statistic to problems in econometrics and the extensions and improvements they have made. My special thanks are due to Professor A. K. Bera whose papers provided me all the information I needed in preparing this chapter.

References

1. Aitchison, J. and Silvey, S. D. (1958). Maximum likelihood estimation of parameters subject to restraints, *Annals of Mathematical Statistics*, **29**, 813–828.

2. Amemiya, T. (1985). *Advanced Econometrics*, Harvard University Press, Boston, Massachusetts.

3. Bartlett, M. S. (1937). Properties of sufficiency and statistical tests, *Proceedings of the Royal Society, Series A*, **160**, 268–282.

4. Bera, A. K. and Bilias, Y. (2001). Rao's score, Neyman's $C(\alpha)$) and Silvey's LM tests: An essay on historical developments and new results, *Journal of Statistical Planning and Inference*, **97**, 9–44.

5. Bera, A. K. and Jarque, C. M. (1981). An efficient large sample test for normality of observations and regression residuals, *Working Papers in Economics and Econometrics*, **40**, The Australian National University, Canberra, Australia.

6. Bera, A. K. and Ullah, A. (1991). Rao's score test in Econometrics, *Journal of Quantitative Economics*, **7**, 189–220.

7. Bickel, P. J. and Doksum, K. A. (2001). *Mathematical Statistics*, Vol. 1, Second Edition, Prentice-Hall, Inglewood Cliffs, New Jersey.

8. Box, G. E. P. and Raminez, J. (1992). Cumulative score charts, *Quality and Reliability Engineering*, **8**, 17–27.

9. Bradley, R. A. (1953). Some statistical methods in taste testing and quality evaluation, *Biometrics*, **9**, 22–38.

10. Breusch, T.S. (1978). Testing for autocorrelation in dynamic linear models, *Australian Economic Papers*, **17**, 334–355.

11. Breusch, T. S. and Pagan, A. R. (1979). A simple test for heteroscedasticity and random coefficient variation, *Econometrics*, **47**, 1287–1294.

12. Breusch, T. S. and Pagan, A. R. (1980). The Lagrange Multiplier test and its application to model specification in econometrics, *Review of Economic Studies*, **47**, 239–253.

13. Byron, R. P. (1968). Methods for estimating demand equations using prior information, A series of experiments with Australian data, *Australian Economic Papers*, **7**, 227–248.

14. Chandra, T. K. and Joshi, S. N. (1983). Comparison of the likelihood ratio, Wald's and Rao's tests, *Sankhyā, Series A*, **45**, 226–246.

15. Chandra, T. K. and Mukherjee, R. (1984). On the optimality of Rao's statistic, *Communications in Statistics—Theory and Methods*, **13**, 1507–1515.

16. Chandra, T. K. and Mukherjee, R. (1985). Comparison of the likelihood ratio, Wald's and Rao's tests, *Sankhyā, Series A*, **47**, 271–284.

17. Chandra, T. K. and Samanta, T. (1988). On second order local comparisons between perturbed maximum likelihood estimators and Rao's statistic as test statistics, *Journal of Multivariate Analysis*, **25**, 201–222.

18. Conniffe, D. (1990). Testing hypotheses with estimated scores, *Biometrika*, **77**, 97–106.

19. Davies, R. B. (1977). Hypothesis testing when a nuisance parameter is present only under the alternative, *Biometrika*, **64**, 247-254.

20. Davies, R. B. (1987). Hypothesis testing when a nuisance parameter is present only under the alternative, *Biometrika*, **74**, 33–43.

21. Dean, C. and Lawless, J. F. (1989). Tests for detecting over dispersion in Poisson regression models, *Journal of the American Statistical Association*, **84**, 467–472.

22. Engle, R. F. (1984). Wald, likelihood ratio and Lagrangian Multiplier test in econometrics, In *Handbook of Econometrics*, Vol. 2 (Eds., Z. Griliches and M. Intriligator), North-Holland Science Publishers, Amsterdam.

23. Fears, T. R., Benichow, J. and Gail, M. H. (1996). A remainder of the fallibility of the Wald statistic, *The American Statistician*, **50**, 226–227.

24. Ghosh, J. K. (1991). Higher order asymptotics for the likelihood ratio, Rao's and Wald's test, *Statistics & Probability Letters*, **12**, 505–509.

25. Ghosh, J. K. and Mukherjee, R. (2001). Test statistic arising from quasi-likelihood: Bartlett adjustment and higher-order power, *Journal of Statistical Planning and Inference*, **97**, 45–55.

26. Godfrey, L. G. (1978a). Testing against general autoregressive and moving average models when the regression include lagged dependent variables, *Econometrica*, **46**, 227–236.

27. Godfrey, L. G. (1978b). Testing for higher order serial correlation in regression equations when the regression includes lagged dependent variables, *Econometrica*, **46**, 1303–1310.

28. Godfrey, L. G. (1988). *Misspecification Tests in Econometrics*, Cambridge University Press, London, England.

29. Godfrey, L. G. and Orme, C. D. (2001). On improving the robustness and reliability of Rao's score test, *Journal of Statistical Planning and Inference*, **97**, 153–176.

30. Godfrey, L. G. and Wickens, M. R. (1981). Testing linear and log-linear regression for functional form, *Review of Economic Studies*, **48**, 487–496.

31. Green, W. H. (1990). *Econometric Analysis*, Macmillan Publishers Limited, Hampshire, England.

32. Gregory, A. W. and Veal, M. R. (1985). Formulating Wald tests of nonlinear restrictions, *Econometrica*, **53**, 1465–1468.

33. Hall, W. J. and Mathiason, D. J. (1990). On large sample estimation and testing in parameter models, *International Statistical Review*, **58**, 77–97.

34. Harris, P. (1985). An asymptotic expansion for the null distribution of the efficient score statistics, *Biometrika*, **72**, 653–659.

35. Harvey, A. (1990). *The Econometric Analysis of Time Series*, Philip Allan, Oxford, England.

36. Hauck, W. W. and Donner, A. (1977). Wald test as applied to hypotheses in logit analysis, *Journal of the American Statistical Association*, **72**, 851–853.

37. *Journal of Statistical Planning and Inference*, 2001, Vol. **97**, No. 1, Special Issue on Rao's Score Test (12 papers, pp. 1–200).

38. Judge, G. G., Griffiths, W. E., Hill, R. C., Lütkepohl, H. and Lee, T. C. (1985). *The Theory and Practice of Econometrics*, Second edition, John Wiley & Sons, New York.

39. Kmenta, J. (1986). *Elements of Econometrics*, Second edition, Macmillan Publishers Limited, Hampshire, England.

40. Kramer, W. and Sonnberger, H. (1986). *The Linear Regression Model Test*, Physica-Verlag, Heidelberg, Germany.

41. Le Cam, L. (1990). On the standard asymptotic confidence ellipsoids of Wald, *International Statistical Review*, **58**, 129–152.

42. Lehmann, E. (1999). *Theory of Asymptotic Inference*, Springer-Verlag, New York.

43. Li, Bing (2001). Sensitivity of Rao's score test, the Wald test and the likelihood ratio test to nuisance parameters, *Journal of Statistical Planning and Inference*, **97**, 57–66.

44. Maddala, G. S. (1988). *Introduction to Econometrics*, Macmillan Publishers Limited, Hampshire, England.

45. Mantel, N. (1987). Understanding Wald's test for exponential families, *The American Statistician*, **41**, 147–149.

46. Mukherjee, R. (1990). Comparison of tests in the multiparameter case I: Second order power, *Journal of Multivariate Analysis*, **33**, 17–30.

47. Mukherjee, R. (1993). Rao's score test: Recent asymptotic results, In *Handbook of Statistics 11* (Eds., G. S. Maddala, C. R. Rao and H. Vinod), pp. 363–379, North- Holland Science Publishers, Amsterdam.

48. Neyman, J. (1959). Optimal asymptotic test of composite statistical hypothesis, In *Probability and Statistics* (Ed., U. Grenander), John Wiley & Sons, New York.

49. Neyman, J, (1979). $C(\alpha)$ tests and their uses, *Sankhyā, Series A*, **41**, 1–21.

50. Neyman, J. and Pearson, E. S. (1928). On the use and interpretation of certain test criteria, *Biometrika*, **20A**, 175–240, 263–294.

51. Pawitan, Y. (2000). A remainder of the fallibility of the Wald statistic: Likelihood explanation, *The American Statistician*, **54**, 54–56.

52. Peers, H. W. (1971). Likelihood ratio and associated test criteria, *Biometrika*, **58**, 577–587.

53. Rao, C. R. (1948). Large sample tests of statistical hypotheses concerning several parameters with application to problems of estimation, *Proceedings of the Cambridge Philosophical Society*, **44**, 50–57.

54. Rao, C. R. (1950). Methods of scoring linkage data giving the simultaneous segregation of three factors, *Heredity*, **4**, 37–59.

55. Rao, C .R. (1951). Sequential tests of null hypotheses, *Sankhyā*, **10**, 361–370.

56. Rao, C. R. (1961). A study of large sample test criteria through properties of efficient estimates. *Sankhyā, Series A*, **23**, 25–40.

57. Rao, C. R. (1973). *Linear Statistical Inference and its Applications*, Second edition, John Wiley & Sons, New York.

58. Rao, C. R. and Poti, S. J. (1946). On locally most powerful tests when alternatives are one-sided, *Sankhyā*, **7**, 439–440.

59. Sen, P. K. (1997). Introduction to Rao (1948). In *Breakthroughs in Statistics* (Eds., S. Kotz and N. L. Johnson), Vol. III, Springer-Verlag, New York.

60. Serfling, R. J. (1980). *Approximation Theorems of Mathematical Statistics*, John Wiley & Sons, New York.

61. Silvey, S. D. (1959). The Lagrangian multiplier test, *Annals of Mathematical Statistics*, **30**, 389–407.

62. Spanos, A. (1986). *Statistical Foundations of Econometric Modeling*, Cambridge University Press, London, England.

63. Taniguchi, M. (2001). On large deviation asymptotics of some tests in time series, *Journal of Statistical Planning and Inference*, **97**, 191–200.

64. Taniguchi, M. (1988). Asymptotic expansions of the distribution of some statistics for Gaussian ARMA process, *Journal of Multivariate Analysis*, **27**, 494–511.

65. Taniguchi, M. (1991). *Higher Order Asymptotic Theory for Time Series Analysis*, Lecture Notes in Statistics, Vol. 68, Springer-Verlag, New York.

66. Terril, G. R. (2001). The gradient statistic, *Personal communication.*

67. Tu, D., Chen, J. and Shi, P. (2004). A Bartlett type correction for Rao Score test in Cox regression model, *Technical Report.*

68. Vaeth, M. (1985). On the use of Wald's test in exponential families, *International Statistical Review*, **53**, 199–214.

69. Wald, A. (1943). Tests of statistical hypotheses concerning several parameters when the number of observations is large, *Transactions of the Americal Mathematical Soceity*, **54**, 426–482.

70. White, H. (1982). Maximum likelihood estimation of misspecified models, *Econometrica*, **50**, 1–25.

71. White, H. (1984). *Asymptotic Theory for Econometricians*, Academic Press, New York.

2

EM Algorithm and Optimal Censoring Schemes for Progressively Type-II Censored Bivariate Normal Data

N. Balakrishnan and J.-A. Kim

Department of Mathematics and Statistics, McMaster University, Hamilton, Ontario, Canada

Abstract: The EM algorithm is used to find the maximum likelihood estimates (MLEs) of the parameters of a bivariate normal distribution based on progressively Type-II right censored samples. The asymptotic variances and covariances of the MLEs are derived, using the missing information principle, from the Fisher information matrix as well as from the partially observed information matrix. Optimal censoring schemes are then investigated with respect to minimum trace of the variance-covariance matrix of the MLEs and also with respect to the maximum information about ρ.

Keywords and phrases: EM algorithm, maximum likelihood estimates, concomitants of order statistics, progressive type-II right censoring, asymptotic variances, missing information principle, optimal censoring scheme

2.1 Introduction

In many life-testing experiments, the experimenter may not observe all failure times either unintentionally or intentionally. For example, some of the experimental units may break accidentally, or subjects in clinical trials may drop out for personal reasons in the middle of the trial. In some life-testing studies involving expensive units, it will be beneficial if some of the units placed on the test could be removed early on from the test so that those units could be used for other tests as well. In some clinical trials, duration of survival after a treatment may be many years and the experimenter may terminate the study prior to observing the durations of survival for all the individuals in the trial. Censored data arises in all these situations wherein the experimenter does not obtain complete information for all the units or individuals under study.

Different types of censoring arise based on how the data are collected from the life-testing experiment. Let us consider a life-testing experiment wherein n items are placed on test. Suppose the experiment has to be terminated at a prefixed time, say T. Then one can only obtain failure times which are less than or equal to T, and the data so obtained are called *Type-I censored data.* Instead of prefixing the total time of the experiment, the experimenter may wish to discontinue the experiment after the first r failures are observed. In such a situation, the data are said to be *Type-II censored.* A generalization of this Type-II censoring is called *Progressive Type-II censoring*, which arises as follows. Of the n items placed on a life-test, suppose R_1 functioning items are randomly removed from the test right after the first failure. Similarly, immediately after observing the second failure, R_2 items are randomly removed from the remaining $n - R_1 - 2$ items on the test, and so on until each item is taken care of either due to its failure or due to its removal from the test. The data obtained in this manner are said to be progressively Type-II censored data. Inference under Type-I and Type-II censoring for various parametric families of distributions have been discussed by Nelson (1982), Cohen and Whitten (1988), Balakrishnan and Cohen (1991), and Cohen (1991). Inference for Weibull and exponential distributions under progressive Type-II censoring have been discussed by Mann (1969, 1971), Viveros and Balakrishnan (1994), and Ng, Chan and Balakrishnan (2002). Further references and details on progressive censoring can be found in the book by Balakrishnan and Aggarwala (2000).

Let $(X_1, Y_1), \ldots, (X_n, Y_n)$ be a random sample from a bivariate normal density function, $\phi_\theta(x, y)$, where $\theta = (\mu_X, \sigma_X, \mu_Y, \sigma_Y, \rho)$, (μ_X, μ_Y) are the means, (σ_X, σ_Y) are the standard deviations, and ρ is the correlation coefficient between X_i and Y_i. Suppose X_1, X_2, ..., X_n are the life-times of the n units placed on a life-test, and Y_1, Y_2, ..., Y_n are the corresponding covariates. Prior to the experiment, a number $m < n$ is fixed as the number of complete failures to be observed and the progressive censoring scheme $(R_1, R_2, \ldots, R_m)$ with $R_j \geq 0$ and $\sum_{j=1}^{m} R_j + m = n$ is also pre-specified. During the experiment, immediately after the jth failure is observed, R_j functioning items are randomly removed from the test. The m complete (ordered) life-times thus observed are denoted by $X_{j:m:n}^{(R_1,\ldots,R_m)}$, $j = 1, 2, \ldots, m$. For convenience, sometimes the progressive censoring scheme will be omitted in the notation of the $X_{j:m:n}$s. These completely observed failure times are referred to as progressively Type-II right censored order statistics; see Balakrishnan and Aggarwala (2000). Let $Y_{[j:m:n]} = Y_i$ if $X_{j:m:n} = X_i$ for $j = 1, ..., m$. Then, one can have the concomitants (David, 1973) $Y_{[j:m:n]}$, $j = 1, ..., m$, of the progressively Type-II right censored order statistics, which are also called the induced order statistics [Bhattacharya (1974)]. The exact and asymptotic distribution theory of $Y_{[i:n]}$, concomitants of the usual order statistics from some bivariate distributions, were studied by Yang (1977) and the asymptotic distribution of the concomitants from a bivariate normal distribution were derived by David and Galambos (1974). David and Nagaraja (1998) provided a detailed review of developments on concomitants of order statistics including the role of concomitants in the estimation of regression parameters and correlation coefficient and the analysis of censored bivariate data. Based on a Type-II right censored bivariate normal sample, Harrell and Sen (1979) derived the maximum likelihood estimators (MLEs) of the parameters and their asymptotic variance-covariance matrix. Balakrishnan and Kim (2005) derived the MLEs of the parameters and their asymptotic variance-covariance matrix based on a progressively Type-II right censored bivariate normal data.

The Expectation-Maximization (EM) algorithm [Dempster et al. (1977), McLachlan and Krishnan (1997)] is a useful tool to estimate the parameters of the distribution based on an incomplete data, especially when the estimation based on complete data is relatively easy. In the EM algorithm for incomplete data problems, the parameters are estimated after filling in initial values for the missing data. Then the initial values for missing data are updated by their expected values using the initially estimated parameters. The parameters are then re-estimated, and so on, proceeding iteratively until convergence. The progressively censored data mentioned above can be viewed as an incomplete data and the EM algorithm can then be applied to obtain the MLEs. It should be mentioned that the EM algorithm for progressively Type-II right censored univariate data has been discussed recently by Ng et al. (2002). These authors also used the Missing Information Principle [Orchard and Woodbury (1970), Louis (1982), Tanner (1993)] to derive the asymptotic variances and covariance of the MLEs when the EM algorithm is used for progressively Type-II right censored univariate data. The EM algorithm for Type-II right censored bivariate data has been discussed recently by Balakrishnan and Kim (2004).

In Section 2.2, conditional distributions of censored data given observed data are determined for progressively Type-II right censored samples from a bivariate distribution using which conditional expectations of censored data given observed data are derived for the case of the bivariate normal distribution. In Section 2.3, the EM algorithm for finding the MLE of θ of a bivariate normal distribution is discussed. In Section 2.4, asymptotic variances and covariances of the MLEs are derived using the missing information principle from the Fisher information matrix as well as from the partially observed information matrix. An illustrative example is presented in Section 2.5. Optimal progressive censoring schemes with respect to minimum trace of the variance-covariance matrix of the MLEs and also with respect to maximum information about ρ are examined and presented in Section 2.6 with some comments.

2.2 Conditional Distributions of Concomitants of Order Statistics

For convenience, the following notation will be adopted throughout this chapter:

$$\begin{aligned} F_Z(z) &\quad - \quad \text{cdf of a random variable } Z, \\ f_Z(z) &\quad - \quad \text{pdf of a random variable } Z, \\ f_{Y|X}(y|x) &\quad - \quad \text{conditional pdf of } Y \text{given} X = x, \\ L_Z &\quad - \quad \text{likelihood function based on the random variable} Z. \end{aligned}$$

Let

$$X_{obs} = (X_{1:m:n}, X_{2:m:n}, \ldots, X_{m:m:n})^T$$

and

$$Y_{obs} = (Y_{[1:m:n]}, Y_{[2:m:n]}, \ldots, Y_{[m:m:n]})^T$$

be the observed data, and

$$X_{cen} = (X_{(1)}, X_{(2)}, \ldots, X_{(m)})^T$$

and

$$Y_{cen} = (Y_{(1)}, Y_{(2)}, \ldots, Y_{(m)})^T$$

be the censored data, where $X_{(j)}$ and $Y_{(j)}$ are $1 \times R_j$ vectors with $X_{(j)} = (X_{j1}, X_{j2}, \ldots, X_{jR_j})$ and $Y_{(j)} = (Y_{j1}, Y_{j2}, \ldots, Y_{jR_j})$ for $j = 1, \ldots, m$. Combine (X_{obs}, Y_{obs}) and (X_{cen}, Y_{cen}) to form $(\mathbf{X}, \mathbf{Y})$ which is the complete data, where $\mathbf{X} = (X_{obs}^T, X_{cen}^T)$ and $\mathbf{Y} = (Y_{obs}^T, Y_{cen}^T)$. The joint density of $(\mathbf{X}, \mathbf{Y})$ is given by

$$\begin{aligned} & f_{(\mathbf{X},\mathbf{Y})}(\mathbf{x}, \mathbf{y}) \\ &= C \prod_{j=1}^{m} \prod_{k=1}^{R_j} f_{Y|X}(y_{[j:m:n]}|x_{j:m:n}) f_X(x_{j:m:n}) f_{Y|X}(y_{jk}|x_{jk}) f_X(x_{jk}), \end{aligned} \tag{2.1}$$

where $C = n(n - R_1 - 1) \cdots (n - R_1 - R_2 - \cdots - R_{m-1} - m + 1)$. The joint density of (X_{obs}, Y_{obs}) can be written as

$$\begin{aligned} & f_{(X_{obs}, Y_{obs})}(x_{obs}, y_{obs}) \\ &= C \prod_{j=1}^{m} f_{Y|X}(y_{[j:m:n]}|x_{j:m:n}) f_X(x_{j:m:n}) [1 - F_X(x_{j:m:n})]^{R_j}, \\ & \qquad x_{1:m:n} \le \cdots \le x_{m:m:n}, \quad -\infty < y_{[j:m:n]} < \infty. \end{aligned} \tag{2.2}$$

From (2.1) and (2.2), the conditional joint distribution of censored data, given observed data, can be written as

$$\begin{aligned} f_{(X_{cen},Y_{cen})|(X_{obs},Y_{obs})}&(x_{cen}, y_{cen}|x_{obs}, y_{obs}) \\ &= \prod_{j=1}^{m}\prod_{k=1}^{R_j} f_{Y|X}(y_{jk}|x_{jk})\frac{f_X(x_{jk})}{[1-F_X(x_{j:m:n})]}, \\ &\quad x_{j:m:n} \le x_{jk} < +\infty, \quad -\infty < y_{jk} < +\infty. \end{aligned} \tag{2.3}$$

Therefore, the density function of (X_{jk}, Y_{jk}), given (X_{obs}, Y_{obs}), $j = 1, 2, \ldots, m$ and $k = 1, 2, \ldots, R_j$, is

$$\begin{aligned} f_{(X_{jk},Y_{jk})|(X_{obs},Y_{obs})}&(x_{jk}, y_{jk}|x_{obs}, y_{obs}) \\ &= f_{(X_{jk},Y_{jk})|X_{j:m:n}}(x_{jk}, y_{jk}|X_{j:m:n} = x_{j:m:n}) \\ &= f_{Y|X}(y_{jk}|x_{jk})\frac{f_X(x_{jk})}{[1-F_X(x_{j:m:n})]}, \\ &\quad x_{j:m:n} \le x_{jk} < +\infty, \quad -\infty < y_{jk} < +\infty. \end{aligned} \tag{2.4}$$

From (2.4), we can derive the first and second moments of Y_{jk} and the product moment of X_{jk} and Y_{jk}, given (X_{obs}, Y_{obs}), when $(\mathbf{X}, \mathbf{Y})$ are from a bivariate normal distribution with parameter $\theta = (\mu_X, \sigma_X, \mu_Y, \sigma_Y, \rho)$, as follows:

$$E[Y_{jk}|X_{j:m:n} = x_{j:m:n}] = \mu_Y + \rho\sigma_Y Q_j, \tag{2.5}$$

$$E[Y_{jk}^2|X_{j:m:n} = x_{j:m:n}] = \sigma_Y^2(1 + \rho^2 z_j Q_j) + 2\rho\sigma_Y\mu_Y Q_j + \mu_Y^2, \tag{2.6}$$

$$\begin{aligned} E[X_{jk}Y_{jk}&|X_{j:m:n} = x_{j:m:n}] \\ &= \mu_Y(\mu_X + \sigma_X Q_j) + \rho\sigma_Y(\mu_X Q_j + \sigma_X + \sigma_X z_j Q_j), \end{aligned} \tag{2.7}$$

where $z_j = \frac{x_{j:m:n} - \mu_X}{\sigma_X}$ and $Q_j = \frac{\phi(z_j)}{1-\Phi(z_j)}$ with $\phi(\cdot)$ and $\Phi(\cdot)$ denoting the standard normal pdf and cdf, respectively.

2.3 The EM Algorithm

The progressively Type-II right censored bivariate normal data can be viewed as an incomplete data and the EM algorithm can then be applied to determine the MLE of the parameter θ. The log-likelihood function $l(\theta; \mathbf{x}, \mathbf{y})$ based on the complete data $(\mathbf{X}, \mathbf{Y})$ is

$$\begin{aligned}
l(\theta;\mathbf{x},\mathbf{y}) &= \text{constant} - n\log\sigma_X - n\log\sigma_Y - \frac{n}{2}\log(1-\rho^2)\\
&\quad -\frac{1}{2(1-\rho^2)}\sum_{j=1}^{n}\left\{\left(\frac{x_j-\mu_X}{\sigma_X}\right)^2 - 2\rho\left(\frac{x_j-\mu_X}{\sigma_X}\right)\left(\frac{y_j-\mu_Y}{\sigma_Y}\right)+\left(\frac{y_j-\mu_Y}{\sigma_Y}\right)^2\right\}\\
&= \text{constant} - n\log\sigma_X - n\log\sigma_Y - \frac{n}{2}\log(1-\rho^2)\\
&\quad -\frac{1}{2(1-\rho^2)}\sum_{j=1}^{m}\left\{\left(\frac{x_{j:m:n}-\mu_X}{\sigma_X}\right)^2 - 2\rho\left(\frac{x_{j:m:n}-\mu_X}{\sigma_X}\right)\left(\frac{-y_{[j:m:n]}-\mu_Y}{\sigma_Y}\right)\right.\\
&\quad \left. +\left(\frac{y_{[j:m:n]}-\mu_Y}{\sigma_Y}\right)^2\right\}\\
&\quad -\frac{1}{2(1-\rho^2)}\sum_{j=1}^{m}\sum_{k=1}^{R_j}\left\{\left(\frac{x_{jk}-\mu_X}{\sigma_X}\right)^2 - 2\rho\left(\frac{x_{jk}-\mu_X}{\sigma_X}\right)\left(\frac{y_{jk}-\mu_Y}{\sigma_Y}\right)\right.\\
&\quad \left. +\left(\frac{y_{jk}-\mu_Y}{\sigma_Y}\right)^2\right\},
\end{aligned}$$

where (x_{jk}, y_{jk}), $j = 1, \ldots, m$, $k = 1, \ldots, R_j$, are the censored data. Based on the complete data, it is well-known that the MLE of $\theta = (\mu_X, \sigma_X, \mu_Y, \sigma_Y, \rho)$ is given by [see Kotz, Balakrishnan and Johnson (2000, p. 294)]

$$\hat{\mu}_X = \frac{1}{n}\left[\sum_{j=1}^{m} X_{j:m:n} + \sum_{j=1}^{m}\sum_{k=1}^{R_j} X_{jk}\right], \tag{2.8}$$

$$\hat{\sigma}_X = \left[\frac{1}{n}\left\{\sum_{j=1}^{m}(X_{j:m:n}-\hat{\mu}_X)^2 + \sum_{j=1}^{m}\sum_{k=1}^{R_j}(X_{jk}-\hat{\mu}_X)^2\right\}\right]^{\frac{1}{2}}, \tag{2.9}$$

$$\hat{\mu}_Y = \frac{1}{n}\left[\sum_{j=1}^{m} Y_{[j:m:n]} + \sum_{j=1}^{m}\sum_{k=1}^{R_j} Y_{jk}\right], \tag{2.10}$$

$$\hat{\sigma}_Y = \left[\frac{1}{n}\left\{\sum_{j=1}^{m}(Y_{[j:m:n]}-\hat{\mu}_Y)^2 + \sum_{j=1}^{m}\sum_{k=1}^{R_j}(Y_{jk}-\hat{\mu}_Y)^2\right\}\right]^{\frac{1}{2}}, \tag{2.11}$$

$$\hat{\rho} = \frac{\frac{1}{n}\left\{\sum_{j=1}^{m}(X_{j:m:n}-\hat{\mu}_X)(Y_{[j:m:n]}-\hat{\mu}_Y)+\sum_{j=1}^{m}\sum_{k=1}^{R_j}(X_{jk}-\hat{\mu}_X)(Y_{jk}-\hat{\mu}_Y)\right\}}{\hat{\sigma}_X\hat{\sigma}_Y}. \tag{2.12}$$

Hence, in the $(h+1)$th iteration of the EM algorithm, the values of $\hat{\mu}_X^{(h+1)}$, $\hat{\sigma}_X^{(h+1)}$, $\hat{\mu}_Y^{(h+1)}$, $\hat{\sigma}_Y^{(h+1)}$ and $\hat{\rho}^{(h+1)}$ are calculated as follows:

$$\hat{\mu}_X^{(h+1)} = \frac{1}{n}\left[\sum_{j=1}^{m} x_{j:m:n} + \sum_{j=1}^{m} R_j E(X_{jk}|X_{jk} \geq x_{j:m:n}, \hat{\mu}_X^{(h)}, \hat{\sigma}_X^{(h)})\right], \tag{2.13}$$

$$\hat{\sigma}_X^{(h+1)} = \Big[\tfrac{1}{n}\Big\{\textstyle\sum_{j=1}^m x_{j:m:n}^2 + \sum_{j=1}^m R_j E(X_{jk}^2|X_{jk} \geq x_{j:m:n}, \hat{\mu}_X^{(h+1)}, \hat{\sigma}_X^{(h)})\Big\} \\ \qquad - (\hat{\mu}_X^{(h+1)})^2\Big]^{\frac{1}{2}}, \tag{2.14}$$

$$\hat{\mu}_Y^{(h+1)} = \tfrac{1}{n}\Big[\textstyle\sum_{j=1}^m y_{[j:m:n]} \\ + \textstyle\sum_{j=1}^m R_j E(Y_{jk}|X_{j:m:n} = x_{j:m:n}, \hat{\mu}_X^{(h+1)}, \hat{\sigma}_X^{(h+1)}, \hat{\mu}_Y^{(h)}, \hat{\sigma}_Y^{(h)}, \hat{\rho}^{(h)})\Big], \tag{2.15}$$

$$\begin{aligned} \hat{\sigma}_Y^{(h+1)} &= \Big[\tfrac{1}{n}\Big\{\textstyle\sum_{j=1}^m y_{[j:m:n]}^2 \\ &\quad + \textstyle\sum_{j=1}^m R_j E(Y_{jk}^2|X_{j:m:n} = x_{j:m:n}, \hat{\mu}_X^{(h+1)}, \hat{\sigma}_X^{(h+1)}, \hat{\mu}_Y^{(h+1)}, \hat{\sigma}_Y^{(h)}, \hat{\rho}^{(h)})\Big\} \\ &\quad -(\hat{\mu}_Y^{(h+1)})^2\Big]^{\frac{1}{2}}, \end{aligned} \tag{2.16}$$

$$\hat{\rho}^{(h+1)} = \frac{A - n\hat{\mu}_X^{(h+1)}\hat{\mu}_Y^{(h+1)}}{BC}, \tag{2.17}$$

where

$$\begin{aligned} A &= \sum_{j=1}^m x_{j:m:n} y_{[j:m:n]} \\ &\quad + \sum_{j=1}^m R_j E\left[X_{jk}Y_{jk}|(X_{obs}, Y_{obs}), \hat{\mu}_X^{(h+1)}, \hat{\sigma}_X^{(h+1)}, \hat{\mu}_Y^{(h+1)}, \hat{\sigma}_Y^{(h+1)}, \hat{\rho}^{(h)}\right], \\ B &= \left\{\sum_{j=1}^m x_{j:m:n}^2 + \sum_{j=1}^m R_j E[X_{jk}^2|X_{jk} \geq x_{j:m:n}, \hat{\mu}_X^{(h+1)}, \hat{\sigma}_X^{(h+1)}] - n(\hat{\mu}_X^{(h+1)})^2\right\}^{\frac{1}{2}}, \end{aligned}$$

and

$$\begin{aligned} C &= \Big\{\sum_{j=1}^m y_{[j:m:n]}^2 \\ &\quad + \sum_{j=1}^m R_j E[Y_{jk}^2|X_{j:m:n} = x_{j:m:n}, \hat{\mu}_X^{(h+1)}, \hat{\sigma}_X^{(h+1)}, \hat{\mu}_Y^{(h+1)}, \hat{\sigma}_Y^{(h+1)}, \hat{\rho}^{(h)}] \\ &\quad -n(\hat{\mu}_Y^{(h+1)})^2\Big\}^{\frac{1}{2}}. \end{aligned}$$

The first and second moments of X_{jk}, given $X_{jk} \geq x_{j:m:n}$, are given by [see Cohen (1991, p. 10)]

$$E(X_{jk}|X_{jk} \geq x_{j:m:n}, \hat{\mu}_X, \hat{\sigma}_X) = \hat{\sigma}_X Q_j + \hat{\mu}_X, \tag{2.18}$$

$$E(X_{jk}^2|X_{jk} \geq x_{j:m:n}, \hat{\mu}_X, \hat{\sigma}_X) = \hat{\sigma}_X^2(1 + z_j Q_j) + 2\hat{\sigma}_X\hat{\mu}_X Q_j + \hat{\mu}_X^2, \tag{2.19}$$

where z_j and Q_j are as defined earlier (with μ_X and σ_X replaced by $\hat{\mu}_X$ and $\hat{\sigma}_X$, respectively). Now, by using Eqs. (2.5), (2.6), (2.7), (2.18) and (2.19) in Eqs. (2.13)–(2.17), we can find the MLE of θ via the EM algorithm.

By the fact that $(\hat{\mu}_X, \hat{\sigma}_X)$ is the MLE of (μ_X, σ_X) based on X_{obs} alone [see Balakrishnan and Kim (2005)], the EM algorithm can be separated into two parts, with one iterating for $(\hat{\mu}_X, \hat{\sigma}_X)$ and the other for $(\hat{\mu}_Y, \hat{\sigma}_Y, \hat{\rho})$, which results in reducing the computational time.

2.4 Asymptotic Variances and Covariances

The Fisher information of complete data, $I_{comp}(\theta)$, for a bivariate normal distribution is known to be [see Kotz, Balakrishnan and Johnson (2000, p. 294)]

$$\begin{aligned} I_{comp}(\theta) & \\ = & -E\left[\frac{\partial^2 l(\theta; \mathbf{X}, \mathbf{Y})}{\partial\theta^2}\right] \\ = & \ n\begin{pmatrix} \frac{1}{\sigma_X^2(1-\rho^2)} & 0 & -\frac{\rho}{\sigma_X\sigma_Y(1-\rho^2)} & 0 & 0 \\ & \frac{2-\rho^2}{\sigma_X^2(1-\rho^2)} & 0 & -\frac{\rho^2}{\sigma_X\sigma_Y(1-\rho^2)} & -\frac{\rho}{\sigma_X(1-\rho^2)} \\ & & \frac{1}{\sigma_Y^2(1-\rho^2)} & 0 & 0 \\ & & & \frac{2-\rho^2}{\sigma_Y^2(1-\rho^2)} & -\frac{\rho}{\sigma_Y(1-\rho^2)} \\ & & & & \frac{1+\rho^2}{(1-\rho^2)^2} \end{pmatrix}. \end{aligned} \tag{2.20}$$

The log-likelihood function for (X_{jk}, Y_{jk}) which is the censored data at the time of the jth failure, given the observed data, can be derived from (2.4) to be

$$\begin{aligned} & l_{miss,j}(\theta, (X_{jk}, Y_{jk}) | X_{j:m:n} = x_{j:m:n}) \\ & = \text{constant} - \log\sigma_Y - \frac{1}{2}\log(1-\rho^2) \\ & -\frac{1}{2\sigma_Y^2(1-\rho^2)}\left\{y_{jk} - \mu_Y - \frac{\rho\sigma_Y}{\sigma_X}(x_{jk} - \mu_X)\right\}^2 \\ & -\log\sigma_X - \frac{1}{2\sigma_X^2}(x_{jk} - \mu_X)^2 - \log\{1 - \Phi(z_j)\}, \\ & \qquad x_{jk} \geq x_{j:m:n}, \ -\infty < y_{jk} < \infty. \end{aligned} \tag{2.21}$$

Then, the missing data information, $I_{miss,j}(\theta)$, in one observation (X_{jk}, Y_{jk}), is given by

$$I_{miss,j}(\theta) = E_{X_{j:m:n}}\left[E\left(-\frac{\partial^2 l_{miss,j}(\theta, (X_{jk}, Y_{jk}) | X_{j:m:n} = x_{j:m:n})}{\partial\theta^2}\right)\right]$$

$$= E_{X_{j:m:n}} \begin{pmatrix} I^{*,11}_{miss,j} & I^{*,12}_{miss,j} & I^{*,13}_{miss,j} & I^{*,14}_{miss,j} & I^{*,15}_{miss,j} \\ & I^{*,22}_{miss,j} & I^{*,23}_{miss,j} & I^{*,24}_{miss,j} & I^{*,25}_{miss,j} \\ & & I^{*,33}_{miss,j} & I^{*,34}_{miss,j} & I^{*,35}_{miss,j} \\ & & & I^{*,44}_{miss,j} & I^{*,45}_{miss,j} \\ & & & & I^{*,55}_{miss,j} \end{pmatrix} \quad (2.22)$$

$$= E_{X_{j:m:n}} \left[I^{*}_{miss,j} \right], \quad (2.23)$$

where

$$I^{*,11}_{miss,j} = \frac{\rho^2}{\sigma_X^2(1-\rho^2)} + \frac{1+z_jQ_j-Q_j^2}{\sigma_X^2}, \quad I^{*,12}_{miss,j} = \frac{\rho^2 Q_j}{\sigma_X^2(1-\rho^2)} + \frac{Q_j(1+z_j^2-z_jQ_j)}{\sigma_X^2},$$

$$I^{*,13}_{miss,j} = -\frac{\rho}{\sigma_X\sigma_Y(1-\rho^2)}, \quad I^{*,14}_{miss,j} = -\frac{\rho^2 Q_j}{\sigma_X\sigma_Y(1-\rho^2)},$$

$$I^{*,15}_{miss,j} = -\frac{\rho Q_j}{\sigma_X(1-\rho^2)}, \quad I^{*,22}_{miss,j} = \frac{\rho^2(1+z_jQ_j)}{\sigma_X^2(1-\rho^2)} + \frac{2+z_jQ_j+z_j^3Q_j-z_j^2Q_j^2}{\sigma_X^2},$$

$$I^{*,23}_{miss,j} = -\frac{\rho Q_j}{\sigma_X\sigma_Y(1-\rho^2)}, \quad I^{*,24}_{miss,j} = -\frac{\rho^2(1+z_jQ_j)}{\sigma_X\sigma_Y(1-\rho^2)},$$

$$I^{*,25}_{miss,j} = -\frac{\rho(1+z_jQ_j)}{\sigma_X(1-\rho^2)}, \quad I^{*,33}_{miss,j} = \frac{1}{\sigma_Y^2(1-\rho^2)},$$

$$I^{*,34}_{miss,j} = \frac{\rho Q_j}{\sigma_Y^2(1-\rho^2)}, \quad I^{*,35}_{miss,j} = \frac{Q_j}{\sigma_Y(1-\rho^2)},$$

$$I^{*,44}_{miss,j} = \frac{2}{\sigma_Y^2} + \frac{\rho^2(1+z_jQ_j)}{\sigma_Y^2(1-\rho^2)}, \quad I^{*,45}_{miss,j} = -\frac{\rho(1-z_jQ_j)}{\sigma_Y(1-\rho^2)},$$

$$I^{*,55}_{miss,j} = \frac{2\rho^2}{(1-\rho^2)^2} + \frac{1+z_jQ_j}{(1-\rho^2)}.$$

Let $\psi_j^{(abcd)}$ be defined by

$$\psi_j^{(abcd)} = E_{X_{j:m:n}} \left[\frac{\{\phi(Z_j)\}^a Z_j^d}{\{\Phi(Z_j)\}^b \{1-\Phi(Z_j)\}^c} \right].$$

By taking expectation of $I^{*,kl}_{miss,j}$ with respect to the jth progressively censored order statistic $X_{j:m:n}$, we can obtain the information for the missing data as

$$I_{miss,j}(\theta) = \begin{pmatrix} I^{11}_{miss,j} & I^{12}_{miss,j} & I^{13}_{miss,j} & I^{14}_{miss,j} & I^{15}_{miss,j} \\ & I^{22}_{miss,j} & I^{23}_{miss,j} & I^{24}_{miss,j} & I^{25}_{miss,j} \\ & & I^{33}_{miss,j} & I^{34}_{miss,j} & I^{35}_{miss,j} \\ & & & I^{44,j}_{miss} & I^{45,j}_{miss} \\ & & & & I^{55,j}_{miss} \end{pmatrix}, \quad (2.24)$$

where

$$I^{11}_{miss,j} = \frac{\rho^2}{\sigma_X^2(1-\rho^2)} + \frac{1+\psi_j^{(1011)}-\psi_j^{(2020)}}{\sigma_X^2}, \quad I^{12}_{miss,j} = \frac{\rho^2\psi_j^{(1010)}}{\sigma_X^2(1-\rho^2)} + \frac{(\psi_j^{(1010)}+\psi_j^{(1012)}-\psi_j^{(2021)})}{\sigma_X^2},$$

$$I^{13}_{miss,j} = -\frac{\rho}{\sigma_X\sigma_Y(1-\rho^2)}, \quad I^{14}_{miss,j} = -\frac{\rho^2\psi_j^{(1010)}}{\sigma_X\sigma_Y(1-\rho^2)},$$

$$I^{15}_{miss,j} = -\frac{\rho\psi_j^{(1010)}}{\sigma_X(1-\rho^2)}, \quad I^{22}_{miss,j} = \frac{\rho^2(1+\psi_j^{(1011)})}{\sigma_X^2(1-\rho^2)} + \frac{2+\psi_j^{(1011)}+\psi_j^{(1013)}-\psi_j^{(2022)}}{\sigma_X^2},$$

$$I^{23}_{miss,j} = -\frac{\rho\psi_j^{(1010)}}{\sigma_X\sigma_Y(1-\rho^2)}, \quad I^{24}_{miss,j} = -\frac{\rho^2(1+\psi_j^{(1011)})}{\sigma_X\sigma_Y(1-\rho^2)},$$

$$I^{25}_{miss,j} = -\frac{\rho(1+\psi_j^{(1011)})}{\sigma_X(1-\rho^2)}, \quad I^{33}_{miss,j} = \frac{1}{\sigma_Y^2(1-\rho^2)},$$

$$I^{34}_{miss,j} = \frac{\rho\psi_j^{(1010)}}{\sigma_Y^2(1-\rho^2)}, \quad I^{35}_{miss,j} = \frac{\psi_j^{(1010)}}{\sigma_Y(1-\rho^2)},$$

$$I^{44}_{miss,j} = \frac{2}{\sigma_Y^2} + \frac{\rho^2(1+\psi_j^{(1011)})}{\sigma_Y^2(1-\rho^2)}, \quad I^{45}_{miss,j} = -\frac{\rho(1-\psi_j^{(1011)})}{\sigma_Y(1-\rho^2)},$$

$$I^{55}_{miss,j} = \frac{2\rho^2}{(1-\rho^2)^2} + \frac{1+\psi_j^{(1011)}}{(1-\rho^2)}.$$

$\psi_j^{(abcd)}$ can be expressed in terms of moments of the smallest usual order statistic from different sample sizes using the density function of $Z_j = \frac{X_{j:m:n}-\mu_X}{\sigma_X}$ given by Balakrishnan *et al.* (2002) as

$$f_{Z_j}(z_j) = c' \sum_{i=0}^{j-1} c_{i,j-1}(R_1+1,\ldots,R_{j-1}+1) f_Z(z_j)\{1-F_Z(z_j)\}^{R_i''-1}, \quad -\infty < z_j < \infty,$$

where

$$c' = n(n-R_1-1)\cdots(n-R_1-\cdots-R_{j-1}-j+1),$$

$$R_i'' = (R_j^*+1) + \sum_{k=j-i}^{j-1}(R_k+1), \quad R_j^* = n-j-R_1-\cdots-R_{j-1},$$

$$c_{i,r}(\mathbf{a}_r) = \frac{(-1)^i}{\left\{\prod_{l=1}^{i}\sum_{k=r-i+1}^{r-i+l} a_k\right\}\left\{\prod_{l=1}^{r-i}\sum_{k=l}^{r-i} a_k\right\}} \quad \text{for} \quad \mathbf{a}_r = (a_1, a_2, \ldots, a_r).$$

To determine $\psi_j^{(1010)} = E\left[\frac{\phi(Z_j)}{1-\Phi(Z_j)}\right]$, for example, using the above density function of Z_j, we need to find an integral of the form $\int_{-\infty}^{\infty}\{\phi(x)\}^2\{1-\Phi(x)\}^a dx$ for $a > 0$ (an integer), where $\phi(\cdot)$ and $\Phi(\cdot)$ are the standard normal pdf and cdf, respectively. Balakrishnan and Kim (2005) have expressed that

$$\int_{-\infty}^{\infty}\{\phi(x)\}^2\{1-\Phi(x)\}^a dx = -\frac{1}{(a+1)(a+2)}E[Z_{1:a+2}], \tag{2.25}$$

where $Z_{1:a+2}$ is the smallest order statistic in a sample of size $a+2$ from the standard normal distribution. Therefore, $\psi_j^{(1010)}$ can be expressed in terms of expected values of the smallest usual order statistic from different sample sizes from the standard normal distribution. Similarly, $\psi_j^{(1011)}$, $\psi_j^{(1012)}$, $\psi_j^{(1013)}$, $\psi_j^{(2020)}$, $\psi_j^{(2021)}$ and $\psi_j^{(2022)}$ can all be expressed in terms of the first, second, third and fourth moments of the smallest usual order statistic from different sample sizes from the standard normal distribution, with the use of the following formulas:

$$\int_{-\infty}^{\infty} x\,\{\phi(x)\}^2\,\{1-\Phi(x)\}^a dx = \frac{1}{(a+1)(a+2)}\left\{1-E[Z_{1:a+2}^2]\right\}, \tag{2.26}$$

$$\int_{-\infty}^{\infty} x^2\,\{\phi(x)\}^2\,\{1-\Phi(x)\}^a dx = \frac{1}{(a+1)(a+2)}\left\{2E[Z_{1:a+2}]-E[Z_{1:a+2}^3]\right\}, \tag{2.27}$$

$$\int_{-\infty}^{\infty} x^3\,\{\phi(x)\}^2\,\{1-\Phi(x)\}^a dx = \frac{1}{(a+1)(a+2)}\left\{3E[Z_{1:a+2}^2]-E[Z_{1:a+2}^4]\right\}, \tag{2.28}$$

$$\int_{-\infty}^{\infty} \{\phi(x)\}^3\,\{1-\Phi(x)\}^a dx = -\frac{2}{(a+1)(a+2)(a+3)}\left\{1-E[Z_{1:a+3}^2]\right\}, \tag{2.29}$$

$$\int_{-\infty}^{\infty} x\,\{\phi(x)\}^3\,\{1-\Phi(x)\}^a dx = -\frac{1}{(a+1)(a+2)(a+3)}\left\{5E[Z_{1:a+3}]-2E[Z_{1:a+3}^3]\right\}, \tag{2.30}$$

$$\int_{-\infty}^{\infty} x^2\,\{\phi(x)\}^3\,\{1-\Phi(x)\}^a dx = \frac{1}{(a+1)(a+2)(a+3)}\left\{2-8E[Z_{1:a+3}^2]+2E[Z_{1:a+3}^4]\right\}. \tag{2.31}$$

From (2.20) and (2.24), we can obtain the Fisher information matrix for the observed data as

$$I = I_{comp} - \sum_{j=1}^{m} R_j I_{miss,j}. \tag{2.32}$$

By inverting the Fisher information matrix above, we can obtain the asymptotic variance-covariance matrix of the MLE of θ as

$$V(\hat{\theta}) = \begin{pmatrix} V_{11} & V_{12} & V_{13} & V_{14} & V_{15} \\ & V_{22} & V_{23} & V_{24} & V_{25} \\ & & V_{33} & V_{34} & V_{35} \\ & & & V_{44} & V_{45} \\ & & & & V_{55} \end{pmatrix}, \tag{2.33}$$

Table 2.1: Simulated progressively Type-II censored samples from a bivariate normal distribution

R=(25.0,. . . ,0)				
(13.9534, 5.2440)	(21.1502.10.5074)	(23.6538, 7.1682)	(27.5243.10.1348)	(29.1558, 9.2772)
(18.7286.10.4529)	(21.4592.10.4308)	(23.7090.10.1993)	(28.1226, 9.3579)	(29.8880, 9.8495)
(18.7648, 5.8012)	(23.1105. 9.0242)	(25.9404.11.9220)	(28.3778,11.4975)	(31.5465,10.6816)
(19.5882, 5.3824)	(23.2390. 7.9312)	(26.6553,10.4963)	(28.6203. 8.2270)	(34.8827,12.1457)
(20.6655, 7.7746)	(23.4543,10.8688)	(27.4103. 8.9235)	(28.7939,11.2662)	(35.8427,10.9435)
R=(5.5.5.5.5.0,. . . ,0)				
(13.9534, 5.2440)	(19.7357. 6.0263)	(22.5961, 8.5164)	(25.9404.11.9220)	(29.1832, 8.8002)
(16.6970, 4.8205)	(20.6655. 7.7746)	(23.0072, 9.7782)	(27.4103. 8.9235)	(29.8880. 9.8495)
(18.7286.10.4529)	(21.1502,10.5074)	(23.1105, 9.0242)	(27.5243.10.1348)	(30.5462, 9.2462)
(18.7648. 5.8012)	(21.6218.11.0130)	(24.0908, 9.4917)	(28.1226. 9.3579)	(31.5465,10.6816)
(19.5882, 5.3824)	(22.4708. 8.1816)	(24.3373, 8.0105)	(28.7939,11.2662)	(32.4081,10.6757)

where the V-matrix is the inverse of the I-matrix evaluated at $\theta = \hat{\theta}$.

Alternatively, the information matrix for the observed data can also be obtained directly from the likelihood function of the observed data in (2.2), and the corresponding expressions can be found in Balakrishnan and Kim (2005). It can be shown that

$$\textstyle\sum_{j=1}^{m} R_j \psi_j^{(1010)} = -\sum_{j=1}^{m} E[Z_j], \qquad \sum_{j=1}^{m} R_j \psi_j^{(1011)} = m - \sum_{j=1}^{m} E[Z_j^2], \tag{2.34}$$

where $Z_j = \frac{X_{j:m:n} - \mu_X}{\sigma_X}$; see the Appendix for a proof. Using the relationships in (2.34), it can be shown that the information matrix presented by Balakrishnan and Kim (2005) and the information matrix derived in (2.32) by the use of the missing information principle are exactly the same. If $I^*_{miss,j}$ is used in (2.32) instead of $I_{miss,j}$, then the partially observed information matrix $I^* = I_{comp} - \sum_{j=1}^{m} R_j I^*_{miss,j}$ is obtained. Note that in this case there is no need to calculate $\psi_j^{(abcd)}$ values [as required in the computation of I in (2.32)].

2.5 Illustrative Example

To illustrate the proposed EM algorithm for the computation of the MLE of θ, a progressively Type-II right censored bivariate normal data with $n = 50$, $m = 25$ and $\theta = (25, 5, 10, 2, 0.6)$ were generated. Progressive censoring schemes $(25, 0, \ldots, 0)$ and $(5, 5, 5, 5, 5, 0, \ldots, 0)$ were applied. The samples obtained are as follows.

A computer program has been written in FORTRAN to compute the MLE of θ via the EM algorithm. Sample means, standard deviations and the sample correlation coefficient are used as initial estimates for the parameters and the level of accuracy of 1×10^{-8} was used in the EM algorithm. After 35 iterations, the estimates of (μ_X, σ_X) converged to $\mu_{X(\infty)} = 25.4999$ and $\sigma_{X(\infty)} = 4.9449$ for the data with the progressive censoring scheme $(25, 0, \ldots, 0)$. The EM algorithm for (μ_Y, σ_Y, ρ), using $\mu_{X(\infty)}$ and $\sigma_{X(\infty)}$, yielded $\mu_{Y(\infty)} = 9.4513$, $\sigma_{Y(\infty)} = 1.8946$ and $\rho_{(\infty)} = 0.6200$ after 24 iterations. The information matrices evaluated at $\hat{\theta}$ were

$$I_{comp}(\hat{\theta}) = \begin{pmatrix} 3.3218 & 0 & -5.3754 & 0 & 0 \\ & 5.3666 & 0 & -3.3328 & -10.1843 \\ & & 22.6278 & 0 & 0 \\ & & & 36.5571 & -26.5806 \\ & & & & 182.6682 \end{pmatrix},$$

$$\sum_{j=1}^{m} R_j I_{miss,j}(\hat{\theta}) = \begin{pmatrix} 1.5680 & 0.2577 & -2.6877 & -0.0765 & -0.2337 \\ & 2.1723 & -0.1234 & -1.5211 & -4.6480 \\ & & 11.3139 & 0.3220 & 0.9839 \\ & & & 17.8992 & -14.4495 \\ & & & & 87.7917 \end{pmatrix},$$

and

$$\begin{aligned} I(\hat{\theta}) &= I_{comp}(\hat{\theta}) - \sum_{j=1}^{m} R_j I_{miss,j}(\hat{\theta}) \\ &= \begin{pmatrix} 1.7538 & -0.2577 & -2.6877 & 0.0765 & 0.2337 \\ & 3.1944 & 0.1234 & -1.8118 & -5.5363 \\ & & 11.3139 & -0.3220 & -0.9839 \\ & & & 18.6579 & -12.1310 \\ & & & & 94.8765 \end{pmatrix}. \end{aligned}$$

From this, we obtain the asymptotic variance-covariance matrix of $\hat{\theta} = (\hat{\mu}_X, \hat{\sigma}_X, \hat{\mu}_Y, \hat{\sigma}_Y, \hat{\rho})$ as

$$I^{-1}(\hat{\theta}) = \begin{pmatrix} 0.9137 & 0.0835 & 0.2171 & 0.0123 & 0.0064 \\ & 0.4076 & 0.0198 & 0.0600 & 0.0315 \\ & & 0.1401 & 0.0052 & 0.0027 \\ & & & 0.0674 & 0.0121 \\ & & & & 0.0139 \end{pmatrix}.$$

On the other hand, we obtain the asymptotic variance-covariance matrix of $\hat{\theta}$ from the partially observed information matrix I^* as

$$(I^*)^{-1}(\hat{\theta}) = \begin{pmatrix} 0.9315 & 0.0667 & 0.2213 & 0.0098 & 0.0051 \\ & 0.4121 & 0.0158 & 0.0607 & 0.0318 \\ & & 0.1410 & 0.0037 & 0.0019 \\ & & & 0.0682 & 0.0126 \\ & & & & 0.0142 \end{pmatrix}.$$

Note that there are slight differences between $I^{-1}(\hat{\theta})$ and $(I^*)^{-1}(\hat{\theta})$, but the advantage of the use of $(I^*)^{-1}(\hat{\theta})$ is that it did not require the determination of $\psi_j^{(abcd)}$ values.

For the progressively Type-II right censored sample with the censoring scheme $R = (5,5,5,5,5,0,\ldots,0)$, the estimates of (μ_X, σ_X) converged to $\mu_{X(\infty)} = 24.6950$ and $\sigma_{X(\infty)} = 4.4439$ after 73 EM-iterations. The EM algorithm for (μ_Y, σ_Y, ρ), using $\mu_{X(\infty)}$ and $\sigma_{X(\infty)}$, yielded $\mu_{Y(\infty)} = 9.0029$, $\sigma_{Y(\infty)} = 1.9092$ and $\rho_{(\infty)} = 0.6293$ after 36 iterations. The information matrices evaluated at $\hat{\theta}$ were

$$I_{comp}(\hat{\theta}) = \begin{pmatrix} 4.1923 & 0 & -6.1412 & 0 & 0 \\ & 6.7242 & 0 & -3.8649 & -11.7248 \\ & & 22.7133 & 0 & 0 \\ & & & 36.4305 & -27.2909 \\ & & & & 191.3843 \end{pmatrix},$$

$$\sum_{j=1}^{m} R_j I_{miss,j}(\hat{\theta}) = \begin{pmatrix} 1.8300 & 0.6586 & -3.0706 & -0.2666 & -0.8087 \\ & 2.4282 & -0.4236 & -1.5798 & -4.7926 \\ & & 11.3566 & 0.9859 & 2.9909 \\ & & & 17.3944 & -16.1355 \\ & & & & 88.1383 \end{pmatrix},$$

and

$$\begin{aligned} I(\hat{\theta}) &= I_{comp}(\hat{\theta}) - \sum_{j=1}^{m} R_j I_{miss,j}(\hat{\theta}) \\ &= \begin{pmatrix} 2.3623 & -0.6586 & -3.0706 & 0.2666 & 0.8087 \\ & 4.2959 & 0.4236 & -2.2851 & -6.9322 \\ & & 11.3566 & -0.9859 & -2.9909 \\ & & & 19.0360 & -11.1555 \\ & & & & 103.2460 \end{pmatrix}. \end{aligned}$$

From this, we obtain the asymptotic variance-covariance matrix of $\hat{\theta} = (\hat{\mu}_X, \hat{\sigma}_X, \hat{\mu}_Y, \hat{\sigma}_Y, \hat{\rho})$ as

$$I^{-1}(\hat{\theta}) = \begin{pmatrix} 0.6931 & 0.1138 & 0.1874 & 0.0194 & 0.0097 \\ & 0.3204 & 0.0308 & 0.0545 & 0.0274 \\ & & 0.1402 & 0.0118 & 0.0059 \\ & & & 0.0658 & 0.0110 \\ & & & & 0.0128 \end{pmatrix}.$$

If we use the partially observed information matrix I^* instead of I, we obtain the asymptotic variance-covariance matrix of $\hat{\theta}$, $(I^*)^{-1}$, as

$$(I^*)^{-1}(\hat{\theta}) = \begin{pmatrix} 0.6920 & 0.1159 & 0.1871 & 0.0197 & 0.0099 \\ & 0.3195 & 0.0313 & 0.0544 & 0.0273 \\ & & 0.1401 & 0.0119 & 0.0060 \\ & & & 0.0656 & 0.0109 \\ & & & & 0.0127 \end{pmatrix}.$$

Once again, the asymptotic variance-covariance matrices of $\hat{\theta}$ from the Fisher information matrix I and the partially observed information matrix I^* are nearly the same.

2.6 Optimal Censoring Scheme

In this section, we discuss optimal censoring schemes in terms of minimum trace of the variance-covariance matrix of the MLEs and also maximum information about ρ.

From Eqs. (2.32) and (2.34), the Fisher information for ρ can be determined as

$$I(\rho) = \frac{2m\rho^2}{(1-\rho^2)^2} + \frac{\sum_{i=1}^{m} E[Z_j^2]}{1-\rho^2}, \tag{2.35}$$

where, as before, $Z_j = \frac{X_{j:m:n} - \mu_X}{\sigma_X}$. The Fisher information about ρ will be affected by the progressive censoring scheme only through the term $E[Z_j^2]$ and monotonically increases as ρ increases for a fixed progressive censoring scheme. The Fisher information about ρ was calculated for various n, m, and ρ values and for all the possible progressive censoring schemes for each n and m chosen. The results for the best, worst, left and right censoring schemes are presented in Table 2.2 for $n = 10$ and $n = 20$ along with the relative efficiency of the censoring scheme compared to the best progressive censoring scheme (one which gives the most information about ρ among all the possible progressive censoring schemes). When ρ is large, say 0.9, the relative efficiency of the worst censoring scheme is at least 96% for all n and m considered, indicating that the censoring scheme does not seem to matter much in this case. However, the relative efficiency of the worst censoring scheme decreases as ρ decreases, and the loss could be as high as 27.8% in this case. Further, the progressive censoring scheme $(n-m, 0, \ldots, 0)$ is one of the best progressive censoring schemes in almost all the cases. Efficiency of the progressive censoring scheme $(n-m, 0, \ldots, 0)$ does decrease by 0.46-8.15% as ρ decreases and censoring proportion increases. On the other hand, the progressive censoring scheme $(0, \ldots, 0, n-m)$ seems to be nearly the worst censoring scheme in most cases.

In order to investigate the performance of the two extreme censoring schemes further, the trace of the variance-covariance matrix [obtained by inverting the Fisher information matrix, I, in (20)] was determined for each of the n, m, and ρ values. The results for the best, worst (right), and left censoring schemes are presented in Table 2.3 along with their relative efficiencies compared to the best progressive censoring scheme (one which gives the least trace value among all possible progressive censoring schemes). It seems that heavy censoring at the beginning of the experiment is more efficient than heavy censoring towards the end of the experiment. The progressive censoring scheme $(0, \ldots, 0, n-m)$ turns out to be the worst censoring scheme in the sense that it gives the largest trace value of the variance-covariance matrix for all combinations of n and m considered. Efficiency loss in using the censoring scheme $(0, \ldots, 0, n-m)$ increases as ρ decreases.

Therefore, from both these considerations, it is observed that Type-II right censoring is not to be recommended in general, and that the progressive censoring scheme $(n-m, 0, \ldots, 0)$ is nearly as efficient as the best progressive censoring scheme for each combination of n and m considered. So, if the censoring can be implemented by the experimenter, then the scheme $(n-m, 0, \cdots, 0)$ will be the best choice for inferential purposes.

Table 2.2: Information about ρ

n	m	Censoring Scheme	$\rho = 0.9$ (1)	$\rho = 0.9$ (2)	$\rho = 0.7$ (1)	$\rho = 0.7$ (2)	$\rho = 0.5$ (1)	$\rho = 0.5$ (2)	$\rho = 0.3$ (1)	$\rho = 0.3$ (2)	$\rho = 0.1$ (1)	$\rho = 0.1$ (2)	$\sum_{i=1}^{m} E[Z_i^2]$
10	9	0,1,7*0†	452.5158	1.0000	52.0300	1.0000	20.3216	1.0000	12.1114	1.0000	9.5182	1.0000	9.2412
		8*0,1	442.2355	0.9773	48.2000	0.9264	17.7172	0.8718	9.9650	0.8228	7.5452	0.7927	7.2879
		1,8*0	452.2478	0.9994	51.9301	0.9981	20.2536	0.9967	12.0554	0.9954	9.4667	0.9946	9.1902
		8*0,1	442.2355	0.9773	48.2000	0.9264	17.7172	0.8718	9.9650	0.8228	7.5452	0.7927	7.2879
	8	0,2,6*0	403.6466	1.0000	46.7743	1.0000	18.4209	1.0000	11.0602	1.0000	8.7313	1.0000	8.4823
		7*0,2	390.9536	0.9686	42.0455	0.8989	15.2053	0.8254	8.4100	0.7604	6.2952	0.7210	6.0707
		2,7*0	403.1105	0.9987	46.5745	0.9957	18.2851	0.9926	10.9482	0.9899	8.6284	0.9882	8.3805
		7*0,2	390.9536	0.9686	42.0455	0.8989	15.2053	0.8254	8.4100	0.7604	6.2952	0.7210	6.0707
	7	0,3,5*0	354.7774	1.0000	41.5186	1.0000	16.5202	1.0000	10.0089	1.0000	7.9444	1.0000	7.7235
		6*0,3	342.8918	0.9665	37.0906	0.8933	13.5092	0.8177	7.5273	0.7521	5.6633	0.7129	5.4652
		3,6*0	353.9732	0.9977	41.2190	0.9928	16.3165	0.9877	9.8410	0.9832	7.7900	0.9806	7.5707
		6*0,3	342.8918	0.9665	37.0906	0.8933	13.5092	0.8177	7.5273	0.7521	5.6633	0.7129	5.4652
	6	0,4,4*0	305.9082	1.0000	36.2629	1.0000	14.6196	1.0000	8.9577	1.0000	7.1575	1.0000	6.9647
		1,4*0,3	295.8817	0.9672	32.5275	0.8970	12.0795	0.8263	6.8642	0.7663	5.2332	0.7311	5.0596
		4,5*0	304.8359	0.9965	35.8634	0.9890	14.3479	0.9814	8.7338	0.9750	6.9517	0.9712	6.7609
		5*0,4	296.4421	0.9691	32.7363	0.9027	12.2215	0.8360	6.9812	0.7794	5.3407	0.7462	5.1661
	5	0,5,3*0,	257.0390	1.0000	31.0072	1.0000	12.7189	1.0000	7.9064	1.0000	6.3706	1.0000	6.2058
		3,3*0,2	248.7282	0.9677	27.9110	0.9001	10.6135	0.8345	6.1712	0.7805	4.7756	0.7496	4.6268
		5,4*0	255.6987	0.9948	30.5079	0.9839	12.3793	0.9733	7.6266	0.9646	6.1133	0.9596	5.9512
		4*0,5	250.6925	0.9753	28.6428	0.9237	11.1111	0.8736	6.5813	0.8324	5.1525	0.8088	5.0000
20	18	3*0,2,14*0	905.2949	1.0000	104.1580	1.0000	40.7098	1.0000	24.2778	1.0000	19.0869	1.0000	18.5324
		17*0,2	881.9453	0.9742	95.4591	0.9165	34.7946	0.8547	19.4026	0.7992	14.6056	0.7652	14.0959
		2,17*0	904.0239	0.9986	103.6845	0.9955	40.3878	0.9921	24.0124	0.9891	18.8429	0.9872	18.2909
		17*0,2	881.9453	0.9742	95.4591	0.9165	34.7946	0.8547	19.4026	0.7992	14.6056	0.7652	14.0959
	16	3*0,4,12*0	807.8198	1.0000	93.7447	1.0000	36.9752	1.0000	22.2303	1.0000	17.5636	1.0000	17.0647
		15*0,4	779.8012	0.9653	83.3064	0.8887	29.8771	0.8080	16.3802	0.7368	12.1863	0.6938	11.7412
		4,15*0	805.2777	0.9969	92.7977	0.9899	36.3312	0.9826	21.6995	0.9761	17.0757	0.9722	16.5817
		15*0,4	779.8012	0.9653	83.3064	0.8887	29.8771	0.8080	16.3802	0.7368	12.1863	0.6938	11.7412
	14	3*0,6,10*0	710.3446	1.0000	83.3314	1.0000	33.2405	1.0000	20.1827	1.0000	16.0403	1.0000	15.5971
		1,12*0,5	683.9983	0.9629	73.5161	0.8822	26.5661	0.7992	14.6818	0.7274	10.9839	0.6848	10.5912
		6,13*0	706.5316	0.9946	81.9109	0.9830	32.2745	0.9709	19.3866	0.9606	15.3085	0.9544	14.8726
		13*0,6	684.3390	0.9634	73.6430	0.8837	26.6524	0.8018	14.7530	0.7810	11.0493	0.6888	10.6560
	12	3*0,8,8*0	612.8695	1.0000	72.9181	1.0000	29.5059	1.0000	18.1352	1.0000	14.5170	1.0000	14.1294
		4,10*0,4	588.9958	0.9610	64.0240	0.8780	23.4579	0.7950	13.1506	0.7251	9.9352	0.6844	9.5934
		8,11*0	607.7854	0.9917	71.0240	0.9740	28.2179	0.9563	17.0737	0.9415	13.5413	0.9328	13.1634
		11*0,8	592.1544	0.9662	65.2007	0.8942	24.2581	0.8221	13.8101	0.7615	10.5414	0.7261	10.1936
	10	3*0,10,6*0	515.3943	1.0000	62.5048	1.0000	25.7712	1.0000	16.0877	1.0000	12.9937	1.0000	12.6618
		6,8*0,4	494.0588	0.9586	54.5563	0.8728	20.3662	0.7903	11.6330	0.7231	8.8990	0.6849	8.6080
		10,9*0	509.0392	0.9877	60.1372	0.9621	24.1613	0.9375	14.7608	0.9175	11.7741	0.9061	11.4543
		9*0,10	501.3850	0.9728	57.2857	0.9165	22.2222	0.8623	13.1627	0.8182	10.3051	0.7931	10.0000

†: For simplicity of notation, censoring scheme 0, 1, 0, 0, 0, 0, 0, 0, 0 will be denoted as 0,1,7*0.

(1): Information, (2): Relative efficiency

Table 2.3: Trace of variance-covariance matrix

		$\rho = 0.9$			$\rho = 0.7$			$\rho = 0.5$			$\rho = 0.3$			$\rho = 0.1$		
n	m	(1)	(2)	(3)	(1)	(2)	(3)	(1)	(2)	(3)	(1)	(2)	(3)	(1)	(2)	(3)
10	9	3*0,1.5*0	0.3254	1.0000	2*0,1.6*0	0.3522	1.0000	2*0,1.6*0	0.3868	1.0000	2*0,1.6*0	0.4168	1.0000	2*0,1.6*0	0.4338	1.0000
		8*0,1	0.3315	0.9816	8*0,1	0.3667	0.9605	8*0,1	0.4100	0.9432	8*0,1	0.4484	0.9294	8*0,1	0.4706	0.9217
		1,8*0	0.3292	0.9886	1,8*0	0.3550	0.9922	1,8*0	0.3890	0.9942	1,8*0	0.4187	0.9954	1,8*0	0.4356	0.9960
	8	2*0,2.5*0	0.3541	1.0000	2*0,2.5*0	0.3860	1.0000	0,2.6*0	0.4260	1.0000	0,2.6*0	0.4598	1.0000	0,2.6*0	0.4789	1.0000
		7*0,2	0.3696	0.9582	7*0,2	0.4223	0.9140	7*0,2	0.4829	0.8822	7*0,2	0.5361	0.8576	7*0,2	0.5671	0.8445
		2,7*0	0.3605	0.9823	2,7*0	0.3906	0.9880	2,7*0	0.4294	0.9919	2,7*0	0.4629	0.9932	2,7*0	0.4819	0.9937
	7	2*0,3.4*0	0.3923	1.0000	0,3.5*0	0.4306	1.0000	0,3.5*0	0.4767	1.0000	0,3.5*0	0.5158	1.0000	0,3.5*0	0.5378	1.0000
		6*0,3	0.4241	0.9250	6*0,3	0.5028	0.8565	6*0,3	0.5880	0.8107	6*0,3	0.6619	0.7793	6*0,3	0.7049	0.7630
		3,6*0	0.4001	0.9434	3,6*0	0.4359	0.9879	3,6*0	0.4808	0.9914	3,6*0	0.5193	0.9934	3,6*0	0.5410	0.9942
	6	0,4.4*0	0.4450	1.0000	0,4.4*0	0.4913	1.0000	0,4.4*0	0.5467	1.0000	0,4.4*0	0.5932	1.0000	0,4.4*0	0.6192	1.0000
		5*0,4	0.5057	0.8800	5*0,4	0.6247	0.7865	5*0,4	0.7470	0.7319	5*0,4	0.8513	0.6968	5*0,4	0.9116	0.6792
		4,5*0	0.4523	0.9837	4,5*0	0.4958	0.9909	4,5*0	0.5489	0.9960	4,5*0	0.5939	0.9988	4,5*0	0.6192	1.0000
	5	0,5.3*0	0.5214	1.0000	5,4*0	0.5798	1.0000	5,4*0	0.6445	1.0000	5,4*0	0.6986	1.0000	5,4*0	0.7288	1.0000
		4*0,5	0.6341	0.8222	4*0,5	0.8191	0.7078	4*0,5	1.0005	0.6442	4*0,5	1.1520	0.6064	4*0,5	1.2390	0.5882
		5,4*0	0.5253	0.9926	5,4*0	0.5798	1.0000	5,4*0	0.6445	1.0000	5,4*0	0.6986	1.0000	5,4*0	0.7288	1.0000
20	18	7*0,2.10*0	0.1626	1.0000	5*0,2.12*0	0.1760	1.0000	4*0,2.13*0	0.1932	1.0000	4*0,2.13*0	0.2081	1.0000	4*0,2.13*0	0.2166	1.0000
		17*0,2	0.1664	0.9775	17*0,2	0.1847	0.9527	17*0,2	0.2071	0.9329	17*0,2	0.2270	0.9169	17*0,2	0.2386	0.9080
		2,17*0	0.1654	0.9830	2,17*0	0.1783	0.9871	2,17*0	0.1952	0.9898	2,17*0	0.2100	0.9910	2,17*0	0.2185	0.9915
	16	5*0,4,10*0)	0.1768	1.0000	4*0,4,11*0	0.1927	1.0000	4*0,4,11*0	0.2125	1.0000	3*0,4,12*0	0.2294	1.0000	3*0,4,12*0	0.2389	1.0000
		15*0,4	0.1861	0.9499	15*0,4	0.2141	0.8998	15*0,4	0.2460	0.8639	15*0,4	0.2740	0.8371	15*0,4	0.2904	0.8227
		4,15*0	0.1821	0.9709	4,15*0	0.1970	0.9782	4,15*0	0.2162	0.9828	4,15*0	0.2329	0.9848	4,15*0	0.2425	0.9855
	14	4*0,6,9*0	0.1957	1.0000	3*0,6,10*0	0.2147	1.0000	3*0,6,10*0	0.2377	1.0000	2*0,6,11*0	0.2572	1.0000	2*0,6,11*0	0.2680	1.0000
		13*0,6	0.2147	0.9117	13*0,6	0.2573	0.8346	13*0,6	0.3029	0.7849	13*0,6	0.3424	0.7510	13*0,6	0.3655	0.7333
		6,13*0	0.2030	0.9642	6,13*0	0.2206	0.9735	6,13*0	0.2428	0.9791	6,13*0	0.2619	0.9819	6,13*0	0.2728	0.9826
	12	3*0,8,8*0	0.2218	1.0000	2*0,8,9*0	0.2448	1.0000	2*0,8,9*0	0.2718	1.0000	2*0,8,9*0	0.2944	1.0000	2*0,8,9*0	0.3071	1.0000
		11*0,8	0.2581	0.8597	11*0,8	0.3237	0.7561	11*0,8	0.3905	0.6961	11*0,8	0.4473	0.6581	11*0,8	0.4803	0.6393
		8,11*0	0.2303	0.9634	8,11*0	0.2514	0.9736	8,11*0	0.2776	0.9790	8,11*0	0.3000	0.9815	8,11*0	0.3126	0.9824
	10	2*0,10,7*0	0.2593	1.0000	2*0,10,7*0	0.2877	1.0000	0,10,8*0	0.3202	1.0000	0,10,8*0	0.3471	1.0000	0,10,8*0	0.3620	1.0000
		9*0,10	0.3276	0.7914	9*0,10	0.4321	0.6659	9*0,10	0.5334	0.6004	9*0,10	0.6179	0.5617	9*0,10	0.6665	0.5432
		10,9*0	0.2676	0.9689	10,9*0	0.2939	0.9792	10,9*0	0.3256	0.9835	10,9*0	0.3523	0.9850	10,9*0	0.3674	0.9855

(1): Censoring scheme, (2): Trace of the variance-covariance matrix, (3): Relative efficiency

Appendix

Here, as presented earlier in (2.34), we will show that

$$\begin{aligned}\sum_{j=1}^{m} R_j \psi_j^{(1010)} &= -\sum_{j=1}^{m} E[Z_j],\\ \sum_{j=1}^{m} R_j \psi_j^{(1011)} &= m - \sum_{j=1}^{m} E[Z_j^2],\end{aligned}$$

where $Z_j = \frac{X_{j:m:n} - \mu_X}{\sigma_X}$.

PROOF. First, using the joint density function of $Z_{1:m:n}, Z_{2:m:n}, \ldots, Z_{j:m:n}$ given by Balakrishnan and Aggarwala (2000, p. 13) as

$$\begin{aligned}&f_{Z_{1:m:n},Z_{2:m:n},\ldots,Z_{j:m:n}}(z_1, z_2, \ldots, z_j)\\ &\quad = n(n - R_1 - 1)(n - R_1 - R_2 - 2)\cdots(n - R_1 - \cdots - R_{j-1} - j + 1)\\ &\qquad \times \prod_{i=1}^{j-1} \phi(z_i)\{1 - \Phi(z_i)\}^{R_i} \phi(z_j)\{1 - \Phi(z_j)\}^{n-R_1-\cdots-R_{j-1}-j},\\ &\qquad\quad -\infty < z_1 \le z_2 \le \cdots \le z_j < \infty,\end{aligned}$$

we can write

$$\begin{aligned}R_j \psi_j^{(1010)} &= R_j E\left[\frac{\phi(Z_j)}{1 - \Phi(Z_j)}\right]\\ &= R_j \int_{-\infty}^{\infty} \int_{z_1}^{\infty} \cdots \int_{z_{j-1}}^{\infty} n(n - R_1 - 1)(n - R_1 - R_2 - 2)\\ &\quad \times \cdots (n - R_1 - \cdots - R_{j-1} - j + 1)\\ &\quad \times \prod_{i=1}^{j-1} \phi(z_i)\{1 - \Phi(z_i)\}^{R_i} \{\phi(z_j)\}^2\\ &\quad \times \{1 - \Phi(z_j)\}^{n-R_1-\cdots-R_{j-1}-j-1} dz_j \cdots dz_1. \qquad (2.36)\end{aligned}$$

Let $IT_j = \int_{z_{j-1}}^{\infty} \{\phi(z_j)\}^2 \{1 - \Phi(z_j)\}^{n-R_1-\cdots-R_{j-1}-j-1} dz_j$. Then,

$$\begin{aligned}IT_j &= -\frac{1}{(n - R_1 - \cdots - R_{j-1} - j)} \int_{z_{j-1}}^{\infty} \phi(z_j) \frac{d}{dz_j} \{1 - \Phi(z_j)\}^{n-R_1-\cdots-R_{j-1}-j}\\ &= -\frac{1}{(n - R_1 - \cdots - R_{j-1} - j)} \left[\phi(z_j)\{1 - \Phi(z_j)\}^{n-R_1-\cdots-R_{j-1}-j} \Big|_{z_j=z_{j-1}}^{z_j=\infty} \right.\\ &\qquad \left. + \int_{z_{j-1}}^{\infty} z_j \phi(z_j)\{1 - \Phi(z_j)\}^{n-R_1-\cdots-R_{j-1}-j} dz_j \right]\\ &= \frac{1}{(n - R_1 - \cdots - R_{j-1} - j)}\end{aligned}$$

$$\times \Big[\phi(z_{j-1}) \{1-\Phi(z_{j-1})\}^{n-R_1-\cdots-R_{j-1}-j} - \int_{z_{j-1}}^{\infty} z_j \phi(z_j) \{1-\Phi(z_j)\}^{n-R_1-\cdots-R_{j-1}-j} \, dz_j \Big].$$

Hence, (2.36) becomes

$$\begin{aligned} R_j \psi_j^{(1010)} &= \frac{R_j(n-R_1-\cdots-R_{j-1}-j+1)}{(n-R_1-\cdots-R_{j-1}-j)} \\ &\quad \times \int_{-\infty}^{\infty}\int_{z_1}^{\infty}\cdots\int_{z_{j-2}}^{\infty} n(n-R_1-1)\cdots(n-R_1-\cdots-R_{j-2}-j+2) \\ &\quad \times \prod_{i=1}^{j-2} \phi(z_i)\{1-\Phi(z_i)\}^{R_i} \{\phi(z_{j-1})\}^2 \{1-\Phi(z_{j-1})\}^{n-R_1-\cdots-R_{j-2}-j} \\ &\qquad dz_{j-1}\cdots dz_1 \\ &\quad - \frac{R_j}{(n-R_1-\cdots-R_{j-1}-j)} E(Z_j). \end{aligned}$$

Integrating by parts

$$IT_{j-1} = \int_{z_{j-2}}^{\infty} \{\phi(z_{j-1})\}^2 \{1-\Phi(z_{j-1})\}^{n-R_1-\cdots-R_{j-2}-j} \, dz_{j-1},$$

we get $R_j \psi_j^{(1010)}$ as

$$\begin{aligned} & R_j \psi_j^{(1010)} \\ &= \frac{R_j(n-R_1-\cdots-R_{j-1}-j+1)(n-R_1-\cdots-R_{j-2}-j+2)}{(n-R_1-\cdots-R_{j-1}-j)(n-R_1-\cdots-R_{j-2}-j+1)} \\ &\quad \times \int_{-\infty}^{\infty}\int_{z_1}^{\infty}\cdots\int_{z_{j-3}}^{\infty} n(n-R_1-1)(n-R_1-R_2-2) \\ &\quad \times\cdots(n-R_1-\cdots-R_{j-3}-j+3) \\ &\quad \times \prod_{i=1}^{j-3} \phi(z_i)\{1-\Phi(z_i)\}^{R_i} \{\phi(z_{j-2})\}^2 \\ &\quad \times \{1-\Phi(z_{j-2})\}^{n-R_1-\cdots-R_{j-3}-j+1} dz_{j-2}\cdots dz_1 \\ &\quad - \frac{R_j(n-R_1-\cdots-R_{j-1}-j+1)}{(n-R_1-\cdots-R_{j-1}-j)(n-R_1-\cdots-R_{j-2}-j+1)} E(Z_{j-1}) \\ &\quad - \frac{R_j}{(n-R_1-\cdots-R_{j-1}-j)} E(Z_j). \end{aligned}$$

By repeating this process, $R_j \psi_j^{(1010)}$ can finally be written as

$$\begin{aligned} & R_j \psi_j^{(1010)} = \\ & - \frac{R_j(n-R_1-\cdots-R_{j-1}-j+1)(n-R_1-\cdots-R_{j-2}-j+2)\cdots(n-R_1-1)}{(n-R_1-\cdots-R_{j-1}-j)(n-R_1-\cdots-R_{j-2}-j+1)\cdots(n-R_1-2)(n-1)} E(Z_1) \end{aligned}$$

$$
\begin{aligned}
&- \frac{R_j(n-R_1-\cdots-R_{j-1}-j+1)(n-R_1-\cdots-R_{j-2}-j+2)\cdots(n-R_1-R_2-2)}{(n-R_1-\cdots-R_{j-1}-j)(n-R_1-\cdots-R_{j-2}-j+1)\cdots(n-R_1-2)} \\
&\qquad \times E(Z_2) \\
&- \frac{R_j(n-R_1-\cdots-R_{j-1}-j+1)(n-R_1-\cdots-R_{j-2}-j+2)\cdots(n-R_1-R_2-R_3-3)}{(n-R_1-\cdots-R_{j-1}-j)(n-R_1-\cdots-R_{j-2}-j+1)\cdots(n-R_1-R_2-3)} \\
&\qquad \times E(Z_3) \\
&- \cdots \\
&- \frac{R_j(n-R_1-\cdots-R_{j-1}-j+1)(n-R_1-\cdots-R_{j-2}-j+2)}{(n-R_1-\cdots-R_{j-1}-j)(n-R_1-\cdots-R_{j-2}-j+1)(n-R_1-\cdots-R_{j-3}-j+2)} \\
&\qquad \times E(Z_{j-2}) \\
&- \frac{R_j(n-R_1-\cdots-R_{j-1}-j+1)}{(n-R_1-\cdots-R_{j-1}-j)(n-R_1-\cdots-R_{j-2}-j+1)} E(Z_{j-1}) \\
&- \frac{R_j}{(n-R_1-\cdots-R_{j-1}-j)} E(Z_j).
\end{aligned}
$$

Therefore, the summation of $R_j\psi_j^{(1010)}$ for $j = 1, \ldots, m$ will end up as a linear combination of $E(Z_j)$ for $j = 1, \ldots, m$, and so all we need to do is show that the coefficient of $E(Z_j)$ is -1 for $j = 1, \ldots, m$. The term $E(Z_m)$ comes only from $R_m\psi_m^{(1010)}$ and its coefficient is $-\frac{R_m}{(n-R_1-\cdots-R_{m-1}-m)} = -1$ since $n - m = R_1 + R_2 + \cdots + R_m$. In general, the coefficient of $E(Z_j)$ is given by

$$
\begin{aligned}
&- \frac{R_j}{(n-R_1-\cdots-R_{j-1}-j)} \\
&-[R_{j+1}(n-R_1-\cdots-R_j-j)] \times [(n-R_1-\cdots-R_j-j-1)(n-R_1-\cdots-R_{j-1}-j)]^{-1} \\
&-[R_{j+2}(n-R_1-\cdots-R_{j+1}-j-1)(n-R_1-\cdots-R_j-j)] \\
&\quad \times [(n-R_1-\cdots-R_{j+1}-j-2)(n-R_1-\cdots-R_j-j-1)(n-R_1-\cdots-R_{j-1}-j)]^{-1} \\
&- \cdots \\
&-[R_{m-1}(n-R_1-\cdots-R_{m-2}-m+2)\cdots(n-R_1-\cdots-R_{j+1}-j-1)(n-R_1-\cdots-R_j-j)] \\
&\quad \times [(n-R_1-\cdots-R_{m-2}-m+1)(n-R_1-\cdots-R_{m-3}-m+2)\cdots \\
&\quad\quad \cdots(n-R_1-\cdots-R_j-j-1)(n-R_1-\cdots-R_{j-1}-j)]^{-1} \\
&-[R_m(n-R_1-\cdots-R_{m-1}-m+1)(n-R_1-\cdots-R_{m-2}-m+2)\cdots(n-R_1-\cdots-R_j-j)] \\
&\quad \times [(n-R_1-\cdots-R_{m-1}-m)(n-R_1-\cdots-R_{m-2}-m+1)\cdots \\
&\quad\quad \cdots(n-R_1-\cdots-R_j-j-1)(n-R_1-\cdots-R_{j-1}-j)]^{-1}, \qquad (2.37)
\end{aligned}
$$

which is simply -1 by using the relationship $n - R_1 - R_2 - \cdots - R_{m-1} - m = R_m$.

Next, let us write

$$
\begin{aligned}
R_j\psi_j^{(1011)} &= R_j E\left[\frac{Z_j\phi(Z_j)}{1-\Phi(Z_j)}\right] \\
&= R_j \int_{-\infty}^{\infty}\int_{z_1}^{\infty}\cdots\int_{z_{j-1}}^{\infty} n(n-R_1-1)(n-R_1-R_2-2) \\
&\quad \times \cdots (n-R_1-\cdots-R_{j-1}-j+1) \\
&\quad \times \prod_{i=1}^{j-1} \phi(z_i)\{1-\Phi(z_i)\}^{R_i} z_j\{\phi(z_j)\}^2 \\
&\quad \times \{1-\Phi(z_j)\}^{n-R_1-\cdots-R_{j-1}-j-1} dz_j \cdots dz_1. \qquad (2.38)
\end{aligned}
$$

Let $IT_j^* = \int_{z_{j-1}}^{\infty} z_j \{\phi(z_j)\}^2 \{1-\Phi(z_j)\}^{n-R_1-\cdots-R_{j-1}-j-1} dz_j$. Then,

$$\begin{aligned}
IT_j^* &= -\frac{1}{(n-R_1-\cdots-R_{j-1}-j)} \\
&\quad \times \int_{z_{j-1}}^{\infty} z_j\phi(z_j)\frac{d}{dz_j}\{1-\Phi(z_j)\}^{n-R_1-\cdots-R_{j-1}-j} \\
&= -\frac{1}{(n-R_1-\cdots-R_{j-1}-j)} \\
&\quad \times \Bigg[z_j\phi(z_j)\{1-\Phi(z_j)\}^{n-R_1-\cdots-R_{j-1}-j}\Big|_{z_j=z_{j-1}}^{z_j=\infty} \\
&\qquad - \int_{z_{j-1}}^{\infty} \left\{\phi(z_j) - z_j^2\phi(z_j)\right\}\{1-\Phi(z_j)\}^{n-R_1-\cdots-R_{j-1}-j} dz_j \Bigg] \\
&= \frac{1}{(n-R_1-\cdots-R_{j-1}-j)} \\
&\quad \times \Bigg[z_{j-1}\phi(z_{j-1})\{1-\Phi(z_{j-1})\}^{n-R_1-\cdots-R_{j-1}-j} \\
&\qquad + \int_{z_{j-1}}^{\infty} \phi(z_j)\{1-\Phi(z_j)\}^{n-R_1-\cdots-R_{j-1}-j} dz_j \\
&\qquad - \int_{z_{j-1}}^{\infty} z_j^2\phi(z_j)\{1-\Phi(z_j)\}^{n-R_1-\cdots-R_{j-1}-j} dz_j \Bigg].
\end{aligned}$$

Eq. (2.38) then becomes

$$\begin{aligned}
R_j\psi_j^{(1011)} &= \frac{R_j(n-R_1-\cdots-R_{j-1}-j+1)}{(n-R_1-\cdots-R_{j-1}-j)} \\
&\quad \times \int_{-\infty}^{\infty}\int_{z_1}^{\infty}\cdots\int_{z_{j-2}}^{\infty} n(n-R_1-1)\cdots(n-R_1-\cdots-R_{j-2}-j+2) \\
&\quad \times \prod_{i=1}^{j-2}\phi(z_i)\{1-\Phi(z_i)\}^{R_i} z_{j-1}\{\phi(z_{j-1})\}^2 \\
&\qquad \times \{1-\Phi(z_{j-1})\}^{n-R_1-\cdots-R_{j-2}-j} dz_{j-1}\cdots dz_1. \\
&\quad + \frac{R_j}{(n-R_1-\cdots-R_{j-1}-j)}\left\{1-E(Z_j^2)\right\}.
\end{aligned}$$

Integrating by parts $IT_{j-1}^*, \cdots, IT_1^*$, we obtain $R_j\psi_j^{(1011)}$ as

$$\begin{aligned}
&R_j\psi_j^{(1011)} \\
&= [R_j(n-R_1-\cdots-R_{j-1}-j+1)(n-R_1-\cdots-R_{j-2}-j+2)\cdots(n-R_1-1) \\
&\qquad \times\{1-E(Z_1^2)\}] \\
&\quad \times((n-R_1-\cdots-R_{j-1}-j)(n-R_1-\cdots-R_{j-2}-j+1)\cdots(n-R_1-2)(n-1))^{-1} \\
&+ [R_j(n-R_1-\cdots-R_{j-1}-j+1)(n-R_1-\cdots-R_{j-2}-j+2)\cdots(n-R_1-R_2-2) \\
&\qquad \times\{1-E(Z_2^2)\}] \\
&\quad \times((n-R_1-\cdots-R_{j-1}-j)(n-R_1-\cdots-R_{j-2}-j+1)\cdots(n-R_1-2))^{-1}
\end{aligned}$$

$$
\begin{aligned}
+ \quad & [R_j(n - R_1 - \cdots - R_{j-1} - j + 1)(n - R_1 - \cdots - R_{j-2} - j + 2) \cdots \\
& \qquad \cdots (n - R_1 - R_2 - R_3 - 3)]\{1 - E(Z_3^2)\} \\
& \times((n - R_1 - \cdots - R_{j-1} - j)(n - R_1 - \cdots - R_{j-2} - j + 1) \cdots (n - R_1 - R_2 - 3))^{-1} \\
+ \quad & \cdots \\
+ \quad & [R_j(n - R_1 - \cdots - R_{j-1} - j + 1)(n - R_1 - \cdots - R_{j-2} - j + 2)\{1 - E(Z_{j-2}^2)\}] \\
& \times((n - R_1 - \cdots - R_{j-1} - j)(n - R_1 - \cdots - R_{j-2} - j + 1) \\
& \qquad \times(n - R_1 - \cdots - R_{j-3} - j + 2))^{-1} \\
+ \quad & [R_j(n - R_1 - \cdots - R_{j-1} - j + 1)\{1 - E(Z_{j-1}^2)\}] \\
& \times((n - R_1 - \cdots - R_{j-1} - j)(n - R_1 - \cdots - R_{j-2} - j + 1))^{-1} \\
+ \quad & \frac{R_j}{(n - R_1 - \cdots - R_{j-1} - j)}\{1 - E(Z_j^2)\}.
\end{aligned}
$$

So, the summation of $R_j\psi_j^{(1011)}$ for $j = 1, \ldots, m$ will end up as a linear combination of $1 - E(Z_j^2)$ for $j = 1, \ldots, m$. The collection of coefficients for $1 - E(Z_j^2)$ is the same as in (2.37) but with a positive sign, which reduces to 1. This completes the proof. ■

References

1. Balakrishnan, N. and Aggarwala, R. (2000). *Progressive Censoring: Theory, Methods and Applications*, Birkhäuser, Boston.

2. Balakrishnan, N., Childs, A. and Chandrasekar. B. (2002). An efficient computational methods for moments of order statistics under progressive censoring, *Statistics & Probability Letters*, **60**, 359–365.

3. Balakrishnan, N. and Cohen, A. C. (1991). *Order Statistics and Inference: Estimation Methods*, Academic Press, Boston.

4. Balakrishnan, N. and Kim, J.-A. (2004). EM algorithm for Type-II right censored bivariate normal data, In *Advances in Parametric and Semi-Parametric Inference with Applications in Reliability, Survival Analysis and Quality of Life* (Eds., M. S. Nikulin, N. Balakrishnan, M. Mesbah and N. Limnios), pp. 177–210, Birkhäuser, Boston.

5. Balakrishnan, N. and Kim, J.-A (2005). Point and interval estimation for bivariate normal distribution based on progressively Type-II censored data, In *Communications in Statistics—Theory and Methods*, Special Issue in Memory of Milton Sobel (Eds., S. Panchapakesan and P. Chen) (to appear).

6. Bhattacharya, P. K. (1974). Convergence of sample paths of normalized sums of induced order statistics, *Annals of Statistics*, **2**, 1034–1039.

7. Cohen, A. C. (1991). *Truncated and Censored Samples: Theory and Applications*, Marcel Dekker, New York.

8. Cohen, A. C. and Whitten, B. J. (1988). *Parameter Estimation in Reliability and Life Span Models*, Marcel Dekker, New York.

9. David, H. A. (1973). Concomitants of order statistics, *Bulletin of the International Statistical Institute*, **45**, 295–300.

10. David, H. A. and Galambos, J. (1974). The asymptotic theory of concomitants of order statistics, *Journal of Applied Probability*, **11**, 762–770.

11. David, H. A. and Nagaraja, H. N. (1998). Concomitants of order statistics, In *Handbook of Statistics 16: Order Statistics: Theory and Methods* (Eds., N. Balakrishnan and C. R. Rao), pp. 487–513, North-Holland Academic Publishers, Amsterdam.

12. Dempster, A. P., Laird, N. M. and Rubin, D. B. (1977). Maximum likelihood from incomplete data via the EM algorithm (with discussions), *Journal of the Royal Statistical Society, Series B*, **39**, 1–38.

13. Harrell, F. E. and Sen, P. K. (1979). Statistical inference for censored bivariate normal distributions based on induced order statistics, *Biometrika*, **66**, 293–298.

14. Joshi, P. C. and Balakrishnan, N. (1981). An identity for the moments of normal order statistics with applications, *Scandinavian Actuarial Journal*, 203–213.

15. Kotz, S., Balakrishnan, N. and Johnson, N. L. (2000). *Continuous Multivariate Distributions*, Vol. 1, Second edition, John Wiley & Sons, New York.

16. Louis, T. A. (1982). Finding the observed information matrix when using the EM algorithm, *Journal of the Royal Statistical Society, Series B*, **44**, 226–233.

17. Mann, N. R. (1969). Exact three-order-statistic confidence bounds on reliable life for a Weibull model with progressive censoring, *Journal of the American Statistical Association*, **64**, 306–315.

18. Mann, N. R. (1971). Best linear invariant estimation for Weibull parameters under progressive censoring, *Technometrics*, **13**, 521–533.

19. McLachlan, G. J. and Krishnan, T. (1997). *The EM Algorithm and Extensions*, John Wiley & Sons, New York.

20. Nelson, W. (1982). *Applied Life Data Analysis*, John Wiley & Sons, New York.

21. Ng, H. K. T., Chan, P. S. and Balakrishnan, N. (2002). Estimation of parameters from progressively censored data using EM algorithm, *Computational Statistics & Data Analysis*, **39**, 371–386.

22. Orchard, T. and Woodbury, M. A. (1970). A missing information principle: Theory and applications, In *Proceedings of the Sixth Berkeley Symposium on Mathematical Statistics and Probability*, Vol. 1, pp. 697–715.

23. Tanner, M. A. (1993). *Tools for Statistical Inference: Observed Data and Data Augmentation Methods*, Second edition, Springer-Verlag, New York.

24. Viveros, R. and Balakrishnan, N. (1994). Interval estimation of parameters of life from progressively censored data, *Technometrics*, **36**, 84–91.

25. Yang, S. S. (1977). General distribution theory of the concomitants of order statistics, *Annals of Statistics*, **5**, 996–1002.

3

Inference Guided Data Exploration

Greg Yothers and Allan R. Sampson

National Surgical Adjuvant Breast and Bowel Project (NSABP), Pittsburgh, Pennsylvania U.S.A.
Department of Statistics, University of Pittsburgh, Pennsylvania U.S.A.

Abstract: We consider comparing two treatments using a given hypothesis test on the full sample and on all possible subsets, and we separately consider restricting the subsets considered to be those defined based on half-intervals of a covariate. Rather than treating this as a family of hypothesis tests, we instead choose the minimum p-value from the group of hypothesis tests as our test statistic. Simulation is employed to find an approximate critical value to control the type I error for our novel test statistic. These techniques may be used as a rule of thumb for judging the potential significance of a result after a "fishing expedition" has been caried out on a dataset, i.e., a large number of tests of hypothesis were performed on subsets of the data or a subset was selected after inspecting the data. When the technique is restricted to subsets defined based on half-intervals of a covariate, it may be useful as a planned methodology for analyzing an experiment.

Keywords and phrases: Multiple subset testing, selecting population, fishing expedition

3.1 Introduction

For a variety of reasons, researchers are often tempted to look at numerous subset analyses of their data. One situation that commonly motivates this is in non-significant critical scientific studies where the study sponsors can feel compelled to find subpopulations where the treatment under study had an effect. This situation is sometimes refered to as a "fishing expedition". Clearly, formal control of the type I error is difficult when the same hypothesis is tested repeatedly on subsets of the original data. Given the ad hoc nature of the hypotheses tested in this context, formal inference is not really possible, but some sort of "rule of thumb" for deciding whether the most extreme p-value observed during a fishing expedition is likely to occur purely by chance would be quite useful in practice. A related problem arises when the number of formal tests of hypothesis is small, but the subsets which are tested are selected after inspecting plots or summary measures of the data or the raw data itself.

To further motivate our concern about this problem, we note that small biotechnology and pharmaceutical companies which are not well capitalized sometimes have their corporate success depending on the outcome of a single study of their experimental treatment in a particular population. If this study has a negative result, the company may very well want to identify, if possible, a subset of the study population for which their treatment is effective. One goal in doing this is to atempt to secure additional funding for a new study, and perhaps more importantly, to allow their company to continue to exist.

We first consider the problem where the company, in effect, can be viewed as considering all subsets where a treatment comparison is possible. Our rationale for this is that the company may have done much data snooping and considered the effects of numerous variables. In theory, if we could identify all search strategies that were used, we could restrict our subset considerations. Barring this, our conservative approach is to base the first part of our approach on all subsets. In the second part of our approach, we do consider restrictions on the subsets of interest.

The group of hypothesis tests is not viewed as a family of tests. Instead, we choose as our test statistic the minimum p-value from all tests. Our aim is the determination of a critical value to control the type I error for this test statistic. Analogous to the Scheffé procedure, which gives one critical value which allows for comparison of all contrasts in the analysis of variance, we would like to derive a critical p-value which would allow hypothesis testing on all possible pairs of subsets while controlling the type I error at level α.

Let S_t and S_c denote the sets of individuals who received treatment and control, respectively. With samples of size n_t from treatment and n_c from control, the number of possible pairs of unique, unordered, and non-empty subsets is $(2^{n_t} - 1)(2^{n_c} - 1)$. Let S_{ti} and S_{cj} denote the ith and jth subsets from treatment and control. We define P_{ij} to be the p-value from a test of hypothesis on the responses associated with the pair of subsets S_{ti} and S_{cj}. Depending on the test statistic used, the cardinalities of S_{ti} and S_{cj}, denoted by $|S_{ti}|$ and $|S_{cj}|$, typically must be at least 2 for P_{ij} to be defined. The most extreme p-value (P^*) obtained by testing all possible pairs of subsets where each treatment has at least 2 observations would be:

$$P^* = \inf \{P_{ij},\ (i,j) :\ |S_{ti}| \geq 2 \text{ and } |S_{cj}| \geq 2\}. \tag{3.1}$$

Given the large number $(2^{n_t} - 1 - n_t)(2^{n_c} - 1 - n_c)$ of possible pairs of subsets which could be compared in evaluating (3.1), one may wish to consider some restriction criteria. If the universe of subsets for hypothesis testing is restricted, then we would expect to see less extreme p-values than if all possible subsets are considered. One natural restriction criteria would be to increase the cardinality of the sets in (3.1) to a number larger than two.

Our first approach, which considers all pairs of subsets, may find a significant result corresponding to a pair of subsets, but there may be no meaningful way of defining the corresponding populations. To circumvent this, as well as the very small critical p-values, our second restriction criterion chooses subsets of the original sample based on half-intervals of a relevant covariate. For example, if the pertinant covariate is age, one would consider pairs of all subjects in treatment and control older (or younger) than a varying threshold of age.

To amplify on the method based on half-intervals, consider a study concerning treatments for wound healing, where the healing rate may be correlated with initial wound area. It may be that the difference in mean response between treatment and control also depends on initial wound area, where larger wounds heal faster when treated, but treatment has little or no effect on smaller wounds. The choice of inclusion/exclusion criteria in a study in this situation can be problematic. Choosing eligibility criteria which are too liberal with respect to the covariate may include patients where the treatment effect is negligible making it difficult to prove the treatment superior to control for the entire study population. Conversely, choosing eligibility criteria which are too conservative with respect to the covariate may lead to denying therapy to patients who may benefit when the treatment is finally approved. A useful design would include all patients who could potentially benefit from treatment, but the analysis would determine the largest subset with respect to the covariate where treatment is significantly better than control. Such a method would allow the target population for the treatment to be determined by the data rather than being chosen somewhat arbitrarily prior to the trial. We note that there are a number of other approaches that can be used in the half-intervals setting [see Yothers (2003) or Fleiss (1986) for linear responses].

For both all subsets and half-intervals, the Bonferroni procedure clearly would be extremely conservative due to the large number of comparisons. More specialized multiple comparison procedures tend to be targeted at the problem of comparing multiple treatments on the same endpoint or the problem of making the same treatment comparison on multiple endpoints and hence do not apply here [Miller (1981) and Hsu (1996)].

There are parallels in our considerations to a problem in spatial statistics, considered by Worsley (1992), who addressed the problem of finding regions of brain activity using PET scans. Such experiments may include 10^5 voxels or brain scan sub-regions, so the problem of determining which voxels are active is difficult. His approach is not really applicable to our setting due to a lack in our case of an analog for the "distance" between non-overlapping subsets, and dealing with the problem of overlapping subsets. Koziol and Wu (1996) consider a problem where the response is binary and the probability of success is monotonically related to the value of the covariate. They approach finding the region of the covariate where the probability of success differs for values of the covariate less than a threshold value.

The problem of comparing all pairs of subsets of at least a minimum subset sample size is considered in Section 3.2 for normal data with both known and unknown variances. In Section 3.3 we restrict our interest to subsets formed by considering only those observations where a covariate falls in a half interval and consider the two sample t-test, the two sample Wilcoxon test, and an analysis of covariance model. An example is provided in Section 3.4 and we conclude in Section 3.5 with a discussion.

3.2 Inference Guided Data Exploration Restricted by Minimum Subset Sample Size

Define $P^*_{n_{min}}$, the most extreme p-value observed after testing all pairs of subsets with cardinality at least n_{min}, as follows:

$$P^*_{n_{min}} = \inf \{P_{ij},\ (i,j):\ |S_{ti}| \geq n_{min} \text{ and } |S_{cj}| \geq n_{min}\}. \tag{3.2}$$

The number of pairs of subsets is:

$$N_{pairs} = \left[\sum_{r=n_{min}}^{n_t} \binom{n_t}{r} \times \sum_{r=n_{min}}^{n_c} \binom{n_c}{r} \right]. \tag{3.3}$$

A critical value for $P^*_{n_{min}}$ which controls the type I error at level α can be estimated via simulation as follows: (1) sample responses for S_t and S_c from a specified null distribution; (2) calculate $P^*_{n_{min}}$ for the sample; (3) repeat steps 1 and 2 N times; and (4) estimate the critical value as the $(N\ \alpha)$th order statistic from the N values of $P^*_{n_{min}}$.

This procedure, while computationally intensive, is feasible for small sample sizes. Obviously the estimated critical value depends on the model and test statistic, the sample sizes for treatment and control (n_t and n_c), and the minimum subset size (n_{min}).

3.2.1 Two-sided Z-test assuming known variance

As a first step, we consider applying the method of inference guided data exploration restricted by minimum subset sample size to the known variance normal setting with the two sample Z-test as our statistic. For the Z-test, the p-value does not depend on the underlying sample sizes used in the calculation, so that one can directly determine a critical Z-score for our method (the t statistic, for example, has a p-value which depends on the degrees of freedom i.e., sample sizes).

Observe that the Z-statistic depends only on the difference in means and the sub-sample sizes. If we consider all pairs of sub-samples of size n_i from treatment and n_j from control, then the most extreme Z-score for this combination of sub-sample sizes comes from the pair having the largest difference in means. This corresponds to comparing the largest n_i order statistics from treatment with the smallest n_j order statistics from control or vice versa. We generically refer to these two combinations as tests comparing maximal order statistics to minimal order statistics. Clearly we need only consider the $2(n_t - n_{min} + 1)(n_c - n_{min} + 1) - 1$ comparisons where maximal order statistics are compared to minimal order statistics as opposed to the large number in (3.3).

For this setting our simulation results are presented in Table 3.1 in the columns headed Z-test. Due to invariance, the simulations ($N = 10{,}000$) for both the treatment and control samples are drawn from the same null distribution $N(0, 1)$. The table lists the relevant parameters (n_t, n_c, and n_{min}), the critical value for the test statistic to control the type I error at level 0.05 ($|Z|$), and the p-value associated with the critical value (p($|Z|$)).

Note that critical values can be quite extreme when a large number of pairs of subsets are considered. For instance, for $n_t = 40$, $n_c = 40$, and $n_{min} = 20$, we would conceptually consider $[\sum_{r=20}^{40} \binom{40}{r}]^2 \approx 3.83 \times 10^{23}$ comparisons, although our shortcut allows us to calculate the critical value with only $2(n_t - n_{min} + 1)(n_c - n_{min} + 1) - 1 = 881$ comparisons. The critical value for this entry is 6.5656 as compared to 1.96 for a critical value for a single comparison. Thus the evidence must be particularly striking in such a fishing expedition to demonstrate a significant difference. Generally, holding other parameters fixed, the critical p-value tends to decrease (become more extreme) as n_t or n_c increases and increase as n_{min} increases.

3.2.2 Two-sided t-test

The preceding shortcut method does not apply for the two-sided t-statistic. Unlike the Z-statistic, the variance is based on the treatment and control samples under consideration, so that the most extreme statistic for a given pair of sub-sample sizes may not necessarily come from the pair having the largest separation in means. Table 3.1, in the columns headed t-test, provides the results of simulations applying the inference guided data exploration restricted by minimum subset sample size method to normal independent identically distributed samples and using the two-sample two-sided t-test. Due to the large amount of computation time required to run the simulations for larger sample sizes, these tabulations are not as extensive as those for the Z-test. The critical value must be determined based on the p-value for the t-test due to the varying degrees of freedom. The Z-score corresponding to the critical p-value is tabulated only

Table 3.1: Minimum subset size, $\alpha = 0.05$, $N = 10{,}000$

			Z-test		t-test		t-test, upper-bound	
n_t	n_c	n_{min}	$\lvert Z\rvert$	p($\lvert Z\rvert$)	p	Z(p)	p	Z(p)
3	3	2	2.6111	9.025E-03	4.629E-03	2.8318	5.848E-03	2.7562
3	3	3	1.9595	5.005E-02	5.322E-02	1.9331	5.322E-02	1.9331
4	2	2	2.4516	1.422E-02	8.010E-03	2.6517	8.923E-03	2.6150
4	3	2	2.8055	5.023E-03	1.905E-03	3.1047	2.807E-03	2.9882
4	4	2	2.9926	2.766E-03	8.358E-04	3.3407	1.472E-03	3.1802
4	4	4	1.9664	4.926E-02	5.133E-02	1.9487	5.133E-02	1.9487
5	2	2	2.6002	9.316E-03	3.816E-03	2.8930	5.087E-03	2.8015
5	3	2	2.9464	3.215E-03	9.524E-04	3.3042	1.574E-03	3.1606
5	4	2	3.1406	1.686E-03	3.785E-04	3.5547	7.333E-04	3.3768
5	5	2	3.2925	9.929E-04	1.798E-04	3.7458	4.130E-04	3.5316
5	5	4	2.6354	8.403E-03	4.945E-03	2.8106	5.268E-03	2.7902
5	5	5	1.9643	4.950E-02	4.701E-02	1.9862	4.701E-02	1.9862
6	6	2	3.5435	3.948E-04	4.736E-05	4.0683	1.248E-04	3.8365
6	6	3	3.3580	7.852E-04	1.598E-04	3.7753	2.923E-04	3.6220
6	6	4	3.0965	1.958E-03	6.872E-04	3.3946	8.794E-04	3.3265
7	3	2	3.1417	1.680E-03	2.871E-04	3.6267	6.157E-04	3.4246
7	7	4	3.4196	6.272E-04			2.154E-04	3.7002
8	8	2	3.9164	8.989E-05			2.034E-05	4.2612
8	8	4	3.7251	1.953E-04			4.823E-05	4.0640
8	8	8	1.9830	4.737E-02			4.966E-02	1.9629
10	10	2	4.2311	2.326E-05			3.501E-06	4.6390
11	11	2	4.3919	1.123E-05			1.671E-06	4.7897
12	12	2	4.5035	6.685E-06			6.249E-07	4.9834
13	13	2	4.6436	3.423E-06			3.196E-07	5.1115
14	14	2	4.7962	1.617E-06			1.510E-07	5.2513
16	16	14	3.0158	2.563E-03	8.143E-04	3.3479	8.143E-04	3.3479
20	20	10	5.1021	3.360E-07			1.051E-08	5.7224
30	30	15	5.8718	4.312E-07			1.304E-11	6.7682
40	40	20	6.5466	5.886E-11			2.309E-14	7.6321
50	40	35	4.8161	1.464E-06			1.393E-08	5.6743

for convenience of interpretation.

As expected, the p-values (and associated Z-scores) are slightly more extreme for the t-test than the corresponding values for the Z-test. For example, the entry corresponding to $n_t = 16$, $n_c = 16$, and $n_{min} = 14$ has a critical p-value of 8.143×10^{-4} for the t-test, but a less extreme critical p-value of 2.563×10^{-3} for the Z-test. The large amount of computer time required to simulate the critical p-values for the relatively small sample sizes considered for the t-test suggests that this approach is impractical if not impossible for moderate to large sample sizes.

3.2.3 Two-sided t-test, upper bound

Due to the computational requirements to implement the t-test, we examined the consequences of applying the shortcut used for the Z-test to the t-test (i.e., only compare the most extreme n_i order statistics from treatment with the most extreme n_j order statistics from control for each pair of sub-sample sizes). Our simulations show that this approach for a given data set often misses the most extreme p-value obtained by considering all pairs of subsets but overall produces critical p-values which are not terribly liberal. In the columns headed t-test upbound, Table 3.1 provides these resulting critical p-values for the two-sample two-sided t-test. Of necessity, these p-values of course are at least as large as the true t-test extreme p-values. Thus the critical values are upper-bounds.

To examine the applicability of this upper-bound and the Z-test approximation for the critical p-values of the t-test, we compare in Table 3.2 select entries of Table 3.1. For both the t-upper and Z approximations, we list the percentile from the empirical distribution of the minimum p-value of the t-test corresponding the 0.05 level critical p-values from the approximations. Use of the t-upper approximation results in type I error rates ranging from near nominal ($n_t = 5$, $n_c = 5$, and $n_{min} = 4$; $n_t = 6$, $n_c = 6$, and $n_{min} = 4$; and $n_t = 16$, $n_c = 16$, and $n_{min} = 14$) to roughly a doubling of the nominal 0.05 level ($n_t = 6$, $n_c = 6$, and $n_{min} = 2$). The exact nominal level for the entry $n_t = 16$, $n_c = 16$, and $n_{min} = 14$, an apparent coincidence, indicates that for the particular simulations carried out, the fifth percentile of the minimum p-value was the same to the number of decimal places recorded for both the shortcut method and the exhaustive search method. Table 3.2 also allows comparison of the potential methods of approximating the critical p-value for the t-test, the Z-test and the upper bound for the t-test. The Z aproximation is inferior to the t-upper approximation for all entries in Table 3.2. In Table 3.1 we see that, for nearly every combination of parameters, the critical p-value for the Z-test is larger than the necessarily liberal critical value of the t-upper approximation. The approach of using the shortcut approximation for the critical p-value using the t-test, while being slightly liberal, would seem to be a better approximation than using the critical value from the Z-test. Also, since the direct approach employed for the t-test is computationally intractable for even moderate sample sizes, we would recommend this approximation for the critical p-value of the t-test when using the method of inference guided data exploration restricted by minimum subset sample size.

Table 3.2: Selected type I error rates for the 0.05 level t-upper and Z approximations to critical p-value for the t-test

n_t	n_c	n_{min}	t-upper	Z
4	4	2	0.0798	0.1306
5	5	2	0.0972	0.1785
5	5	4	0.0527	0.0777
6	6	2	0.1015	0.2152
6	6	3	0.0770	0.1508
6	6	4	0.0595	0.1061
7	3	2	0.0869	0.1704
16	16	14	0.0500	0.1065

3.3 Inference Guided Data Exploration Restricted by Half-Intervals of a Covariate

Suppose now that we observe a covariate along with the response in our experiment, where one has reason to believe that the size of the treatment effect could be related to the value of the covariate. The notions of our multiple subsets procedure is then useful to find the range of the covariate where there is an effect.

Let k index the order statistics of the observed covariate X, $X_{(1)}, \dots, X_{(n_t+n_c)}$, where here we have combined the covariates of the treatment and control groups. Now, for convenience and without loss of generality, let the sets of individuals receiving treatment and control be denoted by $S_t = \{1, \dots, n_t\}$ and $S_c = \{n_t + 1, \dots, n_t + n_c\}$. Define the subsets of individuals where the covariate is no larger than the kth order statistic of the covariates as:

$$S_t^{(k)} = \left\{i : i \leq n_t,\ x_i \leq X_{(k)}\right\}, \tag{3.4}$$

$$S_c^{(k)} = \left\{i : n_t < i \leq n_t + n_c,\ x_i \leq X_{(k)}\right\}, \tag{3.5}$$

where x_i is the value of the covariate associated with the ith individual.

We define P_k to be the p-value from a test of hypothesis on the responses associated with the pair of subsets $S_t^{(k)}$ and $S_c^{(k)}$. Define $P^*_{n_{min},X}$, the most extreme p-value observed after testing all pairs of subsets defined by half-intervals of the covariate and with cardinality at least n_{min}, as follows:

$$P^*_{n_{min},X} = \inf\left\{P_k,\ k:\ \left|S_t^{(k)}\right| \geq n_{min} \text{ and } \left|S_c^{(k)}\right| \geq n_{min}\right\}. \tag{3.6}$$

The number of pairs to be compared in (3.6) is at most $(n_t + n_c - 2n_{min} + 2)$.

A simulation procedure like that in Section 3.2 can again be used under certain assumptions to estimate a critical value for $P^*_{n_{min},X}$ which controls the type I error at level α.

3.3.1 Two-sided t-test

We first apply the method of inference guided data exploration restricted by half-intervals of a covariate using the two-sample t-test. Assume that the data consist of a response, Y and a covariate, X which are jointly distributed as bivariate normal with correlation ρ. Both the treatment and control samples are drawn from:

$$\begin{pmatrix} Y \\ X \end{pmatrix} \sim BVN\left[\begin{pmatrix} 0 \\ 0 \end{pmatrix}, \begin{pmatrix} 1 & \rho \\ \rho & 1 \end{pmatrix}\right]. \tag{3.7}$$

By invariance, the tabulated values correspond to any bivariate normal distribution with correlation ρ. Independent identically distributed samples are generated and the two-sample two-sided t-test is used for the test statistic. Table 3.3 lists the relevant parameters (ρ, n_t, n_c, and n_{min}). The critical p-value, p, for the test to control the type I error at level 0.05 and the Z-score associated with the critical p-value, Z(p), are given in the columns headed t-test. The number of simulations is 10,000.

In contrast to Table 3.1, the comparable entries for critical p-values are not nearly as extreme in Table 3.3. There are two reasons for this. First of all, we have restricted our attention to pairs of subsets formed by half-intervals of a covariate which leads to fewer pairs of subsets to be compared per iteration. For instance the entry for $n_t = 10$, $n_c = 10$, and $n_{min} = 2$, would consider $2(n_t - n_{min} + 1)(n_c - n_{min} + 1) - 1 = 161$ comparisons per iteration in Table 3.1 (using the shortcut for the Z and t statistics respectively), while in this table we would only consider at most $(n_t + n_c - 2n_{min} + 2) = 18$ comparisons per iteration. Secondly, the shortcut for the Z and t statistics only considers pairs of subsets where maximal order statistics are compared to minimal order statistics, while in this table the subsets considered are more "typical". More specifically, consider the case of $\rho = 0$. This implies the response and covariate are independent under the assumption of a bivariate normal distribution. Now, selecting a subset of responses based on the covariate is like selecting a subset

Table 3.3: : Half-intervals, $\alpha = 0.05$, $N = 10{,}000$

				t-test		ANCOVA t-test	
ρ	n_t	n_c	n_{min}	p	Z(p)	p	Z(p)
0	4	4	2	1.945E-02	2.3369	1.677E-02	2.3916
0.25	4	4	2	1.768E-02	2.3722	1.736E-02	2.3790
0.5	4	4	2	1.998E-02	2.3267	1.677E-02	2.3917
0.75	4	4	2	1.828E-02	2.3599	1.754E-02	2.3752
0.99	4	4	2			1.590E-02	2.4111
1	4	4	2	1.658E-02	2.3959	*	*
0	10	10	5	1.897E-02	2.3461	1.874E-02	2.3507
0.25	10	10	5	1.776E-02	2.3707	1.945E-02	2.3368
0.5	10	10	5	1.656E-02	2.3963	1.699E-02	2.3870
0.75	10	10	5	1.489E-02	2.4351	1.803E-02	2.3651
0.99	10	10	5			1.665E-02	2.3944
1	10	10	5	1.136E-02	2.5313	*	*
0	20	20	8	1.307E-02	2.4818	1.381E-02	2.4622
0.25	20	20	8	1.175E-02	2.5194	1.330E-02	2.4756
0.5	20	20	8	1.223E-02	2.5053	1.135E-02	2.5316
0.75	20	20	8	1.022E-02	2.5683	1.349E-02	2.4706
0.99	20	20	8			1.238E-02	2.5011
1	20	20	8	7.782E-03	2.6614	*	*
0	50	50	20	1.227E-02	2.5041	1.201E-02	2.5119
0.25	50	50	20	1.065E-02	2.5540	1.129E-02	2.5336
0.5	50	50	20	9.637E-03	2.5886	1.171E-02	2.5207
0.75	50	50	20	8.994E-03	2.6123	1.163E-02	2.5230
0.99	50	50	20			1.279E-02	2.4896
1	50	50	20	6.775E-03	2.7077	*	*
0	80	80	30	1.084E-02	2.5480	1.123E-02	2.5355
0.25	80	80	30	1.108E-02	2.5401	1.100E-02	2.5426
0.5	80	80	30	9.396E-03	2.5973	1.090E-02	2.5460
0.75	80	80	30	8.049E-03	2.6500	9.993E-03	2.5761
0.99	80	80	30			1.025E-02	2.5672
1	80	80	30	5.562E-03	2.7725	*	*
0	60	60	40	1.912E-02	2.3432	1.933E-02	2.3391
0.25	60	60	40	1.832E-02	2.3591	1.849E-02	2.3557
0.5	60	60	40	1.684E-02	2.3902	1.885E-02	2.3484
0.75	60	60	40	1.400E-02	2.4572	1.886E-02	2.3482
0.99	60	60	40			1.919E-02	2.3419
1	60	60	40	9.906E-03	2.5791	*	*

* The ANCOVA model could not be fit with ρ=1 due to zero variance.

at random. In the case of $\rho = 1$, we would compare minimal order statistics to minimal order statistics in (3.4) or maximal order statistics to maximal order statistics (when $x_i \geq X_{(k)}$ in (3.4)). This is because restricting $x_i \leq X_{(k)}$ with $\rho = 1$ causes us to examine the minimal order statistics of the response. For either extreme and for intermediate values of the correlation, the pairs of subsets considered will tend to produce less extreme p-values than would be expected by comparing maximal order statistics to minimal order statistics as we do in Table 3.1.

The critical p-value depends on the joint distribution of the response and the covariate. Further simulations with a fixed set of covariates and the responses generated from the conditional distribution of the response given the covariate revealed that the critical p-value also depends on the set of observed covariates. Thus, the results for the t-test in Table 3.3 do not apply to an analysis conditional on the observed covariates, but rather apply to the case of the response and covariate being jointly random. When analyzing a dataset where accuracy is critical and where covariates are fixed, we would recommend that simulations be carried out using the fixed covariates and generating the responses from the assumed null conditional distribution.

When $\rho = 1$, the value of the response is determined by the value of the covariate. These entries are identical to what would be obtained by selecting subsets based on the order statistics of the response rather than the covariate. When the other parameters are fixed, the entries for $\rho = 1$ tend to produce the most extreme critical p-values when compared to other values of ρ. The few exceptions are likely due to sampling variation. For $\rho = 1$, outliers from one tail of the distribution of the response for each treatment will be included with certainty in the sub-samples. These outliers will be more and more influential as the sub-sample size decreases. To the extent that one treatment group has more outliers than the other; this may explain some of the most extreme p-values. We suspect the critical p-values for $\rho = 1$ are the most extreme because, for values of ρ less than one, outliers have some probability of being included in sub-samples, but are not included with certainty. This heuristic argument leads us to conjecture that the most extreme critical p-value for any joint distribution between the response and the covariate where the response has a marginal normal distribution corresponds to the case of $\rho = 1$. We would recommend that the critical p-value for $\rho = 1$ be used as a conservative estimate of the critical p-value whenever the joint distribution of the response and the covariate is unknown and the marginal distribution of the response can be assumed to be approximately normal. The empirical evidence supports this recommendation when the response and covariate are assumed to have a bivariate normal distribution.

3.3.2 ANCOVA *t*-test

If there is a relationship between the expected value of the response and the value of the covariate, a suitable model can exploit this relationship to reduce the variance of the estimator of the mean response and provide a more powerful test. Our half-intervals methodology is applicable when the ANCOVA model $Y = \beta_0 + \beta_1 \, I_{[Treatment]} + \beta_2 X + \varepsilon$ is assumed to hold for at least a portion of the range of the covariate defined by a half-interval, where we suppose parallel slopes.

Table 3.3, in the columns headed ANCOVA *t*-test, shows the results of simulations applying the method of inference guided data exploration restricted by half-intervals of a covariate. The data consist of a response and a covariate which are jointly distributed as bivariate normal with correlation ρ. If Y and X are the response and covariate, respectively, then both the treatment and control samples are drawn from the bivariate normal null distribution (3.7).

Independent identically distributed samples were generated and the *t*-test that the coefficient of the treatment indicator (β_1) is equal to zero in the ANCOVA model, was used for the test statistic. The alternative again is two-sided. This model apparently controls for the correlation between response and covariate so that it may be seen that the critical p-value no longer depends on the correlation. The small variation seen in the critical p-values when the other parameters are fixed appears due to sampling variation.

In Section 3.3.1, we previously pointed out that the method restricted by half-intervals of a covariate produced critical p-values appropriate for an analysis where the response and covariate are jointly random and that, conditional on the observed set of covariates, the critical p-value depends on the covariates. In the current setting, we ran further simulations to explore whether the critical p-value for a fixed set of covariates would still depend on the covariates. Twenty-five simulations were run, where for each simulation, the set of covariates was fixed at a single realization of $X_i \sim$ i.i.d. $N(0, 1)$ and the Y_j's were then generated from the model $Y_j = \beta X_j + \varepsilon_j$ where $\varepsilon_j \sim$ i.i.d. $N(0, 1)$. We used parameters $n_t = 50$, $n_c = 50$, $n_{min} = 20$, $\beta = .5$, and $N = 10{,}000$ iterations. The twenty-five simulated critical p-values were all consistent with the entries for ANCOVA in Table 3.3 corresponding to parameters $n_t = 50, n_c = 50$, and $n_{min} = 20$. This is suggestive that the critical p-values for ANCOVA in Table 3.3 are appropriate for an analysis conditional on the observed covariate when the dependence between the response and the covariate is properly modeled.

3.4 Application

To illustrate the application of our methods, we use data from the National Surgical Adjuvant Breast and Bowel Project (NSABP) [Mamounas (1997)]. The NSABP is a National Cancer Institute funded cooperative group which performs clinical trials to study breast and colorectal cancer. For this study, patients are randomized into three groups. The control group (arm 1) receives 4 cycles of Adriamycin Cyclophosphamide (AC) chemotherapy given preoperatively along with 5 years of Tamoxifen beginning concurrently with AC. The two experimental groups receive the same AC chemotherapy and Tamoxifen followed by 4 cycles of Taxotere (T) given preoperatively (arm 2) or postoperatively (arm 3). Study inclusion/exclusion criteria are described in Mamounas (1997).

In the present example, we are concerned with the change in tumor size between baseline (randomization) and surgery. We compare arms 1 and 3 combined, both of which received 4 cycles of AC prior to surgery with arm 2 where patients received 4 cycles of AC followed by 4 cycles of T prior to surgery. Our response variable is proportional reduction in tumor size defined as (baseline tumor size - tumor size at surgery)/baseline tumor size. Our analysis is restricted to patients whose baseline tumor size is at least 1.0 centimeter and for whom the tumor size at surgery is non-missing. A tumor size of zero at surgery is allowable and corresponds to a complete preoperative response to therapy: such patients will have a proportional reduction in tumor size value of 1.0. Our interest concerns the effect of a patient's age on her tumor response measured by its proportional reduction, and whether or not this effect is comparable between arms 1/3 and arm 2.

Our previous experience with this dataset suggests the effect of Taxotere in helping to reduce tumor size is more pronounced in younger women. To examine this, we apply the method of inference guided data exploration by half-intervals of a covariate to this dataset using proportional reduction in tumor size as the response, age as the covariate with subsets formed by including all women no older than a series of decreasing age thresholds, and the two sample t-test as the test of hypothesis. The control group (arms 1/3) has 1,117 observations, and the group treated with Taxotere preoperatively (arm 2) has 512 observations. Our half-intervals procedure in ths example does not consider any subsets where either treatment arm has less than 50 observations, i.e., n_{min}=50 in (3.6). A conservative estimate of the critical p-value was obtained by simulation using bivariate normal data and parameters $\alpha = 0.05$, $\rho = 1$, $n_t = 512$, $n_c = 1{,}117$, $n_{min} = 50$, and $N = 10{,}000$ iterations by the same methodology used in Section 3.3.1 to produce Table 3.3. Since we do not know the true relationship between age and proportional reduction in tumor size, we used a correlation of 1 as this produced the most extreme critical p-values. The simulation yielded an estimated critical p-value of 0.00267.

To implement our approach, we performed a series of two-sample t-tests where we compared the control and treatment groups on subsets of the data beginning with the full sample comparison, then including women no older than the second highest age, and continuing in this fashion until the final comparison which included women no older than 36 years of age. The next comparison would have included women no older than 35, but was not performed since the treatment group would have had less than $n_{min} = 50$ observations. The results of these hypothesis tests are displayed in Table 3.4 where we observe that the minimum p-value obtained from this series of tests was p = 0.0000785 and corresponded to a subset defined as all women no older than 53 years of age. Since the observed minimum p-value (0.0000785) was less than the critical p-value (0.00267), we reject the null hypothesis of no difference in proportional reduction in tumor size between the treatments and suggest that among women no older than 53 years of age the reduction is greater in women receiving Taxotere than in women not receiving Taxotere at the $\alpha = 0.05$ level.

3.5 Summary and Discussion

The method of inference guided data exploration restricted by minimum subset sample size, considered in Section 3.2, provides a critical p-value for the minimum p-value after testing all possible pairs of subsets at least as large as the minimum sample size criteria. Direct application of the method can be computationally prohibitive for all but the smallest sample sizes as we observed in Section 3.2.2 where we applied the method using the two-sample t-test as our test of hypothesis. The shortcut method of only considering pairs of subsets where maximal order statistics are compared to minimal order statistics makes the method computationally feasible, but this approach only yields exact critical p-values for certain hypothesis tests such as the two-sample Z-test considered in Section 3.2.1. The shortcut method is also useful for finding an approximate critical p-value for other tests of hypothesis such as the two-sample t-test considered in Section 3.2.3. We found this approximate critical p-value to be a better approximation for the two-sample t-test than using the critical p-value from the two-sample Z-test.

In applying the method restricted by minimum subset sample size, one tests the null hypothesis of no difference between treatments in any pair of subsets considered versus the alternative that the treatments differ in at least one of the pairs of subsets considered. Concluding the alternative does not lead to a generalizable result, but may lead to further research to identify the factors related to the variation in the effect of treatment. When considering only the

Table 3.4: P-values of the t-test comparing all patients no older than age 36

Maximum Age	N control, in subset	N treatment, in subset	P-value
79	1,117	512	0.0003732
77	1,116	512	0.0003625
76	1,114	512	0.0003719
74	1,114	511	0.0003946
73	1,114	509	0.0004917
72	1,110	507	0.0004783
71	1,106	506	0.0005853
70	1,101	504	0.0004348
69	1,091	501	0.0004135
68	1,083	498	0.0004609
67	1,079	495	0.0004810
66	1,076	495	0.0004105
65	1,072	488	0.0004817
64	1,054	483	0.0004134
63	1,040	473	0.0001582
62	1,026	465	0.0003010
61	1,009	460	0.0001957
60	988	454	0.0001452
59	969	447	0.0001578
58	941	435	0.0002719
57	912	422	0.0004558
56	887	411	0.0009168
55	856	399	0.0008892
54	829	377	0.0015703
53	**801**	**361**	**0.0000785**
52	773	344	0.0000807
51	739	333	0.0001388
50	684	307	0.0001041
49	627	292	0.0003315
48	588	273	0.0008473
47	540	245	0.0020021
46	486	225	0.0024372
45	440	199	0.0032663
44	396	179	0.0026331
43	365	162	0.0028408
42	327	139	0.0052053
41	297	122	0.0102753

Table 3.4 (continued):

Maximum Age	N control, in subset	N treatment, in subset	P-value
40	260	102	0.0024475
39	225	91	0.0050965
38	189	75	0.0059949
37	155	65	0.0102552
36	130	58	0.0330878

sample under study and not attempting to generalize the results, any pair of subsets tested which produce a p-value less than the critical p-value is identified as a pair of subsets where the treatments differ. This methodology will be useful to provide a "rule of thumb" for a statistician to judge potential significance when encountering a researcher who has already been on a fishing expedition in his or her data.

The method of inference guided data exploration restricted by half-intervals of a covariate, considered in Section 3.3, provides a critical p-value for the minimum p-value after testing multiple pairs of subsets defined by including all observations where the covariate is at least as large as (not larger than) a series of threshold values. The half-intervals restriction provides critical p-values which are much less extreme than those provided when the minimum sample size restriction is used. For half-intervals, the critical p-value depends on the relationship between the response and the covariate (correlation for a linear relationship), but, as seen in Section 3.3.2, the dependence can be removed when the relationship is suitably modeled (ANCOVA model for a linear relationship). In practice, the method could be employed by using the critical p-value from the worst case correlation ($\rho = 1$) which would make the method slightly conservative. The results in Sections 3.3.1 (t-test) only apply to the case where the response and covariate are jointly random. Conditional on a fixed set of covariates, the critical p-value depends on the covariates observed.

We feel the methodology employing the half-intervals restriction considers the types of subsets a researcher would more commonly be interested in considering. For example, if the researcher believes that the effect of treatment may not be as strong in patients with very small values of a specific covariate, then this method could be employed to guard against inclusion criteria which were too liberal and possibly included patients where the effect of treatment was negligible. We again test the null hypothesis of no difference between treatments in any pair of subsets considered versus the alternative that the treatments differ in at least one of the pairs of subsets considered, however we are more comfortable in generalizing our results beyond the sample under study because the significant subsets are now identifiable based on the value of a covariate exceeding some threshold. We believe the method employing the half-intervals restriction will be used to provide a "rule of thumb" for judging significance in situations where a researcher has already examined subsets with respect to a covariate and may also be used as a planned methodology when the aim of the research is primarily exploratory in nature.

Acknowledgments. The work of the second author was supported in part by a grants from the NSF (DMS00 72207) and NIH (5 P50 MH 45156-14). We wish to thank Professor Stewart Anderson Ph.D. of the University of Pittsburgh's Department of Biostatistics and the NSABP for providing the tumor response data used in our application.

References

1. Fleiss, J. L. (1986). *The Design and Analysis of Clinical Experiments*, John Wiley & Sons, New York.

2. Hsu, J. C. (1996). *Multiple Comparisons*, Chapman and Hall, New York.

3. Koziol, J. A. and Wu, S. H. (1996). Changepoint statistics for assessing a treatment-vovariate interaction, *Biometrics*, **52**, 1147–1152.

4. Mamounas, E. P. (1997). NSABP Protocol B-27: Preoperative doxorubicin plus cyclophosphamide followed by preoperative or postoperative docetaxel, *Oncology*, **11** (Suppl. No. 6), 37–40.

5. Miller R. G. (1981). *Simultaneous Statistical Inference*, Springer-Verlag, New York.

6. Potthoff, R. F. (1964). On the Johnson-Neyman technique and some extensions thereof, *Psychometrika*, **29**, 241–256.

7. Worsley, K. J. (1992). A three dimensional statistical analysis for CBF activation studies in human brain, *Journal of Cerebral Blood Flow and Metabolism*, **12**, 900–918.

8. Yothers, G. (2003). *Methodologies for Identifying Subsets of the Population Where Two Treatments Differ*, Ph.D. Dissertation, University of Pittsburgh, Pittsburgh, Pennsylvania.

4

Discriminating Between Normal and Laplace Distributions

Debasis Kundu

Department of Mathematics, Indian Institute of Technology Kanpur, India

Abstract: Both normal and Laplace distributions can be used to analyze symmetric data. In this chapter, we consider the logarithm of the ratio of the maximized likelihoods to discriminate between these two distributions. We obtain the asymptotic distributions of the test statistics and it is observed that they are independent of the unknown parameters. When the underlying distribution is normal, the asymptotic distribution works quite well even when the sample size is small. But when the underlying distribution is Laplace, the asymptotic distribution does not work well for small sample sizes. In this case, we propose a bias corrected asymptotic distribution which works well even for small sample sizes. Based on the asymptotic distributions, minimum sample size needed to discriminate between these two distributions is obtained for a given probability of correct selection. Monte Carlo simulations are performed to examine how the asymptotic results work for small sample sizes and two data sets are analyzed for illustrative purposes.

Keywords and phrases: Asymptotic distributions, likelihood ratio tests, probability of correct selection, location scale family

4.1 Introduction

Suppose an experimenter has n observations and the elementary data analysis, say a histogram, stem and leaf plot, or the box plot, suggests that they have come from a symmetric distribution. The experimenter wants to determine which of normal or Laplace distributions fits the data better.

It is well-known that the normal distribution is used to analyze symmetric data with short tails, while the Laplace distribution is used for data with long tails. Although these two distributions may provide similar fit for moderate sample sizes, it is still desirable to choose the better fitting model since the inference procedures often involve tail probabilities, where the distributional assumption becomes critical. Hence, it is important to make the best possible decision based on the available data.

For a given data, determining whether it follows one of the two given distribution functions is a well-known problem. Discriminating between any two general probability distribution functions was studied by Cox (1961, 1962), Chambers and Cox (1967), Atkinson (1969, 1970), Dyer (1973), and Chen (1980). Dumonceaux and Antle (1973) discussed the problem of discriminating between the log-normal and Weibull distributions, while Dumonceaux, Antle and Haas (1973) considered the problem of discriminating between any two distribution functions from location-scale family. In both these articles, the authors proposed test statistics and computed the critical values based on Monte Carlo simulations. Fearn and Nebenzahl (1991) and Bain and Engelhardt (1980) examined the problem of discriminating between the gamma and Weibull distributions. Wiens (1999), Kim, Sun and Tsutakawa (2002), Firth (1988), and Kundu and Manglick (2004) all discussed different aspects of discriminating between the log-normal and gamma distributions.

In this chapter, we discriminate between the normal and Laplace distributions using the ratio of the maximized likelihoods (RML). It should be mentioned that Dumonceaux, Antle and Haas (1973) also used a statistic equivalent to the RML for discriminating between two distribution functions. They did not study any distributional property of the proposed statistic. Here, using the approach of White (1982a,b), we obtain the asymptotic distribution of the logarithm of RML. It is observed that this asymptotic distribution is normal and does not depend on the parameters of the underlying distribution function. Numerical study indicates that when the underlying distribution is normal, the asymptotic distribution works quite well even for small sample sizes but this is not true when the underlying distribution is Laplace. In the latter case, we propose a bias corrected asymptotic distribution which works quite well for small sample sizes as well.

The asymptotic distribution can be used to compute the probability of correct selection (PCS). We also obtain the minimum sample size necessary to discriminate between the two distribution functions for a given PCS.

The rest of this chapter is organized as follows. At the end of this section, we provide a list of notation. We briefly describe the likelihood ratio method in Section 4.2. Asymptotic distributions of the logarithm of RML statistics when the true model is normal and Laplace are obtained in Section 4.3. Sample size determination is discussed in Section 4.4. Some numerical experiments are performed in Section 4.5 and two real data sets are analyzed in Section 4.6. Finally, we draw some conclusions in Section 4.7.

We use the following notation for the rest of this chapter. The density function of a normal random variable, with location parameter $-\infty < \mu < \infty$ and scale parameter $\sigma > 0$, will be denoted by

$$f_N(x; \mu, \sigma) = \frac{1}{\sqrt{2\pi}\sigma} e^{-\frac{(x-\mu)^2}{2\sigma^2}}, \quad -\infty < x < \infty. \tag{4.1}$$

A normal distribution with mean μ and variance σ^2 will be denoted by $N(\mu, \sigma^2)$. The density function of a Laplace distribution, with location parameter $-\infty < \eta < \infty$ and scale parameter $\theta > 0$, will be denoted by

$$f_L(x; \eta, \theta) = \frac{1}{2\theta} e^{-\frac{|x-\eta|}{\theta}}, \quad -\infty < x < \infty. \tag{4.2}$$

A Laplace distribution with location η and scale θ will be denoted by $L(\eta, \theta)$.

In addition, almost sure convergence will be denoted by *a.s.* For any Borel measurable function $h(\cdot)$, $E_N(h(U))$ and $V_N(h(U))$ will denote the mean and variance of $h(U)$ under the assumption that U follows $N(\mu, \sigma^2)$. Similarly, we denote $E_L(h(U))$ and $V_L(h(U))$ for the mean and variance of $h(U)$ when U follows $L(\eta, \theta)$. Moreover, if $g(\cdot)$ and $h(\cdot)$ are two Borel measurable functions, we define $cov_N(g(U), h(U)) = E_N(g(U)h(U)) - E_N(g(U))E_N(h(U))$ and $cov_L(g(U), h(U)) = E_L(g(U)h(U)) - E_L(g(U))E_L(h(U))$. Finally, we define

$$\begin{aligned} \text{median}\{a_1, \ldots, a_{2m+1}\} &= a_{m+1}, \\ \text{median}\{a_1, \ldots, a_{2m}\} &= \frac{a_m + a_{m+1}}{2} \end{aligned}$$

where $a_1 < a_2 < \cdots < a_{2m+1}$ are the ordered values.

4.2 Ratio of Maximized Likelihood

Suppose we have a sample $X_1, \ldots, X_n$ from one of the two distribution functions. The likelihood functions, assuming that the data follow $N(\mu, \sigma^2)$ and $L(\eta, \theta)$, are

$$l_N(\mu, \sigma) = \prod_{i=1}^{n} f_N(X_i; \mu, \sigma)$$

and

$$l_L(\eta, \theta) = \prod_{i=1}^{n} f_L(X_i; \eta, \theta),$$

respectively. The logarithm of RML is defined as

$$T = \ln \left[\frac{l_N(\hat{\mu}, \hat{\sigma})}{l_L(\hat{\eta}, \hat{\theta})} \right]. \tag{4.3}$$

Here, $(\hat{\mu}, \hat{\sigma})$ and $(\hat{\eta}, \hat{\theta})$ are the maximum likelihood estimators (MLEs) of (μ, σ) and (η, θ), respectively, based on the sample $X_1, \ldots, X_n$. So, T can be written as

$$T = \frac{n}{2}\ln 2 - \frac{n}{2}\ln \pi + n \ln \hat{\theta} - n \ln \hat{\sigma} + \frac{n}{2}, \tag{4.4}$$

where

$$\begin{aligned} \hat{\mu} &= \frac{1}{n}\sum_{i=1}^{n} X_i, \qquad \hat{\sigma}^2 = \frac{1}{n}\sum_{i=1}^{n}(X_i - \hat{\mu})^2, \\ \hat{\eta} &= \text{median}\{X_1, \ldots, X_n\} \quad \text{and} \quad \hat{\theta} = \frac{1}{n}\sum_{i=1}^{n}|X_i - \hat{\eta}|. \end{aligned} \tag{4.5}$$

The following discrimination procedure can then be used. Choose the normal distribution if the test statistic $T > 0$, otherwise choose the Laplace distribution as the preferred model. Note that if the true model is $N(\mu, \sigma^2)$, then the distribution of T is independent of μ and σ. Similarly, if the true model is $L(\eta, \theta)$, then the distribution of T is independent of η and θ.

4.3 Asymptotic Properties of the Logarithm of RML

In this section, we derive the asymptotic distributions of T when the true model is normal and Laplace.

Case 1: The Data Follow Normal Distribution

In this case, when the true model is $N(\mu, \sigma^2)$, we have the following result.

Theorem 4.3.1 *Under the assumption that the data follow $N(\mu, \sigma^2)$, the distribution of T is asymptotically normal with mean $E_N(T)$ and $V_N(T)$.*

To prove Theorem 4.3.1, we need the following lemma.

Lemma 4.3.1 *Suppose the data follow $N(\mu, \sigma^2)$, then as $n \to \infty$, we have*
(1) $\hat{\eta} \to \tilde{\eta}$ *a.s., and* $\hat{\theta} \to \tilde{\theta}$ *a.s., where*

$$E_N\left(\ln(f_L(X; \tilde{\eta}, \tilde{\theta}))\right) = \max_{\eta,\theta} E_N\left(\ln(f_L(X; \eta, \theta))\right).$$

Note that $\tilde{\eta}$ and $\tilde{\theta}$ may depend on μ and σ, but we do not make it explicit for brevity.

(2) Let us denote

$$T^* = \ln\left[\frac{l_N(\mu,\theta)}{l_L(\tilde{\eta},\tilde{\theta})}\right]. \tag{4.6}$$

Then, $n^{-\frac{1}{2}}(T - E_{LN}(T))$ *is asymptotically equivalent to* $n^{-\frac{1}{2}}(T^* - E_{LN}(T^*))$.

PROOF OF LEMMA 4.3.1. The proof follows using arguments similar to those for Theorem 1 of White (1982b). ∎

PROOF OF THEOREM 4.3.1. Using the Central Limit Theorem (CLT), it can be easily seen that $n^{-\frac{1}{2}}(T^* - E_{LN}(T^*))$ is asymptotically normally distributed. Therefore, the proof immediately follows from Part (2) of Lemma 4.3.1 and the CLT. ∎

Comments: It should be mentioned here that the distance between $L(\tilde{\eta},\tilde{\theta})$ and $N(\mu,\sigma)$ is minimum in terms of the Kullback-Liebler information measure.

Now we compute $\tilde{\eta}$, $\tilde{\theta}$, $E_N(T)$ and $V_N(T)$. Note that

$$E_N\left(\ln(f_L(X;\eta,\theta))\right) = -\ln 2 - \ln\theta - E\left|\frac{X-\eta}{\theta}\right|.$$

Since X follows $N(\mu,\sigma^2)$, it is clear that

$$\tilde{\eta} = \mu \quad \text{and} \quad \tilde{\theta} = E|X-\mu| = \sigma\sqrt{\frac{2}{\pi}}. \tag{4.7}$$

Now, we provide expressions for $E_N(T)$ and $V_N(T)$. Note that $\lim_{n\to\infty}\frac{E_N(T)}{n}$ and $\lim_{n\to\infty}\frac{V_N(T)}{n}$ exist. Let us denote $\lim_{n\to\infty}\frac{E_N(T)}{n} = AM_N$ and $\lim_{n\to\infty}\frac{V_N(T)}{n} = AV_N$. Since the distribution of T is independent of μ and σ, for the derivation of $E_N(T)$ and $V_N(T)$, without loss of any generality, we take $\mu = 0$ and $\sigma = 1$, i.e., X follows $N(0,1)$. Therefore, for large n,

$$\begin{aligned}
\frac{E_N(T)}{n} &\approx AM_N = E_N\left[\ln f_N(X;0,1) - \ln f_L(X;\tilde{\eta},\tilde{\theta})\right] \\
&= E_N\left[-\frac{1}{2}\ln(2\pi) - \frac{X^2}{2} + \ln 2 + \ln\tilde{\theta} + \left|\frac{X-\tilde{\eta}}{\tilde{\theta}}\right|\right] \\
&= -\frac{1}{2}\ln(2\pi) - E_N\left(\frac{X^2}{2}\right) + \ln 2 + \ln\tilde{\theta} + \sqrt{\frac{\pi}{2}}E_N|X| \\
&= \ln 2 + \frac{1}{2} - \ln\pi = 0.0484172.
\end{aligned} \tag{4.8}$$

Further, we have

$$\begin{aligned}\frac{V_N(T)}{n} &\approx AV_N = V_N\left(\ln f_N(X;0,1) - \ln f_L(X;\tilde{\eta},\tilde{\theta})\right)\\ &= V_N\left(-\frac{1}{2}X^2 + \sqrt{\frac{\pi}{2}}|X|\right)\\ &= \frac{1}{4}V_N(X^2) + \frac{\pi}{2}V_N(|X|) - \sqrt{\frac{\pi}{2}}cov_N(X^2,|X|)\\ &= \frac{1}{2} + \frac{\pi}{2}\left(1-\frac{2}{\pi}\right) - 1 = 0.0707963. \end{aligned} \quad (4.9)$$

Case 2: The Data Follow Laplace Distribution

In this case, when the true model is $L(\eta,\theta)$, we have the following result.

Theorem 4.3.2 *Under the assumption that the data follow $L(\eta,\theta)$, the distribution of T is asymptotically normal with mean $E_L(T)$ and variance $V_L(T)$.*

To prove Theorem 4.3.2, we need Lemma 4.3.2, which is similar to Lemma 4.3.1.

Lemma 4.3.2 *Suppose the data follow $L(\eta,\theta)$, then as $n \to \infty$, we have (1) $\hat{\mu} \to \tilde{\mu}$ a.s., $\hat{\sigma} \to \tilde{\sigma}$ a.s, where*

$$E_L\left(\ln(f_N(X;\tilde{\mu},\tilde{\sigma}))\right) = \max_{\mu,\sigma} E_L\left(\ln(f_N(X;\mu,\sigma))\right).$$

Here again, $\tilde{\mu}$ and $\tilde{\sigma}$ may depend on η and θ, but we do not make it explicit for brevity.

(2) If we denote

$$T_* = \ln\left[\frac{l_N(\tilde{\mu},\tilde{\sigma})}{l_L(\eta,\theta)}\right],$$

then $n^{-\frac{1}{2}}[T - E_L(T)]$ is asymptotically equivalent to $n^{-\frac{1}{2}}[T_* - E_L(T_*)]$.

PROOF OF LEMMA 4.3.2. It follows from Theorem 1 of White (1982b). ■

PROOF OF THEOREM 4.3.2. The proof is similar to that of Theorem 4.3.1. ■

Now, we explain how $\tilde{\mu}$, $\tilde{\sigma}$, $E_L(T)$ and $V_L(T)$ can all be obtained. Consider

$$E_L\left[\ln(f_N(X;\mu,\sigma))\right] = E_L\left[-\frac{1}{2}\ln(2\pi) - \ln\sigma - \frac{(X-\mu)^2}{2\sigma^2}\right].$$

Since X follows $L(\eta,\theta)$, we readily have

$$\tilde{\mu} = \eta \quad \text{and} \quad \tilde{\sigma} = \sqrt{2}\,\theta. \quad (4.10)$$

Now, we provide expressions for $E_L(T)$ and $V_L(T)$. As before, $\lim_{n\to\infty} \frac{E_L(T)}{n}$ and $\lim_{n\to\infty} \frac{V_L(T)}{n}$ exist and we denote $\lim_{n\to\infty} \frac{E_L(T)}{n} = AM_L$ and $\lim_{n\to\infty} \frac{V_L(T)}{n} = AV_L$. As mentioned before, the distribution of T is independent of η and θ and so we take $\eta = 0$ and $\theta = 1$, without loss of any generality, for the calculations of AM_L and AV_L. Hence, for large n,

$$\begin{aligned}\frac{E_L(T)}{n} &\approx AM_L = E_L\left[\ln(f_N(X;\tilde{\mu},\tilde{\sigma})) - \ln(f_L(X:0,1))\right]\\ &= E_L\left[-\frac{1}{2}\ln(2\pi) - \ln\tilde{\sigma} - \frac{(X-\tilde{\mu})^2}{2\tilde{\sigma}^2} + \ln 2 + |X|\right]\\ &= -\frac{1}{2}\ln(2\pi) - \frac{1}{2}\ln 2 - \frac{1}{2} + \ln 2 + 1 = -0.0723649 \qquad (4.11)\end{aligned}$$

and, further,,

$$\begin{aligned}\frac{V_L(T)}{n} &\approx AV_L = V_L\left[\ln(f_N(X;\tilde{\mu},\tilde{\sigma})) - \ln(f_L(X;0,1))\right]\\ &= V_L\left[-\frac{X^2}{4} + |X|\right] = \frac{1}{16}V_L(X^2) + V_L(|X|) - \frac{1}{2}cov_L(X^2,|X|)\\ &= 0.25. \qquad (4.12)\end{aligned}$$

4.4 Determination of Sample Size and Testing

4.4.1 Minimum sample size determination

In this subsection, we propose a method of determining the minimum sample size needed to discriminate between the normal and Laplace distributions for a given probability of correct selection (PCS). It is expected that the user specifies the PCS before hand.

First we consider Case 1, i.e., the data follow $N(\mu,\sigma)$. Since T is asymptotically normally distributed with mean $E_N(T)$ and variance $V_N(T)$, the PCS is

$$PCS = P[T > 0] \approx \Phi\left(\frac{E_N(T)}{\sqrt{V_N(T)}}\right) = \Phi\left(\frac{n \times AM_N}{\sqrt{n \times AV_N}}\right), \qquad (4.13)$$

where Φ is the distribution function of the standard normal distribution. Now, for determining the minimum sample size required to achieve at least p^* protection level (viz., the PCS), we solve

$$\Phi\left(\frac{n \times AM_N}{\sqrt{n \times AV_N}}\right) = p^* \qquad (4.14)$$

to obtain

$$n = \frac{z_{p^*}^2 AV_N}{(AM_N)^2}, \tag{4.15}$$

where z_{p^*} is the $100p^*$ percentile point of the standard normal distribution.

Next, for Case 2, i.e., when the data follow $L(\eta, \theta)$, we similarly obtain

$$n = \frac{z_{p^*}^2 AV_L}{(AM_L)^2}. \tag{4.16}$$

Therefore, to achieve overall p^* protection level, we need at least

$$n = \max\left\{\frac{AV_L}{(AM_L)^2}, \frac{AV_L}{(AM_L)^2}\right\} z_{p^*}^2 = \max\{30.2, 47.7\} z_{p^*}^2 = 47.7\ z_{p^*}^2. \tag{4.17}$$

4.4.2 Testing of hypotheses

Cox (1961, 1962) and Dumonceaux and Antle (1973) considered the above discrimination problem as the following two testing of hypotheses problems:

$$\text{Problem 1:} \quad H_0 : \text{Normal} \quad vs. \ H_1 : \text{Laplace}, \tag{4.18}$$

$$\text{Problem 2:} \quad H_0 : \text{Laplace} \quad vs. \ H_1 : \text{Normal}. \tag{4.19}$$

Dumonceaux and Antle (1973) provided the exact critical regions and the powers of the likelihood ratio tests based on Monte Carlo simulations. The asymptotic results derived in the last section can be used for testing the above two hypotheses as follows:

Test 1: For Problem 1: Reject the null hypothesis H_0 at α % level of significance if $T < n \times 0.0484172 - z_\alpha \times \sqrt{n \times 0.0707963}$, and accept H_0 otherwise.

Test 2: For Problem 2: Reject the null hypothesis H_0 at α % level of significance if $T > -n \times 0.0723649 + z_\alpha \times \sqrt{n \times 0.25}$, and accept H_0 otherwise.

4.5 Numerical Experiments

In this section, we present some numerical results performed to check how the asymptotic results work for different sample sizes. All computations were performed at the Indian Institute of Technology Kanpur, using Pentium IV processor. We used the random deviate generator RAN2 of Press et al. (1993) and all the programs were written in FORTRAN, which are available from the author. We computed the PCS based on simulations and also based on the asymptotic normality results derived in Section 4.3. Since the distribution (numerical value) of T is independent of the location and scale parameters, we took the location and scale parameters to be zero and one, respectively, in all cases. We considered sample sizes n = 20, 40, 60, 80 and 100.

Table 4.1: The probability of correct selection based on Monte Carlo simulations (MC) with 10,000 replications and also based on the asymptotic results (AR) when the true model is normal

n	20	40	60	80	100
MC	0.80	0.87	0.92	0.95	0.96
AR	0.79	0.88	0.92	0.95	0.96

First, we take the case when the true model is normal. In this case, we generated a random sample of size n from $N(0,1)$, computed T, and checked whether T is positive or negative. We replicated this process 10,000 times and obtained an estimate of PCS. These results are presented in Table 4.1. We obtained similar results when the true model is Laplace, and these results are presented in Table 4.2.

It is quite clear from Tables 4.1 and 4.2 that as sample size increases the PCS increases as one would expect. The asymptotic results in Table 4.1 are quite close to the corresponding simulated results. This is not the case for Table 4.2 wherein it is observed that for small sample sizes the asymptotic results do not match well with the simulated results. This is so because T is biased in this case. So, we propose the following biased corrected version for small sample sizes. We suggest to use $AM_L = -0.0723649 + \frac{0.64}{n}$ and $AV_L = 0.25 - \frac{2.09}{n}$. The results, based on the biased corrected version, are also provided in Table 4.2. The bias values were obtained by regressing the simulated mean and variance of T with the theoretical mean and variance, respectively, over a wide range of sample sizes. Based on the bias correction, the value of n in (4.16) can be

Table 4.2: The probability of correct selection based on Monte Carlo simulations (MC), the asymptotic results (AS) and bias corrected asymptotic results (BC) with 10,000 replications when the true model is Laplace

n	20	40	60	80	100
MC	0.65	0.79	0.86	0.90	0.93
AS	0.78	0.86	0.91	0.94	0.96
BC	0.68	0.79	0.85	0.89	0.92

modified to

$$n = \frac{-(1.28AM_L - z_{p^*}^2 AV_L) + \sqrt{(1.28AM_L - z_{p^*}^2 AV_L)^2 - 4AM_L^2(0.64^2 + 2.09z_{p^*}^2)}}{2AM_L^2},$$

and therefore, (4.17) and Test 2 can also be modified accordingly.

Now we consider the discrimination problem as a testing of hypothesis problem as posed in the last section. Let us define the rejection regions as $\{T < 0\}$ and $\{T > 0\}$ for Problems 1 and 2, respectively. It is then immediate that P[Type I error] = 1 − PCS. From Tables 4.1 and 4.2, it is clear that P[Type I error] varies between 0.20 to 0.04 as the sample size varies between 20 to 100 when the true model is normal, and when the true model is Laplace it varies between 0.35 to 0.07. Similarly, in these two cases, the power of the tests vary between 0.65 to 0.93 and 0.80 to 0.96, respectively, as sample size varies between 20 to 100.

4.6 Data Analysis

In this section, we analyze two data sets and use our method to discriminate between the two distribution functions.

Data Set 1: The first data set is as follows [Lawless (1982, p. 228)]. The data are the number of million revolutions before failure for each of the 23 ball bearings in the life tests, and they are: 17.88, 28.92, 33.00, 41.52, 42.12, 45.60, 48.80, 51.84, 51.96, 54.12, 55.56, 67.80, 68.44, 68.64, 68.88, 84.12, 93.12, 98.64, 105.12, 105.84, 127.92, 128.04, 173.40.

Several authors have analyzed this data set using gamma, Weibull, log-normal, and generalized exponential models. We make the log transformation

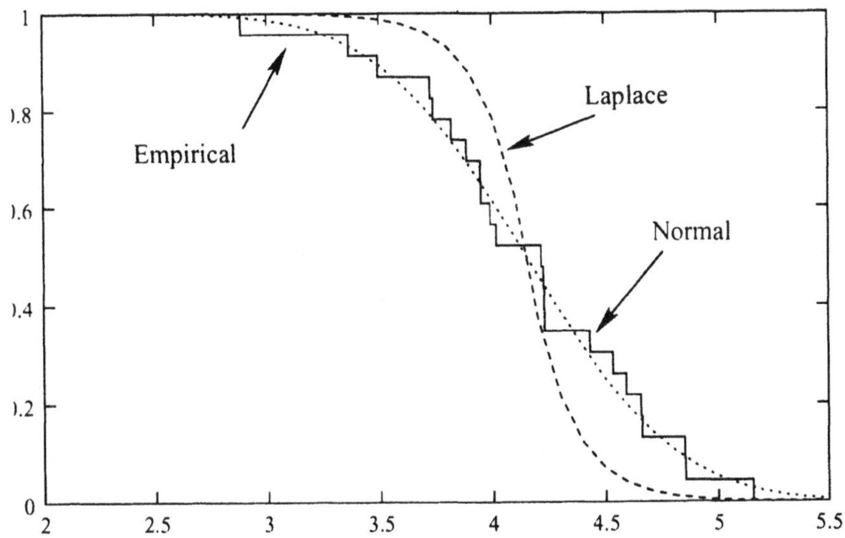

Figure 4.1: Empirical survival function, fitted normal and fitted Laplace survival functions for data set 1.

of the data and try to fit the normal and Laplace distributions. When we use the normal model, the MLEs of the parameters turn out to be $\hat{\mu} = 4.1506$ and $\hat{\sigma} = 0.5215$, and the corresponding log-likelihood (LL) value is -17.6615. The Kolmogorov-Smirnov (K-S) distance between the data and the fitted normal distribution function is 0.0899 and the corresponding p-value is 0.99. Similarly, when we fit the Laplace model, the MLEs turn out to be $\hat{\eta} = 4.1506$ and $\hat{\theta} = 0.4200$, and the corresponding LL value is -18.9899. The non-parametric survival function, fitted normal survival function, and fitted Laplace survival function are all presented in Figure 4.1.

The K-S distance between the data and the fitted Laplace distribution function is 0.2473 and the corresponding p-value is 0.12. We also present the observed and expected frequencies for different intervals in Table 4.3.

The χ^2 values are 0.579 and 2.089 for the normal and Laplace distributions, respectively. For data set 1, the K-S distances, χ^2 values and Figure 4.1 all indicate that for the transformed data, the normal distribution is a better fit compared to the Laplace distribution.

The logarithm of RML, i.e., $T = 1.3284 > 0$. This suggests the normal model as a better fit for the data. From (4.13), it is clear that if the data follow normal distribution, then based on a sample of size 23, PCS = 0.81 and if the data follow Laplace distribution, then the bias corrected PCS = 0.71. Hence, in this case, the PCS is at least $\min\{0.71, 0.81\} = 0.71$. Based on the assumption that the data follow normal distribution, the p-value = 0.57. Similarly, based

Table 4.3: The observed and the expected frequencies for the normal and Laplace distributions for data set 1

Intervals	Observed	Normal	Laplace
$< \ln 35$	3	2.92	2.79
$\ln 35$ - $\ln 55$	7	6.04	5.38
$\ln 55$ - $\ln 80$	5	6.48	8.20
$\ln 80$ - $\ln 100$	3	3.15	2.73
$> \ln 100$	5	4.41	3.90

on the assumption that the data follow Laplace distribution, the p-value = 0.11. Comparison of the p-values also suggests the normal distribution to be better than the Laplace distribution. Therefore, in this case, the LL values, χ^2 values, K-S distances and the proposed method all lead to the normal model and the probability of correct selection is at least 71%. If we consider the two testing of hypotheses problems in (4.18) and (4.19), then based on the data, we note that we can not reject the null hypothesis in both cases even at 10% level of significance.

Data Set 2: Now consider the following data set simulated from the Laplace distribution, viz. $L(0,1)$: -1.28, 0.36, -1.29, -0.80, 0.28, -0.06, -1.53, 0.28, -0.54, 0.17, 0.59, 6.22, 2.41, 0.33, -1.51, 0.25, 2.33, 2.81, -0.92, 2.12, -1.01, 1.35, -0.37, -0.39, -4.39, -2.39, 0.97, -0.58, -2.24, -0.05.

Using the normal model, the MLEs of the parameters are $\hat{\mu} = 0.0373$ and $\hat{\sigma} = 1.8906$, and the corresponding log-likelihood (LL) value is -61.6744. The Kolmogorov-Smirnov (K-S) distance between the data and the fitted normal distribution function is 0.1656 and the corresponding p-value is 0.3833. Similarly, using the Laplace model, the MLEs are $\hat{\eta} = 0.0373$ are $\hat{\theta} = 1.3298$, and the corresponding LL value is -59.3458. The non-parametric, the fitted normal, and the fitted Laplace survival functions are all presented in Figure 4.2.

The K-S distance between the data and the fitted Laplace distribution function is 0.1499 and the corresponding p-value is 0.5100. The observed and expected frequencies are presented in Table 4.4.

The χ^2 values are 1.356 and 1.035 for normal and Laplace distributions, respectively. For data set 2, K-S distances, χ^2 values and Figure 4.2 all indicate that the Laplace distribution is a better fit than the normal distribution.

The logarithm of RML, i.e., $T = -2.3286 < 0$. Hence, the proposed method suggests the Laplace model over the normal. From (4.13), it is clear that if the data follow normal distribution, then based on a sample of size 30, PCS = 0.84 and if the data follow Laplace distribution, then the bias corrected PCS = 0.74. Hence, in this case, the PCS is at least $\min\{0.74, 0.84\} = 0.74$. Based on

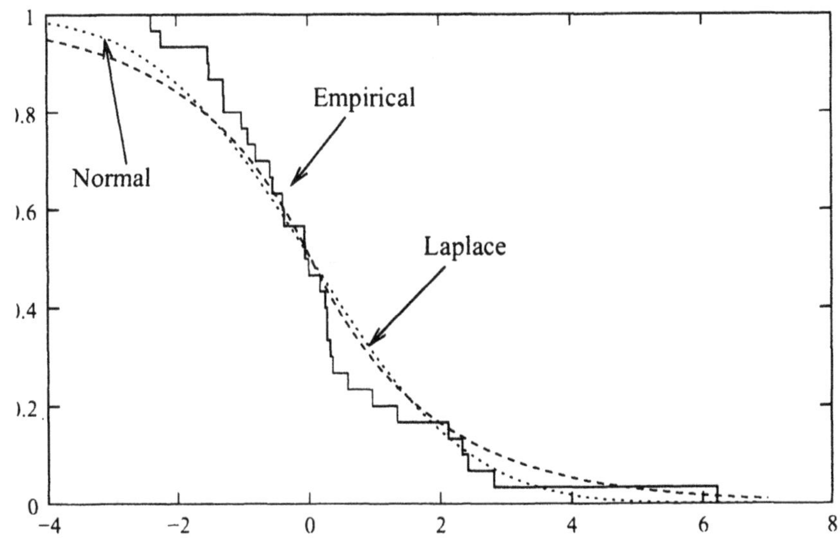

Figure 4.2: Empirical survival function, fitted normal and fitted Laplace survival functions for data set 2.

the assumption that the data follow normal distribution, the p-value < 0.005. Similarly, based on the assumption that the data follow Laplace distribution, the p-value > 0.62. Comparing the two p-values, it is clear that the Laplace distribution is preferable over the normal. Therefore, in this case, the LL values, χ^2 values, K-S distances and the proposed method all lead to the Laplace model and the probability of correct selection is at least 74%. If we consider the two testing of hypotheses problems in (4.18) and (4.19), we reject the null hypothesis for the problem in (4.18) for any $\alpha < 0.005$.

4.7 Conclusions

In this chapter, we consider the problem of discrimination between the normal and Laplace families. We consider the statistic based on the ratio of maximized likelihoods and obtain the asymptotic distributions of the test statistics under the two true models. It is observed that the asymptotic distributions are asymptotically normal and that they are independent of the parameters of the true model. This method can, in fact, be used for discriminating between any two members of the location–scale family. The exact mean and variance of the corresponding normal distribution needs to be derived in each case. Finally,

Table 4.4: The observed and the expected frequencies for the normal and Laplace distributions for data set 2

Intervals	Observed	Normal	Laplace
< -3.0	1	1.64	1.55
-3.0 – -1.5	4	4.42	3.22
-1.5 – 0.0	11	8.70	9.90
0.0 – 1.5	9	8.71	10.34
1.5 – 3.0	4	4.78	3.37
> 3.0	1	1.75	1.62

we should mention that for a given data set, it may happen that neither of the two distribution functions provide good fit, which should be clear from the K-S values and also from the χ^2 values. In such cases, some other distributions should be considered for the data at hand.

References

1. Atkinson, A. (1969). A test of discriminating between models, *Biometrika*, **56**, 337–341.

2. Atkinson, A. (1970). A method for discriminating between models (with discussions), *Journal of the Royal Statistical Society, Series B*, **32**, 323–353.

3. Bain, L. J. and Engelhardt, M. (1980). Probability of correct selection of Weibull versus gamma based on likelihood ratio, *Communications in Statistics—Theory and Methods*, **9**, 375–381.

4. Chambers, E. A. and Cox, D. R. (1967). Discriminating between alternative binary response models, *Biometrika*, **54**, 573–578.

5. Chen, W. W. (1980). On the tests of separate families of hypotheses with small sample size, *Journal of Statistical Computation and Simulation*, **2**, 183–187.

6. Cox, D. R. (1961). Tests of separate families of hypotheses, In *Proceedings of the Fourth Berkeley Symposium in Mathematical Statistics and Probability*, pp. 105–123.

7. Cox, D. R. (1962). Further results on tests of separate families of hypotheses, *Journal of the Royal Statistical Society, Series B*, **24**, 406–424.

8. Dumonceaux, R., Antle, C. E. and Haas, G. (1973). Likelihood ratio test for discriminating between two models with unknown location and scale parameters, *Technometrics*, **15**, 19–31.

9. Dumonceaux, R. and Antle, C. E. (1973). Discriminating between the log-normal and Weibull distributions, *Technometrics*, **15**, 923–926.

10. Dyer, A. R. (1973). Discrimination procedure for separate families of hypotheses, *Journal of the American Statistical Association*, **68**, 970–974.

11. Fearn, D. H. and Nebenzahl, E. (1991). On the maximum likelihood ratio method of deciding between the Weibull and Gamma distributions, *Communications in Statistics—Theory and Methods*, **20**, 579–593.

12. Firth, D. (1988). Multiplicative errors: log-normal or gamma?, *Journal of the Royal Statistical Society, Series B*, **50**, 266–268.

13. Kim, H., Sun, D. and Tsutakawa, R. K. (2002). Lognormal vs. gamma: Extra variations, *Biometrical Journal*, **44**, 305–323.

14. Kundu, D. and Manglick, A. (2004). Discriminating between the log-normal and gamma distributions, *Journal of Applied Statistical Sciences* (to appear).

15. Lawless, J. F. (1982). *Statistical Models & Methods for Lifetime Data*, John Wiley & Sons, New York.

16. Linhardt, H. and Zucchini, W. (1986). *Model Selection*, John Wiley & Sons, New York.

17. Press, W. H., Flannery, B. P., Teukolsky, S. A. and Vetterling, W. T. (1993). *Numerical Recipes: The Art of Scientific Computing*, Cambridge University Press, London, England.

18. White, H. (1982a). Maximum likelihood estimation of mis-specified models, *Econometrica*, **50**, 1–25.

19. White, H. (1982b). Regularity conditions for Cox's test of non-nested hypotheses, *Journal of Econometrics*, **19**, 301–318.

20. Wiens, B. L. (1999). When log-normal and gamma models give different results: a case study, *The American Statistician*, **53**, 89–93.

5

A Simple Classification Rule for Directional Data

Ashis SenGupta[1] **and Supratik Roy**[2]

[1]*Applied Statistics Division, Indian Statistical Institute, Kolkata, India*
[2]*Department of Statistics, University College of Cork, Cork, Republic of Ireland*

Abstract: An intuitive and geometrically motivated chord-length based discriminant statistic is proposed for the classification of a new observation into one of two circular populations when training samples are available from each of them. Assuming that each of the two underlying populations is von Mises, the exact distribution of this statistic is indicated and its relationship to Fisher's discrimination and Cox's Logistic discrimination rules are discussed. The performance of this rule is presented and compared with Fisher's rule in terms of exact error probabilities and apparent error rates. This new rule is illustrated by a real-life data set.

Keywords and phrases: Apparent error rate, classification rule, directional data, logistic discrimination

5.1 Introduction

Consider the problem of classifying a new observation into one of two distinct circular populations. For an introduction to analysis of circular or directional data, see e.g., Mardia (1972), Jammalamadaka and SenGupta (2001). Suppose we have observations as directional data from these two (identifiable) populations given as θ_{ij}, $j = 1, \ldots, n_i$, $i = 1, 2$. We will utilize these observations as training samples to provide estimates of parameters of the above two populations as needed. Let a new observation be denoted by θ. Denote the sample mean directions by $\overline{\theta}_i$, $i = 1, 2$.

Morris and Laycock (1974) have discussed the usual Fisher's discrimination rule for the von Mises or circular normal (CN) populations when the parameters may possibly be unknown. El Khattabi and Streit (1996) have illustrated the use of classical Bayes rule with offset normal distribution on the circle and some other distributions on the sphere. Note that such parametric rules become quite cumbersome even for applications, when invoked for other popular circular distributions, e.g., the family of symmetric wrapped stable distributions [SenGupta and Pal (2001)]. In a somewhat more extended context, Collett and Lewis (1981) have discussed the problem of discriminating between the von Mises and wrapped normal distributions given a set of data assumed to be coming from one (unknown) of these two populations. These works however apply the standard linear techniques and do not address the peculiarity and the distinctive features of directional data.

In Section 5.2 we introduce a very simple and elegant chord-based discrimination rule which is intuitively appealing and geometrically motivated specifically for circular data and which may be used for arbitrary circular distributions with possibly unknown functional forms. The basic idea used here is to find out the average "distance" (in an appropriate sense) from the new observation to the observations in the two known groups. If the distance from one group is less than from the other, then the new observation is classified as belonging to the "closer" population. Though this approach may be used for any circular distribution, here we illustrate it by the von Mises populations. Next we recall that in a linear setup, for the univariate or multivariate normal distributions the Fisher type discrimination rule, which coincides with the Logistic Discrimination (LGD) rule of Cox (1966) with same variances and the Quadratic LGD rule [see e.g., Anderson (1975)] if variances are different, can be viewed as a quadratic distance function i.e., with variance-covariance matrix playing the role of the metric tensor. In Section 5.3 we show that a similar phenomenon holds for the class of directional distributions also. In Section 5.4 we discuss the exact distribution of the discriminant statistic and note how one can compute the threshold value numerically. Section 5.5 presents a study on the efficiency of the chord-based rule and compares it with Fisher's rule in terms of their Apparent Error Rates (AERs). Finally in Section 5.6, the new rule is illustrated by a real-life data set.

5.2 Construction of the Rule

5.2.1 A distance measure

The simplest distance that can be used for circular data is the arc-length, which in the case of the unit circle is equivalent to the radian measure subtended at the center of the circle, i.e., the value of the observation in radians.

But to be a proper distance on the circle, the distance measure must be rotationally invariant, both in terms of magnitude as well as in the sense of rotation. Thus if we have to consider the arc-length in terms of radian measure, we have to transform it in a suitable way, i.e., take absolute value of the difference in angles, modulo 2π. We may also have to consider the minimum of the two arc-lengths into which two points on the circle divides a circle.

These problems do not arise if instead of the arc-length we consider the length of the chord cut off by the two points on the circle. This is always non-negative, invariant under rotation, both in magnitude and displacement. As we shall see, this particular form has also other attractive properties due to its similarity to known descriptive measures, e.g., circular variance [Mardia (1972, p. 21)].

We observe that though the use of chord length as a descriptive measure is quite natural and may have been in use for long, the approach in the following section seems to be the maiden attempt in this direction.

5.2.2 Average distance of a point from a group

Let two points on the unit circle be denoted by θ_1, θ_2. Then the square of the chord-length between the two is given by $2\,(1 - \cos(\theta_1 - \theta_2))$. Based on this we take the distance measure as

$$d_{ij} \;=\; 1 \;-\; \cos(\theta_i - \theta_j). \tag{5.1}$$

Note that d_{ij} has the following properties : It is always non-negative, symmetric in its indices and is invariant under rotation. A measure of deviation between two points on a circle, e.g., two circular observations, the true mean direction and its estimator [SenGupta and Maitra (1998)], etc. may thus be based on it.

The average distance $d_i(\theta)$ of θ from the group i, is given by

$$d_i(\theta) \;=\; 1 \;-\; \frac{1}{n_i}\sum_j \cos(\theta_{ij} - \theta). \tag{5.2}$$

Note that this is similar to the sample circular variance with a shift in the mean direction. Let

$$\overline{C}_i \;=\; \frac{1}{n_i}\sum_j \cos(\theta_{ij}),\; \overline{S}_i \;=\; \frac{1}{n_i}\sum_j \sin(\theta_{ij}),\; \overline{R}_i = \sqrt{\overline{C}_i^2 + \overline{S}_i^2},\; \tan(\overline{\theta}_i) = \frac{\overline{S}_i}{\overline{C}_i}.$$

5.2.3 The chord-based rule

Let the new observation to classify be θ. Let d_{0i} be the distance of θ from $\bar{\theta}_i$, the circular mean for group i, $i = 1, 2$. Define $D(\theta) \;=\; d_{01}(\theta) - d_{02}(\theta)$. Suppose that prior probabilities for the two populations are equal and let c be a real constant. The classification rule is then given as follows:

$$\text{If}\quad D(\theta) < c \quad \text{assign } \theta \text{ to population 1,}$$

$$\text{and assign } \theta \text{ to population 2 otherwise.} \tag{5.3}$$

Now

$$D(\theta) = (\cos(\bar{\theta}_2) - \cos(\bar{\theta}_1))\cos\theta + (\sin(\bar{\theta}_2) - \sin(\bar{\theta}_1))\sin\theta. \quad (5.4)$$

Let

$$\tan(\theta_0) = \frac{\sin(\bar{\theta}_2) - \sin(\bar{\theta}_1)}{\cos(\bar{\theta}_2) - \cos(\bar{\theta}_1)}. \quad (5.5)$$

Note that $P(\bar{\theta}_1 = \bar{\theta}_2) = 0$, assuming that we are dealing with underlying continuous distributions and hence θ_0 is well defined (with probability 1).

Then (5.4) can be written as

$$D(\theta) = \sqrt{2 - 2\cos(\bar{\theta}_1 - \bar{\theta}_2)}\ \cos(\theta - \theta_0). \quad (5.6)$$

Note that by (5.5), there will be two solutions for θ_0.

The classification rule in (5.3) now reduces to an equivalent but a very simple form as

$$\text{If } \cos(\theta - \theta_0) > K \text{ assign } \theta \text{ to population 1,}$$
$$\text{and assign } \theta \text{ to population 2 otherwise,} \quad (5.7)$$

where K is an appropriate constant.

Remarks.

1. The direction θ_0 is orthogonal to the bisector of $\bar{\theta}_1$ and $\bar{\theta}_2$.

2. As is often done for the sake of simplicity of the classification rule [see, e.g., Rao (1973, p. 575, Eq. (8e.1.8))], we can take $K = 0$. The rule as given by equation (5.7) then simply partitions the circle into sectors of width 180°. In this case, explicitly, the sectors can be specified as one semicircle having θ_0 as its midpoint, and the complementary arc. Note that if the sample mean directions are equal, unequal circular variances have no effect on the rule. In this case θ_0 is simply the mean direction itself. However, when the sample mean directions are not equal, the circular variances do affect θ_0. It is obvious that the rule can be modified trivially to cover the case of unequal prior probabilities also. Finally, in case specified misclassification probabilities are to be maintained, K can be suitably determined by using the distribution of $\cos(\theta - \theta_0)$ as discussed in Section 5.4.

5.2.4 An extension of the chord-based rule

Let $V_1 = d_1(\bar{\theta}_1), V_2 = d_2(\bar{\theta}_2)$, i.e., V_i is the average intragroup "distance" from each other for the observations in group or sample i. Note that V_i is nothing but the sample "circular variance" for sample i, see, e.g., Mardia (1972, p. 21). Define the intra-group average d_{ii} from the sample mean direction as

$$d_{ii} = 1 - \frac{1}{n_i}\sum \cos(\theta_{ij} - \bar{\theta}_i). \quad (5.8)$$

Then $d_{ii} = 1 - \bar{R}_i = V_i$. Take constants $\alpha_i > 0, i = 1, 2$, and β, and define

$$D_1(\theta) = \alpha_1 \left(d_1(\theta) - \frac{d_{11} + d_{22}}{2} \right) - \alpha_2 \left(d_2(\theta) - \frac{d_{11} + d_{22}}{2} \right) + \beta.$$

The classification rule is given by

$$\text{If} \quad D_1(\theta) < 0 \quad \text{assign } \theta \text{ to population 1}$$

$$\text{and assign } \theta \text{ to population 2 otherwise.} \qquad (5.9)$$

Now $D_1(\theta)$ reduces to

$$\{\alpha_2 \cos(\bar{\theta}_2) - \alpha_1 \cos(\bar{\theta}_1)\} \cos\theta + \{\alpha_2 \sin(\bar{\theta}_2) - \alpha_1 \sin(\bar{\theta}_1)\} \sin\theta$$

$$+\frac{1}{2}(\alpha_1 - \alpha_2)(\bar{R}_1 + \bar{R}_2). \qquad (5.10)$$

Let

$$\tan(\theta_0) = \frac{\alpha_2 \sin(\bar{\theta}_2). - \alpha_1 \sin(\bar{\theta}_1)}{\alpha_2 \cos(\bar{\theta}_2) - \alpha_1 \cos(\bar{\theta}_1)}. \qquad (5.11)$$

Then by (5.11), there will be two solutions for θ_0. However, as is done [see, e.g., Jammalamadaka and SenGupta (2001)] for defining $\bar{\theta}$, θ_0 also may be defined uniquely by taking the quadrant specific arc-tan function by interpreting the numerator and denominator of the ratio in the right-hand side of (5.11) accordingly.

5.3 Relationship of Chord-based Rule with Other Rules

5.3.1 Fisher's rule

Assume that the underlying populations are in the CN family, i.e., $CN(\mu_i, \kappa_i)$, $i = 1, 2$. Recall that the density function corresponding to $CN(\mu, \kappa)$ is given by

$$f(\theta; \mu, \kappa) = \frac{1}{2\pi I_0(\kappa)} \exp[\kappa \cos(\theta - \mu)], \quad 0 \le \mu < 2\pi, \kappa > 0.$$

Note that, given the parameters, the standard Fisher type ('maximum likelihood') function would have the form

$$-\ln I_0(\kappa_1) + \ln I_0(\kappa_2) + \{\kappa_1 \cos(\mu_1) - \kappa_2 \cos(\mu_2)\} \cos(\theta)$$

$$+\{\kappa_1 \sin(\mu_1) - \kappa_2 \sin(\mu_2)\} \sin(\theta) + \beta \qquad (5.12)$$

$$= -\ln I_0(\kappa_1) + \ln I_0(\kappa_2) + \sqrt{\kappa_1^2 + \kappa_2^2 - 2\kappa_1\kappa_2 \cos(\mu_1 - \mu_2)}$$

$$\times \cos\left(\theta - \tan^{-1}\frac{\kappa_2 \sin\mu_2 - \kappa_1 \sin\mu_1}{\kappa_2 \cos\mu_2 - \kappa_1 \cos\mu_1}\right) + \beta. \quad (5.13)$$

Putting $\alpha_i = \kappa_i$ in (5.10) and observing that

$$\frac{d}{d\kappa} \ln I_0(\kappa) = A(\kappa),$$

we have

$$\ln I_0(\kappa) = \int A(\kappa) d\kappa.$$

Recall from the ML estimation of κ for the CN population, that $\bar{R}$ is asymptotically $A(\kappa)$. Note also that $\kappa\bar{R}^2/2$ approximates the integral $\int A(\kappa)d\kappa$ by a triangle. Note also that

$$\frac{d}{d\kappa} A(\kappa) = 1 - A^2(\kappa) - \frac{A(\kappa)}{\kappa}$$

and hence that for small change in κ, the order of change in $A(\kappa)$ is less than that of κ. Therefore,

$$\ln I_0(\kappa_1) - \ln I_0(\kappa_2) = \int_{\kappa_1}^{\kappa_2} A(\kappa)d\kappa \sim \frac{1}{2}[A(\kappa_1) + A(\kappa_2)](\kappa_2 - \kappa_1),$$

by the trapezoidal rule. Note that, if κ_1 and κ_2 are very close to each other, asymptotically (5.10) approximates (strongly converges to) the corresponding portion of (5.12). The equivalence between Fisher's rule and our rule then becomes clear. Thus although we have kept the rule flexible by introducing the constants α_is, a recommended choice in case von Mises distributions seem to be the underlying populations, is that which is found by substituting the pairs

$$\hat{\kappa}_i = A^{-1}(\bar{R}_i), \ i = 1, 2.$$

5.3.2 Cox's logistic discrimination rule

In the above discussion, the modified rule can easily be identified as a semiparametric rule which approaches the Fisher type rule (ratio of densities) when the underlying populations are circular normals and they are close to each other in terms of population parameters.

Since LGD models the ratio of densities in the case when the log ratio is linear in the underlying random variable, observe that LGD cannot be directly applied to discriminate between two von Mises populations. However, note that a simple generalization of LGD can still be used in such cases, since the log-ratio in this case is linear on the sine and cosine transformations of θ. This also bypasses the rather computationally tricky problem of having to estimate κ,

$A(\kappa)$ and their logarithms, as the constant term in the expression of the LGD subsumes all the Bessel function terms. This can also be approached through the method of Generalized Pseudo Maximum Likelihood estimation [see Roy (1999)]. Also, the rule based on chord lengths as given above, however then need not assume independence of the linear components as done for the LGD rule.

5.4 Exact Distribution of $D(\theta)$

The distribution of $\bar{\theta}$ conditional on $R = r$ is von Mises with mean direction μ and concentration parameter κr. The joint distribution of $C = R\cos\bar{\theta}$, $S = R\sin\bar{\theta}$ is given by

$$f(C, S) = \frac{1}{I_0^n(\kappa)} e^{\kappa(\cos(\mu)C + \sin(\mu)S)} \phi_n(C^2 + S^2). \tag{5.14}$$

Here ϕ_n is the density of R^2 when $\theta_1, \ldots, \theta_n$ is a random sample from a circular uniform distribution. The joint distribution of $U = \alpha_1\cos(\bar{\theta}_1) - \alpha_2\cos(\bar{\theta}_2)$, $V = \alpha_1\sin(\bar{\theta}_1) - \alpha_2\sin(\bar{\theta}_2)$, given $R_1 = r_1, R_2 = r_2, \alpha_1, \alpha_2$, is given by

$$\begin{aligned} f(U, V) &= \frac{e^{\kappa_1 r_1 \cos(\mu_1)\frac{U}{\alpha_1} + \kappa_2 r_2 \cos(\mu_2)\frac{V}{\alpha_2}}}{(2\pi)^2 I_0(\kappa_1 r_1) I_0(\kappa_2 r_2)} \\ &\quad \times \int_{\bar{\theta}_2} \exp\left\{ (\kappa_1 r_1 \cos(\mu_1)\frac{\alpha_2}{\alpha_1} + \kappa_2 r_2 \cos(\mu_2))\cos(\bar{\theta}_2) \right. \\ &\qquad \left. + (\kappa_1 r_1 \sin(\mu_1)\frac{\alpha_2}{\alpha_1} + \kappa_2 r_2 \sin(\mu_2))\sin(\bar{\theta}_2) \right\} d\bar{\theta}_2. \end{aligned}$$

Combining this with (5.14), we have the joint distribution of (C, S, U, V) (where $\cos(\theta) = C$ and $\sin(\theta) = S$), given $R_1 = r_1, R_2 = r_2, \alpha_1, \alpha_2$, to be

$$\begin{aligned} f(C, S, U, V) &= \frac{e^{\kappa_1 r_1 \cos(\mu_1)\frac{U}{\alpha_1} + \kappa_2 r_2 \cos(\mu_2)\frac{V}{\alpha_2} + \kappa\cos(\theta - \mu)}}{(2\pi)^3 I_0(\kappa) I_0(\kappa_1 r_1) I_0(\kappa_2 r_2)} \\ &\quad \times \phi_1(1) \int_{\bar{\theta}_2} e^{(\kappa_1 r_1 \cos(\mu_1)\frac{\alpha_2}{\alpha_1} + \kappa_2 r_2 \cos(\mu_2))\cos(\bar{\theta}_2)} \\ &\quad \times e^{(\kappa_1 r_1 \sin(\mu_1)\frac{\alpha_2}{\alpha_1} + \kappa_2 r_2 \sin(\mu_2))\sin(\bar{\theta}_2)} d\bar{\theta}_2. \end{aligned} \tag{5.15}$$

To get the distribution of the statistic, the conditional density in (5.15) multiplied by the joint density $h_{n_1}(R_1)h_{n_2}(R_2)$ [for the definition of $h_n(R)$, see e.g., Mardia (1972, p. 94)] has to be integrated over regions of the form $d = aCU + (1-a)SV$. This fact may be used in invoking numerical integration to obtain the constant K of Section 5.2 when specified misclassification probabilities are to be met.

Table 5.1: Comparison of Fisher's and chord-based rules

$n_1 = n_2 = 10, \mu_1 = 0, \kappa_1 = 0.10$

$\kappa_2 = 0.10$

μ_2	$\hat{\mu_1}$	$\hat{\mu_2}$	R_1	R_2	ERR_1	AER_1	ERR_2	AER_2
0.1	0.05	0.12	0.09	0.09	0.15	0.19	0.13	0.25
0.2	0.05	0.17	0.091	0.089	0.14	0.16	0.14	0.25
0.3	0.05	0.26	0.091	0.09	0.137	0.15	0.146	0.24
0.4	0.055	0.36	0.092	0.091	0.135	0.148	0.145	0.24
0.5	0.05	0.48	0.093	0.092	0.13	0.14	0.148	0.21

$\kappa_2 = 0.20$

μ_2	$\hat{\mu_1}$	$\hat{\mu_2}$	R_1	R_2	ERR_1	AER_1	ERR_2	AER_2
0.1	0.05	0.12	0.09	0.18	0.15	0.17	0.25	0.3
0.2	0.05	0.17	0.091	0.19	0.14	0.16	0.22	0.26
0.3	0.05	0.26	0.091	0.19	0.132	0.145	0.2	0.25
0.4	0.055	0.36	0.092	0.192	0.129	0.14	0.19	0.24
0.5	0.05	0.48	0.093	0.192	0.126	0.134	0.18	0.22

$\kappa_2 = 0.30$

μ_2	$\hat{\mu_1}$	$\hat{\mu_2}$	R_1	R_2	ERR_1	AER_1	ERR_2	AER_2
0.1	0.05	0.12	0.09	0.28	0.16	0.17	0.25	0.3
0.2	0.05	0.17	0.091	0.289	0.14	0.16	0.22	0.26
0.3	0.05	0.26	0.091	0.29	0.132	0.145	0.2	0.25
0.4	0.055	0.36	0.092	0.291	0.129	0.14	0.19	0.24
0.5	0.05	0.48	0.093	0.292	0.126	0.134	0.18	0.22

5.5 Efficiency of the Rule

As is apparent, closed form expressions for error probabilities do not exist and the actual values have to be numerically computed for each pair of training samples.

Table 5.1 presents and compares the performances of the Fisher's and the chord-based rules, where we have taken $\mu_1 = 0$ (angles are given in radians) without loss of generality. ERR_1 denotes the calculated error probability from the exact distribution of the modified statistic as given above, ERR_2 the calculated error probability from the Fisher type (ratio of densities) discrimination rule, AER_1 the apparent error rate from the modified statistic as given above, and AER_2 the apparent error rate from the Fisher type discrimination rule. It is clear that our proposed rule outperforms Fisher's rule in terms of both ERR and AER over all the parameter combinations considered.

5.6 A Real-life Example

We now consider the data on pigeon-homing, as referred to in Mardia (1972, pp. 156–157), in which the internal clocks of 10 birds were reset by 6 hours clockwise while the clocks of 9 birds were left unaltered. Assuming that the underlying distributions are von Mises with equal concentration parameters [as in Mardia (1972, p. 157)], we classify each observation in the two samples on the basis of the remaining observations, by comparing the average chord-length distance from each group.

The result shows that the apparent error rate (AER) is 0.0 for the control group, 0.25 for the experimental group and 0.117 for the combined sample.

The AERs or the sample misclassification probabilities show that the rule correctly classifies all the observations in the control group, and 75% in the experimental group.

References

1. Anderson, J. A. (1975). Quadratic logistic discrimination, *Biometrika*, **62**, 149–154.

2. Collett, D. and Lewis, T. (1981). Discriminating between the von Mises and wrapped normal distributions, *Australian Journal of Statistics,* **23**, 71–79.

3. Cox, D. R. (1966). Some procedures associated with the logistic qualitative response curve, In *Research Papers in Statistics* (Ed., F. N. David), pp. 55-71, John Wiley & Sons, New York.

4. El Khattabi, S. and Streit, F. (1996). Identification analysis in directional statistics, *Computational Statistics & Data Analysis,* **23**, 45–63.

5. Jammalamadaka, S. R. and SenGupta, A. (2001). *Topics in Circular Statistics,* World Scientific Publishers, New Jersey.

6. Mardia, K. V. (1972). *Statistics of Directional Data,* Academic Press, London.

7. Morris, J. E. and Laycock, P. J. (1974). Discriminant analysis of directional data, *Biometrika,* **61**, 335–341.

8. Rao, C. R. (1973). *Linear Statistical Inference and Its Applications,* Second edition, John Wiley & Sons, New York.

9. Roy, S. (1999). *On Logistic and Some New Discrimination Rules: Characterizations, Inference and Applications*, Ph.D. Thesis, Indian Statistical Institute, Kolkata, India.

10. SenGupta, A. and Maitra, R. (1998). On best equivariance and admissibility of simultaneous MLE for mean direction vectors of several Langevin distributions, *Annals of the Institute of Statistical Mathematics,* **50**, 715–727.

11. SenGupta, A. and Pal, C. (2001). On optimal tests for isotropy against the symmetric wrapped stable-circular uniform mixture family, *Journal of Applied Statistics,* **28**, 129–143.

PART II

Ranking and Selection

6

On Some Ranking and Selection Procedures for MANOVA Models with Applications

Deng-Yuan Huang[1] **and Ren-Fen Lee**[2]

[1]*Department of Business Administration, Fu-Jen Catholic University, Taipei, Taiwan, Republic of China*
[2]*National Kaohsiung University of Applied Sciences, Kaohsiung, Taiwan, Republic of China*

Abstract: The interaction of product or service attributes may influence the preferences of consumers. If the numbers of stimuli (profiles) are large, then the combinations of the levels present difficulty for consumers to rank order in terms of their preference. In this chapter, our purpose is to propose a ranking and selection procedure to select the most preferred attributes in such situations. We use the orthogonal array design to build a suitable MANOVA model in conjoint analysis. A split-plot experiment is conducted and the method of moments is used to obtain unbiased estimators of the parameters. The fitness of the MANOVA models is also studied. Some examples based on a preference study of a household electrical appliance are discussed.

Keywords and phrases: Ranking and selection, conjoint analysis, split-plot experiment, interactions, MANOVA model, utility, Spearman's rank correlation

6.1 Introduction

The conjoint analysis is an efficient tool in marketing research. Consumers are often asked to make a choice among the combinations of the levels of multi-attribute products or services. The interactions among attributes may influence the resulting choice. The main purpose of this paper is to propose a ranking and selection procedure to select the most preferred attributes. For this purpose, we build a MANOVA model based on the orthogonal array design and the split-plot experiment with interactions. The moment method of estimation is used to obtain an unbiased estimator of the part-utility. Then, we can overcome the difficulties in ranking a large number of profiles. The most favored product attributes and the most desirable levels of these product attributes are then selected. We also study the fitness of the MANOVA models. The methods introduced in the chapter are illustrated by examples.

6.2 MANOVA Model with Equal Factor Levels

Let us consider the example of an appliance maker who wants to understand what factor levels are cared most about by consumers when buying a refrigerator. Assume that there are three main attributes (brand, volume and price) that the consumers care about for buying a refrigerator. We consider each of these attributes having three levels. The attributes and the corresponding levels are listed as follows:

Factor	Level		
A. Brand	0. Taiwanese	1. American	2. Other Asians
B. Volume	0. Under 200 liters	1. 200-400 liters	2. Above 400 liters
C. Price	0. Under NT$20000	1. NT$20000-NT$30000	2. Above NT$30000

For the three attributes each with three levels, twenty-seven ($3 \times 3 \times 3$) combinations (profiles, stimuli) can be formed. To avoid the difficulty in ranking too many total number of profiles , we adopt an orthogonal fractional design as shown in Table 6.1, in which only eighteen stimuli are needed to be ranked. There are three steps in the split-plot experiment.

Step 1: We choose a factor as the main attribute.

Step 2: Fixing one level of the main attribute, we rank order the level combinations of the other attributes (called subplot treatments).

Step 3: Fixing another level of the main attribute, we rank order the subplot treatments which may not be the same subplot treatments as those obtained in Step 2. Repeat this process until all levels of the main attribute have been considered.

In the split-plot experiment, the order in which the main attribute levels are considered is randomly decided.

Table 6.1 is an orthogonal array which is produced by the statistical software SPSS [SPSS (1994, pp. 1–36)]. The main attribute in Table 6.1 is brand. Under different levels of brand, the sets of subplot treatments are different. We want to evaluate the interaction between volume and price in the subplot treatments of different sets. In Table 6.2, the code numbers are adapted from Table 6.1.

Table 6.1: Questionnaire

1. If the brand of the refrigerator is Taiwanese, please rank order or assign scores to the set of combinations, where the lower the value, the greater the preference. (The value '1' is the most preferred, and the value '6' is the least preferred.)		
□	Volume: under 200 liters	Price: under NT$20000
□	Volume: under 200 liters	Price: above NT$30000
□	Volume: above 400 liters	Price: NT$20000-NT$30000
□	Volume: above 400 liters	Price: under NT$20000
□	Volume: 200-400 liters	Price: NT$20000-NT$30000
□	Volume: 200-400 liters	Price: above NT$30000
2. If the brand of the refrigerator is American, please rank order or assign scores to the set of combinations, where the lower the value, the greater the preference. (The value '1' is the most preferred, and the value '6' is the least preferred.)		
□	Volume: under 200 liters	Price: NT$20000-NT$30000
□	Volume: 200-400 liters	Price: under NT$20000
□	Volume: above 400 liters	Price: above NT$30000
□	Volume: under 200 liters	Price: above NT$30000
□	Volume: 200-400 liters	Price: NT$20000-NT$30000
□	Volume: above 400 liters	Price: under NT$20000
3. If the brand of the refrigerator is Other Asians, please rank order or assign scores to the set of combinations, where the lower the value, the greater the preference. (The value '1' is the most preferred, and the value '6' is the least preferred.)		
□	Volume: above 400 liters	Price: above NT$30000
□	Volume: 200-400 liters	Price: above NT$30000
□	Volume: under 200 liters	Price: NT$20000-NT$30000
□	Volume: above 400 liters	Price: NT$20000-NT$30000
□	Volume: under 200 liters	Price: under NT$20000
□	Volume: 200-400 liters	Price: under NT$20000

Table 6.2: Orthogonal design

Stimulus no. (1)	A. Brand (2)	B. Volume (3)	C. Price (4)	BC (5)	BC^2 (6)
1	[0] Taiwanese	[0] under 200 liters	[0] under NT$20000	[0]	[0]
2	[0] Taiwanese	[0] under 200 liters	[2] above NT$30000	[2]	[1]
3	[0] Taiwanese	[2] above 400 liters	[1] NT$20000-NT$30000	[0]	[1]
4	[0] Taiwanese	[2] above 400 liters	[0] under NT$20000	[2]	[2]
5	[0] Taiwanese	[1] 200-400 liters	[1] NT$20000-NT$30000	[2]	[0]
6	[0] Taiwanese	[1] 200-400 liters	[2] above NT$30000	[0]	[2]
7	[1] American	[0] under 200 liters	[1] NT$20000-NT$30000	[1]	[2]
8	[1] American	[1] 200-400 liters	[0] under NT$20000	[1]	[1]
9	[1] American	[2] above 400 liters	[2] above NT$30000	[1]	[0]
10	[1] American	[0] under 200 liters	[2] above NT$30000	[2]	[1]
11	[1] American	[1] 200-400 liters	[1] NT$20000-NT$30000	[2]	[0]
12	[1] American	[2] above 400 liters	[0] under NT$20000	[2]	[2]
13	[2] Other Asians	[2] above 400 liters	[2] above NT$30000	[1]	[0]
14	[2] Other Asians	[1] 200-400 liters	[2] above NT$30000	[0]	[2]
15	[2] Other Asians	[0] under 200 liters	[1] NT$20000-NT$30000	[1]	[2]
16	[2] Other Asians	[2] above 400 liters	[1] NT$20000-NT$30000	[0]	[1]
17	[2] Other Asians	[0] under 200 liters	[0] under NT$20000	[0]	[0]
18	[2] Other Asians	[1] 200-400 liters	[0] under NT$20000	[1]	[1]

Footnote: The number in square brackets [ ] is the factor level.

Table 6.3: Ranks assigned to the stimuli listed in Table 6.1

r_1	r_2	r_3	r_4	r_5	r_6	r_7	r_8	r_9	r_{10}	r_{11}	r_{12}	r_{13}	r_{14}	r_{15}	r_{16}	r_{17}	r_{18}
3	6	1	5	2	4	4	3	1	5	2	6	4	6	5	1	2	3

Ranking the combinations of the factor levels

Let r_m represent the ranking result given by the responder to the mth stimulus in Table 6.1. Since under each brand, six subplot treatments are compared, the smallest value of r_m is 1 and the largest value is 6. The larger the value of r_m is, the more one dislikes this subplot treatment. Table 6.3 shows the ranks given by responder A to the stimuli in Table 6.1.

According to Table 6.3, for the Taiwanese brand, we find that stimulus 3 – the refrigerator with volume above 400 liters and price between NT$20000 and NT$30000 – is responder A's favorite.

Since each factor has two degrees of freedom, each two-factor interaction has 4 degrees of freedom. The interaction of volume and price is denoted as $B \times C$. It is easy to partition the two-factor interaction ($B \times C$) into two orthogonal two-degrees-of-freedom components (BC and BC^2). Each component, having no physical interpretation, has two degrees of freedom [Montgomery (1991, pp. 387–438)].

The component BC is constructed by using the equation $L_1 = Z_2 + Z_3$, where Z_2 is the level of the factor volume and Z_3 is the level of the factor price in a particular subplot treatment. The level of BC is given by the value of $L_1 (\text{mod } 3)$. That is, L_1 can take only values 0, 1 and 2. The component BC^2 is constructed by using the equation $L_2 = Z_2 + 2Z_3$, and its level is given by the value of $L_2 (\text{mod } 3)$. For example, for the second stimulus in Table 6.2 with $Z_2 = 0$ and $Z_3 = 2$ we get $Z_2 + Z_3 = 2$ and $Z_2 + 2Z_3 = 1$. So, the level of BC is 2 and the level of BC^2 is 1, which are shown in the 5th and 6th columns of Table 6.2, respectively. The other corresponding levels of BC and BC^2 for all the other stimuli are also listed in these two columns of Table 6.2.

The relations between the rankings of stimuli given by a responder and the levels of factors can be represented by the following MANOVA model.

$$\mathbf{Y}_{18\times 1} = X_{18\times 11}\boldsymbol{\beta}_{11\times 1} + \boldsymbol{\varepsilon}_{18\times 1}, \tag{6.1}$$

where $\mathbf{Y}' = (Y_{000}, Y_{002}, \ldots, Y_{210})$. In Y_{ijk} the subscripts i, j, k denote, respectively, the levels of brand, volume, and price. We let $Y_{ijk} = 7 - r_m$ to convert the rankings for convenience of interpretation. The greater the value of Y_{ijk} is, the higher is a preference for the combination of levels. Then the converted data collected from responder A is

$$\begin{aligned}(&Y_{000}, Y_{002}, Y_{021}, Y_{020}, Y_{011}, Y_{012}, Y_{101}, Y_{110}, Y_{122}, Y_{102}, Y_{111}, Y_{120}, Y_{222}, Y_{212},\\ &Y_{201}, Y_{221}, Y_{200}, Y_{210})\\ &= (4, 1, 6, 2, 5, 3, 3, 4, 6, 2, 5, 1, 3, 1, 2, 6, 5, 4).\end{aligned}$$

The error term

$$\begin{aligned}\boldsymbol{\varepsilon}_{18\times 1} &= (\varepsilon_0^{(1)} + \varepsilon_1^{(2)}, \varepsilon_0^{(1)} + \varepsilon_2^{(2)}, \ldots, \varepsilon_1^{(1)} + \varepsilon_7^{(2)}, \varepsilon_1^{(1)} + \varepsilon_8^{(2)}, \ldots, \varepsilon_2^{(1)} + \varepsilon_{14}^{(2)},\\ &\qquad \varepsilon_2^{(1)} + \varepsilon_{15}^{(2)}, \varepsilon_2^{(1)} + \varepsilon_{16}^{(2)}, \varepsilon_2^{(1)} + \varepsilon_{17}^{(2)}, \varepsilon_2^{(1)} + \varepsilon_{18}^{(2)})\\ &= \boldsymbol{\varepsilon}^{(1)} + \boldsymbol{\varepsilon}^{(2)}\end{aligned}$$

satisfies

$$\begin{aligned}&E(\boldsymbol{\varepsilon}^{(1)}) = \mathbf{0},\ E(\boldsymbol{\varepsilon}^{(2)}) = \mathbf{0},\ \text{Cov}(\boldsymbol{\varepsilon}^{(1)}) = \boldsymbol{\Sigma}_1,\ \text{Cov}(\boldsymbol{\varepsilon}^{(2)}) = \boldsymbol{\Sigma}_2,\\ &\text{Cov}(\boldsymbol{\varepsilon}_i^{(1)}, \boldsymbol{\varepsilon}_j^{(2)}) = 0.\end{aligned}$$

In conjoint analysis, the part-utility is what is known as the "effect" in experimental design. In (6.1), $\boldsymbol{\beta}' = [\mu, a_0, a_1, b_0, b_1, c_0, c_1, bc_0, bc_1, bc_0^2, bc_1^2]$, where μ is the overall mean, a_i is the part-utility of the ith level of brand (factor A), b_j is the part-utility of the jth level of volume (factor B), c_k is the part-utility of the kth level of price (factor C), bc_s is the part-utility of the sth level of BC, and bc_u^2 is the part-utility of the uth level of BC^2. They satisfy the following conditions:

$$\begin{aligned}&a_0 + a_1 + a_2 = 0, \quad b_0 + b_1 + b_2 = 0, \quad c_0 + c_1 + c_2 = 0,\\ &bc_0 + bc_1 + bc_2 = 0, \quad \text{and} \quad bc_0^2 + bc_1^2 + bc_2^2 = 0.\end{aligned}$$

The design matrix $X_{18\times 11}$ for the orthogonal array in Table 6.2 is

$$X_{18\times 11} = \begin{bmatrix}
1 & 1 & 0 & 1 & 0 & 1 & 0 & 1 & 0 & 1 & 0 \\
1 & 1 & 0 & 1 & 0 & -1 & -1 & -1 & -1 & 0 & 1 \\
1 & 1 & 0 & -1 & -1 & 0 & 1 & 1 & 0 & 0 & 1 \\
1 & 1 & 0 & -1 & -1 & 1 & 0 & -1 & -1 & -1 & -1 \\
1 & 1 & 0 & 0 & 1 & 0 & 1 & -1 & -1 & 1 & 0 \\
1 & 1 & 0 & 0 & 1 & -1 & -1 & 1 & 0 & -1 & -1 \\
1 & 0 & 1 & 1 & 0 & 0 & 1 & 0 & 1 & -1 & -1 \\
1 & 0 & 1 & 0 & 1 & 1 & 0 & 0 & 1 & 0 & 1 \\
1 & 0 & 1 & -1 & -1 & -1 & -1 & 0 & 1 & 1 & 0 \\
1 & 0 & 1 & 1 & 0 & -1 & -1 & -1 & -1 & 0 & 1 \\
1 & 0 & 1 & 0 & 1 & 0 & 1 & -1 & -1 & 1 & 0 \\
1 & 0 & 1 & -1 & -1 & 1 & 0 & -1 & -1 & -1 & -1 \\
1 & -1 & -1 & -1 & -1 & -1 & -1 & 0 & 1 & 1 & 0 \\
1 & -1 & -1 & 0 & 1 & -1 & -1 & 1 & 0 & -1 & -1 \\
1 & -1 & -1 & 1 & 0 & 0 & 1 & 0 & 1 & -1 & -1 \\
1 & -1 & -1 & -1 & -1 & 0 & 1 & 1 & 0 & 0 & 1 \\
1 & -1 & -1 & 1 & 0 & 1 & 0 & 1 & 0 & 1 & 0 \\
1 & -1 & -1 & 0 & 1 & 1 & 0 & 0 & 1 & 0 & 1
\end{bmatrix}$$

For example, the 15th stimulus in Table 6.2 (Other Asian brands, volume under 200 liters, and price between NT\$20000 and NT\$30000) can be expressed as:

$$\begin{aligned}
Y_{201} &= \mu + a_2 + \varepsilon_2^{(1)} + b_0 + c_1 + bc_1 + bc_2^2 + \varepsilon_{15}^{(2)} \\
&= \mu - a_0 - a_1 + \varepsilon_2^{(1)} + b_0 + c_1 + bc_1 - bc_0^2 - bc_1^2 + \varepsilon_{15}^{(2)}. \qquad (6.2)
\end{aligned}$$

Similarly, the other stimuli can also be expressed as

$$\begin{aligned}
Y_{000} &= \mu + a_0 + \varepsilon_0^{(1)} + b_0 + c_0 + bc_0 + bc_0^2 + \varepsilon_1^{(2)}, \\
\cdots &= \cdots \\
\cdots &= \cdots \\
\cdots &= \cdots \\
Y_{210} &= \mu - a_0 - a_1 + \varepsilon_2^{(1)} + b_1 + c_0 + bc_1 + bc_1^2 + \varepsilon_{18}^{(2)}.
\end{aligned}$$

Let the set of 18 combinations in the orthogonal array (Table 6.2) be

$$R = \left\{ \begin{matrix} (0,0,0), (0,0,2), (0,2,1), (0,2,0), (0,1,1), (0,1,2), (1,0,1), (1,1,0), (1,2,2), \\ (1,0,2), (1,1,1), (1,2,0), (2,2,2), (2,1,2), (2,0,1), (2,2,1), (2,0,0), (2,1,0) \end{matrix} \right\}.$$

Let

$$Y_{...} = \sum_{i=0}^{2} \sum_{\substack{j=0 \\ (i,j,k)\in R}}^{2} \sum_{k=0}^{2} Y_{ijk}, \quad \overline{Y}_{...} = \frac{Y_{...}}{18}, \quad Y_{i..} = \sum_{\substack{j=0 \\ (i,j,k)\in R}}^{2} \sum_{k=0}^{2} Y_{ijk}, \quad \overline{Y}_{i..} = \frac{Y_{i..}}{6},$$

$$Y_{.j.} = \sum_{\substack{i=0 \\ (i,j,k)\in R}}^{2} \sum_{k=0}^{2} Y_{ijk}, \quad \overline{Y}_{.j.} = \frac{Y_{.j.}}{6}, \quad Y_{..k} = \sum_{\substack{i=0 \\ (i,j,k)\in R}}^{2} \sum_{j=0}^{2} Y_{ijk}, \quad \overline{Y}_{..k} = \frac{Y_{..k}}{6}.$$

Since

$$\begin{aligned}
E(Y_{000}) &= \mu + a_0 + b_0 + c_0 + bc_0 + bc_0^2, \\
\cdots &= \cdots \\
\cdots &= \cdots \\
\cdots &= \cdots \\
E(Y_{210}) &= \mu - a_0 - a_1 + b_1 + c_0 + bc_1 + bc_1^2,
\end{aligned}$$

we get

$$E(\overline{Y}_{...}) = \tfrac{1}{18}E(18\mu) = \mu,$$

$$E\left(\frac{(Y_{000}+Y_{021}+Y_{012}+Y_{212}+Y_{221}+Y_{200})-(Y_{002}+Y_{020}+Y_{011}+Y_{222}+Y_{201}+Y_{210})}{9}\right)$$

$$= \tfrac{1}{9}E(9bc_0) = bc_0,$$

$$E\left(\frac{(Y_{000}+Y_{002}+Y_{021}+Y_{020}+Y_{011}+Y_{012})-(Y_{102}+Y_{111}+Y_{120}+Y_{212}+Y_{221}+Y_{290})}{9}\right) = a_0,$$

$$E\left(\frac{(Y_{101}+Y_{110}+Y_{122}+Y_{102}+Y_{111}+Y_{120})-(Y_{002}+Y_{020}+Y_{011}+Y_{222}+Y_{201}+Y_{210})}{9}\right) = a_1,$$

$$E(\overline{Y}_{.0.} - \overline{Y}_{...}) = b_0, \qquad E(\overline{Y}_{.1.} - \overline{Y}_{...}) = b_1,$$

$$E(\overline{Y}_{..0} - \overline{Y}_{...}) = c_0, \qquad E(\overline{Y}_{..1} - \overline{Y}_{...}) = c_1,$$

$$E\left(\frac{(Y_{101}+Y_{110}+Y_{122}+Y_{222}+Y_{201}+Y_{210})-(Y_{102}+Y_{111}+Y_{120}+Y_{212}+Y_{221}+Y_{200})}{9}\right) = bc_1,$$

$$E\left(\frac{(Y_{000}+Y_{011}+Y_{122}+Y_{111}+Y_{222}+Y_{200})}{6} - \overline{Y_{...}}\right) = bc_0^2,$$

$$E\left(\frac{(Y_{002}+Y_{021}+Y_{110}+Y_{102}+Y_{221}+Y_{210})}{6} - \overline{Y_{...}}\right) = bc_1^2.$$

Then, using the method of moments, the unbiased estimate of $\hat{\beta}$ is

$$\hat{\beta} = \begin{bmatrix} \hat{\mu} \\ \hat{a}_0 \\ \hat{a}_1 \\ \hat{b}_0 \\ \hat{b}_1 \\ \hat{c}_0 \\ \hat{c}_1 \\ \widehat{bc}_0 \\ \widehat{bc}_1 \\ \widehat{bc}_0^2 \\ \widehat{bc}_1^2 \end{bmatrix} = \begin{bmatrix} \overline{Y}_{...} \\ \frac{(Y_{000}+Y_{002}+Y_{021}+Y_{020}+Y_{011}+Y_{021})-(Y_{102}+Y_{111}+Y_{120}+Y_{212}+Y_{221}+Y_{200})}{9} \\ \frac{(Y_{101}+Y_{110}+Y_{122}+Y_{102}+Y_{111}+Y_{120})-(Y_{002}+Y_{020}+Y_{011}+Y_{222}+Y_{201}+Y_{210})}{9} \\ \frac{(Y_{000}+Y_{002}+Y_{101}+Y_{102}+Y_{201}+Y_{200})}{6} - \overline{Y}_{...} \\ \frac{(Y_{011}+Y_{012}+Y_{110}+Y_{111}+Y_{212}+Y_{210})}{6} - \overline{Y}_{...} \\ \frac{(Y_{000}+Y_{020}+Y_{110}+Y_{120}+Y_{200}+Y_{210})}{6} - \overline{Y}_{...} \\ \frac{(Y_{021}+Y_{011}+Y_{101}+Y_{111}+Y_{201}+Y_{221})}{6} - \overline{Y}_{...} \\ \frac{(Y_{000}+Y_{021}+Y_{012}+Y_{212}+Y_{221}+Y_{200})-(Y_{002}+Y_{020}+Y_{011}+Y_{222}+Y_{201}+Y_{210})}{9} \\ \frac{(Y_{101}+Y_{110}+Y_{122}+Y_{222}+Y_{201}+Y_{210})-(Y_{102}+Y_{111}+Y_{120}+Y_{212}+Y_{221}+Y_{200})}{9} \\ \frac{(Y_{000}+Y_{011}+Y_{122}+Y_{111}+Y_{222}+Y_{200})}{6} - \overline{Y}_{...} \\ \frac{(Y_{002}+Y_{021}+Y_{110}+Y_{102}+Y_{221}+Y_{210})}{6} - \overline{Y}_{...} \end{bmatrix} \tag{6.3}$$

Now, using the data of responder A in (6.3), we have

$$\begin{aligned}
\hat{\mu} &= 3.5,\ \hat{a}_0 = \frac{1}{9},\ \hat{a}_1 = \frac{4}{9},\ \hat{a}_2 = -\frac{5}{9}, \\
\hat{b}_0 &= -\frac{2}{3},\ \hat{b}_1 = \frac{1}{6},\ \hat{b}_2 = \frac{1}{2}, \\
\hat{c}_0 &= -\frac{1}{6},\ \hat{c}_1 = 1, \hat{c}_2 = -\frac{5}{6}, \\
\widehat{bc}_0 &= \frac{8}{9},\ \widehat{bc}_1 = \frac{2}{9},\ \widehat{bc}_1 = -\frac{10}{9}, \\
\widehat{bc}_0^2 &= \frac{7}{6},\ \widehat{bc}_1^2 = \frac{1}{3},\ \widehat{bc}_2^2 = -\frac{3}{2}.
\end{aligned}$$

The total utility of a stimulus can be expressed as:

The Total Utility = Overall Mean + Part-utility of Brand
+ Part-utility of Volume + Part-utility of Price
+ Part-utility of Interaction of Volume and Price.

The more the total utility is, the more the responder likes the stimulus. Using responder A's data, the total utility of the 2nd stimulus in Table 6.2 is:

Overall Mean + Part-utility (Taiwanese Brand)
+ Part-utility (Volume under 200 liters)
+ Part-utility (Price above NT$30000) + $bc_2 + bc_1^2$.

The estimate of this total utility is

$$\hat{\mu} + \hat{a}_0 + \hat{b}_0 + \hat{c}_2 + \widehat{bc}_2 + \widehat{bc}_1^2 = \frac{7}{2} + \frac{1}{9} + \left(-\frac{2}{3}\right) + \left(-\frac{5}{6}\right) + \left(-\frac{10}{9}\right) + \frac{1}{3} = \frac{4}{3}\ .$$

Similarly, the estimated total utility of the 5th stimulus is $\frac{7}{2} + \frac{1}{9} + \frac{1}{6} + 1 + \left(-\frac{10}{9}\right) + \frac{7}{6} = \frac{29}{6}$. Therefore, responder A prefers the fifth stimulus more than the second.

Fitness of the MANOVA model

The association coefficient Spearman ρ is used to evaluate the consistency between the predicted ranks obtained from the MANOVA model and the observed ranks. Using the responder A's data, the results are listed in Table 6.4.

The Spearman ρ values in Table 6.4 are all positive and close to 1. It indicates that using the MANOVA model to predict someone's preference ranks is suitable. Kendall τ is also useful to evaluate the goodness of fit.

Table 6.4: Responder A's Spearman ρ

Brand	Spearman ρ
Taisanese	0.97
American	0.93
Other Asians	0.94

Selection procedures for the attribute with most relative importance

Since the estimate of part-utility is under the same scale, the relative importance of factors can be computed directly. The larger the part-utility range is, the more important the factor is. The ranges of the part-utility of the levels of the factors for responder A in Table 6.5 are as follows:

the range of all the part-utility of the levels of the volume (B) is

$$\max_{i \neq j}(b_i - b_j) = \hat{b}_2 - \hat{b}_0 = \frac{1}{2} - \left(-\frac{2}{3}\right) = \frac{7}{6}\,,$$

the range of all the part-utility of the levels of the price (C) is

$$\max_{i \neq j}(c_i - c_j) = \hat{c}_1 - \hat{c}_2 = 1 - \left(-\frac{5}{6}\right) = \frac{11}{6}\,.$$

Fixing the level of brand, we compare the relative importance of volume and price. The relative importance score is computed by taking the range of all the part-utilities (volume (b) or price (c)) of a specific factor and dividing it by the sum of all the part utilities (volume (b) and price (c)) of the factors. If we let $R_1 = \max_{i \neq j}(\hat{b}_i - \hat{b}_j)$ and $R_2 = \max_{i \neq j}(\hat{c}_i - \hat{c}_j)$, then the relative importance scores of the attributes are $\frac{R_i}{R_1+R_2} \times 100\%$, $i = 1, 2$.

The selection procedure is: Select the attribute with the largest relative importance score as the most preferred factor.

For responder A:

The importance score of volume is $\frac{7/6}{7/6+11/6} \times 100\% = 38.9\%$.

The importance score of price is $\frac{11/6}{7/6+11/6} \times 100\% = 61.1\%$.

Based on the importance score (61.1%), we select the price as the most preferred factor for responder A. The above results for responder A are summarized in Table 6.5.

There are two special situations about the method we discussed. One is the interaction model without split-plot design. The second is the split-plot design but without interaction terms. These are described below.

Table 6.5: Results for responder A

Factor	High part-utility	Low part-utility	Range	Importance score
Volume	$\frac{1}{2}$	$-\frac{2}{3}$	$\frac{7}{6}$	38.9%
Price	1	$-\frac{5}{6}$	$\frac{11}{6}$	61.1%

(1) The MANOVA model with interaction but without split-plot design

We reconsider the example of the refrigerators. The new questionnaire is shown in Table 6.6.

The difference between Tables 6.1 and 6.6 is that there are eighteen combinations to be ranked in Table 6.6 whereas there are only six combinations to be ranked under each level of the main attribute in Table 6.1. Since the interaction terms are considered, the number of the questions in the present questionnaire is larger, namely 18.

In Table 6.7 that follows, the code numbers are adapted from Table 6.6. The components (BC and BC^2) of the interaction terms are constructed the same way as discussed earlier.

The rank orders (or scores) given by responder B to the stimuli in Table 6.6 are shown in Table 6.8. According to Table 6.8, we find that stimulus 7 – refrigerator with Other Asians brand, volume above 400 liters and price between NT\$20000 and NT\$30000 – is the responder B's favorite.

The relations between the rankings of stimuli given by a responder and the levels of factors can be represented by the following MANOVA model:

$$\mathbf{Y}_{18\times 1} = X_{18\times 11}\boldsymbol{\beta}_{11\times 1} + \boldsymbol{\varepsilon}_{18\times 1}, \tag{6.4}$$

where $\mathbf{Y}'_{18\times 1} = [Y_1, Y_2, \ldots, Y_{18}]$, and $Y_i = 19 - r_i$. The greater the value Y_j is, the more the responder prefers the associated combination. Then the converted data collected from responder B is

$$\begin{aligned}&(Y_1, Y_2, Y_3, Y_4, Y_5, Y_6, Y_7, Y_8, Y_9, Y_{10}, Y_{11}, Y_{12}, Y_{13}, Y_{14}, Y_{15}, Y_{16}, Y_{17}, Y_{18})\\ &= (7, 16, 4, 5, 1, 8, 18, 10, 12, 14, 3, 6, 13, 15, 11, 9, 2, 17).\end{aligned}$$

The error term $\boldsymbol{\varepsilon}' = (\varepsilon_1, \ldots, \varepsilon_{18})$, satisfies $E(\boldsymbol{\varepsilon}) = \mathbf{0}$, $\text{Cov}(\boldsymbol{\varepsilon}) = \mathbf{0}$. In (6.4), $\boldsymbol{\beta}'_{11\times 1} = (\mu, a_0, a_1, b_0, b_1, c_0, c_1, bc_0, bc_1, bc_0^2, bc_1^2)$, the components of which satisfy the conditions: $a_0 + a_1 + a_2 = 0$, $b_0 + b_1 + b_2 = 0$, $c_0 + c_1 + c_2 = 0$,

Table 6.6: Questionnaire

Please rank order or assign scores to the set of combinations, where the lower the value, the greater the preference.

(The value '1' is the most preferred, and the value '18' is the least preferred)

☐	1. Brand: Taiwanese	Volume: under 200 liters	Price: under NT$20000
☐	2. Brand: Other Asian	Volume: above 400 liters	Price: above NT$30000
☐	3. Brand: Other Asians	Volume: 200-400 liters	Price: above NT$30000
☐	4. Brand: American	Volume: under 200 liters	Price: NT$20000-NT$30000
☐	5. Brand: Taiwanese	Volume: under 200 liters	Price: above NT$30000
☐	6. Brand: Other Asians	Volume: under 200 liters	Price: NT$20000-NT$30000
☐	7. Brand: Other Asians	Volume: above 400 liters	Price: NT$20000-NT$30000
☐	8. Brand: Taiwanese	Volume: above 400 liters	Price: NT$20000-NT$30000
☐	9. Brand: American	Volume: 200-400 liters.	Price: under NT$20000
☐	10.Brand: American.	Volume: above 400 liters	Price: above NT$30000
☐	11. Brand: American	Volume: under 200 liters	Price: above NT$30000
☐	12. Brand: Other Asians	Volume: under 200 liters	Price: under NT$20000
☐	13. Brand: American	Volume: 200-400 liters	Price: NT$20000-NT$30000
☐	14. Brand: Other Asians	Volume: 200-400 liters	Price: under NT$20000
☐	15. Brand: Taiwanese	Volume: above 400 liters	Price: under NT$20000
☐	16. Brand: Taiwanese	Volume: 200-400 liters	Price: NT$20000-NT$30000
☐	17. Brand: Taiwanese	Volume: 200-400 liters	Price: above NT$30000
☐	18. Brand: American	Volume: above 400 liters	Price: under NT$20000

Table 6.7: Orthogonal design

Stimulus no. (1)	*A.* Brand (2)	*B.* Volume (3)	*C.* Price (4)	*BC* (5)	BC^2 (6)
1	[0] Taiwanese	[0] under 200 liters	[0] under NT$20000	[0]	[0]
2	[2] Other Asians	[2] above 400 liters	[2] above NT$30000	[1]	[0]
3	[2] Other Asians	[1] 200-400 liters	[2] above NT$30000	[0]	[2]
4	[1] American	[0] under 200 liters	[1] NT$20000-NT$30000	[1]	[2]
5	[0] Taiwanese	[0] under 200 liters	[2] above NT$30000	[2]	[1]
6	[2] Other Asians	[0] under 200 liters	[1] NT$20000-NT$30000	[1]	[2]
7	[2] Other Asians	[2] above 400 liters	[1] NT$20000-NT$30000	[0]	[1]
8	[0] Taiwanese	[2] above 400 liters	[1] NT$20000-NT$30000	[0]	[1]
9	[1] American	[1] 200-400 liters	[0] under NT$20000	[1]	[1]
10	[1] American	[2] above 400 liters	[2] above NT$30000	[1]	[0]
11	[1] American	[0] under 200 liters	[2] above NT$30000	[2]	[1]
12	[2] Other Asians	[0] under 200 liters	[0] under NT$20000	[0]	[0]
13	[1] American	[1] 200-400 liters	[1] NT$20000-NT$30000	[2]	[0]
14	[2] Other Asians	[1] 200-400 liters	[0] under NT$20000	[1]	[1]
15	[0] Taiwanese	[2] above 400 liters	[0] under NT$20000	[2]	[2]
16	[0] Taiwanese	[1] 200-400 liters	[1] NT$20000-NT$30000	[2]	[0]
17	[0] Taiwanese	[1] 200-400 liters	[2] above NT$30000	[0]	[2]
18	[1] American	[2] above 400 liters	[0] under NT$20000	[2]	[2]

Footnote: The number in square brackets [ ] is the factor level.

Table 6.8: Rank order given to stimuli in Table 6.6

r_1	r_2	r_3	r_4	r_5	r_6	r_7	r_8	r_9	r_{10}	r_{11}	r_{12}	r_{13}	r_{14}	r_{15}	r_{16}	r_{17}	r_{18}
12	3	15	14	18	11	1	9	7	5	16	13	6	4	8	10	17	2

$bc_0 + bc_1 + bc_2 = 0$ and $bc_0^2 + bc_1^2 + bc_2^2 = 0$. The design matrix $X_{18\times 11}$ for the orthogonal array in Table 6.7 is

$$X_{18\times 11} = \begin{bmatrix}
1 & 1 & 0 & 1 & 0 & 1 & 0 & 1 & 0 & 1 & 0 \\
1 & -1 & -1 & -1 & -1 & -1 & -1 & 0 & 1 & 1 & 0 \\
1 & -1 & -1 & 0 & 1 & -1 & -1 & 1 & 0 & -1 & -1 \\
1 & 0 & 1 & 1 & 0 & 0 & 1 & 0 & 1 & -1 & -1 \\
1 & 1 & 0 & 1 & 0 & -1 & -1 & -1 & -1 & 0 & 1 \\
1 & -1 & -1 & 1 & 0 & 0 & 1 & 0 & 1 & -1 & -1 \\
1 & -1 & -1 & -1 & -1 & 0 & 1 & 1 & 0 & 0 & 1 \\
1 & 1 & 0 & -1 & -1 & 0 & 1 & 1 & 0 & 0 & 1 \\
1 & 0 & 1 & 0 & 1 & 1 & 0 & 0 & 1 & 0 & 1 \\
1 & 0 & 1 & -1 & -1 & -1 & -1 & 0 & 1 & 1 & 0 \\
1 & 0 & 1 & 1 & 0 & -1 & -1 & -1 & -1 & 0 & 1 \\
1 & -1 & -1 & 1 & 0 & 1 & 0 & 1 & 0 & 1 & 0 \\
1 & 0 & 1 & 0 & 1 & 0 & 1 & -1 & -1 & 1 & 0 \\
1 & -1 & -1 & 0 & 1 & 1 & 0 & 0 & 1 & 0 & 1 \\
1 & 1 & 0 & -1 & -1 & 1 & 0 & -1 & -1 & -1 & -1 \\
1 & 1 & 0 & 0 & 1 & 0 & 1 & -1 & -1 & 1 & 0 \\
1 & 1 & 0 & 0 & 1 & -1 & -1 & 1 & 0 & -1 & -1 \\
1 & 0 & 1 & -1 & -1 & 1 & 0 & -1 & -1 & -1 & -1
\end{bmatrix}.$$

It can be proved that the moment estimators $\hat{\boldsymbol{\beta}}'_{11\times 1}$ are the unbiased estimators of $\boldsymbol{\beta}'_{11\times 1}$, where

$$\hat{\beta} = \begin{bmatrix} \hat{\mu} \\ \hat{a}_0 \\ \hat{a}_1 \\ \hat{b}_0 \\ \hat{b}_1 \\ \hat{c}_0 \\ \hat{c}_1 \\ \widehat{bc}_0 \\ \widehat{bc}_1 \\ \widehat{bc}_0^2 \\ \widehat{bc}_1^2 \end{bmatrix} = \begin{bmatrix}
\frac{T}{18} \\
\frac{(Y_1+Y_5+Y_8+Y_{15}+Y_{16}+Y_{17})-(Y_3+Y_7+Y_{11}+Y_{12}+Y_{13}+Y_{18})}{9} \\
\frac{(Y_4+Y_9+Y_{10}+Y_{11}+Y_{13}+Y_{18})-(Y_2+Y_5+Y_6+Y_{14}+Y_{15}+Y_{16})}{9} \\
\frac{3(Y_1+Y_4+Y_5+Y_6+Y_{11}+Y_{12})-T}{18} \\
\frac{3(Y_3+Y_9+Y_{13}+Y_{14}+Y_{16}+Y_{17})-T}{18} \\
\frac{3(Y_1+Y_9+Y_{12}+Y_{14}+Y_{15}+Y_{18})-T}{18} \\
\frac{3(Y_4+Y_6+Y_7+Y_8+Y_{13}+Y_{16})-T}{18} \\
\frac{(Y_1+Y_3+Y_7+Y_8+Y_{12}+Y_{17})-(Y_2+Y_5+Y_6+Y_{14}+Y_{15}+Y_{16})}{9} \\
\frac{(Y_2+Y_4+Y_6+Y_9+Y_{10}+Y_{14})-(Y_3+Y_7+Y_{11}+Y_{12}+Y_{13}+Y_{18})}{9} \\
\frac{3(Y_1+Y_2+Y_{10}+Y_{12}+Y_{13}+Y_{16})-T}{18} \\
\frac{3(Y_5+Y_7+Y_8+Y_9+Y_{11}+Y_{14})-T}{18}
\end{bmatrix}, \tag{6.5}$$

and

$$T = Y_1 + Y_2 + \cdots + Y_{18}.$$

Also,

$$\hat{a}_2 = -\hat{a}_0 - \hat{a}_1, \quad \hat{b}_2 = -\hat{b}_0 - \hat{b}_1, \quad \hat{c}_2 = -\hat{c}_0 - \hat{c}_1,$$
$$\widehat{bc}_2 = -\widehat{bc}_0 - \widehat{bc}_1, \quad \widehat{bc}_2^2 = -\widehat{bc}_0^2 - \widehat{bc}_1^2.$$

Table 6.9: Results for responder B

Factor	High part-utility	Low part-utility	Range	Importance score
Brand	1.89	−2.33	4.22	23.2%
Volume	4.83	−4.50	9.33	51.2%
Price	1.83	−2.83	4.66	25.6%
Responder B's Spearman $\rho = 0.97$				

Using the data of responder B in (6.5), we have

$$\begin{aligned}&\hat{\mu} = 9.5,\ \hat{a}_0 = -2.33,\ \hat{a}_1 = 0.44,\ \hat{a}_2 = 1.89,\\ &\hat{b}_0 = -4.5,\ \hat{b}_1 = -0.33,\ \hat{b}_2 = 4.83,\\ &\hat{c}_0 = 1.83,\ \hat{c}_1 = 1,\ \hat{c}_2 = -2.83,\\ &\widehat{bc}_0 = -1.44,\ \widehat{bc}_1 = 1,\ \widehat{bc}_2 = 0.44,\\ &\widehat{bc}_0^2 = 1.33,\ \widehat{bc}_1^2 = 0.33.\ \widehat{bc}_2^2 = -1.66.\end{aligned}$$

After computing, the results are as shown in Table 6.9. For responder B, the variable volume is the most important factor.

Except for the error terms, models (6.1) and (6.4) are alike. From (6.3) and (6.5), we know that the estimators of $\boldsymbol{\beta}'_{11\times 1}$ are also the same. However, in the split-plot experiment, the conclusions are made under each level of the main attribute.

(2) The MANOVA model with split-plot design but without interaction terms

Again, we consider the example of the refrigerators. The questionnaire is shown in Table 6.10.

From Tables 6.1 and 6.10, we know that the split-plot design without interaction terms does not need many questions in the questionnaire. Table 6.11 shows the rank orders for responder C.

The relations between the rankings of stimuli given by a responder and the levels of factors can be represented by the following MANOVA model.

$$\mathbf{Y}_{9\times 1} = X_{9\times 7}\boldsymbol{\beta}_{7\times 1} + \boldsymbol{\varepsilon}_{9\times 1}, \tag{6.6}$$

where $\mathbf{Y}'_{9\times 1} = [Y_{012}, Y_{021}, \ldots, Y_{211}]$. In Y_{ijk}, the subscripts i, j, and k stand for the levels of brand, volume, and price level. We use the equation $Y_{ijk} = 4 - r_m$ to convert the rankings (or ratings) for convenience in explanation. The greater the value Y_{ijk} is, the more the responder prefers the associated combination. Then the converted data collected from responder C is

$$(Y_{012}, Y_{021}, Y_{000}, Y_{122}, Y_{110}, Y_{101}, Y_{220}, Y_{202}, Y_{211}) = (3, 1, 2, 3, 2, 1, 3, 2, 1).$$

Table 6.10: Questionnaire

1. If the brand of the refrigerator is Taiwanese, please rank order or assign scores to the set of combinations, where the lower the value, the greater the preference. (The value '1' is the most preferred, and the value '3' is the least preferred.)

□	Volume: 200-400 liters	Price: above NT$30000
□	Volume: above 400 liters	Price: NT$20000-NT$30000
□	Volume: under 200 liters	Price: under NT$20000

2、If the brand of the refrigerator is American, please rank order or assign scores to the set of combinations, where the lower the value, the greater the preference. (The value '1' is the most preferred, and the value '3' is the least preferred.)

□	Volume: above 400 liters	Price: above NT$30000
□	Volume: 200-400 liters	Price: under NT$20000
□	Volume: under 200 liters	Price: NT$20000-NT$30000

3、If the brand of the refrigerator is Other Asians, please rank order or assign scores to the set of combinations, where the lower the value, the greater the preference. (The value '1' is the most preferred, and the value '3' is the least preferred.)

□	Volume: above 400 liters	Price: under NT$20000
□	Volume: under 200 liters	Price: above NT$30000
□	Volume: 200-400 liters	Price: NT$20000-NT$30000

Table 6.11: Rank orders (or scores) for responder C

r_1	r_2	r_3	r_4	r_5	r_6	r_7	r_8	
1	3	2	1	2	3	1	2	3

In (6.6), $\boldsymbol{\beta}'_{7\times 1} = (\mu, a_0, a_1, b_0, b_1, c_0, c_1)$, the components of which satisfy the conditions: $a_0 + a_1 + a_2 = 0$, $b_0 + b_1 + b_2 = 0$, and $c_0 + c_1 + c_2 = 0$. The components of the error vector

$$\boldsymbol{\varepsilon} = \begin{bmatrix} \varepsilon_0^{(1)} + \varepsilon_{012}^{(2)} \\ \varepsilon_0^{(1)} + \varepsilon_{021}^{(2)} \\ \varepsilon_0^{(1)} + \varepsilon_{000}^{(2)} \\ \varepsilon_1^{(1)} + \varepsilon_{122}^{(2)} \\ \varepsilon_1^{(1)} + \varepsilon_{110}^{(2)} \\ \varepsilon_1^{(1)} + \varepsilon_{101}^{(2)} \\ \varepsilon_2^{(1)} + \varepsilon_{220}^{(2)} \\ \varepsilon_2^{(1)} + \varepsilon_{202}^{(2)} \\ \varepsilon_2^{(1)} + \varepsilon_{211}^{(2)} \end{bmatrix} = (\boldsymbol{\varepsilon}^{(1)} + \boldsymbol{\varepsilon}^{(2)})$$

satisfy: $E(\boldsymbol{\varepsilon}^{(1)}) = \mathbf{0}$, $E(\boldsymbol{\varepsilon}^{(2)}) = \mathbf{0}$, $\mathrm{Cov}(\boldsymbol{\varepsilon}^{(1)}) = \Sigma_1$, $\mathrm{Cov}(\boldsymbol{\varepsilon}^{(2)}) = \Sigma_2$, $\mathrm{Cov}(\varepsilon_i^{(1)}, \varepsilon_j^{(2)}) = 0$. The design matrix $X_{9\times 7}$ for the orthogonal array in Table 6.10 is

$$X_{9\times 7} = \begin{bmatrix} 1 & 1 & 0 & 0 & 1 & -1 & -1 \\ 1 & 1 & 0 & -1 & -1 & 0 & 1 \\ 1 & 1 & 0 & 1 & 0 & 1 & 0 \\ 1 & 0 & 1 & -1 & -1 & -1 & -1 \\ 1 & 0 & 1 & 0 & 1 & 1 & 0 \\ 1 & 0 & 1 & 1 & 0 & 0 & 1 \\ 1 & -1 & -1 & -1 & -1 & 1 & 0 \\ 1 & -1 & -1 & 1 & 0 & -1 & -1 \\ 1 & -1 & -1 & 0 & 1 & 0 & 1 \end{bmatrix}.$$

It can be proved that the moment estimator $\hat{\boldsymbol{\beta}}'_{7\times 1}$ is an unbiased estimator of $\boldsymbol{\beta}'_{7\times 1}$, where

$$\hat{\beta} = \begin{bmatrix} \hat{\mu} \\ \hat{a}_1 \\ \hat{a}_2 \\ \hat{b}_1 \\ \hat{b}_2 \\ \hat{c}_1 \\ \hat{c}_2 \end{bmatrix} = \begin{bmatrix} \frac{T}{9} \\ \frac{Y_{012}+Y_{021}+Y_{000}}{3} - \frac{T}{9} \\ \frac{Y_{122}+Y_{110}+Y_{101}}{3} - \frac{T}{9} \\ \frac{Y_{202}+Y_{101}+Y_{000}}{3} - \frac{T}{9} \\ \frac{Y_{012}+Y_{110}+Y_{211}}{3} - \frac{T}{9} \\ \frac{Y_{000}+Y_{110}+Y_{220}}{3} - \frac{T}{9} \\ \frac{Y_{021}+Y_{101}+Y_{211}}{3} - \frac{T}{9} \end{bmatrix}, \tag{6.7}$$

$$T = \sum_i \sum_{\substack{j \\ (i,j,k)\in R}} \sum_k Y_{ijk},\ \hat{a}_3 = -\hat{a}_1 - \hat{a}_2,\ \hat{b}_3 = -\hat{b}_1 - \hat{b}_2,\ \hat{c}_3 = -\hat{c}_1 - \hat{c}_2,$$

and

$$R = \{(0,1,2),(0,2,1),(0,0,0),(1,2,2),(1,1,0),(1,0,1),(2,2,0),\\ (2,0,2),(2,1,1)\}.$$

Using the data of responder C in (6.7), we have

$$\hat{\mu} = 2,\ \hat{a}_1 = 0,\ \hat{a}_2 = 0,\ \hat{a}_3 = 0,$$
$$\hat{b}_1 = -0.33,\ \hat{b}_2 = 0,\ \hat{b}_3 = 0.33,$$
$$\hat{c}_1 = 0.33,\ \hat{c}_2 = -1,\ \hat{c}_3 = 0.67.$$

The corresponding results are shown in Table 6.12. For responder C, the variable price is the most important factor.

Table 6.12: Results for responder C

Factor	High part-utility	Low part-utility	Range	Importance score
Volume	0.33	−0.33	0.66	28.3%
Price	0.67	−1	1.67	71.7%

6.3 MANOVA Model with Unequal Levels

Throughout this section, we will use the carpet cleaner example used by Green and Wind (1973). There are five main attributes (brand, package design, price, a Good Housekeeping seal, and a money-back guarantee) in this study. The attributes and the corresponding levels are listed below.

Factor		Level	
A. Brand	0. K2R	1. Glory	2. Bissell
B. Package	0. A^*	1. B^*	2. C^*
C. Price	0. \$1.19	1. \$1.39	2. \$1.59
D. Seal	0. No	1. Yes	
F. Guarantee	0. No	1. Yes	

There are 108 combinations (profiles, stimuli) of levels to be considered in a complete design. To avoid the difficulty about the total number of profiles becoming too many for responders to rank or to rate, we adopt an orthogonal fractional design and split-plot experiment as shown in Table 6.13.

Table 6.13: Questionnaire

1. If the brand of the cleaner is K2R, please rank order or assign scores to the set of combinations, where the lower the value, the greater the preference. (The value '1' is the most preferred, and the value '6' is the least preferred.)

□	Package: C*	Price: $1.39	Seal :yes	Guarantee: no
□	Package: B*	Price: $1.59	Seal: yes	Guarantee: yes
□	Package: A*	Price: $1.19	Seal: no	Guarantee: no
□	Package: B*	Price: $1.39	Seal: no	Guarantee: no
□	Package: A*	Price: $1.59	Seal: no	Guarantee: no
□	Package: C*	Price: $1.19	Seal: no	Guarantee: yes

2. If the brand of the cleaner is Glory, please rank order or assign scores to the set of combinations, where the lower the value, the greater the preference. (The value '1' is the most preferred, and the value '6' is the least preferred.)

□	Package: C*	Price: $1.59	Seal: no	Guarantee: no
□	Package: C*	Price: $1.19	Seal: no	Guarantee: no
□	Package: A*	Price: $1.39	Seal: yes	Guarantee: no
□	Package: B*	Price: $1.39	Seal: no	Guarantee: yes
□	Package: B*	Price: $1.19	Seal: yes	Guarantee: no
□	Package: A*	Price: $1.59	Seal: no	Guarantee: yes

3. If the brand of the cleaner is Bissell, please rank order or assign scores to the set of combinations, where the lower the value, the greater the preference. (The value '1' is the most preferred, and the value '6' is the least preferred.)

□	Package: A*	Price: $1.39	Seal: no	Guarantee: no
□	Package: C*	Price: $1.39	Seal: no	Guarantee: yes
□	Package: C*	Price: $1.59	Seal: yes	Guarantee: no
□	Package: A*	Price: $1.19	Seal: yes	Guarantee: yes
□	Package: B*	Price: $1.19	Seal: no	Guarantee: no
□	Package: B*	Price: $1.59	Seal: no	Guarantee: no

Table 6.14: Orthogonal design

Stimulus no. (1)	A. Brand (2)	B. Package (3)	C. Price (4)	D. Seal (5)	F. Guarantee (6)
1	[0] K2R	[2] C*	[1] $1.39	[1]yes	[0] no
2	[0] K2R	[1] B*	[2] $1.59	[1] yes	[1]yes
3	[0] K2R	[0] A*	[0] $1.19	[0]no	[0] no
4	[0] K2R	[1] B*	[1] $1.59	[0] no	[0] no
5	[0] K2R	[0] A*	[2] $1.59	[0] no	[0] no
6	[0] K2R	[2] C*	[0] $1.19	[0] no	[1] yes
7	[1] Glory	[2] C*	[2] $1.59	[0] no	[0] no
8	[1] Glory	[2] C*	[0] $1.19	[0] no	[0] no
9	[1] Glory	[0] A*	[1] $1.39	[1] yes	[0] no
10	[1] Glory	[1] B*	[1] $1.39	[0] no	[1] yes
11	[1] Glory	[1] B*	[0] $1.19	[1] yes	[0] no
12	[1] Glory	[0] A*	[2] $1.59	[0] no	[1] yes
13	[2] Bissell	[0] A*	[1] $1.39	[0] no	[0] no
14	[2] Bissell	[2] C*	[1] $1.39	[0] no	[1] yes
15	[2] Bissell	[2] C*	[2] $1.59	[1] yes	[0] no
16	[2] Bissell	[0] A*	[0] $1.19	[1] yes	[1] yes
17	[2] Bissell	[1] B*	[0] $1.19	[0] no	[0] no
18	[2] Bissell	[1] B*	[2] $1.59	[0] no	[0] no

Table 6.13 is an orthogonal array produced by the statistical software SPSS [SPSS (1994, pp. 1–36)]. The main attribute in Table 6.13 is brand. Under different levels of brand, the sets of subplot treatments are different. Because we want to evaluate the interaction between the seal and price, the number of subplot treatments in different sets is large (Here it is six). In Table 6.14, the code numbers are adapted from Table 6.13.

Let r_m represent the ranking result given by the responder to the mth stimulus in Table 6.14. Since under each brand, six subplot treatments are compared, the smallest value of r_m is 1 and the largest value is 6. The larger the value of r_m is, the more the responder dislikes this subplot treatment. The following table shows the ranks (or scores) assigned by responder D to the stimuli in Table 6.13.

Table 6.15: The ranks assigned by responder D to the stimuli in Table 6.13

r_1	r_2	r_3	r_4	r_5	r_6	r_7	r_8	r_9	r_{10}	r_{11}	r_{12}	r_{13}	r_{14}	r_{15}	r_{16}	r_{17}	r_{18}
1	3	4	5	6	2	6	3	2	4	1	5	5	4	3	1	2	6

According to Table 6.15, based on the K2R brand, we find that stimulus 1 – the cleaner with C^* package, \$1.39 price, seal and without money-back guarantee – is responder D's favorite.

The relations between the rankings (or ratings) of stimuli given by a responder and the levels of factors can be represented by the following MANOVA model:

$$\mathbf{Y}_{18\times 1} = X_{18\times 11}\boldsymbol{\beta}_{11\times 1} + \boldsymbol{\varepsilon}_{18\times 1}, \tag{6.8}$$

where $\mathbf{Y}'_{18\times 1} = (Y_1, Y_2, \ldots, Y_{18})$. We use the equation $Y_m = 7 - r_m$ to convert the rankings (or ratings) for convenience in explanation. The greater the value of Y_m, the higher is the preference of the responder. The converted data collected from responder D is

$$\begin{aligned}&(Y_1, Y_2, Y_3, Y_4, Y_5, Y_6, Y_7, Y_8, Y_9, Y_{10}, Y_{11}, Y_{12}, Y_{13}, Y_{14}, Y_{15}, Y_{16}, Y_{17}, Y_{18})\\ &\quad = (6, 4, 3, 2, 1, 5, 1, 4, 5, 3, 6, 2, 2, 3, 4, 6, 5, 1).\end{aligned}$$

In (6.8), $\boldsymbol{\beta}'_{11\times 1} = (\mu, a_0, a_1, b_0, b_1, c_0, c_1, d, f, \gamma, \delta)$, where μ is the overall mean, a_i is the part-utility of the ith level of the brand factor, b_j is the part-utility of the jth level of the package factor, c_k is the part-utility of the kth level of the price factor, d is the part-utility of the 'yes' level of the seal factor, $-d$ is the part-utility of the 'no' level of the seal factor, f is the part-utility of the 'yes' level of the guarantee factor, and $-f$ is the part-utility of the 'no' level of the guarantee factor. Both γ and δ are the part-utility of the interactions between the seal and price. They satisfy the conditions:

$$a_0 + a_1 + a_2 = 0, \quad b_0 + b_1 + b_2 = 0, \quad \text{and} \quad c_0 + c_1 + c_2 = 0.$$

The error term is

$$\begin{aligned}\boldsymbol{\varepsilon}_{18\times 11} &= (\varepsilon_0^{(1)} + \varepsilon_1^{(2)},\ \varepsilon_0^{(1)} + \varepsilon_2^{(2)}, \ldots, \varepsilon_1^{(1)} + \varepsilon_7^{(2)},\ \varepsilon_0^{(1)} + \varepsilon_8^{(2)}, \ldots, \varepsilon_2^{(1)} + \varepsilon_{13}^{(2)},\\ &\qquad \varepsilon_2^{(1)} + \varepsilon_{14}^{(2)},\ \varepsilon_2^{(1)} + \varepsilon_{15}^{(2)},\ \varepsilon_2^{(1)} + \varepsilon_{16}^{(2)},\ \varepsilon_2^{(1)} + \varepsilon_{17}^{(2)},\ \varepsilon_2^{(1)} + \varepsilon_{18}^{(2)})'\\ &= \boldsymbol{\varepsilon}^{(1)} + \boldsymbol{\varepsilon}^{(2)},\end{aligned}$$

and its components satisfy

$$E(\boldsymbol{\varepsilon}^{(1)}) = \mathbf{0},\ E(\boldsymbol{\varepsilon}^{(2)}) = \mathbf{0},\ \mathrm{Cov}(\boldsymbol{\varepsilon}^{(1)}) = \boldsymbol{\Sigma}_1,\ \mathrm{Cov}(\boldsymbol{\varepsilon}^{(2)}) = \boldsymbol{\Sigma}_2, \mathrm{Cov}(\varepsilon_i^{(1)}, \varepsilon_j^{(2)}) = 0.$$

The interaction effects between price and the seal are listed in Table 6.16.

Table 6.16: Interaction effects between price and seal

	Seal	
Price	No	Yes
\$1.19	$-\gamma$	γ
\$1.39	$-\delta$	δ
\$1.59	$\gamma+\delta$	$-\gamma-\delta$

The design matrix $X_{18\times 11}$ for the orthogonal array in Table 6.14 and model (6.8) is

$$X_{18\times 11} = \begin{bmatrix} 1 & 1 & 0 & -1 & -1 & 0 & 1 & 1 & -1 & 0 & 1 \\ 1 & 1 & 0 & 0 & 1 & -1 & -1 & 1 & 1 & -1 & -1 \\ 1 & 1 & 0 & 1 & 0 & 1 & 0 & -1 & -1 & -1 & 0 \\ 1 & 1 & 0 & 0 & 1 & 0 & 1 & -1 & -1 & 0 & -1 \\ 1 & 1 & 0 & 1 & 0 & -1 & -1 & -1 & -1 & 1 & 1 \\ 1 & 1 & 0 & -1 & -1 & 1 & 0 & -1 & 1 & -1 & 0 \\ 1 & 0 & 1 & -1 & -1 & -1 & -1 & -1 & -1 & 1 & 1 \\ 1 & 0 & 1 & -1 & -1 & 1 & 0 & -1 & -1 & -1 & 0 \\ 1 & 0 & 1 & 1 & 0 & 0 & 1 & 1 & -1 & 0 & 1 \\ 1 & 0 & 1 & 0 & 1 & 0 & 1 & -1 & 1 & 0 & -1 \\ 1 & 0 & 1 & 0 & 1 & 1 & 0 & 1 & -1 & 1 & 0 \\ 1 & 0 & 1 & 1 & 0 & -1 & -1 & -1 & 1 & 1 & 1 \\ 1 & -1 & -1 & 1 & 0 & 0 & 1 & -1 & -1 & 0 & -1 \\ 1 & -1 & -1 & -1 & -1 & 0 & 1 & -1 & 1 & 0 & -1 \\ 1 & -1 & -1 & -1 & -1 & -1 & -1 & 1 & -1 & -1 & -1 \\ 1 & -1 & -1 & 1 & 0 & 1 & 0 & 1 & 1 & 1 & 0 \\ 1 & -1 & -1 & 0 & 1 & 1 & 0 & -1 & -1 & -1 & 0 \\ 1 & -1 & -1 & 0 & 1 & -1 & -1 & -1 & -1 & 1 & 1 \end{bmatrix}.$$

The equation (6.8) can also be written as follows:

$$\begin{aligned} Y_i &= \mu + a_0X_{i1} + a_1X_{i2} + b_0X_{i3} + b_1X_{i4} + c_0X_{i5} + c_1X_{i6} + dX_{i7} \\ &\quad + fX_{i8} + \gamma X_{i5}X_{i7} + \delta X_{i6}X_{i7} + \varepsilon_i. \end{aligned} \tag{6.9}$$

We have

$$\begin{aligned} E(Y_1) &= \mu + a_0 - b_0 + b_1 + c_1 + d - f + \delta, \\ E(Y_2) &= \mu + a_0 + b_1 - c_0 - c_1 + d + f - \gamma - \delta, \\ E(Y_3) &= \mu + a_0 + b_0 + c_0 - d - f - \gamma, \end{aligned}$$

$$
\begin{aligned}
E(Y_4) &= \mu + a_0 + b_1 + c_1 - d - f - \delta, \\
E(Y_5) &= \mu + a_0 + b_0 - c_0 - c_1 - d - f + \gamma + \delta, \\
E(Y_6) &= \mu + a_0 - b_0 - b_1 + c_0 - d + f - \gamma, \\
E(Y_7) &= \mu + a_1 - b_0 - b_1 - c_0 - c_1 - d - f + \gamma + \delta, \\
E(Y_8) &= \mu + a_1 - b_0 - b_1 + c_0 - d - f - \gamma, \\
E(Y_9) &= \mu + a_1 + b_0 + c_1 + d - f + \delta, \\
E(Y_{10}) &= \mu + a_1 + b_1 + c_1 - d + f - \delta, \\
E(Y_{11}) &= \mu + a_1 + b_1 + c_0 + d - f + \gamma, \\
E(Y_{12}) &= \mu + a_1 + b_0 - c_0 - c_1 - d + f + \gamma + \delta, \\
E(Y_{13}) &= \mu - a_0 - a_1 + b_0 + c_1 - d - f - \delta, \\
E(Y_{14}) &= \mu - a_0 - a_1 - b_0 - b_1 + c_1 - d + f - \delta, \\
E(Y_{15}) &= \mu - a_0 - a_1 - b_0 - b_1 - c_0 - c_1 + d - f - \gamma - \delta, \\
E(Y_{16}) &= \mu - a_0 - a_1 + b_0 + c_0 + d + f + \gamma, \\
E(Y_{17}) &= \mu - a_0 - a_1 + b_1 + c_0 - d - f - \gamma, \\
E(Y_{18}) &= \mu - a_0 - a_1 + b_1 - c_0 - c_1 - d - f + \gamma + \delta.
\end{aligned}
$$

Let

$$
\begin{aligned}
\mathbf{A}'_{11\times 1} &= [A_1, A_2, \ldots, A_{11}] = X'Y, \\
T_1 &= \frac{A_2}{6} - \frac{A_3}{6} - \frac{A_4}{6} + \frac{A_6}{6} + \frac{A_{10}}{2}, \\
T_2 &= \frac{A_1}{24} - \frac{A_3}{6} - \frac{A_4}{6} + \frac{A_5}{6} + \frac{A_7}{6} + \frac{A_9}{8} + \frac{A_{11}}{2}. \qquad (6.10)
\end{aligned}
$$

Using the method of moments, the estimators of $\boldsymbol{\beta}$ can be obtained as follows:

$$
\begin{aligned}
\hat{d} &= \frac{A_1}{48} + \frac{A_8}{16}, \\
\hat{\delta} &= -\frac{T_1}{5} + \frac{2T_2}{5}, \\
\hat{\gamma} &= \frac{T_1}{4} - \frac{1}{2}\hat{\delta}, \\
\hat{a}_0 &= \frac{A_2}{9} - \frac{A_3}{18} + \frac{1}{3}\hat{\gamma}, \\
\hat{a}_1 &= -\frac{A_2}{18} + \frac{A_3}{9} - \frac{1}{3}\hat{\gamma} - \frac{1}{3}\hat{\delta}, \\
\hat{b}_0 &= \frac{A_4}{9} - \frac{A_5}{18} - \frac{1}{3}\hat{\gamma} - \frac{1}{3}\hat{\delta}, \\
\hat{b}_1 &= -\frac{A_4}{18} + \frac{A_5}{9} + \frac{1}{3}\hat{\delta}, \\
\hat{c}_0 &= \frac{A_6}{9} - \frac{A_7}{18} + \frac{1}{3}\hat{\gamma},
\end{aligned}
$$

$$\begin{aligned}\hat{c}_1 &= -\frac{A_6}{18} + \frac{A_7}{9} + \frac{1}{3}\hat{\delta}, \\ \hat{f} &= \frac{A_1}{48} + \frac{A_9}{16} + \frac{1}{4}\hat{\delta}, \\ \hat{\mu} &= \frac{A_1}{18} + \frac{1}{3}\hat{d} + \frac{1}{3}\hat{f}. \end{aligned} \tag{6.11}$$

Using the data of responder D in (6.10) and (6.11), we have

$$\begin{aligned}\mathbf{A}'_{11\times 1} &= [63, 0, 0, -4, -2, 16, 8, -1, -17, -8, -2]', \\ T_1 &= -0.666667, \quad T_2 = 1.166667, \\ \hat{d} &= 1.25, \quad \hat{\delta} = 0.6, \quad \hat{\gamma} = -0.466667, \\ \hat{a}_0 &= -0.155556, \quad \hat{a}_1 = -0.044444, \\ \hat{b}_0 &= -0.377778, \quad \hat{b}_1 = 0.2, \\ \hat{c}_0 &= 1.177778, \quad \hat{c}_1 = 0.2, \\ \hat{f} &= 0.4, \quad \hat{\mu} = 4.05. \end{aligned}$$

The total utility of a stimulus can be expressed as:

> The Total Utility = Overall Mean + Part-utility of Brand + Part-utility of Package + Part-utility of Price + Part-utility of Seal + Part-utility of Guarantee + Part-utility of Interaction of Price and Seal.

The more the total utility is, the more the responder likes the stimulus. Using responder D's data, the estimated total utility, for example, of the second stimulus in Table 6.14 is:

$$4.05 - 0.155556 + 0.2 - 1.177778 - 0.2 + 1.25 + 0.4 + 0.466667 - 0.6 = 4.23333.$$

The association coefficient, Spearman ρ, is used to evaluate the consistency between the predicted ranks and the observed ranks. The results for responder D's data are given in Table 6.17. The Spearman ρ values in Table 6.17 indicates that it is suitable to use the MANOVA model to predict someone's preference ranks.

Table 6.17: Responder D's Spearman ρ

Brand	Spearman ρ
K2R	1.00
Glory	1.00
Bissell	1.00

Table 6.18: Results for responder D

Factor	High part-utility	Low part-utility	Range	Importance score
Package	0.377778	−0.2	0.577778	8.98%
Price	1.377778	−1.177778	2.555556	39.72%
Seal	1.25	−1.25	2.5	38.86%
Guarantee	0.4	−0.4	0.8	12.44%

From Table 6.18, we see that price is the most important factor to responder D when buying a carpet cleaner.

6.4 Conclusion

In this chapter, orthogonal array designs and split-plot experiments are employed in the conjoint analysis. Since the split-plot experiment is fixed at one level of the main factor at a time, the numbers of subplot treatments are few. Then, it is feasible for responders to rank the subplot treatments. We also discuss how to build a MANOVA model. Using a suitable model, the interaction between factors and the utility of the stimulus can be estimated. Then we can understand the stimulus and the factor that are the most important to the consumer. The predictive ability of the model can also be evaluated by the Spearman ρ or the Kendall τ. Our models and analyses can be easily extended to the case of two or more responders. We can obtain each part-utility as the average of that part-utility for all the responders.

References

1. Green, P. E. and Wind, Y. (1973). *Multivariate Decisions in Marketing: A Measurement Approach*, Dryden Press, Hinsdale, Illinois.

2. Montgomery, D. C. (1991). *Design and Analysis of Experiments*, John Wiley & Sons, New York.

3. SPSS Categories 6.1. (1994). SPSS Inc, USA.

7

A Restricted Subset Selection Rule for Selecting At Least One of the t Best Normal Populations in Terms of Their Means: Common Known Variance Case

Lifang Hsu[1] **and S. Panchapakesan**[2]

[1]*Department of Mathematics, LeMoyne College, Syracuse, NY U.S.A.*
[2]*Department of Mathematics, Southern Illinois University, Carbondale, IL U.S.A.*

Abstract: Consider k (≥ 2) normal populations with unknown means $\mu_1, \ldots,$ μ_k and a common *known* variance σ^2. Let $\mu_{[1]} \leq \cdots \leq \mu_{[k]}$ denote the ordered μ_i. Our goal is to select a non-empty subset of the k populations whose size is at most m ($1 \leq m \leq k - t$) so that at least one of the populations associated with the t ($1 \leq t \leq k - 1$) largest means is included in the selected subset with a minimum guaranteed probability P^*, whenever $\mu_{[k-t+1]} - \mu_{[k-t]} \geq \delta^*$ where P^* and δ^* are specified in advance of the experiment. Santner (1976) proposed and investigated a procedure (R_S) based on samples of size n from each of the populations. We propose and investigate an alternative procedure R_{HP} with the same sampling scheme. We compare our rule with that of a procedure that selects a subset of fixed size m. The special case of $t = 1$ was earlier studied by Gupta and Santner (1973) and Hsu and Panchapakesan (2003).

Keywords and phrases: Selecting normal means, restricted subset size, fixed subset size, efficiency comparisons

7.1 Introduction

Let $\Pi_1, \ldots, \Pi_k$ be k (≥ 2) normal populations with *unknown* means $\mu_1, \ldots, \mu_k$, respectively, and a common *known* variance σ^2. Let $\mu_{[1]} \leq \cdots \leq \mu_{[k]}$ denote the ordered μ_i and let $\Pi_{(i)}$ denote the normal population associated with $\mu_{[i]}$, $i = 1, \ldots, k$. We assume that there is no prior knowledge concerning the correct pairing of the ordered and the unordered μ_i. Let $\delta^* \geq 0$ be a given constant.

Let

$$\Omega = \{\vec{\mu} = (\mu_1, \ldots, \mu_k) \mid -\infty < \mu_i < \infty,\ i = 1, \ldots, k\},$$
$$\Omega(\delta^*) = \{\vec{\mu} \in \Omega \mid \mu_{[k-t+1]} - \mu_{[k-t]} \geq \delta^*\},$$
$$\Omega^0(\delta^*) = \{\vec{\mu} \in \Omega(\delta^*) \mid \mu_{[1]} = \cdots = \mu_{[k-t]} = \mu_{[k-t+1]} - \delta^* = \cdots = \mu_{[k]} - \delta^*\}.$$

Our goal is to select a non-empty subset of the k populations whose size is at most m $(1 \leq m \leq k-t)$ so that at least one of the populations associated with t $(1 \leq t \leq k-1)$ largest means (called the t *best* populations) is included in the selected subset with a minimum guaranteed probability P^* whenever $\vec{\mu} \in \Omega(\delta^*)$. For any rule R, a *correct selection* (CS) is said to occur if the decision results in selection of a subset of size at most m that contains one of the t best populations. Let $P_{\vec{\mu}}(CS \mid R)$ denote the probability of a correct selection (PCS) corresponding to any parametric configuration $\vec{\mu}$. Any *valid* rule R is required to satisfy the condition:

$$P_{\vec{\mu}}(CS \mid R) \geq P^* \quad \text{whenever} \quad \vec{\mu} \in \Omega(\delta^*). \tag{7.1}$$

The region $\Omega(\delta^*)$ over which the minimum PCS is to be guaranteed is known as the *preference-zone*. We propose the following rule based on independent random samples of size n taken from the k populations. Let Y_i denote the sample mean from Π_i and let $Y_{[1]} \leq \cdots \leq Y_{[k]}$ denote the ordered Y_i. The proposed rule is

R_{HP}: Select Π_i if and only if

$$\frac{Y_i - Y_{[k-m]}}{Y_{[k]} - Y_{[k-m]}} > c, \tag{7.2}$$

where $0 \leq c < 1$ is to be chosen so that the probability requirement (7.1) is satisfied. The form of the rule R_{HP} in (7.2) explains the intuitive basis for the rule. However, we prefer to write it equivalently as

R_{HP}: Select Π_i if and only if

$$Y_i > (1-c)Y_{[k-m]} + cY_{[k]}. \tag{7.3}$$

It is obvious that the populations corresponding to $Y_{[1]}, \ldots, Y_{[k-m]}$ will not be selected. Also, the population that gave $Y_{[k]}$ will always be included. Thus the rule R_{HP} selects a non-empty subset of size at most m.

Santner (1976) developed a general theory for the restricted subset selection with the goal of selecting one of the t best populations. Earlier, Santner (1975) developed the theory for the case of $t = 1$. As one of the examples of his procedure, Santner (1976) has considered the above problem of selecting from normal populations. His rule, in the case of $t \geq 1$, is

R_S: Select Π_i if and only if

$$Y_i \geq \max\{Y_{[k-m+1]}, Y_{[k]} - d\sigma/\sqrt{n}\}, \tag{7.4}$$

where $d > 0$ is to be chosen so that the probability requirement (7.1) is satisfied. The special case of R_S for $t = 1$ was earlier investigated by Gupta and Santner (1973). The special case of R_{HP} for $t = 1$ was investigated by Hsu and Panchapakesan (2003).

In his general theory, Santner (1976) considered two types of probability requirements differing in the parametric region over which the PCS is minimized. The requirement (7.1) using the region $\Omega(\delta^*)$ goes with the assumption that $m \leq k - t$; this is referred to as the indifference-zone probability requirement. When $m > k - t$, the probability requirement (7.1) is modified by replacing $\Omega(\delta^*)$ with Ω and is referred to as the subset selection probability requirement. What characterizes the latter case is the fact that a correct selection is always made when we randomly select m of the k populations. We do not consider this case in this chapter; it will be considered elsewhere.

In Section 7.2, we obtain the PCS for the rule R_{HP} and evaluate its infimum over $\Omega(\delta^*)$. We also discuss tables needed for the implementation of R_{HP} and compare its efficiency with respect to a natural rule that selects a subset of fixed size m. Several properties of R_{HP} are given in Section 7.3. The next two sections deal with the expected size of the selected subset (Section 7.4) and some concluding remarks (Section 7.5).

7.2 PCS and Its Infimum over $\Omega(\delta^*)$ for Rule $\mathbf{R_{HP}}$

We first state a lemma of Alam and Rizvi (1966) and Mahamunulu (1967), which is an immediate consequence of a result in Lehmann (1986, p. 116, Problem 15). This lemma concerns a family $\{F(\cdot|\theta)\}$ of cumulative distribution functions (c.d.f.'s) on the real line which is *stochastically increasing* in θ, that is, the distributions are distinct and $\theta_1 < \theta_2$ implies $F(x|\theta_1) \geq F(x|\theta_2)$ for all x.

Lemma 7.2.1. *Let $X_1, \ldots, X_k$ be independent random variables having c.d.f.'s $F(\cdot|\theta_i)$, $i = 1, \ldots, k$, where the family $\{F(\cdot|\theta)\}$ is stochastically increasing in θ. Let $\psi(x_1, \ldots, x_k)$ be non-decreasing (non-increasing) in x_i when all other x_j's are kept fixed. Then $E\{\psi(X_1, \ldots, X_k)\}$ is non-decreasing (non-increasing) in θ_i, when all other θ_j's are kept fixed.*

We now prove the following theorem concerning the infimum of the PCS over $\Omega(\delta^*)$ for the rule R_{HP}.

Theorem 7.2.1. $\inf_{\Omega(\delta^*)} P_{\vec{\mu}}(CS \mid R_{HP}) = \inf_{\Omega^0(\delta^*)} P_{\vec{\mu}}(CS \mid R_{HP})$.

PROOF. Let $Y_{(i)}$ denote the sample mean from $\Pi_{(i)}$, $i = 1, \ldots, k$. Then

$$P_{\vec{\mu}}(CS \mid R_{HP}) = \Pr\{W > (1-c)Y_{[k-m]} + cY_{[k]}\} = E\{\psi(Y_{(1)}, \ldots, Y_{(k)})\},$$

where $W = \max\{Y_{(k-t+1)}, \ldots, Y_{(k)}\}$ and

$$\psi(Y_{(1)}, \ldots, Y_{(k)}) = \begin{cases} 1 & \text{if } W > (1-c)Y_{[k-m]} + cY_{[k]}\,, \\ 0 & \text{otherwise}\,. \end{cases}$$

It can be verified that $\psi(Y_{(1)}, \ldots, Y_{(k)})$ is non-increasing in $Y_{(i)}$, $i = 1, \ldots, k-t$, and non-decreasing in $Y_{(i)}$, $i = k-t+1, \ldots, k$, when all other components of $(Y_{(1)}, \ldots, Y_{(k)})$ are kept fixed. The $Y_{(i)}$ are independent and $Y_{(i)}$ has a normal distribution, $N(\mu_{[i]}, \tau^2)$, with mean $\mu_{[i]}$ and variance $\tau^2 = \sigma^2/n$. So, by Lemma 7.2.1, it follows that $P\{CS \mid R_{HP}\}$ is non-increasing in $\mu_{[i]}$, $i = 1, \ldots, k-t$, and non-decreasing in $\mu_{[i]}$, $i = k-t+1, \ldots, k$, when all other components of $\vec{\mu}$ are kept fixed. Consequently, $P\{CS \mid R_{HP}\}$ is minimized over $\Omega(\delta^*)$ when

$$\mu_{[1]} = \cdots = \mu_{[k-t]} = \mu \text{ (say)}; \; \mu_{[k-t+1]} = \cdots = \mu_{[k]} = \mu + \delta^*. \tag{7.5}$$

This proves the theorem. ■

The configuration (7.5) for which the PCS attains its infimum over the specified region is called a *least favorable configuration* (LFC). We denote this infimum by $P_{LFC}\{CS \mid R_{HP}\}$. Obviously, this infimum is independent of the value of μ and so we can take $\mu = 0$. Thus,

$$P_{LFC}(CS \mid R_{HP}) = \Pr\{W > (1-c)Y_{[k-m]} + cY_{[k]}\}\,, \tag{7.6}$$

where $Y_{(1)}, \ldots, Y_{(k-t)}$ are $N(0, \tau^2)$ and $Y_{(k-t+1)}, \ldots, Y_{(k)}$ are $N(\delta^*, \tau^2)$. We can now write

$$\begin{aligned} P_{LFC}(CS \mid R_{HP}) &= \sum_{i=k-t+1}^{k} \Pr\{W = Y_{(i)} \text{ and } Y_{(i)} > (1-c)Y_{[k-m]} + cY_{[k]}\} \\ &= t \Pr\{W = Y_{(k)} \text{ and } Y_{(k)} > (1-c)Y_{[k-m]} + cY_{[k]}\} \\ &= tQ \text{ (say)}\,. \end{aligned} \tag{7.7}$$

Now, we can write $Q = Q_1 + Q_2 + Q_3$, where

$$Q_1 = \Pr\{W = Y_{(k)} \text{ and } Y_{(k)} = Y_{[k]}\}\,, \tag{7.8}$$

$$Q_2 = \sum_{i=1}^{k-t} \sum_{\substack{j=1 \\ j \neq i}}^{k-t} \Pr\{W = Y_{(k)}, Y_{(i)} = Y_{[k-m]}, Y_{(j)} = Y_{[k]}, Y_{(k)} > (1-c)Y_{(i)} + cY_{(j)}\}\,, \tag{7.9}$$

and

$$Q_3 = \sum_{i=k-t+1}^{k-1} \sum_{j=1}^{k-t} \Pr\{W = Y_{(k)}, Y_{(i)} = Y_{[k-m]}, Y_{(j)} = Y_{[k]}, \\ Y_{(k)} > (1-c)Y_{(i)} + cY_{(j)}\}. \tag{7.10}$$

Let Z_i denote the standardized $Y_{(i)}$, $i = 1, \ldots, k$, so that the Z_i are i.i.d. $N(0,1)$ variables. Also, let $\Delta^* = \sqrt{n}\delta^*/\sigma$. Then, it is easy to see that

$$\begin{aligned} Q_1 &= \Pr\{Z_k + \Delta^* \geq Z_i, \, i = 1, \ldots, k-t; \; Z_k \geq Z_j, \, j = k-t+1, \ldots, k-1\} \\ &= \int_{-\infty}^{\infty} \Phi^{t-1}(y)\Phi^{k-t}(y + \Delta^*)\phi(y)\, dy, \end{aligned} \tag{7.11}$$

where Φ and ϕ denote, respectively, the cumulative distribution function (c.d.f.) and the density function of a standard normal random variable.

Next,

$$\begin{aligned} Q_2 &= (k-t)(k-t-1)\Pr\left\{ \begin{array}{l} W = Y_{(k)}, Y_{(1)} = Y_{[k-m]}, Y_{(k-t)} = Y_{[k]}, \\ Y_{(k-t)} > Y_{(k)} > (1-c)Y_{(1)} + cY_{(k-t)} \end{array} \right\} \\ &= (k-t)(k-t-1) \sum_{\alpha=0}^{t-1} \binom{t-1}{\alpha} \binom{k-t-2}{k-m-1-\alpha} H(\alpha), \end{aligned} \tag{7.12}$$

where α in the summation stands for the number of populations from the set of t best that give rise to $Y_i < Y_{(1)} = Y_{[k-m]}$ and the probability denoted by $H(\alpha)$, after standardizing the Y_i, is given by

$$\begin{aligned} H(\alpha) &= \Pr \left\{ \begin{array}{l} Z_i + \Delta^* \leq Z_1, \; i = k-t+1, \ldots, k-t+\alpha; \\ Z_1 \leq Z_j + \Delta^* \leq Z_k + \Delta^*, \; j = k-t+\alpha+1, \ldots, k-1; \\ Z_r \leq Z_1, \; r = 2, \ldots, k-m-\alpha; \\ Z_1 \leq Z_s \leq Z_{k-t}, \; s = k-m-\alpha+1, \ldots, k-t-1; \\ (1-c)Z_1 + cZ_{k-t} < Z_k + \Delta^* < Z_{k-t} \end{array} \right\} \\ &= \int_{z_1=-\infty}^{\infty} \int_{z_{k-t}=z_1}^{\infty} \int_{z_k=(1-c)z_1+cz_{k-t}-\Delta^*}^{z_{k-t}-\Delta^*} \Phi^{\alpha}(z_1 - \Delta^*)\Phi^{k-m-\alpha-1}(z_1) \\ &\quad \times [\Phi(z_k) - \Phi(z_1 - \Delta^*)]^{t-\alpha-1}[\Phi(z_{k-t}) - \Phi(z_1)]^{m+\alpha-t-1}\phi(z_k)\phi(z_{k-t}) \\ &\quad \times \phi(z_1)\, dz_k\, dz_{k-t}\, dz_1. \end{aligned} \tag{7.13}$$

Proceeding along the same lines, we get

$$Q_3 = (t-1)(k-t)\sum_{\alpha=0}^{t-2} \binom{t-2}{\alpha} \binom{k-t-1}{k-m-1-\alpha} H^*(\alpha), \tag{7.14}$$

where

$$
\begin{aligned}
&H^*(\alpha)\\
&= \Pr\left\{\begin{array}{l}
Z_i \le Z_{k-1},\ i = k-t+1,\ldots,k-t+\alpha;\\
Z_{k-1} \le Z_j \le Z_k,\ j = k-t+\alpha+1,\ldots,k-2;\\
Z_r \le Z_{k-1}+\Delta^*,\ r = 1,\ldots,k-m-\alpha-1;\\
Z_{k-1}+\Delta^* \le Z_s \le Z_{k-t},\ s = k-m-\alpha,\ldots,k-t-1;\\
(1-c)[Z_{k-1}+\Delta^*] + cZ_{k-t} < Z_k+\Delta^* < Z_{k-t}
\end{array}\right\}\\
&= \int_{z_{k-1}=-\infty}^{\infty}\int_{z_{k-t}=z_{k-1}+\Delta^*}^{\infty}\int_{z_k=(1-c)z_{k-1}+c(z_{k-t}-\Delta^*)}^{z_{k-t}-\Delta^*} \Phi^{k-m-\alpha-1}(z_{k-1}+\Delta^*)\\
&\quad\times[\Phi(z_k)-\Phi(z_{k-1})]^{t-\alpha-2}\Phi^{\alpha}(z_{k-1})[\Phi(z_{k-t})-\Phi(z_{k-1}+\Delta^*)]^{m+\alpha-t}\\
&\quad\times\phi(z_{k-1})\phi(z_{k-t})\phi(z_k)\,dz_k\,dz_{k-t}\,dz_{k-1}\,. \qquad (7.15)
\end{aligned}
$$

Combining all the above results, we have

$$P_{LFC}(CS \mid R_{HP}) = t(Q_1+Q_2+Q_3) \qquad (7.16)$$

where Q_1 is given by (7.11), Q_2 is given by (7.12) and (7.13), and Q_3 is given by (7.14) and (7.15).

Remark 7.2.1. In the special case of $t = 1$, we can see that $Q_3 = 0$ and the PCS in (7.16) reduces to the expression obtained for this special case earlier by Hsu and Panchapakesan (2003).

Now,, for convenience, we let $A_c(k,t,m,\Delta^*) = P_{LFC}(CS \mid R_{HP})$. For given k, t, m and Δ^*, as c increases from 0 to 1, it is easy to see that Q_1 remains unchanged, Q_2 decreases from Q_2^0 to zero, and Q_3 decreases from Q_3^0 to zero, where Q_2^0 and Q_3^0 are obtained from the expressions for Q_2 and Q_3 by putting $c = 0$. Consequently, as c increases from 0 to 1, $A_c(k,t,m,\Delta^*)$ decreases from $t(Q_1+Q_2^0+Q_3^0)$ to tQ_1. Thus, for given k, t, m and Δ^*, we can find a unique $c \in (0,1)$ to satisfy the probability requirement (7.1) provided that

$$tQ_1 < P^* < t(Q_1+Q_2^0+Q_3^0)\,. \qquad (7.17)$$

First, note that $Q_1 > 1/k$. Technically, the lower bound for the admissible range for P^* is zero. However, for a meaningful problem, $P^* > t/k$; otherwise, we can randomly select one of the populations and satisfy the probability requirement. Also, we can satisfy the probability requirement (7.1) for any $t/k < P^* \le tQ_1$ just by selecting the population that yielded the largest sample mean.

When $c = 0$, the rule R_{HP} selects the population that yielded the m largest sample means. In other words, R_{HP} in this special case selects a subset of fixed size m. We will refer to it in this case as rule R_{FS}. With the $Y_{(i)}$ as defined for (7.6), $P_{LFC}(CS \mid R_{FS})$ is given by

$$\begin{aligned}
&A_0(k,t,m,\Delta^*) \\
&= 1 - \Pr\{Y_{(i)} \le Y_{[k-m]},\ i = k-t+1,\ldots,k\} \\
&= 1 - \binom{k-t}{m} \Pr\left\{ \begin{matrix} Y_{(i)} < \min(Y_{(1)},\ldots,Y_{(m)}),\ i = k-t+1,\ldots,k; \\ Y_{(j)} < \min(Y_{(1)},\ldots,Y_{(m)}),\ j = m+1,\ldots,k-t \end{matrix} \right\} \\
&= 1 - \binom{k-t}{m} \Pr\left\{ \begin{matrix} Z_i + \Delta^* < U = \min(Z_1,\ldots,Z_m),\ i = k-t+1,\ldots,k; \\ Z_j < U,\ j = m+1,\ldots,k-t \end{matrix} \right\}
\end{aligned}$$

where the Z_i are i.i.d. $N(0,1)$ random variables. Noting that U has density

$$g(u) = m[1 - \Phi(u)]^{m-1}\phi(u), \quad -\infty < u < \infty,$$

we get

$$A_0(k,t,m,\Delta^*) = 1 - \binom{k-t}{m} m \int_{-\infty}^{\infty} \Phi^{k-t-m}(u)\Phi^t(u-\Delta^*)\big[1-\Phi(u)\big]^{m-1}\phi(u)\,du.$$

Now, by writing $\Phi^t(u-\Delta^*)$ as $1-(1-\Phi^t(u-\Delta^*))$ in the above integral and splitting it into two integrals, we can obtain after some simplification,

$$\begin{aligned}
&A_0(k,t,m,\Delta^*) \\
&= m\binom{k-t}{m} \int_{-\infty}^{\infty} \Phi^{k-t-m}(u)\big[1 - \Phi^t(u-\Delta^*)\big]\big[1-\Phi(u)\big]^{m-1}\phi(u)\,du\,.
\end{aligned} \tag{7.18}$$

The upper bound for P^* in (7.17) and the right-hand side of (7.18) are both expressions for $A_0(k,t,m,\Delta^*)$ obtained by different arguments and so are equal. However, we are not successful in showing this directly.

For implementing the rule R_{HP}, given k, t, m, δ^* and P^*, we have many possible pairs of values for n and c such that the probability requirement (7.1) is met. We will later (Section 7.5) discuss some possible criteria that could be used for choosing a pair. Alternatively, given k, t, m, δ^*, P^* and c, we can determine the minimum sample size n needed to satisfy the probability requirement (7.1). For this purpose, given k, t, m, P^* and c, we need the value of Δ^* that satisfies the equation:

$$t(Q_1 + Q_2 + Q_3) = P^*. \tag{7.19}$$

By Lemma 7.2.1, $A_c(k,t,m,\Delta^*)$, which is the left-hand side of (7.19), increases in Δ^* for given k, t, m and c. Therefore, we can find a unique Δ^* satisfying (7.19). Values of Δ^* are given in Table 7.1 for $P^* = 0.900, 0.950, 0.975$; $c = 0.25, 0.50, 0.75$; $k = 3(1)10$, and several selected values of (m,t).

Table 7.1: Values of $\Delta^* = \frac{\sqrt{n}\delta^*}{\sigma}$ for rule R_{HP}

k	m	$c \rightarrow$ / t	$P^* = 0.900$ 0.25	0.50	0.75	$P^* = 0.950$ 0.25	0.50	0.75	$P^* = 0.975$ 0.25	0.50	0.75
5	3	2	0.563	1.000	1.368	0.959	1.388	1.758	1.302	1.722	2.094
6	3	2	0.862	1.227	1.549	1.234	1.599	1.928	1.557	1.920	2.256
	4	2	0.438	0.954	1.407	0.815	1.319	1.778	1.139	1.632	2.098
7	3	2	1.055	1.382	1.678	1.415	1.746	2.050	1.726	2.061	2.373
	4	2	0.718	1.157	1.560	1.074	1.512	1.924	1.381	1.815	2.238
	5	2	0.349	0.922	1.440	0.713	1.273	1.799	1.025	1.574	2.112
	4	3	0.236	0.581	1.033	0.437	0.826	1.363	0.569	1.011	1.651
8	3	2	1.197	1.499	1.778	1.548	1.857	2.145	1.853	2.167	2.464
	4	2	0.904	1.301	1.674	1.249	1.648	2.033	1.543	1.946	2.345
	5	2	0.613	1.110	1.574	0.959	1.452	1.930	1.256	1.745	2.235
	6	2	0.280	0.898	1.465	0.635	1.240	1.819	0.938	1.533	2.127
	4	3	0.488	0.800	1.190	0.707	1.054	1.513	0.857	1.249	1.792
	5	3	0.061	0.492	1.058	0.238	0.726	1.391	0.350	0.904	1.685
9	3	2	1.307	1.594	1.859	1.653	1.947	2.222	1.954	2.252	2.538
	4	2	1.041	1.411	1.763	1.377	1.755	2.121	1.670	2.052	2.427
	5	2	0.793	1.243	1.678	1.130	1.580	2.030	1.416	1.889	2.338
	6	2	0.532	1.074	1.588	0.871	1.408	1.939	1.159	1.700	2.245
	7	2	0.225	0.879	1.490	0.574	1.215	1.838	0.871	1.514	2.142
	4	3	0.658	0.950	1.305	0.887	1.207	1.622	1.048	1.410	1.929
	5	3	0.314	0.704	1.193	0.513	0.946	1.518	0.644	1.134	1.801
	6	3	0.012	0.439	1.090	0.102	0.669	1.426	0.204	0.849	1.727
	5	4	0.336	0.541	0.889	0.518	0.730	1.147	0.630	0.856	1.353

Comparison of $\mathbf{R_{HP}}$ with fixed-size subset selection rule $\mathbf{R_{FS}}$. As we pointed out earlier, our rule R_{HP} with $c = 0$ is the rule R_{FS}. For $t = 1$, the rule R_{FS} is a special case of the fixed-size subset selection rule of Desu and Sobel (1968) who considered selecting a subset of fixed size s that will contain all the t best populations. Now, for our goal, it will be of interest to study how much more we will have to pay in terms of the sample size for using the restricted subset size rule R_{HP} with $0 < c < 1$ in order to have the advantage of not necessarily selecting m populations but achieving the same guaranteed PCS P^* as does R_{FS} with a fixed subset size m.

For such a comparison, we define

$$e(P^*, k, t, m, c) = \frac{n_{HP}}{n_{FS}} \approx \left(\frac{\Delta^*_{HP}(n_{HP})}{\Delta^*_{FS}(n_{FS})}\right)^2, \tag{7.20}$$

where $\Delta^*_{HP}(n_{HP})$ and $\Delta^*_{FS}(n_{FS})$ are the values of Δ^* necessary to satisfy the probability requirement using the rules R_{HP} (with the specified value of c) and R_{FS}, respectively. We first note that values of Δ^*_{HP} are available from Table 7.1. Table 7.2 gives the values of Δ^*_{FS} for $P^* = 0.900, 0.950, 0.975$; $k = 3(1)10, 15$; and selected values of t and m. Table 7.3 gives the values of $e(P^*, k, t, m, c)$ for the same ranges of the arguments as in Table 7.2.

Remark 7.2.2. For Table 7.2, we solve for Δ^* by equating the right-hand side of (7.18) to P^*. It is easy to show that $A_0(k, t, m, \Delta^*)$ is increasing from $1 - \left[\binom{k-t}{m}/\binom{k}{m}\right]$ to 1 as Δ^* increases from zero to infinity. Thus, a solution for Δ^* exists provided that

$$1 - \left[\binom{k-t}{m}\Big/\binom{k}{m}\right] < P^* < 1. \tag{7.21}$$

The lower bound in (7.21) represents the PCS when we randomly select m of the k given populations. For any P^* less than or equal to this lower bound, we can then have PCS $\geq P^*$ by randomly selecting m populations. Such cases are marked by asterisks (*) in Table 7.2. In these cases, randomly choosing a subset of m populations is the best because the probability requirement can be met without taking any observation. So these cases are marked with asterisks also in Table 7.3.

We have seen that $A_c(k, t, m, \Delta^*) = P_{LFC}(CS \mid R_{HP})$ decreases in c for given k, t, m and Δ^*. Also, as mentioned earlier, $A_c(k, t, m, \Delta^*)$ increases in Δ^* for given k, t, m and c. Consequently, for given k, t, m, c and P^*, we have $e(P^*, k, t, m, c) > 1$ and increasing in c for $0 < c < 1$. This fact is borne out by the tabulated values in Table 7.3. We see that (1) $e(P^*, k, t, m, c)$ is increasing in m for given P^*, k, t and c, and (2) $e(P^*, k, t, m, c)$ is decreasing in P^* for given k, t, m and c. If we let P_L denote the lower bound of P^* in (7.21), then we see

Table 7.2: Values of $\Delta^* = \dfrac{\sqrt{n}\delta^*}{\sigma}$ for rule R_{FS}

k	m t	$P^* = 0.900$	$P^* = 0.950$	$P^* = 0.975$
5	3 2	*	0.382	0.714
6	3 2	0.436	0.792	1.099
	4 2	*	0.146	0.463
7	3 2	0.686	1.031	1.328
	4 2	0.206	0.552	0.850
	5 2	*	*	0.286
	4 3	*	*	0.385
8	3 2	0.861	1.199	1.490
	4 2	0.457	0.793	1.082
	5 2	0.038	0.377	0.669
	6 2	*	*	0.151
	4 3	*	0.523	0.671
	5 3	*	*	*
9	3 2	0.995	1.327	1.614
	4 2	0.635	0.964	1.247
	5 2	0.288	0.617	0.901
	6 2	*	0.241	0.528
	7 2	*	*	0.043
	4 3	0.087	0.394	0.658
	5 3	*	*	0.409
	6 3	*	*	*
	5 4	0.270	0.462	0.578

For entries denoted by asterisks, see Remark 7.2.2.

Table 7.3: Values of $e(P^*, k, t, m, c)$ defined in (7.20)

k	m	t	$P^* = 0.900$			$P^* = 0.950$			$P^* = 0.975$		
		$c \rightarrow$	0.25	0.50	0.75	0.25	0.50	0.75	0.25	0.50	0.75
5	3	2	*	*	*	6.30	13.20	21.18	3.33	5.82	8.60
6	3	2	3.91	7.92	12.62	2.43	4.08	5.93	2.01	3.05	4.21
	4	2	*	*	*	31.16	81.62	148.31	6.05	12.42	20.53
7	3	2	2.37	4.06	5.98	1.88	2.87	3.95	1.69	2.41	3.19
	4	2	12.15	31.55	57.35	3.79	7.50	12.15	2.64	4.56	6.93
	5	2	*	*	*	*	*	*	12.84	30.29	54.53
	4	3	*	*	*	*	*	*	2.18	6.90	18.39
8	3	2	1.93	3.03	4.26	1.67	2.40	3.20	1.55	2.12	2.73
	4	2	3.91	8.10	13.42	2.48	4.32	6.57	2.03	3.23	4.70
	5	2	260.23	853.25	1715.00	6.47	14.83	26.21	3.52	6.80	11.16
	6	2	*	*	*	*	*	*	38.59	103.07	198.42
	4	3	*	*	*	1.83	4.06	8.37	1.63	3.46	7.13
9	3	2	1.73	2.57	3.49	1.55	2.15	2.80	1.47	1.95	2.47
	4	2	2.69	4.94	7.71	2.04	3.31	4.84	1.79	2.71	3.79
	5	2	7.58	18.63	33.95	3.35	6.56	10.82	2.47	4.40	6.73
	6	2	*	*	*	13.06	34.13	64.73	4.82	10.37	18.08
	7	2	*	*	*	*	*	*	410.30	1239.69	2481.43
	4	3	57.20	119.24	225.00	5.07	9.38	16.95	2.54	4.59	8.59
	5	3	*	*	*	*	*	*	2.48	7.69	19.39
	5	4	1.55	4.01	10.84	1.26	2.50	6.16	1.19	2.19	5.48

For entries denoted by asterisks, see Remark 7.2.2.

that $e(P^*, k, t, m, c)$ gets enormously large when P^* is close to P_L. One would rather anticipate this in view of the significance of this bound P_L discussed in Remark 7.2.2. When $e(P^*, k, t, m, c)$ is close to 1, it is advantageous to use the rule R_{HP} at a slightly additional cost of sampling compared to R_{FS}. Table 7.3 shows this advantage can be had by taking small values of c especially when P^* is sufficiently larger than P_L.

7.3 Properties of $\mathbf{R_{HP}}$

In this section, we discuss some properties of our rule R_{HP}. For convenience, let

$$P_{\vec{\mu}}(i) \equiv \Pr\{R_{HP} \text{ selects } \Pi_{(i)} \mid \vec{\mu}\} = \Pr\{Y_{(i)} > (1-c)Y_{[k-m]} + cY_{[k]}\}. \quad (7.22)$$

Remark 7.3.1. By taking $t = 1$ in the proof of Theorem 7.2.1, we can see that $P_{\vec{\mu}}(i)$ is non-decreasing in $\mu_{[i]}$ and non-increasing in $\mu_{[j]}$, $j \neq i$, each holding when all other components of $\vec{\mu}$ are kept fixed.

We now define consistency and strong monotonicity of any selection rule R based on samples of common size n from the given populations.

Definition 7.3.1. R is said to be consistent w.r.t. the parametric region Ω' if $\lim_{n\to\infty} \inf_{\Omega'} P(CS \mid R) = 1$.

Definition 7.3.2. R is said to be strongly monotone in $\Pi_{(i)}$ if

$$P_{\vec{\mu}}(i) \begin{cases} \uparrow & \text{in } \mu_{[i]} \text{ when all other components of } \vec{\mu} \text{ are kept fixed}, \\ \downarrow & \text{in } \mu_{[j]},\ j \neq i, \text{ when all other components of } \vec{\mu} \text{ are kept fixed}. \end{cases}$$

Theorem 7.3.1. *For any $i = 1, \ldots, k$, rule R_{HP} is strongly monotone in $\Pi_{(i)}$.*

PROOF. The theorem is a consequence of Remark 7.3.1 and Definition 7.3.2. ■

Theorem 7.3.2. *For every $\delta^* > 0$, rule R_{HP} is consistent w.r.t. $\Omega(\delta^*)$.*

PROOF. As $n \to \infty$, so does Δ^*. We must show that

$$\lim_{\Delta^*\to\infty} t(Q_1 + Q_2 + Q_3) = 1. \quad (7.23)$$

Note that the integrands of the integrals that appear in the expressions for Q_1, Q_2 and Q_3 are bounded and so the dominated convergence theorem applies. We can see that $Q_1 \to 1$, $Q_2 \to 0$ and $Q_3 \to 0$ as $n \to \infty$. Hence the result follows. ■

Remark 7.3.2. Theorem 7.3.2 says that, no matter what $\delta^* > 0$ and $1/k < P^* < 1$ are chosen, rule R_{HP} with any $c \in (0,1)$ can be made to satisfy the probability requirement (7.1) by taking a sufficiently large sample.

Theorem 7.3.3. *For every n and $0 < c < 1$,*

$$\lim_{\delta^* \to \infty} \inf_{\Omega(\delta^*)} P(CS \mid R_{HP}) = 1.$$

PROOF. As $\delta^* \to \infty$, so does Δ^*. The result now follows from (7.23) being true. ∎

Remark 7.3.3. Theorem 7.3.3 states that by taking δ^* sufficiently large we can attain any P^* $(1/k < P^* < 1)$ probability requirement for R_{HP} based on any n and c $(0 < c < 1)$.

Remark 7.3.4. Theorems 7.3.1 through 7.3.3 represent properties parallel to those of R_S proved by Santner (1976).

7.4 Expected Size of the Selected Subset for Rule $\mathbf{R_{HP}}$

Let S denote the number of populations which the rule R_{HP} selects, for any $\vec{\mu} \in \Omega$. It is easy to see that

$$E_{\vec{\mu}}(S \mid R_{HP}) = \sum_{i=1}^{k} \Pr\{\Pi_{(i)} \text{ is selected by } R_{HP}\} = \sum_{i=1}^{k} P_{\vec{\mu}}(i), \tag{7.24}$$

where

$$\begin{aligned} P_{\vec{\mu}}(i) =& \Pr\{Y_{(i)} > (1-c)Y_{[k-m]} + cY_{[k]}\} = \Pr\{Y_{(i)} = Y_{[k]}\} + \\ & \sum_{\substack{l=1 \\ l \neq i}}^{k} \sum_{\substack{j=1 \\ j \neq l,i}}^{k} \Pr\{Y_{(l)} = Y_{[k-m]}, Y_{(j)} = Y_{[k]}, (1-c)Y_{(l)} + cY_{(j)} < Y_{(i)} < Y_{(j)}\} \end{aligned} \tag{7.25}$$

and $Y_{(i)}$ is distributed $N(\mu_{[i]}, \tau^2)$, $i = 1, \ldots, k$. We have not studied the behavior of $E_{\vec{\mu}}(S \mid R_{HP})$ over Ω due to the difficulty in analytically handling the integral expressions involved. We conjecture that the supremum of $E_{\vec{\mu}}(S \mid R_{HP})$ over Ω is attained when $\mu_{[1]} = \cdots = \mu_{[k]}$.

Remark 7.4.1. For the case of $t = 1$ considered earlier by Hsu and Panchapakesan (2003), the selection rule is still R_{HP} defined in (7.3). Therefore, the expression for $P_{\vec{\mu}}(i)$ given in (7.25) is the same for all t $(1 \leq t \leq k-1)$. Consequently, so is the expression for $E_{\vec{\mu}}(S \mid R_{HP})$.

7.5 Concluding Remarks

In this section, we make several comments regarding our rule R_{HP}. Given k, t, m, δ^* and P^*, there will be many choices of c and n satisfying the probability requirement (7.1). From its definition, it is clear that $P_{\vec{\mu}}(i)$ (and consequently $E_{\vec{\mu}}(S \mid R_{HP})$) is decreasing in c which means that the experimenter should seek to use rules with large values of c. On the other hand, for fixed δ^* and P^*, the larger the value of c, the larger Δ^* (and hence n) must be to satisfy the probability requirement (7.1). Hence, the experimenter must decide what trade off among the values of n and c is acceptable. If the cost of sampling is negligible and the emphasis is solely on the expected subset size, then we should let $c \to 1$. If the sampling cost is important and we are satisfied with the subset size, no matter what, being only at most m, then we should let $c = 0$ in which case the rule reduces to R_{FS}.

The problem of selecting at least one of the populations with the t smallest means is formulated in an analogous manner. The populations $\{\Pi_{(1)}, \ldots, \Pi_{(t)}\}$ are now called the t best and the preference-zone is $\Omega(\delta^*) = \{\vec{\mu} \in \Omega \mid \mu_{[t+1]} - \mu_{[t]} \geq \delta^*\}$. Following our earlier notations, let $U_i = -Y_i$, $i = 1, \ldots, k$. Then U_i can be viewed as the mean of a random sample of size n from a $N(\theta_i, \tau^2)$ population where $\theta_i = -\mu_i$. Since $\theta_{[i]} = -\mu_{[k-i+1]}$, $i = 1, \ldots, k$, we can formulate the problem as one of selecting at least one the populations associated with the t largest θ_is with the preference-zone $\Omega(\delta^*) = \{\vec{\theta} = (\theta_1, \ldots, \theta_k) \mid \theta_{[k-t+1]} - \theta_{[k-t]} \geq \delta^*\}$. We can thus apply the rules R_{HP} and R_S based on the U_i. These rules, expressed in terms of the Y_i for selecting one of the populations associated with the t smallest means μ_i, are:

R'_{HP}: Select Π_i if and only if

$$Y_i < cY_{[1]} + (1-c)Y_{[m+1]},$$

where $0 \leq c < 1$ is to be chosen so that the probability requirement (7.1) is satisfied, and

R'_S: Select Π_i if and only if

$$Y_i \leq \min\{Y_{[m]}, Y_{[1]} + d\sigma/\sqrt{n}\},$$

where $d > 0$ is to be chosen so that the probability requirement (7.1) is satisfied. Note that Table 7.1 and Table 1 of Santner (1976) are now applicable to R'_{HP} and R'_S, respectively.

A natural next step in our investigation is to consider the case of unknown common variance σ^2. As well known in the literature, we cannot have a single-stage procedure in this case because the determination of the sample size n

(continuing with equal sample sizes) depends upon the unknown σ. Following the earlier investigations in the case of fixed subset size procedures, one can consider two-stage procedures wherein the first stage samples are used to obtain an estimate of σ. This is an important investigation yet to be carried out even with the Santner type restricted subset size procedures.

Acknowledgment. The authors thank the referee for a critical reading of the manuscript and for several suggestions which have improved the presentation of the article.

References

1. Alam, K. and Rizvi, M. H. (1966). Selection from multivariate normal populations, *Annals of the Institute of Statistical Mathematics*, **18**, 307–318.

2. Desu, M. M. and Sobel, M. (1968). A fixed subset-size approach to the selection problem, *Biometrika*, **55**, 401-410. Corrections and Amendments, *Biometrika*, **55** (1968), 685.

3. Gupta, S. S. and Santner, T. J. (1973). On selection and ranking procedures - a restricted subset selection rule, In *Proceedings of the 39th Session of the International Statistical Institute*, Volume 45, Book I, pp. 478–486, Vienna, Austria.

4. Hsu, L. and Panchapakesan, S. (2003). Selecting the normal population with largest mean: a restricted subset selection rule, In *Statistical Methods and Practice: Recent Advances* (Eds., N. Balakrishnan, N. Kannan and M. R. Srinivasan), Chapter 10, pp. 145–161, Narosa Publishing House, New Delhi, India.

5. Lehmann, E. L. (1986). *Testing of Statistical Hypotheses*, Second edition, John Wiley & Sons, New York.

6. Mahamunulu, D. M. (1967). Some fixed-sample ranking and selection problems, *Annals of Mathematical Statistics*, **38**, 1079–1081.

7. Santner, T. J. (1975). A restricted subset selection approach to ranking and selection problems, *Annals of Statistics*, **3**, 334–349.

8. Santner, T. J. (1976). A generalized goal in restricted subset selection theory, *Sankhyā, Series B*, **38**, 129–143.

8

Selecting the Best Population with Two Controls: An Empirical Bayes Approach

Wen-Tao Huang[1] **and Yao-Tsung Lai**[2]

[1]*Department of Management Sciences and Decision Making, Tamkang University, Tamsui, Taipei, Taiwan, R.O.C.*
[2]*Department of Mathematical Statistics and Actuarial Science, Aletheia University, Tamsui, Taipei, Taiwan, R.O.C.*

Abstract. Consider k $(k \geq 2)$ populations whose means θ_i and variances σ_i^2 are all unknown. For given control values θ_0 and σ_0^2, a qualified subset is defined to be a set that contains every population whose mean is no smaller than θ_0 and whose variance is no larger than σ_0^2. We are interested in selecting the population with the largest mean from those members in the qualified subset if it is non-empty. If this subset is empty, we want no population to be selected. In this chapter we focus on the case of normal populations and consider a Bayes approach. An empirical Bayes selection procedure is derived and is shown to be asymptotically optimal. Analogous methods can be applied to cases other than the normal.

Keywords and phrases: Selection, controls, empirical Bayes, asymptotic optimality

8.1 Introduction

In many practical problems, an experimenter often faces the problem of testing for homogeneity. When the hypothesis of homogeneity is rejected, the experimenter often needs to select one of several categories or treatments under consideration according to his pre-fixed criterion or criteria. This multiple decision problem is considered as a problem of ranking and selection. Ranking and selection theory has been developed over the last fifty years by many researchers, in particular by those of the Purdue School. Readers are referred to Gupta and Panchapakesan (1979) for an extensive bibliography.

In the area of ranking and selection, a large part of the literature is concerned with using a single criterion. For example, a population is considered as the best if it is associated with the largest (or smallest) parameter in a finite set of populations. In many practical situations, on the other hand, this may not satisfy the experimenter's demand. For example, in industrial applications, one not only looks for the product with the largest target value, but also wants to keep the variability of the product for certain measurements under control. Under such circumstances, a single criterion for selection of potential treatments does not meet our requirement. Gupta, Liang and Rau (1994) were the first to consider selecting the best normal population compared with a control using two criteria for selection; however, they relate to the same parameter, namely, the mean. Huang and Lai (1999) have considered the selection of the best normal population compared with two controls. They consider two different quantities, namely, mean and variance. They assume a prior distribution for the mean and permit no perturbation for the variance. In this paper, we allow perturbation for the variance. In a Bayes framework, we develop an empirical Bayes procedure for selecting the best normal population with a normal-gamma prior which, as is well known, is a joint conjugate prior for mean and variance.

In Section 8.2, we formulate the problem and develop the Bayesian setup. In Section 8.3, we propose an empirical Bayes procedure. Section 8.4 deals with the large sample behavior of the proposed empirical Bayes rule. It is shown that the proposed empirical Bayes selection rule is asymptotically optimal.

8.2 Formulation of Problem and a Bayes Selection Rule

Suppose there are k populations (treatments, designs, etc.) $\pi_1, \ldots, \pi_k$ such that π_i has the distribution function $F_i(x)$ whose mean and variance are, respectively, θ_i and σ_i^2, $i = 1, \ldots, k$. We are interested in selecting the population that has the largest mean provided its variance is not large. More exactly, let θ_0 and σ_0^2 be two given control values. We want to choose the population with the largest mean provided that its mean is not less than θ_0 and its associated variance is not larger than σ_0^2. We state this precisely as follows.

Definition 8.2.1 Let $\pi_1, \ldots, \pi_k$ be k populations such that π_i has mean θ_i and variance σ_i^2, $i = 1, \ldots, k$. Let θ_0 and σ_0^2 be two control values (pre-fixed). Define $S = \{\pi_i \mid \sigma_i^2 \leq \sigma_0^2\}$. A population π_i is called σ-qualified, if $\pi_i \in S$. A population π_i is considered as the best, if it simultaneously satisfies the following conditions:

(i) $\pi \in S$,

(ii) $\theta_i \geq \theta_0$ and

(iii) $\theta_i = \max_{\pi_j \in S} \theta_j$.

Let, $\boldsymbol{\theta} = (\theta_1, \ldots, \theta_k)$, $\boldsymbol{\sigma} = (\sigma_1, \ldots, \sigma_k)$ and $\Omega = \{(\theta_i, \sigma_i) | -\infty < \theta_i < +\infty, \sigma_i > 0, i = 1, \ldots, k\}$ be the parameter space. Let $\mathbf{a} = (a_0, a_1, \ldots, a_k)$ denote an action, where $a_i = 0$ or 1; $i = 0, 1, \ldots, k$, and $\sum_{i=0}^{k} a_i = 1$. If $a_i = 1$, for some $i = 1, \ldots, k$, it means that population π_i is selected as the best. When $a_0 = 1$, it means that no population is considered as the best, i.e., none of the k populations satisfies both restrictions (i) and (ii) in Definition 8.2.1. Let $\mathbf{A}$ denote the action space, the set of all possible $\mathbf{a}$.

For the sake of convenience, we consider π_0 as a control population with the parameter vector (θ_0, σ_0). Corresponding $\boldsymbol{\theta} = (\theta_1, \ldots, \theta_k)$ and θ_0, we define $\boldsymbol{\theta}' = (\theta'_0, \theta'_1, \ldots, \theta'_k)$ as follows.

Definition 8.2.2 For a given positive δ and for $i = 0, 1, \ldots, k$, define

$$\theta'_i = \theta_i I_{\{\sigma_i \le \sigma_0\}} + (\theta_0 - \delta) I_{\{\sigma_i > \sigma_0\}}.$$

Accordingly, those populations which do not meet the requirement (i) will also fail to meet the requirement (ii) in Definition 8.2.1 in terms of the transformed parameter θ'_i.

For our decision-theoretic approach, we adopt the loss function which has been used by Huang and Lai (1999).

Definition 8.2.3 For $\delta > 0$, two control values θ_0, σ_0, and parameters $\boldsymbol{\theta}, \boldsymbol{\sigma}$ (equivalently, $\boldsymbol{\theta}', \boldsymbol{\sigma}$), if action $\mathbf{a}$ is taken, a loss $L(\boldsymbol{\theta}, \boldsymbol{\sigma}, \mathbf{a})$ is incurred, defined by

$$\begin{aligned} L(\boldsymbol{\theta}, \boldsymbol{\sigma}, \mathbf{a}) &= L(\boldsymbol{\theta}', \boldsymbol{\sigma}, \mathbf{a}) \\ &= \alpha \left[\theta'_{[k]} - \sum_{i=0}^{k} a_i \theta'_i \right] + (1-\alpha) \sum_{i=0}^{k} \gamma a_i \left(\frac{\sigma_i}{\sigma_0} - 1 \right) I_{\{\sigma_i > \sigma_0\}}, \end{aligned} \tag{8.1}$$

where $\alpha (0 \le \alpha \le 1)$ and $\gamma (> 0)$ are pre-fixed constants, and $\theta'_{[k]} = \max_{0 \le i \le k} \theta'_i$.

The properties of $L(\boldsymbol{\theta}, \boldsymbol{\sigma}, \mathbf{a})$ defined in (8.1) have been discussed in Huang and Lai (1999). In this paper, we consider a Bayes approach for the problem of selecting the best normal population.

Let $X_{i1}, \ldots, X_{iM}$ $(i = 1, \ldots, k)$ be a random sample of size M from the normal population π_i with mean θ_i and variance σ_i^2. The observed value is denoted by $x_{i1}, \ldots, x_{iM}$. Let $\tau_i = 1/\sigma_i^2$, $i = 1, \ldots, k$. It is assumed that (θ_i, τ_i) is a realization of a random vector (Θ_i, T_i) with a normal-gamma prior distribution which is the product of conditional normal distribution $N(\mu_i, [(2\alpha_i - 1)\tau_i]^{-1})$ given τ_i on Θ_i, and a marginal gamma prior distribution $G(\alpha_i, \beta_i)$ on T_i.

For convenience, for $i = 1, \ldots, k$, we denote $\mathbf{x}_i = (x_{i1}, \ldots, x_{iM})$ and $\bar{x}_i = \frac{1}{M}\sum_{j=1}^{M} x_{ij}$. Raiffa and Schlaifer (1961) show that the conditional posterior distribution of Θ_i given $\mathbf{x}_i$ and τ_i is a normal distribution $N(\phi_i(\bar{x}_i), [(2\alpha_i + M-1)\tau_i]^{-1})$, where

$$\phi_i(\bar{x}_i) = E[\Theta_i|\mathbf{x}_i, \tau_i] = \frac{(2\alpha_i - 1)\mu_i + M\bar{x}_i}{2\alpha_i + M - 1}, \tag{8.2}$$

and the marginal posterior distribution of T_i given $\mathbf{x}_i$ is a gamma distribution $G(2\alpha_i + \frac{M}{2} - 1, \eta_i)$, where

$$\eta_i = \beta_i + \frac{(M-1)S^2}{2} + \frac{(2\alpha_i - 1)M(\bar{x}_i - \mu_i)^2}{2(2\alpha_i + M - 1)}. \tag{8.3}$$

The random vectors $(\Theta_1, T_1), \ldots, (\Theta_k, T_k)$ are assumed to be mutually independent.

Let $\mathbf{x} = (\mathbf{x}_1, \ldots, \mathbf{x}_k)$ and let χ be the sample space generated by $\mathbf{x}$. A selection rule $\mathbf{d} = (d_0, d_1, \ldots, d_k)$ is a mapping defined on the sample space χ into the $k+1$ product space $[0,1] \times [0,1] \times \cdots \times [0,1]$ such that $\sum_{i=0}^{k} d_i(\mathbf{x}) = 1$, for all $\mathbf{x} \in \chi$. For every $\mathbf{x} \in \chi$, $d_i(\mathbf{x})$ denotes the probability of selecting population π_i as the best, $i = 1, \ldots, k$; and $d_0(\mathbf{x})$ denotes the probability that none is selected as the best.

For ease of notation, let $\tau_0 = 1/\sigma_0^2$, $\boldsymbol{\tau} = (\tau_1, \ldots, \tau_k)$, $\boldsymbol{\mu} = (\mu_1, \ldots, \mu_k)$, $\boldsymbol{\alpha} = (\alpha_1, \ldots, \alpha_k)$, $\boldsymbol{\beta} = (\beta_1, \ldots, \beta_k)$, $\boldsymbol{\eta} = (\eta_1, \ldots, \eta_k)$, $\boldsymbol{\Theta} = (\Theta_1, \ldots, \Theta_k)$ and $\mathbf{T} = (T_1, \ldots, T_k)$. Let $h(\boldsymbol{\theta}|\mathbf{x}, \boldsymbol{\tau}; \boldsymbol{\mu}, \boldsymbol{\alpha})$ be the joint conditional posterior probability density function of $\boldsymbol{\Theta}$ given $\mathbf{x}$ and $\boldsymbol{\tau}$; and $g(\boldsymbol{\tau}|\mathbf{x}; \boldsymbol{\alpha}, \boldsymbol{\eta})$ be the joint conditional posterior probability density function of $\mathbf{T}$ given $\mathbf{x}$. Let $h_i(\theta_i|\mathbf{x}_i, \tau_i; \mu_i \alpha_i)$ and $g_i(\tau_i|\mathbf{x}_i; \alpha_i \eta_i)$ be the marginally conditional probability density function of Θ_i and T_i, respectively. Under the preceding formulation, the Bayes risk of a selection rule $\mathbf{d}$, denoted by $r(\mathbf{d})$, is given by

$$\begin{aligned}
r(\mathbf{d}) &= E_{\boldsymbol{\tau}} E_{\boldsymbol{\theta}} E_{\mathbf{X}} L(\boldsymbol{\theta}, \boldsymbol{\tau}; \mathbf{d}) \\
&= \alpha \int \int_{\Omega} \int_{\chi} \theta'_{[k]} f(\mathbf{x}|\boldsymbol{\theta}, \boldsymbol{\tau}) h(\mathbf{x}|\boldsymbol{\tau}; \boldsymbol{\mu}, \boldsymbol{\alpha}) g(\boldsymbol{\tau}; \boldsymbol{\alpha}, \boldsymbol{\beta}) d\mathbf{x}\, d\boldsymbol{\theta}\, d\boldsymbol{\tau} \\
&\quad -\alpha \int \int_{\Omega} \int_{\chi} \sum_{i=0}^{k} d_i(\mathbf{x}) \theta'_i f(\mathbf{x}|\boldsymbol{\theta}, \boldsymbol{\tau}) h(\mathbf{x}|\boldsymbol{\tau}; \boldsymbol{\mu}, \boldsymbol{\alpha}) g(\boldsymbol{\tau}; \boldsymbol{\alpha}, \boldsymbol{\beta}) d\mathbf{x}\, d\boldsymbol{\theta}\, d\boldsymbol{\tau} \\
&\quad +(1-\alpha) \int \int_{\Omega} \int_{\chi} \sum_{i=0}^{k} \gamma d_i(\mathbf{x}) \left(\sqrt{\frac{\tau_0}{\tau_i}} - 1 \right) \\
&\qquad \times I_{(\tau_i < \tau_0)} f(\mathbf{x}|\boldsymbol{\theta}, \boldsymbol{\tau}) h(\mathbf{x}|\boldsymbol{\tau}; \boldsymbol{\mu}, \boldsymbol{\alpha}) g(\boldsymbol{\tau}; \boldsymbol{\alpha}, \boldsymbol{\beta}) d\mathbf{x}\, d\boldsymbol{\theta}\, d\boldsymbol{\tau} \\
&= \alpha I_1 - \alpha I_2 + (1 - \alpha I_3), \quad \text{say.}
\end{aligned}$$

Then, straightforward computation yields

$$I_1 = \int \int_{\Omega} \theta'_{[k]} h(\boldsymbol{\theta}|\boldsymbol{\tau}; \boldsymbol{\mu}, \boldsymbol{\alpha}) g(\boldsymbol{\tau}; \boldsymbol{\alpha}, \boldsymbol{\beta}) d\boldsymbol{\theta}\, d\boldsymbol{\tau} = C$$

for some constant C, and

$$\begin{aligned}
I_2 &= \int_\chi \int \int_\Omega \sum_{i=0}^k d_i(\mathbf{x})\theta_i' h(\boldsymbol{\theta}|\mathbf{x},\boldsymbol{\tau};\boldsymbol{\mu},\boldsymbol{\alpha}) g(\boldsymbol{\tau}|\mathbf{x},\boldsymbol{\alpha},\boldsymbol{\eta}) f(\mathbf{x}) d\boldsymbol{\theta}\, d\boldsymbol{\tau}\, d\mathbf{x} \\
&= \int_\chi \sum_{i=0}^k d_i(\mathbf{x}) \left[\int_0^{\tau_0} \int_{-\infty}^{+\infty} (\theta_0 - \delta) h_i(\theta_i|\mathbf{x}_i,\tau_i;\mu_i,\alpha) g_i(\tau_i|\mathbf{x}_i;\alpha_i,\eta_i) d\theta_i\, d\tau_i \right. \\
&\quad \left. + \int_{\tau_0}^{+\infty} \int_{-\infty}^{+\infty} \theta_i h_i(\theta_i|\mathbf{x}_i,\tau_i;\mu_i,\alpha) g_i(\tau_i|\mathbf{x}_i;\alpha_i,\eta_i) d\theta_i\, d\tau_i \right] f(\mathbf{x}) d\mathbf{x} \\
&= \int_\chi \sum_{i=0}^k d_i(\mathbf{x}) \phi_i'(\bar{x}_i) f(\mathbf{x}) d\mathbf{x},
\end{aligned}$$

where

$$\phi_i'(\bar{x}_i) = (\theta_0 - \delta) G_i(\tau_0|\mathbf{x}_i,\alpha_i,\eta_i) + \phi_i(x_i)(1 - G_i(\tau_0|\mathbf{x}_i,\alpha_i,\eta_i)), \tag{8.4}$$

and $G_i(\tau_0|\mathbf{x}_i;\alpha_i,\eta_i)$ is the cumulative probability of gamma distribution $G(2\alpha_i + \frac{M}{2} - 1, \eta_i)$. Finally,

$$\begin{aligned}
I_3 &= \int_\chi \sum_{i=0}^k \gamma d_i(\mathbf{x}) \left[\int_0^{\tau_0} \left(\sqrt{\frac{\tau_0}{\tau_i}} - 1 \right) g_i(\tau_i|\mathbf{x}_i,\alpha_i,\eta_i) d\tau_i \right] f(\mathbf{x}) d\mathbf{x} \\
&= \int_\chi \sum_{i=0}^k \gamma d_i(\mathbf{x}) \left[\frac{\sqrt{\tau_0}\eta_i^{\frac{1}{2}} \Gamma(2\alpha_i + \frac{M}{2} - \frac{3}{2})}{\Gamma(2\alpha_i + \frac{M}{2} - 1)} \right. \\
&\quad \left. \times G_i(\tau_0|\mathbf{x}_i,\alpha_i - \frac{1}{2},\eta_i) - G_i(\tau_0|\mathbf{x}_i,\alpha_i,\eta_i) \right] f(\mathbf{x}) d\mathbf{x}.
\end{aligned}$$

Hence, for some constant C,

$$r(\mathbf{d}) = \alpha C - \int_\chi \sum_{i=0}^k d_i(\mathbf{x}) U_i(\mathbf{x}_i) f(\mathbf{x}) d\mathbf{x}, \tag{8.5}$$

where

$$\begin{aligned}
U_i(\mathbf{x}_i) &= \alpha\phi_i'(\bar{x}_i) + (1-\alpha)\gamma \left[\frac{\sqrt{\tau_0}\eta_i^{\frac{1}{2}} \Gamma(2\alpha_i + \frac{M}{2} - \frac{3}{2})}{\Gamma(2\alpha_i + \frac{M}{2} - 1)} \right. \\
&\quad \left. \times G_i(\tau_0|\mathbf{x}_i,\alpha_i - \frac{1}{2},\eta_i) - G_i(\tau_0|\mathbf{x}_i,\alpha_i,\eta_i) \right].
\end{aligned} \tag{8.6}$$

For each $\mathbf{x} \in \chi$, let

$$Q(\mathbf{x}) = \left\{ i \,\middle|\, U_i(\mathbf{x}_i) = \max_{0 \le j \le k} U_j(x_j), \quad i = 1, \ldots, k \right\}. \tag{8.7}$$

Then, define

$$i^* = i^*(\mathbf{x}) = \begin{cases} 0 & \text{if } Q(\mathbf{x}) = \{0\}, \\ \min\{i | i \in Q(\mathbf{x}),\ i \neq 0\} & \text{otherwise.} \end{cases} \tag{8.8}$$

Then, from (8.6)–(8.8), we derive a Bayes selection rule $\mathbf{d}^B = (d_0^B, d_1^B, \ldots, d_k^B)$ given by

$$d_{i^*}^B(\mathbf{x}) = 1 \quad \text{and} \quad d_j^B(\mathbf{x}) = 0, \quad \text{for } j \neq i^*. \tag{8.9}$$

8.3 An Empirical Bayes Selection Rule

In the problem formulated in Section 8.2, we consider that $\alpha_1, \ldots, \alpha_k$ are known. It is well-known that the exponential distribution is contained in this case. Since $U_i(\mathbf{x}_i)$ still involves the unknown parameters μ_i, β_i, $i = 1, \ldots, k$, the proposed Bayes selection rule $\mathbf{d}^B$ is not applicable. However, based on the past data, these unknown parameters can be estimated and a decision can be made if one more observation is taken. For $i = 1, \ldots, k$, let X_{ijt} ($j = 1, \ldots, M$) denote a sample of size M from π_i with a normal distribution $N(\theta_{it}, \tau_{it}^{-1})$ at time t ($t = 1, \ldots, n$), where (θ_{it}, τ_{it}) is a realization of a random vector (Θ_{it}, T_{it}) which is an independent copy of (Θ_i, T_i) with a normal-gamma distribution described in the preceding section. It is assumed that (Θ_{it}, T_{it}), $i = 1, \ldots, k$, $t = 1, \ldots, n$, are mutually independent. For our convenience, we denote the current random sample $X_{ij(n+1)}$ by X_{ij}, for $j = 1, \ldots, M$, $i = 1, \ldots, k$.

Consistent estimates of the parameters μ_i, β_i have been studied by several authors, for instance, Ghosh and Lahiri (1987) and Ghosh and Meeden (1986).

For each π_i, $i = 1, \ldots, k$, we estimate the unknown parameters μ_i and β_i based on the past data X_{ijt}, $j = 1, \ldots, M$, $t = 1, \ldots, n$. We let

$$\begin{aligned} X_{it} &= \frac{1}{M} \sum_{j=1}^{m} X_{ijt}, \quad X_i(n) = \frac{1}{n} \sum_{t=1}^{n} X_{it}, \\ W_{it}^2 &= \frac{1}{M-1} \sum_{j=1}^{M} (X_{ijt} - X_{it})^2, \text{ and } W_i^2(n) = \frac{1}{n} \sum_{t=1}^{n} W_{it}^2. \end{aligned} \tag{8.10}$$

For ease of notation, we define μ_{in} and β_{in} as estimators of μ_i and β_i, respectively, given by

$$\mu_{in} = X_i(n) \quad \text{and} \quad \beta_{in} = (\alpha_i - 1) W_i^2(n). \tag{8.11}$$

Also, for $i = 1, \ldots, k$, we define

$$\begin{aligned} U_{in}(\mathbf{x}_i) &= \alpha\phi'_{in}(x_i) \\ &\quad +(1-\alpha)\gamma\Bigg[\frac{\sqrt{\tau_0}\eta_{in}^{\frac{1}{2}}\Gamma(2\alpha_i+\frac{M}{2}-\frac{3}{2})}{\Gamma(2\alpha_i+\frac{M}{2}-1)} \\ &\quad \times G_{in}(\tau_0|\mathbf{x}_i,\alpha_i-\frac{1}{2},\eta_{in}) - G_{in}(\tau_0|\mathbf{x}_i,\alpha_i,\eta_{in})\Bigg], \end{aligned} \tag{8.12}$$

where

$$\phi'_{in}(x_i) = (\theta_0-\delta)G_{in}(\tau_0|\mathbf{x}_i,\alpha_i,\eta_{in}) + \phi_{in}(\bar{x}_i)(1-G_{in}(\tau_0|\mathbf{x}_i,\alpha_i,\eta_{in})), \tag{8.13}$$

$$\phi_{in}(\bar{x}_i) = \frac{(2\alpha_i-1)\mu_{in}+M\bar{x}_i}{2\alpha_i+M-1}, \tag{8.14}$$

$$\eta_{in} = \beta_{in} + \frac{(M-1)S^2}{2} + \frac{(2\alpha_i-1)M(\bar{x}_i-\mu_{in})^2}{2(2\alpha_i+M-1)}, \tag{8.15}$$

and

$$G_{in}(\tau_0|\mathbf{x}_i,\alpha_i,\eta_{in}) = G_i(\tau_0|\mathbf{x}_i,\alpha_i,\eta_{in}). \tag{8.16}$$

Note that $U_{0n}(\mathbf{x}_0) = \theta_0$. We consider $U_{in}(\mathbf{x}_i)$ to be an estimator of $U_i(\mathbf{x}_i)$. The properties of the estimators proposed above will be discussed in the following section.

For each $\mathbf{x} \in \chi$, let

$$Q_n(\mathbf{x}) = \left\{ i \,\middle|\, U_{in}(\mathbf{x}_i) = \max_{0\le j\le k} U_{jn}(\mathbf{x}_j)\ ,\ i = 0, 1, \ldots, k \right\}. \tag{8.17}$$

Again, define

$$i_n^* = i_n^*(\mathbf{x}) = \begin{cases} 0 & \text{if } Q_n(\mathbf{x}) = \{0\}, \\ \min\{i|i \in Q_n(\mathbf{x}),\ i \neq 0\} & \text{otherwise.} \end{cases} \tag{8.18}$$

Then, from (8.12), (8.17) and (8.18), we obtain the empirical Bayes selection rule $d^{*n} = (d_0^{*n}, d_1^{*n}, \ldots, d_k^{*n})$ given by

$$d_{i_n^*}^{*n}(\mathbf{x}) = 1 \quad \text{and} \quad d_j^{*n}(\mathbf{x}) = 0, \quad \text{for } j \neq i_n^*. \tag{8.19}$$

8.4 Some Large Sample Properties

In this section, we study the asymptotic optimality of the proposed empirical Bayes selection rule. First, we discuss the consistency of the estimators defined in (8.10) through (8.16) in the case of $M \geq 2$. These results are stated below in Lemma 8.4.1. Since proofs for all parts of Lemma 8.4.1 are routine, we omit them.

Lemma 8.4.1 *(i) $\phi_{in}(\bar{x}_i)$ defined in (8.14), η_{in} defined in (8.15) and G_{in} defined in (8.16), are consistent estimators of $\phi_i(\bar{x}_i)$, η_i and G_i, respectively, $i = 1, \ldots, k$.*

(ii) $U_{in}(\mathbf{x}_i)$ defined in (8.12), is a consistent estimator of $U_i(\mathbf{x}_i)$, $i = 1, \ldots, k$.

Consider an empirical Bayes selection rule $\mathbf{d}^n = (d_0^n, d_1^n, \ldots, d_k^n)$ and denote its Bayes risk by $r(\mathbf{d}^n)$. Since $r(\mathbf{d}^B)$ is the minimum Bayes risk, it implies that $r(\mathbf{d}^n) - r(\mathbf{d}^B) \geq 0$. Thus $E_n[r(\mathbf{d}^n)] - r(\mathbf{d}^B) \geq 0$, where the expectation E_n is taken with respect to the past observations X_{ijt}, $i = 1, \ldots, k$, $j = 1, \ldots, M$, and $t = 1, \ldots, n$, The non-negative difference $E_n[r(\mathbf{d}^n)] - r(\mathbf{d}^B)$ can be used to measure the performance of the selection rule $\mathbf{d}^n$.

Definition 8.4.1 A sequence of empirical Bayes selection rule $\{\mathbf{d}^n\}_{n=1}^{\infty}$ is said to be asymptotically optimal, if $\lim_{n \to \infty}[E_n[r(\mathbf{d}^n)] - r(\mathbf{d}^B)] = 0$.

Theorem 8.4.1 *The empirical Bayes selection rule $d^{*n}(\mathbf{x})$, defined in (8.19) is asymptotically optimal.*

PROOF. By straightforward computation and some simplification, we have

$$
\begin{aligned}
& E_n[r(\mathbf{d}^{*n})] - r(\mathbf{d}^B) \\
&= E_n \int_{\mathcal{X}} [d_{i^*}^B(\mathbf{x}) U_{i^*}(\mathbf{x}_{i^*}) - d_{i_n^*}^{*n}(\mathbf{x}) U_{i_n^*}(\mathbf{x}_{i_n^*})] f(\mathbf{x}) d\mathbf{x} \\
&= \sum_{i=1}^{k} \int_{R^M} P_n \{|U_{in}(\mathbf{x}_i) - U_i(\mathbf{x}_i)| \geq |U_i(\mathbf{x}_i) - \theta_0|\} \, |U_i(\mathbf{x}_i) - \theta_0| \, f_i(\mathbf{x}_i) d\mathbf{x}_i \\
&\quad + \sum_{i=1}^{k} \sum_{j=1}^{k} \int_{R^M} \int_{R^M} \left[P_n \left\{ |U_{in}(\mathbf{x}_i) - U_i(\mathbf{x}_i)| > \frac{1}{2} |U_i(\mathbf{x}_i) - U_j(\mathbf{x}_j)| \right\} \right. \\
&\qquad \left. + P_n \left\{ |U_{jn}(\mathbf{x}_j) - U_j(\mathbf{x}_j)| > \frac{1}{2} |U_i(\mathbf{x}_i) - U_j(\mathbf{x}_j)| \right\} \right] \\
&\qquad \times |U_i(\mathbf{x}_i) - U_j(\mathbf{x}_j)| \, f_i(\mathbf{x}_i) f_j(\mathbf{x}_j) d\mathbf{x}_i \, d\mathbf{x}_j \\
&= I_n + I_n^*, \quad \text{say.}
\end{aligned}
$$

From (8.2) and (8.6), it can be shown that

$$|U_i(\mathbf{x}_i) - \theta_0| \le \frac{M|\bar{x}_i - \mu_i|}{2\alpha_i + M - 1} + b_i(\bar{x}_i - \mu_i)^2 + c_i,$$

where

$$b_i = \frac{\gamma\sqrt{\tau_0}(2\alpha_i - 1)M\Gamma(2\alpha_i + \frac{M}{2} - \frac{3}{2})}{4(2\alpha_i + M - 1)\Gamma(2\alpha_i + \frac{M}{2} - 1)},$$

and

$$c_i = |\theta_0 - \delta| + \gamma\left[\left(\frac{\sqrt{\tau_0}\,\Gamma(2\alpha_i + \frac{M}{2} - \frac{3}{2})}{\beta_i^{\frac{1}{2}}\Gamma(2\alpha_i + \frac{M}{2} - 1)}\right)\left(\frac{2\beta_i + (M-1)S^2}{4}\right) + 1\right] + |\theta_0| + |\mu_i|.$$

It can then be shown that

$$\int_{R^M} |U_i(\mathbf{x}_i) - \theta_0| f_i(\mathbf{x}_i) d\mathbf{x}_i \le c_{\max},$$

where

$$\begin{aligned} c_{\max} &= \max_{1\le i\le k}\left\{\frac{M}{2\alpha_i + M - 1} + \left(\frac{M}{2\alpha_i + M - 1} + b_i\right)\right. \\ &\quad \left.\times\left(\frac{\alpha_i}{M\beta_i} + \frac{\beta_i}{(2\alpha_i - 1)(\alpha_i - 1)}\right) + c_i\right\}. \end{aligned}$$

For any $\varepsilon > 0$, since $U_{in}(\mathbf{x}_i)$ is a consistent estimate for $U_i(\mathbf{x}_i)$, there exists $N_1 > 0$ such that for any $n > N_1$, we have

$$P_n\left\{|U_{in}(\mathbf{x}_i) - U_i(\mathbf{x}_i)| > \frac{\varepsilon}{2k}\right\} < \frac{\varepsilon}{2kc_{\max}}.$$

Let $\chi_i = \left\{\mathbf{x}_i : |U_i(\mathbf{x}_i) - \theta_0| \le \frac{\varepsilon}{2k}\right\}$. We have

$$\begin{aligned} I_1 &= \sum_{i=1}^{k}\int_{\chi_i} P_n\left\{|U_{in}(\mathbf{x}_i) - U_i(\mathbf{x}_i)| \ge |U_i(\mathbf{x}_i) - \theta_0|\right\}|U_i(\mathbf{x}_i) - \theta_0|\, f_i(\mathbf{x}_i)d\mathbf{x}_i \\ &\quad + \sum_{i=1}^{k}\int_{R^M - \chi_i} P_n\left\{|U_{in}(\mathbf{x}_i) - U_i(\mathbf{x}_i)| \ge |U_i(\mathbf{x}_i) - \theta_0|\right\} \\ &\qquad\qquad \times |U_i(\mathbf{x}_i) - \theta_0|\, f_i(\mathbf{x}_i)d\mathbf{x}_i \\ &\le \sum_{i=1}^{k}\int_{\chi_i}\frac{\varepsilon}{2k} f_i(\mathbf{x}_i)d\mathbf{x}_i \\ &\quad + \sum_{i=1}^{k}\int_{R^M - \chi_i} P_n\left\{|U_{in}(\mathbf{x}_i) - U_i(\mathbf{x}_i)| \ge \frac{\varepsilon}{2k}\right\}|U_i(\mathbf{x}_i) - \theta_0|\, f_i(\mathbf{x}_i)d\mathbf{x}_i \\ &\le k\frac{\varepsilon}{2k} + kc_{\max}\frac{\varepsilon}{2kc_{\max}} = \varepsilon. \end{aligned}$$

Hence, $\lim_{n\to+\infty} I_n = 0$. Similarly, $\lim_{n\to+\infty} I_n^* = 0$. This completes the proof. ■

References

1. Ghosh, M. and Lahiri, P. (1987). Robust empirical Bayes estimation of means from stratified samples, *J. Amer. Statist. Assoc.*, **82**, 1153–1162.

2. Ghosh, M. and Meeden, G. (1986). Empirical Bayes estimation in finite population sampling, *J. Amer. Statist. Assoc.*, **81**, 1058–1062.

3. Gupta, S. S., Liang, T. and Rau, R.-B. (1994). Empirical Bayes rules for selecting the best normal population compared with a control, *Statistics and Decisions*, **12**, 125–147.

4. Gupta, S. S. and Panchapakesan, S. (1979). *Multiple Decision Procedures*, John Wiley & Sons, New York; Republished by SIAM, Philadelphia, PA, 2002.

5. Huang, W.-T. and Lai, Y.-T. (1999). Empirical Bayes procedure for selecting the best population with multiple criteria, *Ann. Inst. Statist. Math.*, **51**, 281–299.

6. Raiffa, H. and Schlaifer, R. (1961). *Applied Statistical Decision Theory*, Harvard University Press, Cambridge, MA.

9

Simultaneous Selection of Extreme Populations: Optimal Two-stage Decision Rules

Neeraj Misra and Ishwari D. Dhariyal

Department of Mathematics, Indian Institute of Technology Kanpur, India

Abstract: Consider k (≥ 2) independent populations $\Pi_1, \ldots, \Pi_k$, such that the observations from the population Π_i, $i = 1, \ldots, k$, have a probability density function (pdf) belonging to the exponential family $c(\theta_i)b(x)e^{\theta_i x}$, $x \in \mathbb{R}$ (the real line), $\theta_i \in \Omega \subseteq \mathbb{R}$. Unknown populations associated with $\theta_{[1]} = \min_{1 \leq i \leq k} \theta_i$ and $\theta_{[k]} = \max_{1 \leq i \leq k} \theta_i$ are called the lower extreme population (LEP) and the upper extreme population (UEP), respectively. This chapter deals with optimal decision (or selection) rules for the goal of simultaneously selecting the LEP and the UEP in two stages: screening out non-extreme populations at the first stage and selecting the LEP (the ones associated with $\theta_{[1]}$) and the UEP (those associated with $\theta_{[k]}$) at the second stage. Let $\mathcal{D}_I$ denote the class of permutation invariant two-stage selection rules (ν, η, δ), where at Stage 1, ν and η decide about how many populations and which ones to select, respectively, as contenders for the LEP and the UEP, and at Stage 2, additional samples are taken from the selected populations and δ makes the final decision about the LEP and the UEP from the respective set of these contending populations. After giving the formulation of the problem and defining two-stage selection rules, we derive an essentially complete class of two-stage permutation invariant selection rules under the assumption that the underlying pdfs are unimodal.

Keywords and phrases: Exponential family, lower extreme population, upper extreme population, two-stage selection rule, permutation invariance, monotone likelihood ratio property

9.1 Introduction

Let $X_1, \ldots, X_k$ denote k (≥ 2) independent random variables representing the populations $\Pi_1, \ldots, \Pi_k$, respectively, and suppose that X_i has a probability density function (pdf) $g(\cdot;\theta_i)$, $i = 1,\ldots,k$. Let $\Pi_{(i)}$, $i = 1,\ldots,k$, denote the population associated with $\theta_{[i]}$, the ith smallest of θ_is. Throughout we assume that $\theta_1,\ldots,\theta_k$ are completely unknown and that there is no prior information about which of $\Pi_1,\ldots,\Pi_k$ is $\Pi_{(i)}$, $i = 1,\ldots,k$. Suppose χ is the sample space of the observation vector $\underline{X}$ and let Ω denote the parameter space of $\boldsymbol{\theta} = (\theta_1,\ldots,\theta_k)$. The populations $\Pi_{(1)}$ and $\Pi_{(k)}$, associated with $\theta_{[1]}$ and $\theta_{[k]}$ respectively, are called the lower extreme population (LEP) and the upper extreme population (UEP), respectively. In case of ties for the LEP (or the UEP), we assume that one of the tied populations is arbitrarily tagged as the LEP (or the UEP).

Mishra and Dudewicz (1987) considered the problem of simultaneously selecting two non-empty and random subsets S_L and S_U of $\{1,\ldots,k\}$, where i is associated with Π_i, such that the probability of the event $\{(1) \in S_L \text{ and } (k) \in S_U\}$ is at least P^* $(1/(k(k-1)) < P^* < 1)$, a pre-assigned constant; here (1) is associated with the LEP $\Pi_{(1)}$ and (k) is associated with the UEP $\Pi_{(k)}$. For this goal, they proposed a selection rule under the assumption that the θ_is are the location parameters and the underlying pdfs have the monotone likelihood ratio (MLR) property. Mishra (1986a,b) considered some variations of this problem under different population models. In an extensive decision theoretic study of the problem of simultaneously selecting the LEP and the UEP, Misra and Dhariyal (1994) and Dhariyal and Misra (1994) derived Bayes, minimax and the best permutation invariant selection rules under different loss functions. Recently, Hussein and Panchapakesan (2001) proposed selection rules for simultaneous selection of the LEP and the UEP from k (≥ 2) two parameter exponential populations.

For the goal of selecting the UEP $\Pi_{(k)}$, Gupta and Miescke (1983) proposed a class of two-stage permutation invariant selection rules (ϕ, ψ, δ), where, at Stage 1, ϕ and ψ, respectively, decide about how many and which populations to select, and at Stage 2, after additional samples from the populations selected at Stage 1 have been taken, δ makes a final decision about the UEP from the contending populations selected at Stage 1. Under quite general loss functions and underlying pdfs from unimodal exponential family, Gupta and Miescke (1983) derived an essentially complete class of permutation invariant selection rules. Gupta and Miescke (1984) applied the results of Gupta and Miescke (1983) to a specific loss function under normal probability models.

In this chapter, we extend the results of Gupta and Miescke (1983), derived for the goal of selecting the UEP, to the goal of simultaneously selecting the LEP and the UEP. In proving various results here, we make use of the work by Eaton (1967) which we briefly describe here.

Eaton (1967) considered the general goal of partitioning $\{\Pi_1,\ldots,\Pi_k\}$ into s (≥ 2) disjoint subsets $A_1,\ldots,A_s$, such that A_1 contains populations corresponding to the k_1 largest θ_is, A_2 contains populations corresponding to the next k_2 largest θ_is, $\cdots$, and A_s contains populations corresponding to the k_s smallest θ_is, where $1 \leq k_i < k$ and $\sum_{i=1}^{s} k_i = k$. Here, the action space consists of all partitions $\mathbf{A} = \{A_1, \ldots, A_s\}$ of $\{1, \ldots, k\}$. Let G denote the group of permutations on indices $\{1, \ldots, k\}$. Suppose that $L(\boldsymbol{\theta}, \mathbf{A})$ denotes the loss associated with partition $\mathbf{A}$ when $\boldsymbol{\theta}$ is the true parameter value. For the partitions $\mathbf{A} = \{A_1, \ldots, A_s\}$ and $\mathbf{B} = \{B_1, \ldots, B_s\}$, with $i \in A_\beta \cap B_{\beta+1}, j \in A_{\beta+1} \cap B_\beta$, $A_\beta \bigcup\{j\} = B_\beta \bigcup\{i\}$, $A_{\beta+1} \bigcup\{i\} = B_{\beta+1} \bigcup\{j\}$, $A_l = B_l$, $\forall\, l \neq \beta, \beta+1$ (for some $1 \leq \beta < s$), suppose that $L(\boldsymbol{\theta}, \mathbf{A}) \leq L(\boldsymbol{\theta}, \mathbf{B})$, when $\theta_i \geq \theta_j$, and suppose that $L(\boldsymbol{\theta}, \mathbf{A}) = L(g\boldsymbol{\theta}, g\mathbf{A})$, for all $\boldsymbol{\theta}, \mathbf{A}$ and $g \in G$; here, for a vector $\mathbf{x} = (x_1, \ldots, x_k) \in \mathbb{R}^k$ (the k-dimensional Euclidean space), a nonempty set $B \subseteq \{1, \ldots, k\}$, a partition $\mathbf{A} = \{A_1, \ldots, A_s\}$ of $\{1, \ldots, k\}$ and $g \in G$, $g\mathbf{x} = (x_{g^{-1}1}, \ldots, x_{g^{-1}k})$, $gB = \{gi : \; i \in B\}$ and $g\mathbf{A} = \{gA_1, \ldots, gA_s\}$, where g^{-1} denotes the inverse of permutation g. Under the assumption that the pdf $f(x;\theta)$ has the MLR property and the prior distribution $\tau(\cdot)$ is permutation symmetric, Eaton (1967) established that the natural selection rule which selects the partition $\mathbf{A}^* = \{A_1^*, \ldots, A_s^*\}$, where A_1^* contains the indices of populations corresponding to the k_1 largest X_is, A_2^* contains the indices of populations corresponding to the next k_2 largest X_is, $\cdots$, and A_s^* contains the indices of populations corresponding to the k_s smallest X_is (breaking ties at random), is Bayes under any permutation symmetric prior. This selection rule is further shown to be the best permutation invariant and hence minimax and admissible.

Now assume that the observations from the population Π_i, $i = 1, \ldots, k$ ($k \geq 3$), belong to an exponential family and have a pdf $c(\theta_i)b(x)e^{\theta_i x}$, $x \in \mathbb{R}$ (the real line), $\theta_i \in \Omega \subseteq \mathbb{R}$, with respect to measure μ, which is either the Lebesgue measure on $\mathbb{R}$ or the counting measure on the set of integers $\mathbb{Z}$; here, $c(\cdot)$ and $b(\cdot)$ are nonnegative functions defined on Ω and $\mathbb{R}$, respectively. Further assume that $b(\cdot)$ is log-concave, so that the underlying pdfs are unimodal. We consider the goal of simultaneously selecting the LEP and the UEP in two stages: screening out non-extreme populations at the first stage and selecting two populations associated with $\theta_{[1]}$ and $\theta_{[k]}$ at the second stage. In Section 9.2, we formulate the problem and give the definition of two-stage selection rules. In Sections 9.3 and 9.4, we derive an essentially complete class of two-stage permutation invariant selection rules.

9.2 Formulation of the Problem

For a set A, let $|A|$ denote the cardinality of the set. We will consider the following class of two-stage selection rules: At Stage 1, observe independent observations $X_{i1}, \ldots, X_{in_1}$ from the population Π_i, $i = 1, \ldots, k$, and based on these independent random samples, decide about the number of populations to be selected as contenders for the LEP and the number of populations to be selected as contenders for the UEP, and then, based on this decision, select a disjoint and non-empty subset pair (S_L, S_U) such that $S_L, S_U \subseteq \{1, \ldots, k\}$, where S_L contains the indices of populations to be selected as contenders for the LEP and S_U contains the indices of populations to be selected as contenders for the UEP. If $|S_L| = |S_U| = 1$, then the procedure stops at Stage 1 and claims that the populations with indices in S_L and S_U are the LEP and the UEP, respectively. In all other cases, the procedure proceeds to Stage 2. If, at Stage 1, $|S_L| = 1$ and $2 \leq |S_U| \leq k-1$ ($|S_U| = 1$ and $2 \leq |S_L| \leq k-1$), then the procedure claims that the population with index in S_L (S_U) is the LEP (UEP) and the decision about the UEP (LEP) is made after taking n_2 additional independent observations $Y_{i1}, \ldots, Y_{in_2}$ from each Π_i, $i \in S_U$ ($i \in S_L$). If at Stage 1, $2 \leq |S_L|, |S_U| \leq k-1$, $4 \leq |S_L| + |S_U| \leq k$, $S_L \bigcap S_U = \phi$, then, at Stage 2, n_2 additional independent observations $Y_{i1}, \ldots, Y_{in_2}$ are taken from each Π_i, $i \in S_L \bigcup S_U$, and the final decision about the LEP and the UEP, respectively, is made from the corresponding contending populations having indices in S_L and S_U. Let $U_i = X_{i1} + \cdots + X_{in_1}$ and $V_i = Y_{i1} + \cdots + Y_{in_2}$, $i = 1, \ldots, k$, be sufficient statistics for θ_i at stage 1 and stage 2, respectively, and $f(\cdot\,; \theta_i)$ and $h(\cdot\,; \theta_i)$ denote the pdfs of U_i and V_i, $i = 1, \ldots, k$, respectively, so that

$$f(u; \theta_i) = (c(\theta_i))^{n_1} b_{n_1}(u) e^{\theta_i u}, \quad u \in \mathbb{R}, \ \theta_i \in \Omega, \ i = 1, \ldots, k$$

and

$$h(u; \theta_i) = (c(\theta_i))^{n_2} b_{n_2}(u) e^{\theta_i u}, \quad u \in \mathbb{R}, \ \theta_i \in \Omega, \ i = 1, \ldots, k,$$

where $b_{n_i}(u)$, $i = 1, 2$, denotes the n_i-fold convolution of $b(u)$ with respect to μ. Let $W_i = U_i + V_i$, $i = 1, \ldots, k$, $\mathbf{U} = (U_1, \ldots, U_k)$, $\mathbf{V} = (V_1, \ldots, V_k)$ and $\mathbf{W} = (W_1, \ldots, W_k)$. We will consider only those selection rules which depend on observations only through sufficient statistics $\mathbf{U}$ and $\mathbf{V}$. In order to define a two-stage selection rule, let

$$\nu = \{\nu_{s,t};\ s, t = 1, \ldots, k, \ 2 \leq s + t \leq k\},$$

where, for $s, t = 1, \ldots, k$, $2 \leq s + t \leq k$, $\nu_{s,t}(\cdot)$ is a measurable map from $\mathbb{R}^k$ to $[0, 1]$ with the following interpretation: after observing $\mathbf{U} = \mathbf{u}$, $\nu_{s,t}(\mathbf{u})$ denotes the conditional probability (given $\mathbf{U} = \mathbf{u}$) of selecting s indices in S_L and t indices in S_U at Stage 1. Let

$$\begin{aligned} \eta(\mathbf{u}) &= \{\eta_{s,t}((a_1, a_2)|\mathbf{u}) : \ a_1, a_2 \subseteq \{1, \ldots, k\}, \ a_1, a_2 \neq \phi, \ a_1 \cap a_2 = \phi, \\ &\qquad |a_1| = s, |a_2| = t, \ s, t = 1, \ldots, k, \ 2 \leq s + t \leq k\}, \end{aligned}$$

where

$$\sum_{(a_1, a_2): |a_1| = s, |a_2| = t} \eta_{s,t}((a_1, a_2)|\mathbf{u}) = 1, \ \forall \ \mathbf{u} \in \mathbb{R}^k,$$

and, for $s, t \in \{1, \ldots, k\}$, $a_1, a_2 \subseteq \{1, \ldots, k\}$, $a_1, a_2 \neq \phi$, $a_1 \cap a_2 = \phi$, $\eta_{s,t}((a_1, a_2)|\mathbf{u})$ is a measurable map from $\mathbb{R}^k$ to $[0, 1]$ with the interpretation that $\eta_{s,t}((a_1, a_2)|\mathbf{u})$ denotes the conditional probability (given that $\mathbf{U} = \mathbf{u}$ and given that ν has decided to select s indices in S_L and t indices in S_U) of selecting subset pair (a_1, a_2), with $|a_1| = s$, $|a_2| = t$ at Stage 1.

If $\nu(\mathbf{u})$ decides that $|S_L| = |S_U| = 1$, then the procedure stops at Stage 1, and a final decision is based on

$$\{\eta_{1,1}((\{i\}, \{j\})|\mathbf{u})\,;\; i, j = 1, \ldots, k,\; i \neq j\}\,.$$

In all other cases the procedure proceeds to Stage 2. If at stage 1, η selects $S_L = \{i\}, S_U = \{j_1, \ldots, j_t\}$, with $i \in \{1, \ldots, k\}$, $\{j_1, \ldots, j_t\} \subseteq \{1, \ldots, k\}$, $j_1 < j_2 \ldots < j_t$, $2 \leq t \leq k-1$, $S_L \cap S_U = \phi$ ($S_L = \{i_1, \ldots, i_s\}, S_U = \{j\}$ with $j \in \{1, \ldots, k\}$, $\{i_1, \ldots, i_s\} \subseteq \{1, \ldots, k\}$, $i_1 < i_2 \ldots < i_s$, $2 \leq s \leq k-1$, $S_L \cap S_U = \phi$), then after having observed $V_{j_1} = v_{j_1}, \ldots, V_{j_t} = v_{j_t}$ ($V_{i_1} = v_{i_1}, \ldots, V_{i_s} = v_{i_s}$) at Stage 2, a final decision is based on $\{\delta_{i,j,\{i\},S_U}(\mathbf{u}; v_{j_1}, \ldots, v_{j_t});\ j \in S_U\}$ ($\{\delta_{i,j,S_L,\{j\}}(\mathbf{u}; v_{i_1}, \ldots, v_{i_s});\ i \in S_L\}$). On the other hand if, at Stage 1, η selects $S_L = \{i_1, \ldots, i_s\}$ and $S_U = \{j_1, \ldots, j_t\}$, with $S_L \cap S_U = \phi$, $\{i_1, \ldots, i_s\} \subseteq \{1, \ldots, k\}$, $\{j_1, \ldots, j_t\} \subseteq \{1, \ldots, k\}$, $i_1 < i_2 \ldots < i_s$, $j_1 < j_2 \ldots < j_t$, $s, t \in \{2, \ldots, k-2\}$, $4 \leq s + t \leq k$, then after having observed $V_{i_1} = v_{i_1}, \ldots, V_{i_s} = v_{i_s}; V_{j_1} = v_{j_1}, \ldots, V_{j_t} = v_{j_t}$, the final decision is based on

$$\{\delta_{i,j,S_L,S_U}(\mathbf{u}; v_{i_1}, \ldots, v_{i_s}; v_{j_1}, \ldots, v_{j_t});\ i \in S_L, j \in S_U\}\,,$$

where δ_{i,j,a_1,a_2} is a measurable map from $\mathbb{R}^k \times \mathbb{R}^{|a_1|+|a_2|}$ to $[0, 1]$ with

$$\sum_{i \in a_1} \sum_{j \in a_2} \delta_{i,j,a_1,a_2} \equiv 1$$

and has the following interpretation: $\delta_{i,j,a_1,a_2}, i \in a_1, j \in a_2$, is the conditional probability (given the observation and given that subset pair (a_1, a_2) is selected at Stage 1) of selecting the population Π_i ($i \in a_1$) as the LEP and the population Π_j ($j \in a_2$) as the UEP. Define

$$\begin{aligned} \delta \;=\; & \{\delta_{i,j,a_1,a_2} :\ i \in a_1, j \in a_2,\ a_1, a_2 \neq \phi,\ a_1, a_2 \subseteq \{1, \ldots, k\},\ a_1 \cap a_2 = \phi, \\ & 3 \leq |a_1| + |a_2| \leq k\}\,. \end{aligned}$$

Definition 9.2.1 For ν, η and δ explained above, the triplet (ν, η, δ) will be referred to as a two-stage selection rule.

Let $\mathcal{D}$ denote the class of all two-stage selection rules and let G denote the group of permutations on the indices $\{1, \ldots, k\}$. For $g \in G$, $B = \{i_1, \ldots, i_l\} \subseteq \{1, \ldots, k\}, B \neq \phi$, and $\mathbf{x} \in \mathbb{R}^k$, let $gB = \{gi_1, \ldots, gi_l\}$ denote the image of B under $g \in G$ and let $g\mathbf{x} = (x_{g^{-1}1}, \ldots, x_{g^{-1}l})$, where g^{-1} denotes the inverse of permutation g.

Definition 9.2.2 We say that a two-stage selection rule (ν, η, δ) is permutation invariant if

(i) $\nu_{s,t}(\mathbf{u}) = \nu_{s,t}(g\mathbf{u}), \ \forall g \in G, \ \forall \mathbf{u} \in \mathbb{R}^k, \ \forall s,t \in \{1,\ldots,k\}, \ 2 \le s+t \le k,$

(ii) for every $s,t \in \{1,\ldots,k\}, \ 2 \le s+t \le k, \ a_1, a_2 \subseteq \{1,\ldots,k\}, \ a_1 \cap a_2 = \phi,$ with $|a_1| = s, \ |a_2| = t$, and $\mathbf{u} \in \mathbb{R}^k$,

$$\eta_{s,t}((a_1,a_2)|\mathbf{u}) = \eta_{s,t}((ga_1,ga_2)|g\mathbf{u}), \ \forall g \in G,$$

(iii) for every $a_1 = \{i_1,\ldots,i_s\} \subseteq \{1,\ldots,k\}, \ a_2 = \{j_1,\ldots,j_t\} \subseteq \{1,\ldots,k\},$ with $a_1, a_2 \neq \phi, \ a_1 \bigcap a_2 = \phi, \ i_1 < i_2 < \ldots < i_s, \ j_1 < j_2 < \ldots < j_t, \ ga_1 = \{i'_1,\ldots,i'_s\}, \ ga_2 = \{j'_1,\ldots,j'_t\}, \ i'_1 < i'_2 < \ldots < i'_s, \ j'_1 < j'_2 < \ldots < j'_t$, and every $(v_{i_1},\ldots,v_{i_s},v_{j_1},\ldots,v_{j_t}) \in \mathbb{R}^{s+t}$,

$$\begin{aligned}&\delta_{i,j,a_1,a_2}(\mathbf{u}; v_{gi_1},\ldots,v_{gi_s}; v_{gj_1},\ldots,v_{gj_t})\\ &\quad= \delta_{gi,gj,ga_1,ga_2}(g\mathbf{u}; v_{i'_1},\ldots,v_{i'_s}; v_{j'_1},\ldots,v_{j'_t}),\end{aligned}$$

$\forall \ g \in G,$

Let $\mathcal{D}_\mathcal{I}$ denote the class of all two-stage permutation invariant selection rules. Let $\tau(\cdot)$ be a permutation symmetric prior on the k-dimensional space $\Omega^k \ (= \Omega \times \cdots \times \Omega)$. Let $L(\boldsymbol{\theta}, (a_1,a_2), i, j)$ denote the loss incurred in selecting the subset pair (a_1,a_2) at Stage 1 and then finally selecting populations Π_i and Π_j $(i \in a_1, \ j \in a_2)$ as the LEP and the UEP, respectively. Assume that L is non-negative and, $\forall \ a_1, a_2 \subseteq \{1,\ldots,k\}, \ a_1, a_2 \neq \phi, \ a_1 \bigcap a_2 = \phi, \ i,j \in \{1,\ldots,k\},$ $\boldsymbol{\theta} \in \Omega$, and $\forall \ g \in G$,

$$L(\boldsymbol{\theta}, (a_1,a_2), i, j) = L(g\boldsymbol{\theta}, (ga_1, ga_2), gi, gj). \tag{9.1}$$

In addition to (9.1), suppose that the loss function satisfies:

$$\left.\begin{array}{l}
\text{(a) } L(\boldsymbol{\theta},(\{i\},\{j\}),i,j) \le L(\boldsymbol{\theta},(\{i'\},\{j\}),i',j), \text{ for } \theta_i \le \theta_{i'},\ i \ne j,\ i' \ne j \\
\text{(b) } L(\boldsymbol{\theta},(\{i\},\{j\}),i,j) \le L(\boldsymbol{\theta},(\{i\},\{j'\}),i,j'), \text{ for } \theta_j \ge \theta_{j'},\ i \ne j,\ i \ne j' \\
\text{(c) } L(\boldsymbol{\theta},(\{a_1\},\{a_2\}),i,j) \le L(\boldsymbol{\theta},(\{a_1\},\{a_2\}),i',j), \\
\qquad\qquad \text{for } \theta_i \le \theta_{i'},\ i,i' \in a_1,\ j \in a_2,\ a_1 \cap a_2 = \phi \\
\text{(d) } L(\boldsymbol{\theta},(\{a_1\},\{a_2\}),i,j) \le L(\boldsymbol{\theta},(\{a_1\},\{a_2\}),i,j'), \\
\qquad\qquad \text{for } \theta_j \ge \theta_{j'},\ i \in a_1,\ j,j' \in a_2,\ a_1 \cap a_2 = \phi \\
\text{(e) For } a_1 = \{i_1,\ldots,i_{r-1},i\},\ a_2 = \{j_1,\ldots,j_{s-1},j\},\ \tilde{a}_1 = \{i_1,\ldots,i_{r-1},i'\}, \\
\quad \tilde{a}_2 = \{j_1,\ldots,j_{s-1},j'\}, \text{ with } a_1 \cap a_2 = \phi,\ \tilde{a}_1 \cap \tilde{a}_2 = \phi,\ a_1 \cap \tilde{a}_2 = \phi, \\
\quad \tilde{a}_1 \cap a_2 = \phi,\ q \in a_1 \cap \tilde{a}_1 \text{ and } m \in a_2 \cap \tilde{a}_2, \text{ the following hold:} \\
\quad \text{(i) } L(\boldsymbol{\theta},(\{a_1\},\{a_2\}),q,m) \le L(\boldsymbol{\theta},(\{\tilde{a}_1\},\{a_2\}),q,m), \text{ for } \theta_{i'} \ge \theta_i \\
\quad \text{(ii) } L(\boldsymbol{\theta},(\{a_1\},\{a_2\}),q,m) \le L(\boldsymbol{\theta},(\{a_1\},\{\tilde{a}_2\}),q,m), \text{ for } \theta_j \ge \theta_{j'} \\
\quad \text{(iii) } L(\boldsymbol{\theta},(\{a_1\},\{a_2\}),i,m) \le L(\boldsymbol{\theta},(\{\tilde{a}_1\},\{a_2\}),i',m), \text{ for } \theta_i \le \theta_{i'} \\
\quad \text{(iv) } L(\boldsymbol{\theta},(\{a_1\},\{a_2\}),q,j) \le L(\boldsymbol{\theta},(\{a_1\},\{\tilde{a}_2\}),q,j'), \text{ for } \theta_j \ge \theta_{j'}.
\end{array}\right\} \tag{9.2}$$

Remark 9.2.1 A class of loss functions satisfying (9.1) and (9.2) is

$$\begin{aligned} L(\boldsymbol{\theta},(a_1,a_2),i,j) &= c(|a_1|+|a_2|) + h(\theta_i - \theta_{[1]}, \theta_{[k]} - \theta_j), \\ &\qquad \theta \in \mathbb{R}^k,\ a_1,a_2 \subseteq \{1,\ldots,k\},\ a_1 \cap a_2 = \phi,\ i \in a_1,\ j \in a_2, \end{aligned}$$

where $h(\cdot,\cdot)$ is a non-negative function which is non-decreasing in its arguments, with $h(0,0)=0$, and $c \ge 0$ is the cost for every population that enters Stage 2.

Under the above setup, we will first find the optimal second stage selection rule and then the optimal two-stage selection rule.

9.3 The Optimal Second Stage Selection Rule

For $s, t \subseteq \{1, \ldots, k\}$, $a_1, a_2 \subseteq \{1, \ldots, k\}$, $a_1, a_2 \neq \phi$, $a_1 \cap a_2 = \phi$, with $|a_1| = s, |a_2| = t$, and $\mathbf{u} \in \mathbb{R}^k$, define

$$\eta^*_{s,t}((a_1, a_2)|\mathbf{u}) = \begin{cases} |H_{\mathbf{u}}(s,t)|^{-1}, & \text{if } (a_1, a_2) \in H_{\mathbf{u}}(s,t), \\ 0, & \text{otherwise,} \end{cases}$$

where, for $\mathbf{u} \in \mathbb{R}^k$ and $s, t \in \{1, \ldots, k\}$, and $H_{\mathbf{u}}(s,t)$ is given by

$$\{(a_1, a_2) : |a_1| = s, |a_2| = t, a_1 \cap a_2 = \phi, u_i \leq u_q \leq u_j, \\ \forall i \in a_1, j \in a_2, q \in a_1^c \cap a_2^c\}. .$$

Given that the decision (by ν) of selecting s indices in S_L and t indices in S_U has been made, $\eta^*_{s,t}$ selects the indices of populations yielding the s smallest observed values of U_is in S_L and selects the indices of populations yielding the t largest observed values of U_is (breaking ties at random) in S_U.

Similarly, for $a_1 = \{i_1, \ldots, i_s\}$, $a_2 = \{j_1, \ldots, j_t\}$, with $i_1 < \cdots < i_s$, $j_1 < \cdots < j_t$, $s, t \geq 1$, $a_1 \cap a_2 = \phi$, $\mathbf{u} \in \mathbb{R}^k$, $(v_{i_1}, \ldots, v_{i_s}) \in \mathbb{R}^s$, $(v_{j_1}, \ldots, v_{j_t}) \in \mathbb{R}^t$, $w_l = u_l + v_l$, $l \in a_1 \cup a_2$, define

$$\delta^*_{i,j,a_1,a_2}(\mathbf{u}; v_{i_1}, \ldots, v_{i_s}; v_{j_1}, \ldots, v_{j_t}) \\ = \begin{cases} |C_{s,t,a_1,a_2}(w_{i_1}, \ldots, w_{i_s}; w_{j_1}, \ldots, w_{j_t})|^{-1}, \\ \qquad \text{if } (i,j) \in C_{s,t,a_1,a_2}(w_{i_1}, \ldots, w_{i_s}; w_{j_1}, \ldots, w_{j_t}) \\ 0, \text{ otherwise,} \end{cases} \tag{9.3}$$

where

$$C_{s,t,a_1,a_2}(w_{i_1}, \ldots, w_{i_s}; w_{j_1}, \ldots, w_{j_t}) = \left\{(i,j) : w_i = \min_{q \in a_1} w_q \text{ and } w_j = \max_{q \in a_2} w_q\right\}.$$

Given that the decision (by ν and η) of selecting the subset pair (a_1, a_2), with $|a_1| = s$ and $|a_2| = t$, has been made at Stage 1, δ^* selects the index of the population yielding the smallest value of the W_is in S_L and selects the index of the population yielding the largest value of the W_is in S_U.

Now, the risk of a selection procedure (ν, η, δ) at $\boldsymbol{\theta} \in \Omega$ is given by

$$\begin{aligned}
&R(\boldsymbol{\theta}, (\nu, \eta, \delta)) \\
&= \sum_{i=1}^{k} \sum_{\substack{j=1 \\ i \neq j}}^{k} L(\boldsymbol{\theta}, (\{i\}, \{j\}), i, j) E_{\boldsymbol{\theta}}(\nu_{1,1}(\mathbf{U}) \eta_{1,1}((\{i\}, \{j\}) | \mathbf{U})) \\
&\quad + \sum_{i=1}^{k} \sum_{t=2}^{k-1} E_{\boldsymbol{\theta}} \left(\nu_{1,t}(\mathbf{U}) \sum_{\substack{a_2 : a_2 = \{j_1, \ldots, j_t\} \\ j_q \neq i, q = 1, \ldots, t}} \eta_{1,t}((\{i\}, a_2) | \mathbf{U}) \right. \\
&\quad \left. \times \sum_{j \in a_2} L(\boldsymbol{\theta}, (\{i\}, a_2), i, j) \delta_{i,j,\{i\},a_2}(\mathbf{U}; V_{j_1}, \ldots, V_{j_t}) \right) \\
&\quad + \sum_{s=2}^{k-1} \sum_{j=1}^{k} E_{\boldsymbol{\theta}} \left(\nu_{s,1}(\mathbf{U}) \sum_{\substack{a_1 : a_1 = \{i_1, \ldots, i_s\} \\ i_q \neq j, q = 1, \ldots, s}} \eta_{s,1}((a_1, \{j\}) | \mathbf{U}) \right. \\
&\quad \left. \times \sum_{i \in a_1} L(\boldsymbol{\theta}, (a_1, \{j\}), i, j) \delta_{i,j,a_1,\{j\}}(\mathbf{U}; V_{i_1}, \ldots, V_{i_s}) \right) \\
&\quad + \sum_{s=2}^{k} \sum_{\substack{t=2 \\ s+t \leq k}}^{k} E_{\boldsymbol{\theta}} \left(\nu_{s,t}(\mathbf{U}) \sum_{\substack{(a_1, a_2) : a_1 = \{i_1, \ldots, i_s\} \\ a_2 = \{j_1, \ldots, j_t\}, a_1 \cap a_2 = \phi}} \eta_{s,t}((a_1, a_2) | \mathbf{U}) \right. \\
&\quad \left. \times \sum_{i \in a_1} \sum_{j \in a_2} L(\boldsymbol{\theta}, (a_1, a_2), i, j) \delta_{i,j,a_1,a_2}(\mathbf{U}; V_{i_1}, \ldots, V_{i_s}; V_{j_1}, \ldots, V_{j_t}) \right).
\end{aligned}$$

The Bayes risk for (ν, η, δ) under the prior $\tau(\cdot)$ is given by

$$r(\tau, (\nu, \eta, \delta)) = \int_{\Omega^k} R(\underline{\theta}, (\nu, \eta, \delta)) d\tau(\underline{\theta}). \tag{9.4}$$

Let $r^*_{\mathbf{u}}((\nu, \eta, \delta))$ denote the conditional posterior risk of a selection procedure (ν, η, δ) given $\mathbf{U} = \mathbf{u}$. In order to find a selection rule which minimizes (9.4), it is sufficient to find a selection rule which minimizes $r^*_{\mathbf{u}}((\nu, \eta, \delta))$, for every $\mathbf{u} \in \mathbb{R}^k$. Now,

$$r^*_{\mathbf{u}}((\nu, \eta, \delta)) = \sum_{i=1}^{4} D_i(\mathbf{u}, (\nu, \eta, \delta)), \tag{9.5}$$

where

$$
\begin{aligned}
&D_1(\mathbf{u},(\nu,\eta,\delta)) \\
&= \nu_{1,1}(\mathbf{u})\sum_{i=1}^{k}\sum_{\substack{j=1\\ i\neq j}}^{k}\eta_{1,1}((\{i\},\{j\})|\mathbf{u})E\left(L(\boldsymbol{\theta},(\{i\},\{j\}),i,j)|\mathbf{U}=\mathbf{u}\right),
\end{aligned}
$$

$$
\begin{aligned}
D_2(\mathbf{u},(\nu,\eta,\delta)) &= \sum_{i=1}^{k}\sum_{t=2}^{k-1}\nu_{1,t}(\mathbf{u})\sum_{\substack{a_2:a_2=\{j_1,\ldots,j_t\}\\ j_q\neq i,q=1,\ldots,t}}\eta_{1,t}((\{i\},a_2)|\mathbf{u}) \\
&\quad\times E\left(\sum_{j\in a_2}L(\boldsymbol{\theta},(\{i\},a_2),i,j)\delta_{i,j,\{i\},a_2}(\mathbf{u};V_{j_1},\ldots,V_{j_t})|\mathbf{U}=\mathbf{u}\right), \\
D_3(\mathbf{u},(\nu,\eta,\delta)) &= \sum_{s=2}^{k-1}\sum_{j=1}^{k}\nu_{s,1}(\mathbf{u})\sum_{\substack{a_1:a_1=\{i_1,\ldots,i_s\}\\ i_q\neq j,q=1,\ldots,s}}\eta_{s,1}((a_1,\{j\})|\mathbf{u}) \\
&\quad\times E\left(\sum_{i\in a_1}L(\boldsymbol{\theta},(a_1,\{j\}),i,j)\delta_{i,j,a_1,\{j\}}(\mathbf{u};V_{i_1},\ldots,V_{i_s})|\mathbf{U}=\mathbf{u}\right),
\end{aligned}
$$

$$
\begin{aligned}
&D_4(\mathbf{u},(\nu,\eta,\delta)) \\
&= \sum_{s=2}^{k}\sum_{\substack{t=2\\ s+t\leq k}}^{k}\nu_{s,t}(\mathbf{u})\sum_{\substack{(a_1,a_2):a_1=\{i_1,\ldots,i_s\}\\ a_2=\{j_1,\ldots,j_t\},a_1\cap a_2=\phi}}\eta_{s,t}((a_1,a_2)|\mathbf{u}) \\
&\times E\left(\sum_{i\in a_1}\sum_{j\in a_2}L(\boldsymbol{\theta},(a_1,a_2),i,j)\delta_{i,j,a_1,a_2}(\mathbf{u};V_{i_1},\ldots,V_{i_s};V_{j_1},\ldots,V_{j_t})|\mathbf{U}=\mathbf{u}\right),
\end{aligned}
$$

and the expectations involved in the above expressions are the posterior expectations (of $\boldsymbol{\theta}$ given $\mathbf{U}=\mathbf{u}$).

Since the underlying pdfs have the MLR property and our loss function satisfies (a) and (b) of (9.2), Lemma 4.2 of Eaton (1967) is applicable. Hence,

$$
\begin{aligned}
&D_1(\mathbf{u},(\nu,\eta,\delta)) \\
&= \nu_{1,1}(\mathbf{u})\sum_{i=1}^{k}\sum_{\substack{j=1\\ i\neq j}}^{k}\eta_{1,1}((\{i\},\{j\})|\mathbf{u})E\left(L(\boldsymbol{\theta},(\{i\},\{j\}),i,j)|\mathbf{U}=\mathbf{u}\right) \\
&\geq \nu_{1,1}(\mathbf{u})\sum_{i=1}^{k}\sum_{\substack{j=1\\ i\neq j}}^{k}\eta^*_{1,1}((\{i\},\{j\})|\mathbf{u})E\left(L(\boldsymbol{\theta},(\{i\},\{j\}),i,j)|\mathbf{U}=\mathbf{u}\right).
\end{aligned}
\tag{9.6}
$$

For $a_1 = \{i_1, \ldots, i_s\}$, $a_2 = \{j_1, \ldots, j_t\}$, $a_1 \cap a_2 = \phi$, $|a_1|, |a_2| \geq 2$, on writing

$$\begin{aligned} & D^{\delta}_{s,t}(\mathbf{u}, (a_1, a_2)) \\ &= E(\sum_{i \in a_1} \sum_{j \in a_2} L(\boldsymbol{\theta}, (a_1, a_2), i, j) \delta_{i,j,a_1,a_2}(\mathbf{u}; V_{i_1}, \ldots, V_{i_s}; V_{j_1}, \ldots, V_{j_t}) | \mathbf{U} = \mathbf{u}), \end{aligned}$$

we have

$$\begin{aligned} & D_4(\mathbf{u}, (\nu, \eta, \delta)) \\ &= \sum_{s=2}^{k} \sum_{\substack{t=2 \\ s+t \leq k}}^{k} \nu_{s,t}(\mathbf{u}) \sum_{\substack{(a_1, a_2): a_1 = \{i_1, \ldots, i_s\} \\ a_2 = \{j_1, \ldots, j_t\}, a_1 \cap a_2 = \phi}} \eta_{s,t}((a_1, a_2) | \mathbf{u}) D^{\delta}_{s,t}(\mathbf{u}, (a_1, a_2)). \end{aligned} \tag{9.7}$$

Now

$$\begin{aligned} & D^{\delta}_{s,t}(\mathbf{u}, (a_1, a_2)) \\ &= (m(\mathbf{u}))^{-1} \int_{\mathbb{R}^{|a_1|+|a_2|}} \sum_{i \in a_1} \sum_{j \in a_2} \delta_{i,j,a_1,a_2}(\mathbf{u}; v_{i_1}, \ldots, v_{i_s}; v_{j_1}, \ldots, v_{j_t}) \\ &\times \left\{ \int_{\Omega^k} L(\boldsymbol{\theta}, (a_1, a_2), i, j) \prod_{q=1}^{k} f(u_q; \theta_q) \prod_{q \in a_1 \cup a_2} h(v_q; \theta_q) d\tau(\boldsymbol{\theta}) \right\} \prod_{q \in a_1 \cup a_2} d\mu(v_q), \end{aligned}$$

where

$$m(\mathbf{u}) = \int_{\Omega^k} \left(\prod_{q=1}^{k} f(u_q; \theta_q) \right) d\tau(\boldsymbol{\theta}).$$

Using Fubini's theorem, we get

$$\begin{aligned} & D^{\delta}_{s,t}(\mathbf{u}, (a_1, a_2)) \\ &= (m(\mathbf{u}))^{-1} \int_{\mathbb{R}^k} \sum_{i \in a_1} \sum_{j \in a_2} \delta_{i,j,a_1,a_2}(\mathbf{u}; v_{i_1}, \ldots, v_{i_s}; v_{j_1}, \ldots, v_{j_t}) \\ &\quad \times \left\{ \int_{\Omega^k} L(\boldsymbol{\theta}, (a_1, a_2), i, j) \prod_{q=1}^{k} \{ f(u_q; \theta_q) h(v_q; \theta_q) \} d\tau(\boldsymbol{\theta}) \right\} \prod_{r=1}^{k} d\mu(v_r) \\ &= (\tilde{m}(\mathbf{u}))^{-1} \int_{\mathbb{R}^k} \left(\prod_{q=1}^{k} b_{n_2}(v_q) \right) \sum_{i \in a_1} \sum_{j \in a_2} \delta_{i,j,a_1,a_2}(\mathbf{u}; v_{i_1}, \ldots, v_{i_s}; v_{j_1}, \ldots, v_{j_t}) \\ &\times \left\{ \int_{\Omega^k} L(\boldsymbol{\theta}, (a_1, a_2), i, j) \prod_{q=1}^{k} \left\{ (c(\theta_q))^{n_1+n_2} e^{(u_q+v_q)\theta_q} \right\} d\tau(\boldsymbol{\theta}) \right\} \prod_{r=1}^{k} d\mu(v_r), \end{aligned}$$

where

$$\tilde{m}(\mathbf{u}) = \int_{\Omega^k} \prod_{q=1}^{k} \left\{ (c(\theta_q))^{n_1} e^{u_q \theta_q} \right\} d\tau(\boldsymbol{\theta}).$$

Making the change of variable $u_q + v_q = w_q$, $q = 1, \ldots, k$, in the above integrals, we get

$$\begin{aligned} & D^{\delta}_{s,t}(\mathbf{u}, (a_1, a_2)) \\ &= (\tilde{m}(\mathbf{u}))^{-1} \\ &\times \int_{\mathbb{R}^k} \sum_{i \in a_1} \sum_{j \in a_2} \delta_{i,j,a_1,a_2}(\mathbf{u}; w_{i_1} - u_{i_1}, \ldots, w_{i_s} - u_{i_s}; w_{j_1} - u_{j_1}, \ldots, w_{j_t} - u_{j_t}) \\ &\times \left\{ \int_{\Omega^k} L(\boldsymbol{\theta}, (a_1, a_2), i, j) e^{\sum_{q=1}^{k} \theta_q w_q} d\hat{\tau}(\boldsymbol{\theta}) \right\} \left(\prod_{q=1}^{k} b_{n_2}(w_q - u_q) \right) \prod_{r=1}^{k} d\mu(w_r), \end{aligned}$$

where

$$d\hat{\tau}(\boldsymbol{\theta}) = \left\{ \prod_{q=1}^{k} (c(\theta_q))^{n_1 + n_2} \right\} d\tau(\boldsymbol{\theta}).$$

Note that $d\hat{\tau}(\boldsymbol{\theta})$ is permutation symmetric in $\boldsymbol{\theta}$. For $a_1, a_2 \subseteq \{1, \ldots, k\}, a_1, a_2 \neq \phi, a_1 \cap a_2 = \phi$, define

$$E_1((a_1, a_2), i, j | \mathbf{w}) = \int_{\Omega^k} L(\boldsymbol{\theta}, (a_1, a_2), i, j) e^{\sum_{q=1}^{k} \theta_q w_q} d\hat{\tau}(\boldsymbol{\theta}).$$

Then,

$$\begin{aligned} & D^{\delta}_{s,t}(\mathbf{u}, (a_1, a_2)) \\ &= (\tilde{m}(\mathbf{u}))^{-1} \\ & \int_{\mathbb{R}^k} \sum_{i \in a_1} \sum_{j \in a_2} \delta_{i,j,a_1,a_2}(\mathbf{u}; w_{i_1} - u_{i_1}, \ldots, w_{i_s} - u_{i_s}; w_{j_1} - u_{j_1}, \ldots, w_{j_t} - u_{j_t}) \\ & E_1((a_1, a_2), i, j | \mathbf{w}) \left\{ \prod_{q=1}^{k} b_{n_2}(w_q - u_q) \right\} \prod_{r=1}^{k} d\mu(w_r). \end{aligned} \tag{9.8}$$

We now prove the monotonicity of $E_1((a_1, a_2), i, j | \mathbf{w})$.

Lemma 9.3.1 *Suppose $i, i' \in a_1$ and $j, j' \in a_2$, then for every fixed $\mathbf{w} \in \mathbb{R}^k$,*

(i) $E_1((a_1, a_2), i, j | \mathbf{w}) \geq E_1((a_1, a_2), i', j | \mathbf{w})$, *if* $w_i \geq w_{i'}$,

(ii) $E_1((a_1, a_2), i, j | \mathbf{w}) \geq E_1((a_1, a_2), i, j' | \mathbf{w})$, *if* $w_{j'} \geq w_j$.

PROOF. We prove only the first assertion, as the proof of the second assertion follows exactly on similar lines. For proving (i), consider

$$\begin{aligned}
&E_1((a_1,a_2),i,j|\mathbf{w}) - E_1((a_1,a_2),i',j|\mathbf{w})\\
&= \int_{\{\boldsymbol{\theta}:\theta_i>\theta_{i'}\}} \left[L(\boldsymbol{\theta},(a_1,a_2),i,j) - L(\boldsymbol{\theta},(a_1,a_2),i',j)\right] e^{\sum_{q=1}^k \theta_q w_q} d\hat{\tau}(\boldsymbol{\theta})\\
&+ \int_{\{\boldsymbol{\theta}:\theta_i=\theta_{i'}\}} \left[L(\boldsymbol{\theta},(a_1,a_2),i,j) - L(\boldsymbol{\theta},(a_1,a_2),i',j)\right] e^{\sum_{q=1}^k \theta_q w_q} d\hat{\tau}(\boldsymbol{\theta})\\
&+ \int_{\{\boldsymbol{\theta}:\theta_i<\theta_{i'}\}} \left[L(\boldsymbol{\theta},(a_1,a_2),i,j) - L(\boldsymbol{\theta},(a_1,a_2),i',j)\right] e^{\sum_{q=1}^k \theta_q w_q} d\hat{\tau}(\boldsymbol{\theta}).
\end{aligned}$$

Since $\hat{\tau}(\cdot)$ is permutation symmetric, using (a) and (c) of (9.2), the second integral is seen to be zero. On interchanging the roles of θ_i and $\theta_{i'}$ in the third integral and using the fact that $\hat{\tau}(\cdot)$ is permutation symmetric, we get

$$\begin{aligned}
&E_1((a_1,a_2),i,j|\mathbf{w}) - E_1((a_1,a_2),i',j|\mathbf{w})\\
&= \int_{\{\boldsymbol{\theta}:\theta_i>\theta_{i'}\}} \left[L(\boldsymbol{\theta},(a_1,a_2),i,j) - L(\boldsymbol{\theta},(a_1,a_2),i',j)\right] \prod_{\substack{q=1\\ q\neq i,i'}}^{k} e^{\theta_q w_q}\\
&\quad \times [e^{\theta_i w_i} e^{\theta_{i'} w_{i'}} - e^{\theta_{i'} w_i} e^{\theta_i w_{i'}}] d\hat{\tau}(\boldsymbol{\theta})\\
&\quad \geq 0,
\end{aligned}$$

on using the MLR property and property (c) of (9.2). ∎

Lemma 9.3.2 *For $a_1 \subseteq \{i_1,\ldots,i_s\}$ and $a_2 \subseteq \{j_1,\ldots,j_t\}$, $s,t \geq 2$, $a_1 \cap a_2 = \phi$ and for every δ,*

$$D^{\delta}_{s,t}(\mathbf{u},(a_1,a_2)) \geq D^{\delta^*}_{s,t}(\mathbf{u},(a_1,a_2)), \tag{9.9}$$

where δ^* is as defined by (9.3).

PROOF. On using Lemma 9.3.1 and from the definition of δ^*, it follows that for every fixed $\mathbf{w} \in \mathbb{R}^k$ and $\mathbf{u} \in \mathbb{R}^k$,

$$\begin{aligned}
&\sum_{i\in a_1}\sum_{j\in a_2} \delta_{i,j,a_1,a_2}(\mathbf{u}; w_{i_1}-u_{i_1},\ldots,w_{i_s}-u_{i_s}; w_{j_1}-u_{j_1},\ldots,w_{j_t}-u_{j_t})\\
&\qquad\qquad \times E_1((a_1,a_2),i,j|\mathbf{w})\\
&\geq \sum_{i\in a_1}\sum_{j\in a_2} \delta^*_{i,j,a_1,a_2}(\mathbf{u}; w_{i_1}-u_{i_1},\ldots,w_{i_s}-u_{i_s}; w_{j_1}-u_{j_1},\ldots,w_{j_t}-u_{j_t})\\
&\qquad\qquad \times E_1((a_1,a_2),i,j|\mathbf{w}).
\end{aligned}$$

Now the result follows from (9.8). ∎

For $i, j = 1, \ldots, k$, $a_1 \subseteq \{i_1, \ldots, i_s\}$, $a_2 \subseteq \{j_1, \ldots, j_t\}, 2 \leq s, t \leq k-1$, $\{i\} \nsubseteq a_2, \{j\} \nsubseteq a_1$, let

$$D^{\delta}_{1,t}(\mathbf{u}, (\{i\}, a_2)) = E\left(\sum_{j \in a_2} L(\boldsymbol{\theta}, (\{i\}, a_2), i, j)\delta_{i,j,\{i\},a_2}(\mathbf{u}; V_{j_1}, \ldots, V_{j_t}) \mid \mathbf{U} = \mathbf{u}\right)$$

and

$$D^{\delta}_{s,1}(\mathbf{u}, (a_1, \{j\})) = E\left(\sum_{i \in a_1} L(\boldsymbol{\theta}, (a_1, \{j\}), i, j)\delta_{i,j,a_1,\{j\}}(\mathbf{u}; V_{i_1}, \ldots, V_{i_s}) \mid \mathbf{U} = \mathbf{u}\right).$$

Thus,

$$D_2(\mathbf{u}, (\nu, \eta, \delta)) = \sum_{i=1}^{k} \sum_{t=2}^{k-1} \nu_{1,t}(\mathbf{u}) \sum_{\substack{a_2 : a_2 = \{j_1, \ldots, j_t\} \\ j_q \neq i, q = 1, \ldots, t}} \eta_{1,t}((\{i\}, a_2)|\mathbf{u}) D^{\delta}_{1,t}(\mathbf{u}, (\{i\}, a_2)) \tag{9.10}$$

and

$$D_3(\mathbf{u}, (\nu, \eta, \delta)) = \sum_{s=2}^{k-1} \sum_{j=1}^{k} \nu_{s,1}(\mathbf{u}) \sum_{\substack{a_1 : a_1 = \{i_1, \ldots, i_s\} \\ i_q \neq j, q = 1, \ldots, s}} \eta_{s,1}((a_1, \{j\})|\mathbf{u}) D^{\delta}_{s,1}(\mathbf{u}, (a_1, \{j\})). \tag{9.11}$$

Now,

$$\begin{aligned} & D^{\delta}_{1,t}(\mathbf{u}, (\{i\}, a_2)) \\ & = (\tilde{m}(\mathbf{u}))^{-1} \int_{\mathbb{R}^k} \left\{ \prod_{q=1}^{k} b_{n_2}(w_q - u_q) \right\} \\ & \qquad \sum_{j \in a_2} \delta_{i,j,\{i\},a_2}(\mathbf{u}; w_{j_1} - u_{j_1}, \ldots, w_{j_t} - u_{j_t}) E_1((\{i\}, a_2), i, j|\mathbf{w}) \prod_{r=1}^{k} d\mu(w_r), \end{aligned}$$

and

$$\begin{aligned} & D^{\delta}_{s,1}(\mathbf{u}, (a_1, \{j\})) \\ & = (\tilde{m}(\mathbf{u}))^{-1} \int_{\mathbb{R}^k} \left\{ \prod_{q=1}^{k} b_{n_2}(w_q - u_q) \right\} \\ & \qquad \sum_{i \in a_1} \delta_{i,j,a_1,\{j\}}(\mathbf{u}; w_{i_1} - u_{i_1}, \ldots, w_{i_s} - u_{i_s}) E_1((a_1, \{j\}), i, j|\mathbf{w}) \prod_{r=1}^{k} d\mu(w_r). \end{aligned}$$

The proof of the following lemma is similar to that of Lemma 9.3.2.

Lemma 9.3.3 *For $i, i' \in a_1$, $j, j' \in a_2$ and $a_1 \cap a_2 = \phi$,*

$$\text{(i)}\;\; D^{\delta}_{s,1}(\mathbf{u}, (a_1, \{j\})) \geq D^{\delta^*}_{s,1}(\mathbf{u}, (a_1, \{j\})),\;\; 2 \leq s \leq k-1, \tag{9.12}$$

$$\text{(ii)}\;\; D^{\delta}_{1,t}(\mathbf{u}, (\{i\}, a_2)) \geq D^{\delta^*}_{1,t}(\mathbf{u}, (\{i\}, a_2)),\;\; 2 \leq t \leq k-1. \tag{9.13}$$

The following theorem establishes the optimality of the selection rule δ^*.

Theorem 9.3.1 *Let $\tau(\cdot)$ be a permutation symmetric prior and suppose that the loss function L satisfies (9.1) and (9.2). Then*

$$\text{(i)}\;\; r(\tau, (\nu, \tilde{\eta}, \delta^*)) \leq r(\tau, (\nu, \eta, \delta)),\;\; \forall\; (\nu, \eta, \delta) \in \mathcal{D}$$

and

$$\text{(ii)}\;\; R(\boldsymbol{\theta}, (\nu, \tilde{\eta}, \delta^*)) \leq R(\boldsymbol{\theta}, (\nu, \eta, \delta)),\;\; \forall\; (\nu, \eta, \delta) \in \mathcal{D}_I,$$

$\tilde{\eta}$ is the same as η except that $\tilde{\eta}_{1,1} = \eta^*_{1,1}$.

PROOF. (i) From (9.6) and from the definition of $\tilde{\eta}$, it follows that

$$D_1(\mathbf{u}, (\nu, \eta, \delta)) \geq D_1(\mathbf{u}, (\nu, \tilde{\eta}, \delta)),\;\; \forall\; (\nu, \eta, \delta) \in \mathcal{D}. \tag{9.14}$$

Also, from (9.7) and (9.9), we get

$$D_4(\mathbf{u}, (\nu, \eta, \delta)) \geq D_4(\mathbf{u}, (\nu, \eta, \delta^*)),\;\; \forall\; (\nu, \eta, \delta) \in \mathcal{D}. \tag{9.15}$$

Similarly, from (9.10) and (9.13), we get

$$D_2(\mathbf{u}, (\nu, \eta, \delta)) \geq D_2(\mathbf{u}, (\nu, \eta, \delta^*)),\;\; \forall\; (\nu, \eta, \delta) \in \mathcal{D}, \tag{9.16}$$

and, from (9.11) and (9.12), we get

$$D_3(\mathbf{u}, (\nu, \eta, \delta)) \geq D_3(\mathbf{u}, (\nu, \eta, \delta^*)),\;\; \forall\; (\nu, \eta, \delta) \in \mathcal{D}. \tag{9.17}$$

Now, on using (9.14)–(9.17) in (9.5), we get

$$r^*_{\mathbf{u}}((\nu, \eta, \delta)) \geq r^*_{\mathbf{u}}((\nu, \tilde{\eta}, \delta)),\;\; \forall\; (\nu, \eta, \delta) \in \mathcal{D},$$

which proves the first assertion.

(ii) For $(\nu, \eta, \delta) \in \mathcal{D}_I$, we have

$$R(\boldsymbol{\theta}, (\nu, \eta, \delta)) = R(g\boldsymbol{\theta}, (\nu, \eta, \delta)),\;\; \forall\; g \in G. \tag{9.18}$$

Fix $\boldsymbol{\theta} \in \mathbb{R}^k$ and let τ be a prior that puts equal weight on each of the permutations of $\boldsymbol{\theta}$. Then, on using (i) and (9.18), we get

$$\frac{1}{k!}\sum_{g \in G} R(g\boldsymbol{\theta}, (\nu, \tilde{\eta}, \delta^*)) \;\leq\; \frac{1}{k!}\sum_{g \in G} R(g\boldsymbol{\theta}, (\nu, \eta, \delta)), \forall\; (\nu, \eta, \delta) \in \mathcal{D}_I.$$

This implies $R(\boldsymbol{\theta}, (\nu, \tilde{\eta}, \delta^*)) \;\leq\; R(\boldsymbol{\theta}, (\nu, \eta, \delta)),\;\; \forall\; (\nu, \eta, \delta) \in \mathcal{D}_I$.

■

9.4 The Optimal Two-Stage Selection Rule

For $1 \le s,t \le k-1,\ 3 \le s+t \le k$, we can write

$$
\begin{aligned}
&D^{\delta^*}_{s,t}(\mathbf{u},(a_1,a_2))\\
&= (\tilde{m}(\mathbf{u}))^{-1}\int_{\mathbb{R}^k}\left\{\min_{i\in a_1}\min_{j\in a_2}\int_{\Omega^k} L(\boldsymbol{\theta},(a_1,a_2),i,j)e^{\sum_{q=1}^k \theta_q w_q}d\hat{\tau}(\boldsymbol{\theta})\right\}\\
&\qquad \times\left\{\prod_{q=1}^k b_{n_2}(w_q-u_q)\right\}\prod_{r=1}^k d\mu(w_r)\\
&= (\tilde{m}(\mathbf{u}))^{-1}\int_{\mathbb{R}^k} L^*(\mathbf{w},(a_1,a_2))\left\{\prod_{q=1}^k b_{n_2}(w_q-u_q)\right\}\prod_{r=1}^k d\mu(w_r),
\end{aligned}
$$

where

$$
\begin{aligned}
L^*(\mathbf{w},(a_1,a_2)) &= \min_{i\in a_1}\min_{j\in a_2}\int_{\Omega^k} L(\boldsymbol{\theta},(a_1,a_2),i,j)e^{\sum_{q=1}^k \theta_q w_q}d\hat{\tau}(\boldsymbol{\theta})\\
&= \min_{i\in a_1}\min_{j\in a_2} E_1((a_1,a_2),i,j|\mathbf{w}).
\end{aligned}
$$

The following lemma proves the monotonicity of L^*.

Lemma 9.4.1 *Let* $a_1 = \{i_1,\ldots,i_{r-1},i\}$, $a_2 = \{j_1,\ldots,j_{s-1},j\}$, $\tilde{a}_1 = \{i_1,\ldots,i_{r-1},i'\}$, $\tilde{a}_2 = \{j_1,\ldots,j_{s-1},j'\}$, $|a_1|,|a_2|,|\tilde{a}_1|,|\tilde{a}_2| \ge 1$, $r+s\ge 3$, *with* $a_1\cap a_2=\phi$, $\tilde{a}_1\cap\tilde{a}_2=\phi$, $\tilde{a}_1\cap a_2=\phi$, $a_1\cap\tilde{a}_2=\phi$. *Then*

(i) $L^*(\mathbf{w},(a_1,a_2)) \le L^*(\mathbf{w},(\tilde{a}_1,a_2))$, *if* $w_{i'}\ge w_i$,

(ii) $L^*(\mathbf{w},(a_1,a_2)) \le L^*(\mathbf{w},(a_1,\tilde{a}_2))$, *if* $w_j\ge w_{j'}$.

PROOF. (i) For $q\in\tilde{a}_1\cap a_1$ and $m\in a_2$, consider

$$
\begin{aligned}
&E_1((\tilde{a}_1,a_2),q,m|\mathbf{w}) - E_1((a_1,a_2),q,m|\mathbf{w})\\
&= \int_{\{\boldsymbol{\theta}:\theta_i>\theta_{i'}\}}[L(\boldsymbol{\theta},(\tilde{a}_1,a_2),q,m)-L(\boldsymbol{\theta},(a_1,a_2),q,m)]\,e^{\sum_{q=1}^k \theta_q w_q}d\hat{\tau}(\boldsymbol{\theta})\\
&\quad+\int_{\{\boldsymbol{\theta}:\theta_i=\theta_{i'}\}}[L(\boldsymbol{\theta},(\tilde{a}_1,a_2),q,m)-L(\boldsymbol{\theta},(a_1,a_2),q,m)]\,e^{\sum_{q=1}^k \theta_q w_q}d\hat{\tau}(\boldsymbol{\theta})\\
&\quad+\int_{\{\boldsymbol{\theta}:\theta_i<\theta_{i'}\}}[L(\boldsymbol{\theta},(\tilde{a}_1,a_2),q,m)-L(\boldsymbol{\theta},q,m)]\,e^{\sum_{q=1}^k \theta_q w_q}d\hat{\tau}(\boldsymbol{\theta}).
\end{aligned}
$$

Since $\hat{\tau}(\cdot)$ is permutation symmetric, using property (c) of (9.2), we conclude that the second integral is zero. Since $\hat{\tau}(\cdot)$ is permutation symmetric, on interchanging the roles of θ_i and $\theta_{i'}$ in the third integral. we get

$$\begin{aligned} & E_1((\tilde{a}_1, a_2), q, m|\mathbf{w}) - E_1((a_1, a_2), q, m|\mathbf{w}) \\ = & \int_{\{\boldsymbol{\theta}:\theta_i > \theta_{i'}\}} [L(\boldsymbol{\theta}, (\tilde{a}_1, a_2), q, m) - L(\boldsymbol{\theta}, (a_1, a_2), q, m)] \prod_{\substack{q=1 \\ q \neq i,i'}}^{k} e^{\theta_q w_q} \\ & \times [e^{\theta_i w_i} e^{\theta_{i'} w_{i'}} - e^{\theta_{i'} w_i} e^{\theta_i w_{i'}}] d\hat{\tau}(\boldsymbol{\theta}) \\ \geq & \;\; 0, \text{ if } w_{i'} \geq w_i, \end{aligned}$$

on using the MLR property and property (c)-(i) of (9.2).

Again using the same arguments, we get

$$E_1((\tilde{a}_1, a_2), i', m|\mathbf{w}) - E_1((a_1, a_2), i, m|\mathbf{w}) \geq 0.$$

(ii) Follows in a similar fashion. ∎

Now we present the main theorem.

Theorem 9.4.1 *Let $\tau(\cdot)$ be a permutation symmetric prior and suppose that the loss function L satisfies (9.1) and (9.2). If $b(\cdot)$ is log-concave (so that the underlying pdfs are unimodal), then*

$$(i)\;\; r(\tau, (\nu, \eta^*, \delta^*)) \leq r(\tau, (\nu, \eta, \delta)),\;\; \forall\; (\nu, \eta, \delta) \in \mathcal{D}$$

and

$$(ii)\;\; R(\boldsymbol{\theta}, (\nu, \eta^*, \delta^*)) \leq R(\boldsymbol{\theta}, (\nu, \eta, \delta)),\;\; \forall\; (\nu, \eta, \delta) \in \mathcal{D}_I.$$

PROOF. (i) From Theorem 9.3.1, we have

$$r(\tau, (\nu, \tilde{\eta}, \delta^*)) \leq r(\tau, (\nu, \eta, \delta)),\;\; \forall\; (\nu, \eta, \delta) \in \mathcal{D}. \tag{9.19}$$

For $s, t \geq 1$, $3 \leq s + t \leq k$, consider the problem of partitioning $\{1, \ldots, k\}$ into three (two) disjoint subsets of sizes s, $k - s - t$ and t (s and t) with $s + t < k$ ($s + t = k$). Since log-concavity is preserved under convolutions, it follows that $b_{n_2}(x)$, $x \in \mathbb{R}$ or $x \in \mathbb{Z}$, is log-concave, i.e., $b_{n_2}(w - u)$ has the MLR property in (w, u). Also, using Lemma 4.1, it is easy to see that $L^*(\mathbf{w}, (a_1, a_2))$ has the property of the loss function assumed by Eaton (1967). Hence, for

$1 \leq s, t \leq k,\ 3 \leq s+t \leq k,$

$$
\begin{aligned}
&\sum_{\substack{(a_1,a_2):|a_1|=s,\\|a_2|=t,a_1\cap a_2=\phi}} \eta_{s,t}((a_1,a_2)|\mathbf{u}) D^{\delta^*}_{s,t}(\mathbf{u},(a_1,a_2)) \\
&= (\tilde{m}(\mathbf{u}))^{-1} \sum_{\substack{(a_1,a_2):|a_1|=s,\\|a_2|=t,a_1\cap a_2=\phi}} \eta_{s,t}((a_1,a_2)|\mathbf{u}) \int_{\mathbb{R}^k} L^*(\mathbf{w},(a_1,a_2)) \\
&\qquad \times \left\{\prod_{q=1}^{k} b_{n_2}(w_q-u_q)\right\} \prod_{r=1}^{k} d\mu(w_r) \\
&\geq (\tilde{m}(\mathbf{u}))^{-1} \sum_{\substack{(a_1,a_2):|a_1|=s,\\|a_2|=t,a_1\cap a_2=\phi}} \eta^*_{s,t}((a_1,a_2)|\mathbf{u}) \int_{\mathbb{R}^k} L^*(\mathbf{w},(a_1,a_2)) \\
&\qquad \times \left\{\prod_{q=1}^{k} b_{n_2}(w_q-u_q)\right\} \prod_{r=1}^{k} d\mu(w_r) \\
&= \sum_{\substack{(a_1,a_2):|a_1|=s,\\|a_2|=t,a_1\cap a_2=\phi}} \eta^*_{s,t}((a_1,a_2)|\mathbf{u}) D^{\delta^*}_{s,t}(\mathbf{u},(a_1,a_2)).
\end{aligned} \tag{9.20}
$$

Now the result follows from (9.5), (9.19) and (9.20). ■

Corollary 9.4.1 *Under the assumptions of Theorem 9.3.1, if for a given* ν, (ν, η^*, δ^*) *is minimax in* $\mathcal{D}_I$, *then* (ν, η^*, δ^*) *is minimax in* $\mathcal{D}$.

PROOF. Since the group of permutations G is finite, the result follows from Blackwell and Girshick (1954, Chapter 8). ■

Remark 9.4.1 (i) Under the assumptions of Theorem 9.4.1, the class of selection procedures $(\nu, \eta^*, \delta^*) \in \mathcal{D}_I$ is essentially complete in $\mathcal{D}_I$.

(ii) For finding a Bayes selection rule with respect to a permutation symmetric prior or for finding the best permutation invariant selection rule, it is enough to optimize ν, which decides about how many indices to select in S_L and how many indices to select in S_U at Stage 1.

References

1. Blackwell, D. and Girshick, M. A. (1954). *Theory of Games and Statistical Decisions*, John Wiley & Sons, New York.

2. Dhariyal, I. D. and Misra, N. (1994). Simultaneous selection of extreme populations: A Bayesian approach, *Communications in Statistics—Theory and Methods*, **23**, 1993–2027.

3. Eaton, M. L. (1967). Some optimum properties of ranking procedures. *Annals of Mathematical Statistics*, **38**, 124–137.

4. Gupta, S. S. and Miescke, K. J. (1983). An essentially complete class of two-stage procedures with screening at first stage, *Statistics & Decisions*, **1**, 427–439.

5. Gupta, S. S. and Miescke, K. J. (1984). On two-stage Bayes selection procedures, *Sankhyā, Series B*, **46**, 123–134.

6. Hussein, K. and Panchapakesan, S. (2001). Simultaneous selection of extreme populations from a set of two-parameter exponential distributions, In *Handbook of Statistics 20: Advances in Reliability* (Eds., N. Balakrishnan and C. R. Rao), pp. 813–830, North-Holland Academic Publishers, Amsterdam.

7. Mishra, S. N. (1986a). Simultaneous selection of extreme populations: Scale parameter case, *Progress of Mathematics*, **20**, 91–105.

8. Mishra, S. N. (1986b). Simultaneous selection of extreme populations: Indifference-zone formulation, *American Journal of Mathematical and Management Sciences*, **6**, 131–142.

9. Mishra, S. N. and Dudewicz, E. J. (1987). Simultaneous selection of extreme populations: A subset selection approach, *Biometrical Journal*, **29**, 471–483.

10. Misra, N. and Dhariyal, I. D. (1994). Simultaneous selection of extreme populations: Bayes-P^* and minimax rules, *Communications in Statistics—Theory and Methods*, **23**, 1963–1992.

PART III
Multiple Comparisons and Tests

10

Comparing Variances of Several Measurement Methods Using a Randomized Block Design with Repeat Measurements: A Case Study

Ajit C. Tamhane[1] **and Anthony J. Hayter**[2]

[1]*Department of Statistics, Northwestern University, Evanston, IL U.S.A.*
[2]*School of Industrial and Systems Engineering, Georgia Institute of Technology, Atlanta, GA U.S.A.*

Abstract: In this chapter we consider the problem of comparing variances of several measurement methods in a randomized block design with repeat measurement methods. The analysis is presented in the context of an actual consulting study, which motivated this chapter. We demonstrate how in a practical data analysis, a combination of informal graphical methods and formal inferential methods (multiple comparison methods) are employed to detect outliers, identify patterns and draw conclusions with confidence.

Keywords and phrases: Multiple comparisons of variances, graphical methods, normal theory methods, robust methods, mixed model, audiological measurements

10.1 Introduction

Professor Panchapakesan's contributions have been instrumental for a number of advances in the fields of ranking and selection, and multiple comparison procedures. Practical applications of these methods are still lagging, however. Most published applications use simple one-way layout designs. To spread the applications of these methods, it is important to publish actual case studies involving more complex designs used in practice. This chapter is a contribution in this direction.

We consider the problem of comparing variances of several measurement methods/instruments when repeat measurements are made with each method on a randomly selected sample of subjects/items. Such a design is frequently used in laboratory experiments in physics, chemistry, biology and psychology, to name a few. The purpose here is to illustrate how a combination of graphical and formal inference (in particular, multiple comparison) methods aids in the analysis of data. The following example from our consulting experience is typical of such studies, and will be used to illustrate the various methods.

Example: The *insertion gain* of a hearing aid is defined as the difference between the sound pressure level (SPL) measured at the eardrum of the wearer with the hearing aid in place and the SPL at the eardrum with no aid in place, the stimulus being the same under both conditions. For clinical measurement of insertion gain, the stimulus is presented over a nearby loudspeaker and the response is measured by a probe microphone in the ear canal of the subject. The standard practice was to locate the loudspeaker in the ear-level horizontal plane of the subject. It was claimed that loudspeaker locations above the horizontal plane would yield more precise (less variable) results. A study was conducted at the Department of Communication Studies and Disorders at Northwestern University to check this claim. The study compared the following loudspeaker locations:

- Location 0: 0° azimuth, 0° elevation (Standard/Control),
- Location 1: 45° azimuth, 0° elevation (New),
- Location 2: 0° azimuth, 90° elevation (New),
- Location 3: 45° azimuth, 45° elevation (New).

There were 10 subjects with five replicate measurements of insertion gain (obtained by measuring the SPL with and without the hearing aid in a random order in pairs) at each of the four loudspeaker locations. The order of the locations presented was randomized for each subject. Measurements of insertion gains were made at 6000 Hz frequency. Table 10.1 gives the raw data. The measurements are presented in the order they were taken. The investigator was primarily interested in comparing the within-subject variances for different measurement methods (loudspeaker locations). The locations apparently affect the variability of measurements because the measurement error depends on the angle of incidence of the sound waves at the ear drum.

The outline of this chapter is as follows. Section 10.2 gives graphical analyses for detecting outliers and identifying patterns in the data. Section 10.3 shows how the conventional normal theory and related robust methods can be adapted to compare the variances in the present setting. Section 10.4 applies these methods to the data under study. Section 10.5 gives some concluding remarks.

Table 10.1: Insertion gain (y_{ijk}) in decibels

Methods	Subject									
	1	2	3	4	5	6	7	8	9	10
0	11.4	4.7	31.2	22.5	17.5	16.0	19.5	6.5	22.2	14.0
	11.5	4.5	35.5	19.0	11.5	3.0	14.3	12.0	21.0	7.5
	15.0	8.5	27.2	16.5	13.5	19.7	8.5	8.5	19.5	11.0
	10.0	3.8	31.8	18.6	18.0	1.2	15.0	2.0	18.0	7.5
	9.0	3.3	36.5	18.8	15.5	14.2	12.5	4.5	17.8	9.5
1	9.2	−1.0	28.2	22.8	14.0	14.7	16.0	7.5	16.5	0.0
	8.0	6.3	30.2	17.0	9.2	0.0	9.5	6.5	13.7	−2.5
	15.0	4.0	22.0	14.0	11.0	15.3	8.0	4.5	13.5	4.5
	10.3	2.5	23.8	15.5	16.0	−0.3	9.8	6.0	17.7	5.5
	9.5	4.0	30.8	17.3	13.8	14.5	8.2	4.0	12.8	4.5
2	16.8	16.0	39.5	26.2	20.3	20.5	25.3	13.0	25.2	19.0
	17.7	23.2	41.0	21.2	16.5	5.0	19.0	15.0	20.5	19.0
	23.8	17.0	31.7	23.5	18.5	22.0	17.2	12.5	20.7	17.0
	18.5	15.5	34.2	22.5	23.3	3.5	18.5	11.7	21.3	17.0
	19.2	17.3	45.2	24.3	20.0	18.5	20.8	10.7	18.5	18.0
3	11.3	4.7	28.0	19.3	15.5	12.0	20.0	3.8	13.5	7.0
	9.8	8.3	31.2	14.2	9.2	−2.3	11.5	5.0	10.7	4.5
	13.0	8.0	23.0	15.0	12.2	11.8	8.8	2.3	10.7	9.0
	11.0	6.2	26.5	15.2	15.2	−4.0	1.7	3.0	13.0	7.5
	10.7	7.0	35.7	17.0	14.7	12.0	10.8	1.7	9.7	7.0

10.2 Graphical Analyses and Descriptive Statistics

Before proceeding with detailed analyses it is important to examine the data for any excessive or systematic variability and to determine whether the data on any subjects should be discarded or modified. We first made box plots (shown in Figure 10.1) for each subject of his/her median centered measurements (using the separate median for each set of five subject × method measurements). Note that because of the median centering, four out of the twenty centered measurements for each subject are forced to be zero. From these plots it appears that subjects 3 and 6 have a much higher variability than other subjects. To examine the reasons for this high variability, run charts were made for the two subjects using the methods as labels. These plots are shown in Figures 10.2 and 10.3 for subjects 3 and 6, respectively. We see that there is something unusual going on with these two subjects: for subject 3, the second and fifth measurements are the largest while the remaining three are the smallest for each method. For subject 6, the pattern is more pronounced: measurements form two distinct clusters for each method with the second and the fourth measurements close to zero while the remaining three measurements are much larger, about 10 to 20. Such systematic patterns were not found for any other subjects. It could not be determined why these systematic patterns occurred for subjects 3 and 6, and not for others. Because of these systematic patterns (more than for reasons of high variability), these two subjects were discarded from further analyses. Note that these systematic patterns could not be revealed by the box plots.

In order to compare the within-subject variabilities of the methods, box plots were made of the same median centered measurements but now stratified by the method. Note that this forces eight of the 40 centered measurements (for eight subjects) to be zero. The result is shown in Figure 10.4. This plot provides a preliminary answer to the question under study, namely, an indication that the current method has a higher variability than the new methods, with method 3 having the least variability. We shall investigate this suggestion more fully and determine whether any differences are significant.

Summary statistics for all cells (means $\overline{y}_{ij.}$, variances s_{ij}^2 and logarithms of variances given by $x_{ij} = \ln s_{ij}^2$) as well as the corresponding statistics for the row (method) margins are shown later in Table 10.3. Subjects 3 and 6 are not included in the calculation of the marginal statistics for the methods. These summary statistics generally confirm the findings of the graphical displays.

To study the structure of the cell means, a mixed model analysis of variance (treating the methods fixed and the subjects random) of the data (omitting subjects 3 and 6) is given in Table 10.2. Using the F-tests as guidelines (because of the violation of the homoscedasticity assumption; however, due to the

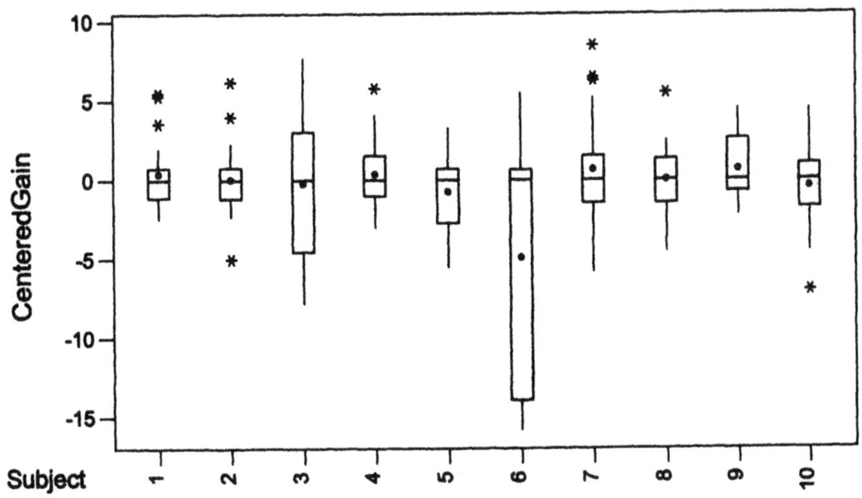

Figure 10.1: Box plots of median-centered insertion gains for subjects

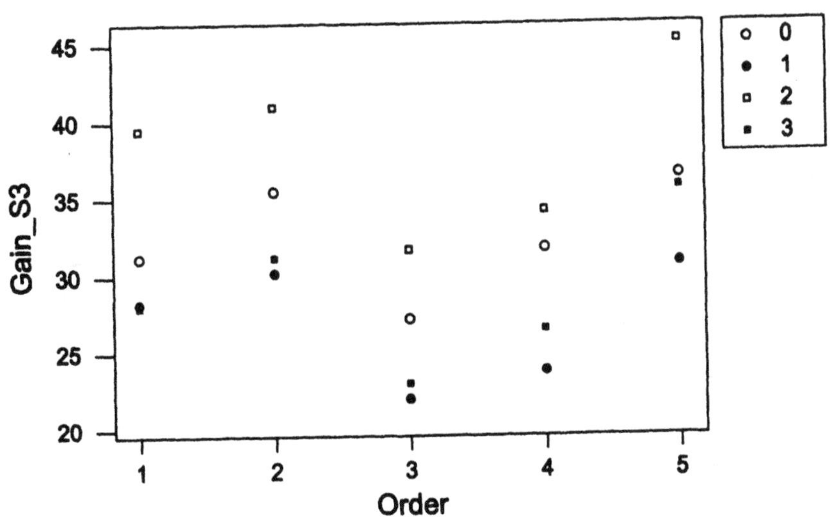

Figure 10.2: Run chart of insertion gains for subject 3

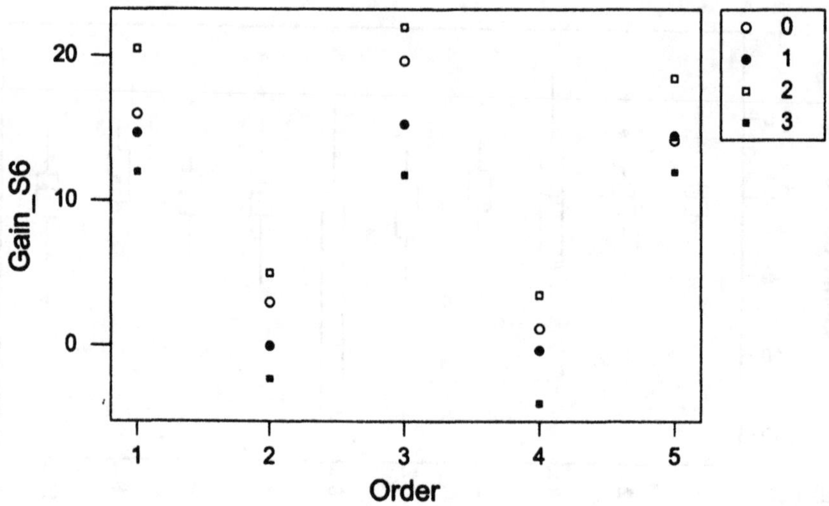

Figure 10.3: Run chart of insertion gains for subject 6

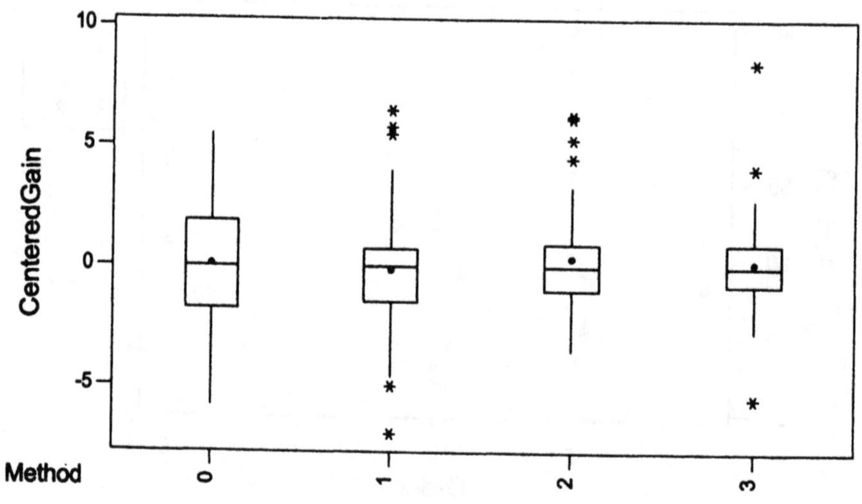

Figure 10.4: Box plots of median-centered insertion gains for methods

Table 10.2: Analysis of variance

Source	SS	DF	MS	F	p-value
Methods	2225.8	3	741.92	35.80	0.000
Subjects	2593.2	7	370.45	17.87	0.000
Methods×Subjects	435.3	21	20.73	3.13	0.000
Error	848.8	128	6.63		
Total	6103.0	159			

balanced design the F-tests should be quite robust), we see that all three effects as significant. However, the main effects of the methods and subjects dominate over their interaction effect. Thus the structure of the cell means appears to be additive. This can be verified by fitting an additive model to the data and studying the residuals.

One could make more plots to detect further quirks and patterns in the data, but one should be cautious of not crossing the boundary between prudently selected graphics and data-dredging. So we will stop here and summarize our findings thus far: subjects 3 and 6 were discarded because they exhibited excessive and, more importantly, systematic variation. After omitting these two subjects, the remaining subjects form a relatively homogeneous group, both with regard to their means and SDs. The data on these subjects indicate that method 3 has less within-subject variability than method 0. Methods differ in their mean levels (with method 2 giving the highest readings and method 1 giving the lowest readings for most subjects) and the structure of the means is roughly additive.

10.3 Formal Statistical Analyses

10.3.1 Model

Let there be $M \geq 2$ new measurement methods, whose variances are to be compared with a control method, labeled 0. Suppose we have available $N \geq 2$ randomly selected subjects and we make $n_{ij} \geq 2$ repeat measurements on the jth subject using the ith method $(0 \leq i \leq M)$. The following mixed-effects model [Scheffé (1959, Ch. 8)] is proposed for the data:

$$y_{ijk} = m_{ij} + e_{ijk} \quad (0 \leq i \leq M, 1 \leq j \leq N, 1 \leq k \leq n_{ij}).$$

Here $(m_{0j}, m_{1j}, \ldots, m_{Mj})$ for $j = 1, \ldots, N$ are assumed to be i.i.d. random vectors from an $(M+1)$-variate distribution (not necessarily normal) with mean vector $(\mu_0, \mu_1, \ldots, \mu_M)$ and covariance matrix $\mathbf{\Sigma} = (\sigma_{ii'})$. Furthermore, for each method i, the measurement errors e_{ijk} are assumed to be i.i.d. (not necessarily normal) with mean 0 and variance σ_i^2, and they are independent of the m_{ij}. Note that in this model any two observations y_{ijk} and $y_{i'j'k'}$ are correlated (assuming a general $\mathbf{\Sigma}$ with nonzero off-diagonal elements) if and only if $j = j'$. All parameters in the model are unknown. The primary interest lies in comparing the imprecisions of the methods as measured by the σ_i^2.

10.3.2 Multiple comparison procedures

Let $\bar{y}_{ij\cdot}$ and the $\bar{e}_{ij\cdot}$ be the cell means of the y_{ijk} and e_{ijk}, respectively. Then the

$$z_{ijk} = y_{ijk} - \bar{y}_{ij\cdot} = e_{ijk} - \bar{e}_{ij\cdot}$$

are distributed independently of the vector $(m_{0j}, m_{1j}, \ldots, m_{Mj})$ with

$$E(z_{ijk}) = 0, \mathrm{Var}(z_{ijk}) = \sigma_i^2\left(\frac{n_{ij}-1}{n_{ij}}\right), \mathrm{Cov}(z_{ijk}, z_{ijk'}) = -\frac{\sigma_i^2}{n_{ij}}. \quad (10.1)$$

Note that the widths of the box plots in Figure 10.4 are roughly proportional to $\mathrm{SD}(z_{ijk}) = \sigma_i\sqrt{(n_{ij}-1)/n_{ij}}$. Furthermore, the z_{ijk} $(1 \leq k \leq n_{ij})$ are mutually independent for $0 \leq i \leq M$ and $1 \leq j \leq N$. Hence for each method i, the within-subject sample variances

$$s_{ij}^2 = \frac{\sum_{k=1}^{n_{ij}} z_{ijk}^2}{n_{ij}-1} \quad (10.2)$$

are mutually independent with

$$E(s_{ij}^2) = \sigma_i^2 \text{ and } \mathrm{Var}(s_{ij}^2) = \sigma_i^4\left(\frac{2}{n_{ij}-1} + \frac{\gamma_{ij}}{n_{ij}}\right),$$

where γ_{ij} is the kurtosis of the distribution of the e_{ijk} [see Bartlett and Kendall (1946)]. Therefore standard procedures based on the independence assumption can be used to make inferences on the σ_i^2 from the s_{ij}^2. Below we briefly discuss such normal theory procedures, and robust procedures that do not assume that the e_{ijk} are normally distributed.

Normal theory procedures

Under the normality assumption, $\gamma_{ij} = 0, s_{ij}^2 \sim \sigma_i^2\chi_{\nu_{ij}}^2/\nu_{ij}$ and the minimum variance unbiased estimate of σ_i^2 is

$$s_i^2 = \frac{\sum_{j=1}^N \nu_{ij} s_{ij}^2}{\sum_{j=1}^N \nu_{ij}} \sim \sigma_i^2\chi_{\nu_i}^2/\nu_i,$$

where $\nu_{ij} = n_{ij} - 1$ and $\nu_i = \sum_{j=1}^{N} \nu_{ij}$. The usual normal theory procedures can be applied to the s_i^2 to make inferences about the σ_i^2. For example, Hartley's (1950) $F_{\max}$ test can be used if all pairwise comparisons between the σ_i^2 are of interest. An illustration of a normal theory procedure for comparing the new method variances σ_i^2 $(1 \le i \le 3)$ with σ_0^2 is given in Section 10.4.

Robust procedures

It is well-known that normal theory procedures for variances are not robust when the assumption of normality is violated. Box (1953) proposed a robust procedure for testing the equality of variances in a one-way layout. This procedure randomly divides each sample into subsamples of equal sizes and calculates an estimate of the treatment variance from each subsample. In the present problem, the necessity of forming random subsamples is obviated because for each method we have N "natural" subsamples, namely $\{y_{ijk},\ 1 \le k \le n_{ij}\}$ for $j = 1, \ldots, N$. Thus Box's procedure can be used as follows: Compute the subsample variances s_{ij}^2 using (10.2) and let

$$x_{ij} = \ln s_{ij}^2.$$

Then the x_{ij} are independent and, for large n_{ij}, are approximately normally distributed with

$$E(x_{ij}) \approx \ln \sigma_i^2 \text{ and } \operatorname{Var}(x_{ij}) \approx \frac{2}{n_{ij} - 1} + \frac{\gamma_{ij}}{n_{ij}};$$

see Bartlett and Kendall (1946). The appropriate procedure depends on the sample sizes n_{ij} and the assumptions we are willing to make about the γ_{ij}. For instance, if $n_{ij} = n$ for all i, j and if we assume $\gamma_{ij} = \gamma$ for all i, j, then the x_{ij} have approximately a constant variance. Therefore the usual ANOVA procedures can be used to compare the $\ln \sigma_i^2$. Multiple comparisons with a control can be carried out using Dunnett's (1955) method. This method is illustrated in the next section.

An alternative to the above procedure is to use a nonparametric procedure. Different approaches are possible, but a simple one is as follows: Compute the Wilcoxon signed rank statistics for comparing the current method with each one of the new methods using the x_{ij} as the data values, and then refer these statistics to the critical point of the null distribution of the maximum of such correlated statistics. This test procedure due to Nemenyi (1963) is described in Section 2.3.2 of Chapter 9 of Hochberg and Tamhane (1987) and is illustrated in the following section. A test based on the signed ranks of $x_{0j} - x_{ij} = \ln(s_{0j}^2 / s_{ij}^2)$

will not only avoid the problems caused by lack of normality of the original data or lack of approximate normality of the x_{ij} caused by insufficiently large n, but will also be resistant to outliers in the orginal data.

10.4 Return to Example

We now return to the example of Section 10.1. Table 10.3 gives the cell variances, s_{ij}^2, and their logarithms, x_{ij}, for the data given in Table 10.1. As noted in Section 10.2, subjects 3 and 6 are deleted as outliers. All subsequent analyses are based on the remaining $N = 8$ subjects.

In this example, one-sided comparisons of new methods with the current method are of interest because new methods would be useful only if they reduced the variability as compared to the current method. To see if it is appropriate to analyze the data under the normality assumptions, we first made normal plots of the residuals z_{ijk} for each method. (Note that this strictly requires that the z_{ijk} be independent, which they are not. However, (10.1) gives the $\text{Corr}(z_{ijk}, z_{ijk'}) = -1/(n-1) = -1/4$, which is fairly small.) The plots, not shown here, do not exhibit any gross violation of normality. Therefore we consider $100(1-\alpha)\%$ simultaneous upper one-sided confidence limits based on normal theory:

$$\frac{\sigma_i^2}{\sigma_0^2} \le \frac{1}{c}\frac{s_i^2}{s_0^2} \quad (i = 1, 2, 3).$$

Here, c is the lower 100α percentage point of the smallest of M random variables, $\chi_{\nu_i}^2/\chi_{\nu_0}^2$ $(1 \le i \le M)$, where the $\chi_{\nu_i}^2$ are independent chi-square random variables with ν_i d.f. In the present example, $M = 3, \nu_i = 32$ and the s_i^2 are the middle entries in the last column of Table 10.3. The values of c have been tabulated by Gupta and Sobel (1962). For $1 - \alpha = 0.90, M = 3$ and common d.f. $= 32$ for all sample variances, we find $c = 0.5314$ from Table 3 in their paper. Therefore the upper confidence limits for σ_i^2/σ_0^2 for $i = 1, 2, 3$ are 1.858, 1.385 and 1.204, respectively. The ordering of these confidence limits are in agreement with Figure 10.4. Thus none of the ratios σ_i^2/σ_0^2 can be demonstrated to be less than 1, and hence none of the methods can be shown to significantly improve on the current method using a familywise type I error rate of $\alpha = 0.10$.

If one does not wish to make the normality assumption, a robust analysis can be based on the transformation $x_{ij} = \ln s_{ij}^2$ discussed in Section 10.3.2. In this case, approximate $100(1-\alpha)\%$ simultaneous upper one-sided confidence limits on $\ln(\sigma_i^2/\sigma_0^2)$ are given by

$$\ln(\sigma_i^2/\sigma_0^2) \le \overline{x}_i - \overline{x}_0 + d\, s_x\sqrt{2/N} \quad (1 \le i \le M),$$

Table 10.3: Cell means $\overline{y}_{ij\cdot}$, cell variances s_{ij}^2 and their logs $x_{ij} = \ln s_{ij}^2$

Method	Subject 1	2	3	4	5	6	7	8	9	10	Average
0	11.40	4.96	32.44	19.08	15.20	10.82	13.96	6.70	19.70	9.90	12.613
	5.15	4.24	13.84	4.67	7.45	67.73	15.92	14.59	3.61	7.40	7.879
	1.639	1.445	2.628	1.541	2.008	4.216	2.768	2.680	1.284	2.001	1.921
1	10.40	3.16	27.00	17.32	12.80	8.84	10.30	5.70	14.84	2.40	9.615
	7.29	7.24	15.37	11.09	7.24	67.40	10.76	2.07	4.54	12.04	7.784
	1.987	1.980	2.732	2.406	1.980	4.211	2.376	0.728	1.513	2.488	1.932
2	19.20	17.80	38.32	23.54	19.72	13.90	20.16	12.58	21.24	18.00	19.030
	7.40	9.67	29.16	3.53	6.25	79.39	9.92	2.59	6.00	1.00	5.795
	2.001	2.269	3.373	1.261	1.833	4.374	2.295	0.952	1.792	0.000	1.550
3	11.16	6.84	28.88	16.14	13.36	5.90	12.56	3.16	11.52	7.00	10.218
	1.37	2.13	23.23	4.16	7.13	68.56	18.58	1.66	2.69	2.63	5.044
	0.315	0.756	3.145	1.426	1.964	4.228	2.922	0.507	0.990	0.967	1.231

The upper entry in each cell is $\overline{y}_{ij\cdot}$, the middle entry is s_{ij}^2 and the lower entry is $x_{ij} = \ln s_{ij}^2$. The entries in the 'Average' column are the corresponding averages, $\overline{y}_{i\cdot\cdot}, s_i^2$ and $\overline{x}_i$, over subjects (excluding subjects 3 and 6).

where d is the upper 100α percentage point of the maximum of M jointly distributed student t random variables with $\nu = (M+1)(N-1)$ d.f. and common associated correlation coefficient $\rho = 1/2$. The values of d have been tabulated by Bechhofer and Dunnett (1988). For $1-\alpha = 0.90, M = 3$ and $\nu = 28$, we find $d = 1.78935$ from Table 1 in their paper. Also, s_x, the pooled sample standard deviation of the x_{ij}, is calculated to be 0.733. Therefore simultaneous 90% upper one-sided confidence limits on $\ln(\sigma_i^2/\sigma_0^2)$ are calculated to be 0.667, 0.285 and -0.034 for $i = 1, 2, 3$, respectively. The corresponding confidence limits on σ_i^2/σ_0^2 are then 1.948, 1.330 and 0.967, respectively. These limits are not too different from the confidence limits calculated from the normal theory, but here we are able to show that method 3 is significantly better than the method 0. (This finding is in accord with the box plots shown in Figure 10.4.) It should be noted, however, that these robust confidence limits are based on approximate normality of the x_{ij} which holds only if the n_{ij} are large, whereas in the present example n_{ij} equal only 5.

If we do not wish to rely on the approximate normality of the x_{ij}, then we can use the nonparametric tests referred to previously. The Wilcoxon signed rank statistics between method 0 and the three new methods (calculated using the x_{ij}) are as follows: $W_1 = 13, W_2 = 22$ and $W_3 = 33$. It is well-known that the joint null distribution of W_1, W_2, W_3 is not distribution-free [see Hochberg and Tamhane (1987, p. 255)], and so the exact critical points of $\max(W_1, W_2, W_3)$ are not available. We can apply the Bonferroni method by comparing the one-sided marginal p-value of W_3 with $0.10/3 = 0.033$. This p-value is found to be 0.0195 from Table H in Lehmann (1975) for $N = 8$. Hence method 3 is shown to be significantly better than method 0 at a type I familywise error rate of $\alpha = 0.10$. Alternatively, we can calculate a large sample approximate critical value of $\max(W_1, W_2, W_3)$ given by [see equation (2.16) of

Hochberg and Tamhane (1987, Ch. 9)]

$$\frac{N(N+1)}{4} + \frac{1}{2} + z_{k,1/2,\alpha}\sqrt{\frac{N(N+1)(2N+1)}{24}},$$

where $z_{k,1/2,\alpha}$ is the upper α critical point of k equicorrelated standard normal variates with common correlation $= 1/2$. For $k = 3$ and $\alpha = 0.10$, we find $z_{k,1/2,\alpha} = 1.7336$ using Dunnett's (1989) program. By substituting $N = 8$, we obtain the desired critical value to be 30.88, which is exceeded by W_3. The other two methods are not significantly better than method 0.

One can use Steel's (1959) sign test procedure above if the assumption of the symmetric distribution of the $x_{ij} - x_{0j}$ is not valid. The three sign statistics are $S_1 = 3, S_2 = 5$ and $S_3 = 7$ (because only one subject, viz. subject 7, has less variability for method 0 than for method 3). Using Table 1 of Rhyne and Steel (1965), we find that the exact multiplicity adjusted p-value of $S_3 = 7$ equals $0.066 < 0.10$. Hence the conclusion drawn using the Wilcoxon signed rank is confirmed. Alternatively, we can calculate a large sample critical value of $\max(S_1, S_2, S_3)$ given by [see equation (2.7) of Hochberg and Tamhane (1987, Ch. 9)]

$$\frac{N}{2} + \frac{1}{2} + z_{k,1/3,\alpha}\frac{\sqrt{N}}{2}.$$

For $k = 3$ and $\alpha = 0.10$, we find $z_{k,1/3,\alpha} = 1.7738$ using Dunnett's (1989) program available at the web site http://lib.stat.cmu.edu/general. By substituting $N = 8$, we obtain the desired critical value to be 7.01, which S_3 just fails to exceed.

10.5 Concluding Remarks

We have given an example of a case study that employed a combination of graphical methods and formal multiple comparison methods to arrive at practically useful conclusions. Graphical examination of the data is helpful in not only revealing the main patterns in the data, but also in identifying outliers that can vitiate the formal analyses. We demonstrated how standard multiple comparison methods for one-way layouts can be readily modified to a more complex (but balanced) design used here.

Based on the graphical displays and formal statistical analyses we can conclude that method 3 is more precise (less variable) than method 0 at $\alpha = 0.10$. It is interesting that only the normal theory method for comparing variances did not support this conclusion. Since this conclusion is based on the data from only eight subjects, it would be desirable to do further experimentation to confirm this finding. The investigator should be advised to find possible reasons for the systematic and excessive variability observed in subjects 3 and 6, and to avoid those sources of variation in future experimentation.

Acknowledgment: The authors wish to thank Mr. Larry Revit, a graduate student in the School of Communication Studies at Northwestern University, for allowing the use of the data that he gathered for his thesis research.

References

1. Bartlett, M. S. and Kendall, D. G. (1946). The statistical analysis of variance-heterogeneity and logarithmic transformation, *Journal of Royal Statistical Society Supplement*, **8**, 128–138.

2. Bechhofer, R. E. and Dunnett, C. W. (1988). Tables of the multivariate Student's t distribution, *Selected Tables in Mathematical Statistics*, **11**, American Mathematical Society, Providence, RI.

3. Box, G. E. P. (1953). Non-normality and tests on variances, *Biometrika*, **40**, 318–335.

4. Dunnett, C. W. (1955). A multiple comparison procedure for comparing several treatments with a control, *Journal of the American Statistical Association*, **50**, 1096–1121.

5. Dunnett, C. W. (1989). Multivariate normal probability integrals with product correlation structure, Algorithm AS251, *Applied Statistics*, **38**, 564–579.

6. Gupta, S. S. and Sobel, M. (1962). On the smallest of several correlated F statistics, *Biometrika*, **49**, 509–523.

7. Hartley, H. O. (1950). On the maximum F-ratio as a shortcut test for heterogeneity of variances, *Biometrika*, **37**, 308–312.

8. Hochberg, Y. and Tamhane, A. C. (1987). *Multiple Comparison Procedures*, John Wiley & Sons, New York.

9. Lehmann, E. L. (1975). *Nonparametrics: Statistical Methods Based on Ranks*, Holden-Day, San Francisco.

10. Nemenyi, P. (1963). Distribution-free multiple comparisons, unpublished doctoral dissertation, Priceton University, Princeton, NJ.

11. Rhyne, A. L. and Steel, R. G. D. (1965). Tables for a treatments versus control multiple comparison test, *Technometrics*, **7**, 293–306.

12. Scheffé, H. (1959). *The Analysis of Variance*, John Wiley & Sons, New York.

13. Steel, R. G. D. (1959). A multiple comparison sign test: Treatments vs. control, *Journal of the American Statistical Association*, **54**, 767–775.

11

Impact of Missing Data and Imputation Methods on Multiple Test Procedures

Alka Indurkhya

Department of Society, Human Development and Health, Harvard University, Boston, MA U.S.A.

Abstract: This chapter illustrates how two common missing data mechanisms (missing completely at random, and missing at random) influence the Type I familywise error rate of the therapeutic window of a drug using multiple test procedures. A therapeutic window is determined by the minimum dose needed to effectively treat a condition or ailment, and the maximum dose that can be safely administered. The effect of multiple imputation procedures for these missing data mechanisms is also assessed. Simulation results suggest that multiple imputations reduce the familywise error rate of the therapeutic window in the presence of missing data.

Keywords and phrases: Multiple comparisons, missing data, multiple imputation, bootstrap, therapeutic window, step-down procedure

11.1 Introduction

Most drugs need to satisfy the minimum effectiveness and maximum safe dose threshold criteria during clinical trials before they are marketed. In order to measure the doses that correspond to the minimum effectiveness and toxicity, respectively we need an experimental design that measures effectiveness and toxicity for a range of doses. The range delimited by the minimum effective dose (MINED) and the maximum safe dose (MAXSD) of a drug is defined as the therapeutic window of the drug. A therapeutic window is determined using two approaches: 1) The MINED and MAXSD may be obtained simultaneously [Thall and Chen (1999) and Tamhane and Logan (2002)] or, 2) The MINED and MAXSD may be obtained individually from the data [Tamhane, Hochberg and Dunnett (1996), Tamhane et al. (2001)]. If the experimental design does not contain data to simultaneously evaluate both safety and efficacy, then the latter approach may be used to obtain the therapeutic window of the drug. However since safety is always a concern in all phases of a clinical trial, data on safety and efficacy are generally available to estimate MINED and MAXSD simultaneously. Contrary to popular belief that randomized controlled clinical trials rarely contain missing data, recent evidence suggests that this is not the case [Shekelle et al. (2003), USFDA (2003)].

A recent review of the literature indicates that most Phase II and III reports from clinical trials do not indicate whether any of the data was missing or censored. In the rare case that they do, no mention is made of what procedures were used to handle such data in order to estimate the therapeutic window. In particular, the impact of missing data on multiple test procedures for estimating the therapeutic window has to date received no attention. This paper attempts to fill this void in the literature by studying the effect of different missing data mechanisms, and multiple imputation procedures on simultaneous estimation of MINED and MAXSD.

11.2 The Therapeutic Window

Following the traditional notation used in estimation of the therapeutic window of a drug, let $i = 0, 1, \ldots, k$ correspond to increasing levels of a drug dose whose therapeutic window is to be estimated. Let dose $i = 0$ represent the control unit, and doses $1 \leq i \leq k$ represent the experimental units. Let the bivariate random variable (X_{ij}, Y_{ij}) represent the observed effectiveness measure x_{ij}, and safety measure y_{ij}, respectively, for subject j treated with dose i. Further, suppose that n_i subjects are treated with dose i and $N = \sum_{i=0}^{k} n_i$ is the total number of subjects.

Let $(X_{ij}, Y_{ij}) \sim N(\boldsymbol{\theta}, \boldsymbol{\Sigma})$ where

$$\boldsymbol{\theta} = \begin{pmatrix} \mu_i \\ \eta_i \end{pmatrix}$$

and

$$\boldsymbol{\Sigma} = \begin{bmatrix} \sigma^2 & \rho\sigma\tau \\ \rho\sigma\tau & \tau^2 \end{bmatrix}.$$

It is customary to assume that values of μ_i that exceed μ_0 indicate high efficacy, while values of η_i that exceed η_0 indicate high toxicity. In order to formally define MINED and MAXSD we need to consider prespecified threshold values for efficacy and toxicity. Let δ_e be the threshold value for efficacy and δ_s be the corresponding threshold value for toxicity. Thus if $\mu_i > \mu_0 + \delta_e$ and $\eta_i < \eta_0 + \delta_s$, then dose i is both effective, and safe.

Definition 11.2.1 The minimum effectiveness dose for a drug is defined as

$$D_e = MINED = \min_{1 \le i \le k} \{\mu_i > \mu_0 + \delta_e\}$$

and the maximum safe dose for a drug is defined as

$$D_s = MAXSD = \max_{1 \le j \le k} \{\eta_j < \eta_0 + \delta_s\}$$

Note that if $D_e < D_s$ in Definition 11.2.1 then $[D_e, D_s]$ is the desired therapeutic window of the drug where all doses greater than D_e must be effective, and all doses less than D_s must be safe. The aim is to estimate D_e and D_s such that for a prespecified Type I familywise error rate α:

$$\Pr(\text{an unsafe dose declared safe or an ineffective dose declared effective}) \le \alpha. \tag{11.1}$$

D_e is said to exceed k if there exists no i such that $\mu_i > \mu_0 + \delta_e$. Similarly, D_s does not exceed 1 if there is no dose j for which $\eta_j < \eta_0 + \delta_s$. Let $\widehat{D_e}$ and $\widehat{D_s}$ denote the estimated dose levels for D_e and D_s, respectively. If $\widehat{D_e} > \widehat{D_s}$, then we conclude that no therapeutic window for the drug can be found; if $\widehat{D_e}$ does not exist, then we estimate that $\widehat{D_e} > k$, and if $\widehat{D_s}$ does not exist, then we conclude that $\widehat{D_s} < 1$.

The null and alternate hypotheses to test for efficacy for different dose levels of a drug are:

$$\begin{aligned} H_i^e : \mu_i &\le \mu_0 + \delta_e, \\ A_i^e : \mu_i &> \mu_0 + \delta_e, \quad i = 1, 2, \ldots, k, \end{aligned} \tag{11.2}$$

and to test for safety of the drug, these are given by:

$$\begin{aligned} H_j^s : \eta_j &\ge \eta_0 + \delta_s, \\ A_j^s : \eta_j &< \eta_0 + \delta_s, \quad j = 1, 2, \ldots, k. \end{aligned} \tag{11.3}$$

Note that the null hypotheses in (11.2) corresponds to the hypotheses that the ith dose is ineffective, while the null hypotheses in (11.3) states that the jth dose is unsafe. As discussed in Tamhane and Logan (2002), the following t-statistics are used to develop the test procedures to test the hypotheses in (11.2) and (11.3), respectively.

$$\begin{aligned} t_i^e &= \frac{\overline{x}_i - \overline{x}_0 - \delta_e}{\hat{\sigma}\sqrt{1/n_i + 1/n_0}} \\ t_j^s &= \frac{\overline{y}_0 - \overline{y}_j + \delta_s}{\hat{\tau}\sqrt{1/n_i + 1/n_0}}, \quad i, j = 1, 2, \ldots, k. \end{aligned} \tag{11.4}$$

Thus it follows from equations (11.2) through (11.4) that the estimates of D_e and D_s are:

$$\begin{aligned} \widehat{D_e} &= \min\{i : H_i^e \text{ is rejected }\}, \\ \widehat{D_s} &= \max\left\{j : H_j^s \text{ is rejected }\right\}. \end{aligned} \tag{11.5}$$

The main assumptions in estimating the therapeutic window are that there exist predetermined efficacy and safety end points for the drug, and that there exists a lower bound for efficacy, and an upper bound for toxicity of the drug. Recall that we have also assumed that the data come from a bivariate normal distribution, and that the variance across different dose levels is the same, namely, σ for effectiveness, and τ for toxicity. The correlation ρ between the toxicity and efficacy measures is also assumed to be constant across all dose levels.

11.3 Bootstrap Procedures to Estimate the Therapeutic Window

It has been recently demonstrated [Tamhane and Logan (2002)] that the bootstrap versions of the procedures based on normal theory have several advantages. First, we can relax the assumption of bivariate normality; second, we do not need to know the correlation coefficient between efficacy and toxicity; and last, significant gains in power are achieved for simultaneous testing of efficacy and toxicity.

The algorithm of the bootstrap SD procedures that has been rigorously tested by Tamhane and Logan (2002) is described in Figure 11.1. The procedure begins after centering the data with respect to the sample means for efficacy and safety, respectively.

Suppose we are to determine the therapeutic window for a drug with three dose levels, i.e., $i = 1, 2, 3$ along with a control level ($i = 0$). We first use the bivariate sample data to compute the t-statistics for each dose level using the formulae in (11.4). Let us denote these by t_i^e, t_j^s for $i, j = 1, 2, 3$.

The BOOTSTRAP FUNCTION in Figure 11.1 starts by generating M random samples of size M^* for each of the dose levels ($i = 0, 1, 2, 3$). The t-statistics for each of the M bootstrap samples are then computed using (11.4). The function BOOTSTRAP STATISTICS returns an M dimensional bivariate array of these t-statistics for dose levels 1 through 3. Note that the control data is also used in (11.4) to obtain the requisite t-statistics at dose levels 1 through 3.

The BOOTSTRAP SD PROCEDURE (see Figure 11.1) estimates the therapeutic window by performing a series of hypothesis tests using the t-statistics from the BOOTSTRAP FUNCTION and the original dataset, for each dose level at a prespecified level of type I error. We denote the t-statistics corresponding to the original dataset by using a single subscript (t_i^e, t_j^s), and those from the bootstrap with a double subscript (t_{ig}^e, t_{jg}^s) for $i = 1, 2, 3$ and $g = 1, 2, \ldots, M$. The hypothesis $H_3^e \cap H_1^s$ for the SD1 version is tested using the empirical p-values. These are (11.6). For the SD2 version, we need to only omit the maximum criteria for the t-statistics to obtain the adjusted SD2 p-values.

$$\begin{aligned} \text{p}_{l,r}^e &= \frac{\sum_{g=1}^M I[\max_{1\le i\le l, r\le j\le k}(t_{ig}^e, t_{jg}^s) \ge \max_{1\le i\le l} t_i^e]}{M}, \\ \text{p}_{l,r}^s &= \frac{\sum_{g=1}^M I[\max_{1\le i\le l, r\le j\le k}(t_{ig}^e, t_{jg}^s) \ge \max_{r\le j\le k} t_j^s]}{M}, \end{aligned} \tag{11.6}$$

where

$$I[\max_{1\le i\le l, r\le j\le k}(t_{ig}^e, t_{jg}^s) \ge \max_{1\le i\le l} t_i^e]$$

is the indicator function of the set $\{\max_{1\le i\le l, r\le j\le k}(t_{ig}^e, t_{jg}^s) \ge \max_{1\le i\le l} t_i^e\}$. Thus it is the number of bootstrap samples for which

$$\max_{1\le i\le l, r\le j\le k}(t_{ig}^e, t_{jg}^s) \ge \max_{1\le i\le l} t_i^e.$$

If both $\text{p}_{3,1}^e$ and $\text{p}_{3,1}^s$ are $\ge \alpha$, then we get the estimated therapeutic window [4, 0], i.e, no therapeutic window exists as $\widehat{D}_e > \widehat{D}_s$.

If both $\text{p}_{3,1}^e$ and $\text{p}_{3,1}^s$ are $< \alpha$, then we reject both H_3^e and H_1^s and test the hypothesis $H_2^e \cap H_2^s$ using the p-values $\text{p}_{2,2}^e$ and $\text{p}_{2.2}^s$. If only $H_2^s < \alpha$, then we test the hypothesis $H_2^e \cap H_3^s$. If both $\text{p}_{2,3}^e$ and $\text{p}_{2,3}^s$ are $\ge \alpha$, then we need to test only the efficacy hypothesis as $\widehat{D}_s > 3$, i.e., all doses are safe. We can test the efficacy hypothesis H_1^e by calculating the empirical p-value in (11.7) given by

$$\text{p}_l^e = \frac{\sum_{g=1}^M I[\max_{1\le i\le l} t_{ig}^e \ge \max_{1\le i\le l} t_i^e]}{M}. \tag{11.7}$$

If p_1^e is $\ge \alpha$, then we have estimated the therapeutic window as [1, 3], i.e:, all doses are effective and safe.

However if only $H_2^e < \alpha$, then we test the hypothesis $H_1^e \cap H_2^s$. If only $\text{p}_{1,2}^e \ge \alpha$, then we need to test only the safety hypothesis as $\widehat{D}_e < 1$, i.e., all doses are effective. We can test the safety hypothesis H_3^s by calculating the empirical p-value in (11.8) given by

$$\text{p}_r^s = \frac{\sum_{g=1}^M I[\max_{r\le j\le k} t_{jg}^s \ge \max_{r\le j\le k} t_j^s]}{M}. \tag{11.8}$$

If p_3^s is $\ge \alpha$, then we have estimated the therapeutic window as [1, 3], i.e.. all doses are effective and safe.

---

function BOOTSTRAP STATISTICS (*sample*) **returns** t-statistics for efficacy and safety

inputs: *sample* is a random sample of size M^* from each dose of the mean centered bivariate dataset

return t^e_{ig}, t^s_{jg} for $i, j = 1, 2, \ldots, k$

---

function BOOTSTRAP SD PROCEDURE (t-*statistics*, α) **returns** therapeutic window $[\widehat{D_e}, \widehat{D_s}]$

inputs:t-*statistics* from original dataset, and from M bootstrap samples of size M^* using function BOOTSTRAP STATISTICS
α is the probability of type I error

initialize $l = k, r = 1$

if $l > 0$ or $r < k+1$ and H^e_i and H^s_j are rejected at level $\alpha$for all $i \geq l+1$, and $j \leq r-1$
then test $H^e_l \cap H^s_r$ for $l \geq 1$ and $r \leq k$

if H^e_l is rejected at level α **then** set $l = l' - 1$ where l' is the index corresponding to the maximum t-statistic D_e value between 1 and l.
else return $\widehat{D_e} = l + 1$
if H^s_r is rejected at level α **then** set $r = r' + 1$ where r' is the index corresponding to the maximum t-statistic value between r and k.
else return $\widehat{D_s} = r - 1$
else if $l = 0$ and $r = k + 1$ **return** $[\widehat{D_e} < 1, \widehat{D_s} > k]$.
else if $r = k + 1, l \geq 1$ **while** H^e_l is rejected at level α and $l > 0$
set $l = l' - 1$where l' is the index corresponding to the maximum t-statistic value between 1 and l.
if H^e_l is not rejected at level α **return** $[\widehat{D_e} = l + 1, \widehat{D_s} > k]$
if $l = 0$ **return** $[\widehat{D_e} < 1, \widehat{D_s} > k]$
else if $l = 0, r \leq k$ **while** H^s_r is rejected at level α and $r < k + 1$
set $r = r' + 1$ where r' is the index corresponding to the maximum t-statistic value between r and k.
if H^s_r is not rejected at level α **return** $[\widehat{D_e} < 1, \widehat{D_s} = r - 1]$
if $r = k + 1$ **return** $[\widehat{D_e} < 1, \widehat{D_s} > k]$ **else return**$[\widehat{D_e} = l + 1, \widehat{D_s} = r - 1]$

Figure 11.1: Algorithm for the SD bootstrap procedure to estimate the therapeutic window

11.4 Missing Data Mechanisms

Whenever we encounter missing data in clinical settings, the mechanism that causes data to be missing [Little and Rubin (1987)] is either (1) Missing completely at random (MCAR), or (2) Missing at random (MAR), or (3) Not missing at random (NMAR). Data are said to be MCAR if the missing values bear no relationship to the variables in the dataset. Data are said to be MAR when the missing values depend on the values of some variables in the dataset, but for the fact that conditional on those values the data are MAR. If the data are neither MCAR nor MAR then they are said to be NMAR. This mechanism requires the missing data to depend on unobserved variables. Since we are assuming that the efficacy and safety data for different dose levels follow a bivariate normal distribution we will focus on only the MCAR and MAR mechanisms.

Many approaches have been investigated in the last two decades to handle missing data [Little and Rubin (1987); Schafer (1997)]. The advantage of using statistical models that predict missing values is that they allow us to incorporate uncertainty in the prediction of missing values. Multiple imputation procedures are becoming the tool of choice in estimating missing values as they include a random component to reflect the fact that these imputed values are estimated. Thus datasets with imputed values are no longer unique due to the random component used in estimating the missing values. Hence with multiple imputation, an incomplete dataset will have the missing values estimated several times. Each imputed dataset is then separately analyzed as if it were complete. The variance of each parameter of interest is now composed of two parts: the estimated variance within each imputed dataset, and the variance across the datasets. The principle of multiple imputation is summarized in Figure 11.2.

11.5 Simulations

We generated 100 observations from a bivariate normal distribution with means $(\mu_i = i, \eta_i = i)$, standard deviations $(\sigma = .5, \tau = .75)$, and correlation $\rho = .5$ for $i = 0, 1, 2, 3, 4$. The threshold constants for efficacy and toxicity were specified as $\delta_e = 1.01$ and $\delta_s = 1.99$ respectively. The true familywise error rate is $\alpha = .05$.

We performed 5000 replications on an IBM workstation using SAS/IML. We simulated data missing completely at random by randomly deleting 5%, 10%,

Step 1 Generate M sets of imputed values $\hat{\mu}_{im}$ for the missing data thereby creating M complete datasets.

Step 2 Obtain the point estimates $(\widehat{\mu}_{im}, \widehat{\eta}_{im})$ and variances $(\widehat{\sigma}^2_m, \widehat{\tau}^2_m)$ of the parameters of interest for $m = 1, 2, \ldots, M$ complete datasets from Step 1.

Step 3 Obtain the adjusted estimates and variances from the imputations in Step 2 by

(a) adjusted parameter estimates $\widehat{\mu}_i = \sum_{m=1}^{M} \widehat{\mu}_{im}/M$ estimates of the parameter value

(b) adjusted parameter variance $\widehat{\sigma}^2 = \frac{\sum_{m=1}^{M} \widehat{\sigma}^2_m}{M} + \left[\frac{M+1}{M(M-1)}\right] \sum_{m=1}^{M} (\widehat{\mu}_{im} - \widehat{\mu}_i)^2$.

Estimates for $\widehat{\eta}_i$ and $\widehat{\tau}^2$ are obtained by following Steps 2 and 3 above by replacing μ with η, and σ with τ respectively.

Figure 11.2: Principle of multiple imputation

and 15% of the observations from (a) efficacy only; (b) safety only; and (c) efficacy and safety. We used the SAS procedure PRQEX3 to estimate the MCAR data. Then 700 bootstrap samples were generated to estimate the therapeutic window using the SD1 and SD2 procedures. Note that the doses here are linear. Next, we simulated MAR data by deleting 5%, 10%, and 15% of the observations from a particular dose for the three situations described above. Again, 700 bootstrap samples were generated to estimate the familywise error rate of the therapeutic window. The NORM program by Schafer (1997) was used to impute data when values were missing at random.

11.6 Results and Discussion

We first present the results for the familywise error rate for (11.1) in Table 11.1 for the different missing data mechanisms, and the bootstrap estimation procedures. The mechanism column indicates the missing data mechanism as applied to the efficacy, or safety, or both components of the dataset. The percent missing column indicates the amount of data that is missing from the dataset. The last column indicates the empirical familywise error rate from the booststrap SD1 (SD2) procedure. Thus if 10% of the data are missing at random from the safety column for a particular dose level, then average Type I familywise error rate from the SD1(SD2) bootstrap procedure is .076(.062) which is a bit higher than the true error rate of .05.

Table 11.1: Comparison of familywise error rate using bootstrap procedures in the presence of missing data

Mechanism	Percent Missing	Error SD1 (SD2)
MCAR [Efficacy]	5%	.034 (.020)
	10%	.048 (.039)
	15%	.062 (.054)
MCAR [Safety]	5%	.030 (.026)
	10%	.046 (.042)
	15%	.060 (.058)
MCAR [Safety and efficacy]	5%	.023 (.020)
	10%	.028 (.034)
	15%	.042 (.044)
MAR [Efficacy]	5%	.044 (.050)
	10%	.067 (.059)
	15%	.083 (.064)
MAR [Safety]	5%	.060 (.046)
	10%	.076 (.062)
	15%	.091 (.078)
MAR [Safety and efficacy]	5%	.043 (.040)
	10%	.091 (.067)
	15%	.112 (.089)

The results show that regardless of the type of missing data mechanism, the familywise error rate increases as the percent of missing data increases. Further, data missing at random has a higher familywise error rate than data missing completely at random. This is not surprising as we may be incorrectly estimating a particular dose if a large percent of missing data happened to be from that particular dose. Note that the error rates for the situation when data are missing completely at random from the concatenated safety and efficacy vector are the lowest. This may be an indication of ample power in the sample due to the reduced probability of eliminating both safety and efficacy data simultaneously from a particular dose response.

Tamhane and Logan (2002) showed that the SD2 bootstrap procedure provided a 5 percent advantage in power for a linear dose configuration as compared to the SD1 bootstrap procedure because it does not use the maximum of t-statistics in computing the adjusted p-values. We find that this advantage is demonstrated in most of the error rate values of Table 11.1.

The results from multiple imputations for the two missing data mechanisms

Table 11.2: Comparison of familywise error rate using bootstrap procedures after imputation of missing data

Mechanism	Percent Missing	error SD1 (SD2)
MCAR [Efficacy]	5%	.032 (.019)
	10%	.034 (.029)
	15%	.036 (.034)
MCAR [Safety]	5%	.030 (.024)
	10%	.032 (.030)
	15%	.041 (.033)
MCAR [Safety and efficacy]	5%	.021 (.020)
	10%	.028 (.024)
	15%	.022 (.024)
MAR [Efficacy]	5%	.024 (.020)
	10%	.027 (.019)
	15%	.033 (.021)
MAR [Safety]	5%	.031 (.020)
	10%	.026 (.022)
	15%	.031 (.018)
MAR [Safety and efficacy]	5%	.033 (.030)
	10%	.031 (.027)
	15%	.032 (.019)

are presented in Table 11.2. The most important difference is that imputation does markedly decrease the familywise error rates for estimating the therapeutic window. The imputations are especially helpful when data is missing at random, i.e., the missing data mechanism depends on a particular dose level. We also assessed the familywise error rate for step dose configuration where $\mu_0 = 0, \mu_1 = 1$ and $\mu_i = 2$ for $i = 2, 3, 4$, and $\eta_0 = 0, \eta_i = 1$ for $i = 1, 2, 3, 4$. The rest of the parameter specifications were identical to the linear dose configuration. The results were similar to those obtained in Tables 11.1 and 11.2 with one important difference: SD1 performed consistently better than SD2, i.e., had lower familywise error rates.

In conclusion, it is important to assess the amount of missing data at each dose level before proceeding to estimate the therapeutic window of a drug. Appropriate multiple imputation methods may be helpful in controlling the familywise error rate during estimation of the therapeutic window of a drug in the presence of missing data.

References

1. Little, R. J. A. and Rubin, D. B. (1987). *Statistical Analysis with Missing Data*, John Wiley & Sons, New York.

2. Schafer, J. L. (1997). *Analysis of Incomplete Multivariate Data*, Chapman and Hall, London.

3. Shekelle, P. G., Hardy, M. L., Maglione, M., Morton S. C. (2003). Efficacy and safety of ephedra and ephedrine for weight loss and athletic performance: A meta-analysis, *Journal of the American Medical Association*, **289**, 1537–1545.

4. Tamhane, A. C. and Logan, B. R. (2002). Multiple test procedures for indentifying the minimum effective and maximum safe doses of a drug, *Journal of the American Statistical Association*, **97**, 293–301.

5. Tamhane, A. C., Hochberg, Y., Dunnett, C. W. (1996). Multiple test procedures for dose finding, *Biometrics*, **52**, 21–37.

6. Tamhane, A. C., Dunnett, C. W., Green, J. W., Wetherington, J. F. (2001). Multiple test procedures for identifying the maximum safe dose, *Journal of the American Statistical Association*, **96**, 835–843.

7. Thall, P. F. and Cheng, S. C. (1999). Treatment comparisons based on two dimensional safety and efficacy alternatives in oncology trials, *Biometrics*, **55**, 746–753.

8. United States Food and Drug Administration. (2003). Ephedra and Ephedrine for Weight Loss and Atheletic Performance Enhancement: Clinical Efficacy and Side Effects. Report Number 76. U.S. Department of Health and Human Services. Agency for Healthcare Research and Quality.

12

Asymptotic Second-order Efficiency for Two-stage Multiple Comparisons with Components of a Linear Function of Mean Vectors

Makoto Aoshima[1] **and Takuya Kushida**[2]

[1] *Institute of Mathematics, University of Tsukuba, Ibaraki, Japan*
[2] *Graduate School of Pure and Applied Sciences, University of Tsukuba, Ibaraki, Japan*

Abstract: We consider an asymptotic second-order efficiency of two-stage procedures for multiple comparisons concerning components of a linear function of mean vectors from $N_p(\boldsymbol{\mu}_i, \boldsymbol{\Sigma}_i)$, $i = 1, ..., k$, where $\boldsymbol{\Sigma}_i$'s are unknown but spherical structures. The sample size is determined by using a two-stage procedure so that simultaneous confidence intervals for component contrasts will cover the true parameters with a guaranteed high accuracy and a specified narrow width. It is shown that no matter how the initial sample size is chosen, the two-stage multiple comparisons procedures do not become asymptotically second-order efficient even under the assumption that a known lower bound is available for a spherical parameter of each $\boldsymbol{\Sigma}_i$. An adjustment of the design constant and a proper choice of the initial sample size that occur in the two-stage procedure are proposed in order to have the asymptotic second-order efficiency. Numerical examples are presented to show how much sample the modified procedure saves in each of the multiple comparisons methods.

Keywords and phrases: Asymptotic consistency, exact consistency, multiple comparisons, sample size, second-order efficiency, sphericity, two-stage procedure

12.1 Introduction

Suppose that there exist k independent p (≥ 2)–variate normal populations $\pi_i : N_p(\boldsymbol{\mu}_i, \boldsymbol{\Sigma}_i)$, $i = 1, ..., k$, where $\boldsymbol{\mu}_i$ is unknown and $\boldsymbol{\Sigma}_i = (\sigma_{(i)rs})$ $(> \mathbf{0})$ is unknown, but has spherical structure, i.e., it holds that

$$\sigma_{(i)rr} + \sigma_{(i)ss} - 2\sigma_{(i)rs} = 2\tau_i^2 \quad (1 \leq r < s \leq p) \tag{12.1}$$

with τ_i (> 0) unknown. A special case of the spherical model is the intraclass correlation model such as $\boldsymbol{\Sigma}_i = \sigma_i^2\{(1-\rho_i)\boldsymbol{I}_p+\rho_i\boldsymbol{J}\}$ for some ρ_i, where $\boldsymbol{J}$ denotes a $p \times p$ matrix of all 1s. Let us consider a linear function of the mean vectors, namely, $\boldsymbol{\xi} = \sum_{i=1}^k b_i\boldsymbol{\mu}_i$, where b_i's are known and nonzero scalars. We are interested in conducting multiple comparisons experiments concerning its correlated components $(\xi_1, ..., \xi_p)$ as p treatments along the lines of Tukey's (1953) method of all pairwise multiple comparisons (MCA), Hsu's (1984) method of multiple comparisons with the best (MCB), and Dunnett's (1955) method of multiple comparisons with a control (MCC). This type of multiple comparisons is motivated by the situation in which the user would be typically interested in both the direction and magnitude of differences concerning the correlated p time components $(\xi_1, ..., \xi_p)$ among several populations. See Hochberg and Tamhane (1987).

Let $\boldsymbol{X}_{i1}, \boldsymbol{X}_{i2}, ...$ be a sequence of i.i.d. random vectors for each π_i. Having recorded $\boldsymbol{X}_{i1}, ..., \boldsymbol{X}_{in_i}$ from each π_i, let us write $\overline{\boldsymbol{X}}_{in_i} = n_i^{-1}\sum_{j=1}^{n_i}\boldsymbol{X}_{ij}$ and define $\boldsymbol{Y}_{\boldsymbol{n}n} = \sum_{i=1}^k b_i\overline{\boldsymbol{X}}_{in_i}$ with $\boldsymbol{n} = (n_1, ..., n_k)$. Then, as for three methods stated above, for specified $d > 0$ (suitably narrow), the following simultaneous confidence intervals are defined by using $\boldsymbol{Y}_{\boldsymbol{n}} = (Y_{1\boldsymbol{n}}, ..., Y_{p\boldsymbol{n}})$:

(MCA) For the $p(p-1)/2$ differences of component effects,

$$R_{\boldsymbol{n}} = \{\boldsymbol{\xi} |\ \xi_r - \xi_s \in [Y_{r\boldsymbol{n}} - Y_{s\boldsymbol{n}} - d,\ Y_{r\boldsymbol{n}} - Y_{s\boldsymbol{n}} + d],\ 1 \le r < s \le p\};$$

(MCB) For comparing each component with the best of the other components when a larger component effect is supposed to be better,

$$R_{\boldsymbol{n}} = \{\boldsymbol{\xi} |\ \xi_r - \max_{s\neq r}\xi_s \in [(Y_{r\boldsymbol{n}} - \max_{s\neq r}Y_{s\boldsymbol{n}} - d)^-,\ (Y_{r\boldsymbol{n}} - \max_{s\neq r}Y_{s\boldsymbol{n}} + d)^+],$$
$$r = 1, ..., p\},$$

where $x^+ = \max\{0, x\}$ and $x^- = \min\{0, x\}$;

(MCC) For comparing each component with a control component,

$$R_{\boldsymbol{n}} = \{\boldsymbol{\xi} |\ \xi_r - \xi_p \in [Y_{r\boldsymbol{n}} - Y_{p\boldsymbol{n}} - d,\ Y_{r\boldsymbol{n}} - Y_{p\boldsymbol{n}} + d],\ r = 1, ..., p-1\},$$

where the component p is supposed to be the control.

Note that MCB implies the inference of both the indifference-zone and the subset selection methodologies for ranking and selection of the best component. See Hsu (1984, 1996) for the details. For each method, we want to construct $R_{\boldsymbol{n}}$ such that

$$P_{\boldsymbol{\theta}}(\boldsymbol{\xi} \in R_{\boldsymbol{n}}) \ge 1 - \alpha \quad \text{for all } \boldsymbol{\theta}, \tag{12.2}$$

where $\boldsymbol{\theta} = (\boldsymbol{\mu}_1, ..., \boldsymbol{\mu}_k, \boldsymbol{\Sigma}_1, ..., \boldsymbol{\Sigma}_k)$ with $\boldsymbol{\Sigma}_i$s satisfying (12.1), and d (> 0) and $\alpha \in (0, 1)$ are both prespecified.

It is shown for MCA and MCC that

$$P_{\boldsymbol{\theta}}(\boldsymbol{\xi} \in R_{\boldsymbol{n}}) = G_p\left(\frac{d^2}{\sum_{i=1}^k b_i^2\tau_i^2/n_i}\right), \tag{12.3}$$

where $G_p(y)$ for $y > 0$ is defined by

$$G_p(y) = p\int_{-\infty}^{\infty}\{\Phi(x) - \Phi(x-\sqrt{y})\}^{p-1}d\Phi(x) \quad \text{for MCA,} \qquad (12.4)$$

$$G_p(y) = \int_{-\infty}^{\infty}\{\Phi(x+\sqrt{y}) - \Phi(x-\sqrt{y})\}^{p-1}d\Phi(x) \quad \text{for MCC} \quad (12.5)$$

with $\Phi(\cdot)$ representing a $N(0,1)$ c.d.f. It is shown for MCB that

$$P_{\boldsymbol{\theta}}(\boldsymbol{\xi} \in R_{\boldsymbol{n}}) \geq G_p\left(\frac{d^2}{\sum_{i=1}^k b_i^2\tau_i^2/n_i}\right), \qquad (12.6)$$

where

$$G_p(y) = \int_{-\infty}^{\infty}\{\Phi(x+\sqrt{y})\}^{p-1}d\Phi(x). \qquad (12.7)$$

[See Aoshima (2001) for the details.] So, the sample sizes $\boldsymbol{n}$ for MCA, MCB and MCC which minimize the sum $\sum_{i=1}^k n_i$ while satisfying requirement (12.2) are given as the smallest integer such that

$$n_i \geq \frac{a}{d^2}|b_i|\tau_i\sum_{j=1}^k|b_j|\tau_j \quad = n_i^\star \text{ (say)} \qquad (12.8)$$

for each π_i, where a (> 0) is the constant such that $G_p(a) = 1-\alpha$ with $G_p(\cdot)$ defined by (12.4), (12.5) or (12.7) accordingly for each method. Since τ_i's are unknown, Aoshima (2001) proposed the following Stein (1945)–type two-stage procedure to determine the sample sizes $\boldsymbol{n}$ for each method.

Two-stage procedure

(T1) First, take a pilot sample $\boldsymbol{X}_{ij} = (X_{ij1},...,X_{ijp})$, $j = 1,...,m$, of size m (≥ 2) and calculate $S_i^2 = \nu^{-1}\sum_{r=1}^p\sum_{j=1}^m(X_{ijr} - \overline{X}_{ij.} - \overline{X}_{i.r} + \overline{X}_{i..})^2$ with $\nu = (p-1)(m-1)$ as an estimate of τ_i^2 for each π_i. Here, $\overline{X}_{ij.} = p^{-1}\sum_{r=1}^p X_{ijr}$, $\overline{X}_{i.r} = m^{-1}\sum_{j=1}^m X_{ijr}$ and $\overline{X}_{i..} = (pm)^{-1}\sum_{r=1}^p\sum_{j=1}^m X_{ijr}$. Define the total sample size of each π_i by

$$N_i = \max\left\{m, \left[\frac{t_\nu}{d^2}|b_i|S_i\sum_{j=1}^k|b_j|S_j\right]+1\right\}, \qquad (12.9)$$

where $[x]$ denotes the largest integer less than x and $t_\nu(> 0)$ is a design constant. Aoshima (2001) suggested that one may determine t_ν as the solution $t = t_\nu$ to the equation

$$k\int_0^\infty G_p(ty/\nu)\,(1-F_\nu(y))^{k-1}\,dF_\nu(y) = 1-\alpha, \qquad (12.10)$$

where $G_p(\cdot)$ is defined by (12.4), (12.5) or (12.7) according to method and $F_r(\cdot)$ denotes a chi–squared distribution function with r degrees of freedom (d.f.).

(T2) Next, take an additional sample $\boldsymbol{X}_{i(m+1)}, ..., \boldsymbol{X}_{iN_i}$ of size $N_i - m$ for each π_i. Let $\boldsymbol{N} = (N_1, ..., N_k)$. By combining the initial sample and the additional sample, calculate $\overline{\boldsymbol{X}}_{iN_i} = N_i^{-1} \sum_{j=1}^{N_i} \boldsymbol{X}_{ij}$ for each π_i. Then, Aoshima (2001) showed that the simultaneous confidence intervals $R_{\boldsymbol{N}}$ based on the components $(Y_{1\boldsymbol{N}}, ..., Y_{p\boldsymbol{N}})$ of $\boldsymbol{Y}_{\boldsymbol{N}} = \sum_{i=1}^{k} b_i \overline{\boldsymbol{X}}_{iN_i}$ satisfies requirement (12.2) *(exact consistency)*.

Here, we consider asymptotic second-order properties of the above two-stage procedure. If it holds, for any $\boldsymbol{\theta}$, that

$$\limsup_{d \to 0} E_{\boldsymbol{\theta}} \left(\sum_{i=1}^{k} N_i - \sum_{i=1}^{k} n_i^\star \right) < \infty,$$

the two-stage procedure is asymptotically second-order efficient. Aoshima (2001) showed that in the case that $k = 1$ if it is possible to assume a known and positive lower bound for the spherical parameter τ, a proper choice of the initial sample size m makes the two-stage procedure asymptotically second-order efficient. However, it will turn out here that when $k \geq 2$, no matter how the initial sample size m is chosen even under the assumption that a known and positive lower bound is available for each τ_i, the two-stage procedure cannot become asymptotically second-order efficient. To overcome this disadvantage, we shall propose an adjustment of the design constant t_ν in (12.9) so that a proper choice of m makes the two-stage procedure asymptotically second-order efficient under the assumption that there exists a known and positive lower bound for each τ_i. Numerical examples are given to illustrate how the proposed modification works well to reduce the sample size required in each multiple comparisons method. Proofs of all the results are postponed to the Appendix.

12.2 Main Results

When $k = 1$, Aoshima (2001) showed that if it is possible to assume a known and positive lower bound for the spherical parameter τ, a proper choice of the initial sample size m makes the two-stage procedure (12.9) with (12.10) asymptotically second-order efficient. The following theorem shows that it cannot possess that property if $k \geq 2$.

Theorem 12.2.1 *When $k \geq 2$, the two-stage procedure (12.9) with (12.10) cannot become asymptotically second-order efficient no matter how the initial sample size m is chosen even under the assumption that a known and positive lower bound is available for the spherical parameter τ_i.*

When $k \geq 2$, instead of using the design constant t_ν given by (12.10), we shall propose an adjustment of t_ν to make the two-stage procedure asymptotically second-order efficient.

Theorem 12.2.2 *(i) For MCA and MCC in the cases that $p = 2, 3$ and for MCB in the case that $p \leq 45$, define the design constant t_ν in (12.9) as the solution $t = t_\nu$ to the equation*

$$k \int_0^\infty G_p(ty/\nu) dF_\nu(y) - (k-1) \int_0^\infty G_p(ty/\nu) dF_{\nu+1}(y) = 1 - \alpha, \qquad (12.11)$$

where $\nu = (p-1)(m-1)$ and $G_p(\cdot)$ is defined by (12.4), (12.5) or (12.7) as appropriate. Then, the two-stage procedure satisfies requirement (12.2) (exact consistency).
(ii) Otherwise, define t_ν by

$$t_\nu = a + \frac{a}{\nu} \left(k - 1 - \frac{a G_p''(a)}{G_p'(a)} \right) \qquad (12.12)$$

with $G_p'(a) = \frac{d}{dy} G_p(y)|_{y=a}$ and so on, where a is the constant such that $G_p(a) = 1 - \alpha$. Then, the two-stage procedure possesses the asymptotic consistency as $\nu \to \infty$, i.e.,

$$P_{\boldsymbol{\theta}}(\boldsymbol{\xi} \in R_{\boldsymbol{N}}) \geq 1 - \alpha + o(\nu^{-1}) \quad \text{for all } \boldsymbol{\theta}. \qquad (12.13)$$

Remark 1 It is easy to see that the right-hand side in (12.24) (in the Appendix) is asymptotically expanded as in (12.27). Then, we notice that the design constant t_ν defined by (12.11) is coincident with that defined by (12.12) up to the order $O(\nu^{-1})$. When $k = 1$, the expression (12.11) is the same as (12.10).

Table 12.1 gives the values of a and $aG_p''(a)/G_p'(a)$ required in the calculation of the formula (12.12) for each method when $p = 2(1)10$ and $\alpha = .10, .05$. When $p = 2, 3$, Table 12.2 gives the value of t_ν for MCA and MCC for $\alpha = .10$, $k = 2(1)5$ and $m = 10(10)50$ by solving (12.11) numerically. When $p = 2(1)5$, Table 12.3 gives the value of t_ν for MCB for $\alpha = .10$, $k = 2(1)5$ and $m = 10(10)50$ by solving (12.11) numerically. In both Tables 12.2 and 12.3, the corresponding value of t_ν computed by (12.12) using Table 12.1 is given within parentheses in each cell.

We observe from Tables 12.2 - 12.3 that when $p = 2, 3$, the expansion formula (12.12) gives a good approximation to the value of t_ν defined as the solution to the equation (12.11) when m or ν is large. As for MCB, especially

Table 12.1: Values of a and $aG''_p(a)/G'_p(a)$ in (12.12)

	MCA		MCB		MCC	
p	a	$\frac{aG''_p(a)}{G'_p(a)}$	a	$\frac{aG''_p(a)}{G'_p(a)}$	a	$\frac{aG''_p(a)}{G'_p(a)}$
			$\alpha = .10$			
2	5.41109	-1.85277	3.28475	-1.32119	5.41109	-1.85277
3	8.42381	-2.29929	4.97379	-1.59689	7.34419	-2.19464
4	10.50049	-2.60623	6.01019	-1.75086	8.50398	-2.38784
5	12.09844	-2.84397	6.75846	-1.85588	9.33375	-2.52026
6	13.40088	-3.03912	7.34385	-1.93475	9.97960	-2.62002
7	14.50161	-3.20509	7.82445	-1.99748	10.50807	-2.69952
8	15.45551	-3.34971	8.23196	-2.04931	10.95510	-2.76533
9	16.29756	-3.47798	8.58558	-2.09332	11.34233	-2.82127
10	17.05150	-3.59328	8.89785	-2.13146	11.68378	-2.86981
			$\alpha = .05$			
2	7.68292	-2.42073	5.41109	-1.85277	7.68292	-2.42073
3	10.98586	-2.98420	7.34466	-2.19793	9.78702	-2.82086
4	13.19985	-3.36242	8.50438	-2.39105	11.03533	-3.04585
5	14.88151	-3.65142	9.33408	-2.52314	11.92449	-3.20027
6	16.24164	-3.88649	9.97988	-2.62258	12.61489	-3.31681
7	17.38518	-4.08507	10.50831	-2.70183	13.17894	-3.40987
8	18.37245	-4.25720	10.95531	-2.76742	13.65556	-3.48702
9	19.24146	-4.40921	11.34251	-2.82319	14.06807	-3.55271
10	20.01779	-4.54539	11.68395	-2.87159	14.43159	-3.60977

Table 12.2: Values of t_ν for MCA and MCC when $\alpha = .10$

p	m	k = 2	3	4	5
MCA	10	7.555 (7.126)	8.393 (7.727)	9.217 (8.329)	10.017 (8.930)
	20	6.312 (6.224)	6.645 (6.508)	6.977 (6.793)	7.307 (7.078)
2	30	5.980 (5.943)	6.187 (6.130)	6.394 (6.317)	6.600 (6.503)
	40	5.827 (5.807)	5.977 (5.946)	6.127 (6.084)	6.277 (6.223)
	50	5.739 (5.726)	5.856 (5.837)	5.974 (5.947)	6.091 (6.057)
	10	10.161 (9.968)	10.715 (10.436)	11.253 (10.904)	11.775 (11.372)
	20	9.197 (9.155)	9.436 (9.377)	9.674 (9.599)	9.908 (9.820)
3	30	8.921 (8.903)	9.073 (9.048)	9.225 (9.193)	9.376 (9.339)
	40	8.790 (8.780)	8.902 (8.888)	9.014 (8.996)	9.125 (9.104)
	50	8.713 (8.707)	8.802 (8.793)	8.890 (8.879)	8.978 (8.965)
MCC	10	7.555 (7.126)	8.393 (7.727)	9.217 (8.329)	10.017 (8.930)
	20	6.312 (6.224)	6.645 (6.508)	6.977 (6.793)	7.307 (7.078)
2	30	5.980 (5.943)	6.187 (6.130)	6.394 (6.317)	6.600 (6.503)
	40	5.827 (5.807)	5.977 (5.946)	6.127 (6.084)	6.277 (6.223)
	50	5.739 (5.726)	5.856 (5.837)	5.974 (5.947)	6.091 (6.057)
	10	8.811 (8.648)	9.293 (9.056)	9.766 (9.464)	10.226 (9.872)
	20	7.997 (7.962)	8.206 (8.155)	8.413 (8.348)	8.619 (8.541)
3	30	7.764 (7.749)	7.897 (7.875)	8.030 (8.002)	8.162 (8.129)
	40	7.653 (7.645)	7.751 (7.739)	7.848 (7.833)	7.946 (7.927)
	50	7.589 (7.584)	7.666 (7.659)	7.743 (7.733)	7.820 (7.808)

Table 12.3: Values of t_ν for MCB when $\alpha = .10$

p	m	k = 2	3	4	5
2	10	4.300 (4.132)	4.805 (4.497)	5.329 (4.862)	5.865 (5.227)
	20	3.721 (3.686)	3.923 (3.859)	4.131 (4.032)	4.343 (4.205)
	30	3.562 (3.548)	3.688 (3.661)	3.816 (3.774)	3.946 (3.887)
	40	3.488 (3.480)	3.579 (3.564)	3.672 (3.649)	3.765 (3.733)
	50	3.445 (3.440)	3.517 (3.507)	3.589 (3.574)	3.662 (3.641)
3	10	5.765 (5.691)	6.091 (5.968)	6.419 (6.244)	6.749 (6.520)
	20	5.330 (5.314)	5.471 (5.445)	5.614 (5.575)	5.757 (5.706)
	30	5.203 (5.196)	5.294 (5.282)	5.384 (5.368)	5.476 (5.454)
	40	5.143 (5.139)	5.209 (5.203)	5.276 (5.267)	5.343 (5.331)
	50	5.108 (5.106)	5.160 (5.156)	5.213 (5.207)	5.265 (5.258)
4	10	6.665 (6.623)	6.914 (6.845)	7.163 (7.068)	7.412 (7.290)
	20	6.310 (6.300)	6.421 (6.406)	6.532 (6.511)	6.644 (6.617)
	30	6.204 (6.200)	6.276 (6.269)	6.347 (6.338)	6.419 (6.407)
	40	6.154 (6.152)	6.206 (6.203)	6.259 (6.254)	6.312 (6.306)
	50	6.124 (6.123)	6.166 (6.164)	6.208 (6.204)	6.249 (6.245)
5	10	7.323 (7.295)	7.527 (7.482)	7.731 (7.670)	7.934 (7.858)
	20	7.019 (7.012)	7.111 (7.101)	7.204 (7.190)	7.296 (7.279)
	30	6.928 (6.925)	6.987 (6.983)	7.047 (7.041)	7.107 (7.100)
	40	6.884 (6.882)	6.928 (6.926)	6.972 (6.969)	7.016 (7.012)
	50	6.858 (6.857)	6.893 (6.891)	6.928 (6.926)	6.963 (6.960)

when $4 \le p \le 45$, the formula (12.12) would give a fairly good approximation to the value of t_ν with sufficient accuracy.

Now, we shall show the asymptotic second-order efficiency of the two-stage procedure (12.9) with (12.11) or (12.12). To derive asymptotic properties up to the second order, it is crucial to assume that there exists a known lower bound $\tau_{i\star}$ (> 0) for τ_i. [See Mukhopadhyay and Duggan (1997).] We assume that

$$\tau_i > \tau_{i\star}, \quad i = 1, ..., k. \tag{12.14}$$

Theorem 12.2.3 *Define the initial sample size m by*

$$m = \max\left\{m_0, \left[\frac{a}{d^2}\min_{1\le i\le k}|b_i|\tau_{i\star}\sum_{j=1}^{k}|b_j|\tau_{j\star}\right] + 1\right\} \tag{12.15}$$

with m_0 (≥ 2) fixed. Then, the two-stage procedure (12.9) with t_ν defined by (12.11) or (12.12) asserts for any $\boldsymbol{\theta}$ satisfying (12.14) that

$$\lim_{d\to 0} E_{\boldsymbol{\theta}}(N_i - n_i^\star) = \frac{\left(2k - 3 - \frac{2aG_p''(a)}{G_p'(a)}\right)\sum_{j=1}^{k}|b_i||b_j|\tau_i\tau_j + b_i^2\tau_i^2}{2(p-1)\min_{1\le i\le k}|b_i|\tau_{i\star}\sum_{j=1}^{k}|b_j|\tau_{j\star}} + \frac{1}{2} \tag{12.16}$$

for each π_i, and hence

$$\lim_{d\to 0} E_{\boldsymbol{\theta}}\left(\sum_{i=1}^{k}N_i - \sum_{i=1}^{k}n_i^\star\right) = \frac{\left(2k - 3 - \frac{2aG_p''(a)}{G_p'(a)}\right)\left(\sum_{i=1}^{k}|b_i|\tau_i\right)^2 + \sum_{i=1}^{k}b_i^2\tau_i^2}{2(p-1)\min_{1\le i\le k}|b_i|\tau_{i\star}\sum_{j=1}^{k}|b_j|\tau_{j\star}} + \frac{k}{2}. \tag{12.17}$$

By using (12.16)–(12.17) for each method, one may estimate the discrepancies between the expected and observed sample sizes. The specification of the lower bound $\tau_{i\star}$s is crucial to reduce the required sample sizes. By observing the value of $aG_p''(a)/G_p'(a)$ tabulated in Table 12.1, we notice that MCB has the most significant improvement when $d \to 0$.

12.3 Moderate Sample Performances

In order to study the performance of the modified two-stage procedure (12.9)–(12.15) with (12.11) or (12.12), we resort to computer simulations. We fix $p = 4$, $k = 2$ and $(b_1, b_2) = (1, -1)$. Independent pseudo-random normal observations from $\pi_i : N_4(\boldsymbol{\mu}_i, \boldsymbol{\Sigma}_i)$, $i = 1, 2$, were generated where $\boldsymbol{\mu}_i$s were fixed as $\boldsymbol{\mu}_1 = (0.1, 0.1, 0.1, 0.1)'$, $\boldsymbol{\mu}_2 = (0.1, 0.2, 0.3, 0.4)'$, and $\boldsymbol{\Sigma}_i$s were fixed as

$\boldsymbol{\Sigma}_1 = \frac{4}{3}(\frac{3}{4}\boldsymbol{I}_4 + \frac{1}{4}\boldsymbol{J})$, $\boldsymbol{\Sigma}_2 = \frac{1}{3}(\frac{3}{4}\boldsymbol{I}_4 + \frac{1}{4}\boldsymbol{J})$ with $\boldsymbol{J}$ a 4×4 matrix of all 1s. That is, $\tau_1 = 1$ and $\tau_2 = 0.5$.

Tables 12.4, 12.5 and 12.6 present simulated results for MCA, MCB and MCC, respectively. Our goal is to construct 90% fixed–width simultaneous confidence intervals for the components of $\boldsymbol{\xi} = \boldsymbol{\mu}_1 - \boldsymbol{\mu}_2$. In other words, we have $\alpha = .10$. Then, from Table 12.1, $a = 10.50049$ (MCA), $a = 6.01019$ (MCB) and $a = 8.50398$ (MCC). Let $n^\star$ $(= n_1^\star + n_2^\star)$ be 100, 200 or 300, whereas with a fixed value of $n^\star$ one easily obtains from (12.8) that d= .486, .344 or .281 for Table 12.4, d= .368, .260 or .212 for Table 12.5 and d= .437, .309, or .253 for Table 12.6.

In Tables 12.4–12.6, the findings obtained by averaging the outcomes from 10,000 (= R, say) replications are summarized in each situation. Under a fixed scenario, suppose that the rth replication ends with $N_i = n_{ir}$ $(i = 1, 2)$ observations and the corresponding fixed–width simultaneous confidence intervals $R\boldsymbol{n}_r$ based on $\boldsymbol{n}_r = (n_{1r}, n_{2r})$ for $r = 1, ..., R$. Now, $\overline{n}_i = R^{-1}\sum_{r=1}^R n_{ir}$ which estimates $n_i^\star$ with its estimated standard error $s(\overline{n}_i)$ where $s^2(\overline{n}_i) = (R^2 - R)^{-1}\sum_{r=1}^R(n_{ir} - \overline{n}_i)^2$, $i = 1, 2$. Then, $\overline{n}$ $(= \overline{n}_1 + \overline{n}_2)$ estimates the total fixed–sample–size $n^\star$ $(= n_1^\star + n_2^\star)$ with its estimated standard error $s(\overline{n})$, computed analogously. In the end of the rth replication, we also check whether $\boldsymbol{\xi}$ belongs to the constructed simultaneous confidence intervals $R\boldsymbol{n}_r$ and define p_r=1 (or 0) according as $\boldsymbol{\xi}$ does (or does not) belong to $R\boldsymbol{n}_r$, $r = 1, ..., R$. Let $\overline{p} = R^{-1}\sum_{r=1}^R p_r$ which estimates the target coverage probability, having its estimated standard error $s(\overline{p})$ where $s^2(\overline{p}) = R^{-1}\overline{p}(1 - \overline{p})$.

We studied the performance of the modified two-stage procedure (12.9)–(12.15) with (12.12) for MCA and MCC or with (12.11) for MCB, under the additional assumption that we knew some positive lower bounds for the unknown τ_i's. We considered two situations: With the previous setup of $\boldsymbol{\Sigma}_i$s, we had fixed two separate cases for Tables 12.4–12.6: $(\tau_{1\star}, \tau_{2\star}) = (0.6, 0.3)$ and $(\tau_{1\star}, \tau_{2\star}) = (0.9, 0.45)$. For the first choice, the known lower bounds are far off from the assumed true values whereas in the case of the second choice, the known lower bounds are much closer to the true entries.

Let us explain, for example, the entries from the third block in the first part of Table 12.4 (MCA). We consider $n^\star = 300$ and hence $d = .281$, $n_1^\star = 200$, $n_2^\star = 100$ from (12.8). One obtains $m = 36$ using (12.15) (having $m_0 = 10$, for example) and thus $t_\nu = 10.861$ using (12.12). Then, we also gave the corresponding value due to Aoshima's procedure (12.9) with (12.10) within the parenthesis. With the initial sample of size $m = 36$, one obtains $t_\nu = 11.591$ using (12.10) in Aoshima's procedure. Next, from 10,000 independent simulations, we observed $\overline{n}_1 = 206.98$ (220.86), $s(\overline{n}_1) = .241$ (.257), $\overline{n}_2 = 103.52$ (110.45), $s(\overline{n}_2) = .105$ (.112) and $\overline{n} = 310.50$ (331.31), $s(\overline{n}) = .316$ (.337). Also, we had $\overline{p} = .9066$ (.9235), $s(\overline{p}) = .00291$ (.00266) and $\overline{n}_1 - n_1^\star = 6.98$ (20.86), $\overline{n}_2 - n_2^\star = 3.52$ (10.45), $\overline{n} - n^\star = 10.50$ (31.31). The values within parentheses are those corresponding values due to Aoshima's procedure (12.9) with (12.10) as before.

One has $C_1 = 6.87$, $C_2 = 3.53$ ($C = C_1 + C_2 = 10.40$) which express the values of the right side in (12.16) for $i = 1, 2$. We note that $\overline{n}_i$ and $\overline{n}$ are closer to $n_i^\star$ and $n^\star$, respectively, and that their estimated standard errors have gone down compared with the entries within parentheses for Aoshima's procedure. Theorem 12.2.3 indicates that one may expect $\overline{n}_i - n_i^\star$ to fall in the vicinity of the value of C_i, $i = 1, 2$. Our experimental evidence favors that sentiment. One will observe that the values of $E_{\boldsymbol{\theta}}(N_i - n_i^\star)$ are approximated fairly well by these asymptotic values for small d. We clearly observe that the extent of over-sampling has reduced when we moved from the first set of $\tau_{i\star}$s to the next in Table 12.4 as expected in Theorem 12.2.3. The specification of the lower bounds is crucial to reduce the sample size from each π_i. Throughout, by considering both $\overline{p}$ and $s(\overline{p})$ values, we may claim that the modified procedure (12.9)–(12.15) with (12.11) or (12.12) seems to satisfy requirement (12.2) well for both small and moderate values of $n^\star$ and $n_i^\star$s under consideration.

12.4 Concluding Remarks

The asymptotic expressions from Theorem 12.2.3 provide useful guidelines by explaining the discrepancies between the expected and observed results, even when the total fixed-sample-size $n^\star$ is not very large. We have reduced the amount of over-sampling significantly by considering an adjustment of the design constant in Theorem 12.2.2. Section 12.3 includes encouraging findings based on extensive computer simulations.

To achieve the goal described above, we have simply used a common number for the initial sample size of all π_i's by taking $\min_{1 \le i \le k} |b_i| \tau_{i\star}$ in (12.15). However, once we focus on the reduction of the sample size, that choice of the initial sample size might not be always an optimal one. The initial sample size should be determined for each π_i by reflecting on each prior information through $\tau_{i\star}$ and may be a different number for each π_i. See Aoshima and Miyajima (2001) for such an attempt.

Appendix

The following lemma is crucial for the proof of Theorem 12.2.1.

Lemma 12.4.1 *The design constant t_ν in (12.9) with (12.10) is asymptotically*

Table 12.4: Simulated results for MCA

		d	m	t_ν	$\overline{n}$	$s(\overline{n})$	$\overline{p}$	$s(\overline{p})$	$\overline{n}-n^\star$
		$\tau_{1\star}=0.6,\ \tau_{2\star}=0.3,\ C=10.40,\ C_1=6.87,\ C_2=3.53$							
$n^\star$	100	.486		11.648	110.79	.200	.9147	.00279	10.79
				(12.932)	(122.89)	(.222)	(.9329)	(.00250)	(22.89)
$n_1^\star$	66.67		12		73.83	.153			7.17
					(81.92)	(.169)			(15.25)
$n_2^\star$	33.33		12		36.96	.067			3.62
					(40.97)	(.074)			(7.64)
$n^\star$	200	.344		11.049	210.47	.266	.9087	.00288	10.47
				(11.938)	(227.32)	(.287)	(.9228)	(.00267)	(27.32)
$n_1^\star$	133.33		24		140.23	.203			6.90
					(151.48)	(.219)			(18.15)
$n_2^\star$	66.67		24		70.23	.088			3.57
					(75.84)	(.096)			(9.17)
$n^\star$	300	.281		10.861	310.50	.316	.9066	.00291	10.50
				(11.591)	(331.31)	(.337)	(.9235)	(.00266)	(31.31)
$n_1^\star$	200.00		36		206.98	.241			6.98
					(220.86)	(.257)			(20.86)
$n_2^\star$	100.00		36		103.52	.105			3.52
					(110.45)	(.112)			(10.45)
		$\tau_{1\star}=0.9,\ \tau_{2\star}=0.45,\ C=5.18,\ C_1=3.33,\ C_2=1.85$							
$n^\star$	100	.486		10.986	105.21	.124	.9042	.00294	5.21
				(11.824)	(113.14)	(.133)	(.9233)	(.00266)	(13.14)
$n_1^\star$	66.67		27		70.00	.094			3.33
					(75.30)	(.102)			(8.63)
$n_2^\star$	33.33		27		35.22	.041			1.88
					(37.84)	(.044)			(4.51)
$n^\star$	200	.344		10.739	205.31	.171	.9004	.00299	5.31
				(11.341)	(216.76)	(.181)	(.9155)	(.00278)	(16.76)
$n_1^\star$	133.33		54		136.79	.130			3.46
					(144.44)	(.137)			(11.10)
$n_2^\star$	66.67		54		68.52	.057			1.85
					(72.33)	(.060)			(5.66)
$n^\star$	300	.281		10.658	305.33	.206	.9027	.00296	5.33
				(11.155)	(319.50)	(.216)	(.9138)	(.00281)	(19.50)
$n_1^\star$	200.00		81		203.47	.157			3.47
					(212.93)	(.164)			(12.93)
$n_2^\star$	100.00		81		101.86	.069			1.86
					(106.58)	(.072)			(6.58)

Table 12.5: Simulated results for MCB

		d	m	t_ν	$\overline{n}$	$s(\overline{n})$	$\overline{p}$	$s(\overline{p})$	$\overline{n}-n^\star$
		$\tau_{1\star}=0.6,\ \tau_{2\star}=0.3,\ C=8.02,\ C_1=5.29,\ C_2=2.74$							
$n^\star$	100	.368		6.540	108.83	.198	.9355	.00246	8.83
				(7.257)	(120.66)	(.220)	(.9453)	(.00227)	(20.66)
$n_1^\star$	66.67		12		72.55	.151			5.88
					(80.45)	(.167)			(13.79)
$n_2^\star$	33.33		12		36.28	.066			2.95
					(40.21)	(.073)			(6.87)
$n^\star$	200	.260		6.256	208.50	.264	.9227	.00267	8.50
				(6.771)	(225.56)	(.286)	(.9337)	(.00249)	(25.56)
$n_1^\star$	133.33		24		139.04	.201			5.71
					(150.43)	(.218)			(17.10)
$n_2^\star$	66.67		24		69.45	.088			2.79
					(75.12)	(.095)			(8.46)
$n^\star$	300	.212		6.170	308.47	.317	.9181	.00274	8.47
				(6.595)	(329.63)	(.338)	(.9291)	(.00257)	(29.63)
$n_1^\star$	200.00		36		205.72	.240			5.72
					(219.84)	(.257)			(19.84)
$n_2^\star$	100.00		36		102.75	.106			2.75
					(109.79)	(.113)			(9.79)
		$\tau_{1\star}=0.9,\ \tau_{2\star}=0.45,\ C=4.12,\ C_1=2.63,\ C_2=1.49$							
$n^\star$	100	.368		6.227	104.33	.124	.9339	.00248	4.33
				(6.714)	(112.39)	(.134)	(.9422)	(.00233)	(12.39)
$n_1^\star$	66.67		27		69.47	.094			2.81
					(74.87)	(.102)			(8.20)
$n_2^\star$	33.33		27		34.85	.041			1.52
					(37.52)	(.044)			(4.18)
$n^\star$	200	.260		6.115	204.35	.170	.9213	.00269	4.35
				(6.466)	(215.99)	(.180)	(.9279)	(.00259)	(15.99)
$n_1^\star$	133.33		54		136.18	.130			2.85
					(143.96)	(.137)			(10.63)
$n_2^\star$	66.67		54		68.16	.056			1.50
					(72.04)	(.060)			(5.37)
$n^\star$	300	.212		6.080	303.97	.208	.9189	.00273	3.97
				(6.369)	(318.39)	(.218)	(.9268)	(.00260)	(18.39)
$n_1^\star$	200.00		81		202.55	.158			2.55
					(212.18)	(.165)			(12.18)
$n_2^\star$	100.00		81		101.41	.069			1.41
					(106.22)	(.073)			(6.22)

We observed that the fixed-sample procedure (12.8) of size $n^\star = 100$, with (12.7) guaranteeing the inequality (12.6), had $\overline{p} = .9272$ ($s(\overline{p}) = .00260$). When $n^\star = 200$ or 300, it was $\overline{p} = .9175$ ($s(\overline{p}) = .00275$) or $\overline{p} = .9143$ ($s(\overline{p}) = .00280$).

Table 12.6: Simulated results for MCC

		d	m	t_ν	$\overline{n}$	$s(\overline{n})$	$\overline{p}$	$s(\overline{p})$	$\overline{n}-n^\star$
$\tau_{1\star}=0.6,\ \tau_{2\star}=0.3,\ C=9.79,\ C_1=6.47,\ C_2=3.33$									
$n^\star$	100	.437		9.377	110.64	.203	.9136	.00281	10.64
				(10.423)	(122.88)	(.226)	(.9316)	(.00252)	(22.88)
$n_1^\star$	66.67		12		73.75	.154			7.09
					(81.93)	(.171)			(15.26)
$n_2^\star$	33.33		12		36.89	.068			3.56
					(40.95)	(.076)			(7.62)
$n^\star$	200	.309		8.922	210.31	.266	.9074	.00290	10.31
				(9.646)	(227.31)	(.287)	(.9243)	(.00265)	(27.31)
$n_1^\star$	133.33		24		140.18	.202			6.85
					(151.52)	(.218)			(18.19)
$n_2^\star$	66.67		24		70.14	.089			3.47
					(75.79)	(.096)			(9.13)
$n^\star$	300	.253		8.778	310.45	.317	.9049	.00293	10.45
				(9.373)	(331.41)	(.339)	(.9208)	(.00270)	(31.41)
$n_1^\star$	200.00		36		206.95	.243			6.95
					(220.93)	(.259)			(20.93)
$n_2^\star$	100.00		36		103.50	.105			3.50
					(110.48)	(.112)			(10.48)
$\tau_{1\star}=0.9,\ \tau_{2\star}=0.45,\ C=4.91,\ C_1=3.15,\ C_2=1.76$									
$n^\star$	100	.437		8.873	105.14	.124	.9092	.00287	5.14
				(9.557)	(113.14)	(.134)	(.9260)	(.00262)	(13.14)
$n_1^\star$	66.67		27		69.98	.095			3.32
					(75.33)	(.102)			(8.66)
$n_2^\star$	33.33		27		35.16	.041			1.83
					(37.81)	(.045)			(4.48)
$n^\star$	200	.309		8.685	205.04	.169	.9049	.00293	5.04
				(9.175)	(216.55)	(.179)	(.9160)	(.00277)	(16.55)
$n_1^\star$	133.33		54		136.58	.130			3.24
					(144.25)	(.137)			(10.92)
$n_2^\star$	66.67		54		68.46	.056			1.79
					(72.29)	(.059)			(5.63)
$n^\star$	300	.253		8.624	304.84	.203	.9029	.00296	4.84
				(9.029)	(319.11)	(.213)	(.9163)	(.00277)	(19.11)
$n_1^\star$	200.00		81		203.11	.155			3.11
					(212.62)	(.162)			(12.62)
$n_2^\star$	100.00		81		101.73	.068			1.73
					(106.49)	(.071)			(6.49)

expanded as $\nu \to \infty$ as follows:

$$t_\nu = a - \frac{\sqrt{2}a_1 a}{\sqrt{\nu}} + O(\nu^{-1}), \tag{12.18}$$

where $\nu = (p-1)(m-1)$, a is the constant such that $G_p(a) = 1 - \alpha$ for each $G_p(\cdot)$ defined by (12.4), (12.5) or (12.7), and $a_1 = E\left(\min_{1\le i\le k} Z_i\right)$ with Z_is being i.i.d. $N(0,1)$ random variables.

Remark 2 Note that $a_1 < 0$ when $k \geq 2$.

PROOF OF LEMMA 12.4.1. We note that (12.10) can be rewritten as

$$E\left\{G_p\left(\frac{t}{\nu}\min_{1\le i\le k} W_i^2\right)\right\} = 1 - \alpha, \tag{12.19}$$

where W_i^2s are i.i.d. chi–squared random variables with ν d.f. By using the technique similar to Lemma 1 (ii) in Aoshima and Takada (2004), the left-hand side in (12.19) is asymptotically expanded as

$$G_p(t) + \frac{\sqrt{2}a_1 G_p'(t)t}{\sqrt{\nu}} + O(\nu^{-1}). \tag{12.20}$$

Then, the expression (12.18) is obtained in a straightforward manner. ∎

PROOF OF THEOREM 12.2.1. Once we obtain the design constant t_ν expanded in the form (12.18), the proof is essentially same as that of Theorem 1 in Aoshima and Takada (2004). ∎

The following lemma is used in the proof of Theorem 12.2.2.

Lemma 12.4.2 *(i) For MCA, MCB and MCC, $-G_p(y)$ is convex for $y > 0$ when $p = 2, 3$;*
(ii) For MCA and MCC, $-G_p(y)$ is not always convex for $y > 0$ when $p \geq 4$, where $G_p(\cdot)$ is defined by (12.4), (12.5) or (12.7) according to the method.

Remark 3 As for MCB with $p \geq 4$, we numerically observed that $-G_p(y)$ is convex for $y > 0$ unless $p \geq 46$.

PROOF OF LEMMA 12.4.2. (i) When $p = 2$, we see that $G_2''(y) < 0,\ y > 0$ for each method since

$$G_2''(y) = -\frac{e^{-y/4}}{4\sqrt{\pi y}}\left(\frac{1}{2} + \frac{1}{y}\right) \quad \text{(for MCA and MCC);}$$

$$G_2''(y) = -\frac{e^{-y/4}}{8\sqrt{\pi y}}\left(\frac{1}{2} + \frac{1}{y}\right) \quad \text{(for MCB).}$$

It shows that $-G_2(y)$ for each method is convex for $y > 0$. When $p \geq 3$, we have that

$$\begin{aligned}
G_p''(y) &= \frac{p(p-1)(p-2)}{4y} \int_{-\infty}^{\infty} \{\Phi(x) - \Phi(x-\sqrt{y})\}^{p-3} \{\phi(x-\sqrt{y})\}^2 d\Phi(x) \\
&\quad - \frac{p(p-1)}{4\sqrt{y}} \left(1 + \frac{1}{y}\right) \int_{-\infty}^{\infty} \{\Phi(x) - \Phi(x-\sqrt{y})\}^{p-2} \phi(x-\sqrt{y}) d\Phi(x) \\
&\quad + \frac{p(p-1)}{4y} \int_{-\infty}^{\infty} \{\Phi(x) - \Phi(x-\sqrt{y})\}^{p-2} x\phi(x-\sqrt{y}) d\Phi(x) \quad \text{(for MCA)}; \\
G_p''(y) &= \frac{(p-1)(p-2)}{4y} \int_{-\infty}^{\infty} \{\Phi(x+\sqrt{y})\}^{p-3} \{\phi(x+\sqrt{y})\}^2 d\Phi(x) \\
&\quad - \frac{(p-1)}{4\sqrt{y}} \left(1 + \frac{1}{y}\right) \int_{-\infty}^{\infty} \{\Phi(x+\sqrt{y})\}^{p-2} \phi(x+\sqrt{y}) d\Phi(x) \\
&\quad - \frac{(p-1)}{4y} \int_{-\infty}^{\infty} \{\Phi(x+\sqrt{y})\}^{p-2} x\phi(x+\sqrt{y}) d\Phi(x) \quad \text{(for MCB)}; \\
G_p''(y) &= \frac{(p-1)(p-2)}{4y} \int_{-\infty}^{\infty} \{\Phi(x+\sqrt{y}) - \Phi(x-\sqrt{y})\}^{p-3} \\
&\qquad \{\phi(x+\sqrt{y}) + \phi(x-\sqrt{y})\}^2 d\Phi(x) \\
&\quad - \frac{(p-1)}{4\sqrt{y}} \left(1 + \frac{1}{y}\right) \int_{-\infty}^{\infty} \{\Phi(x+\sqrt{y}) - \Phi(x-\sqrt{y})\}^{p-2} \\
&\qquad \{\phi(x+\sqrt{y}) + \phi(x-\sqrt{y})\} d\Phi(x) \\
&\quad - \frac{(p-1)}{4y} \int_{-\infty}^{\infty} \{\Phi(x+\sqrt{y}) - \Phi(x-\sqrt{y})\}^{p-2} x \\
&\qquad \{\phi(x+\sqrt{y}) - \phi(x-\sqrt{y})\} d\Phi(x) \quad \text{(for MCC)},
\end{aligned}$$

where $\phi(\cdot)$ denotes a $N(0,1)$ density function. To evaluate $G_p''(y)$ when $p = 3$, the following formulae are useful:

$$\begin{aligned}
&(1) \int_{-\infty}^{\infty} \{\phi(x+\sqrt{y})\}^2 d\Phi(x) = \frac{e^{-y/3}}{2\sqrt{3}\pi}, \\
&(2) \int_{-\infty}^{\infty} \{\phi(x-\sqrt{y})\}^2 d\Phi(x) = \frac{e^{-y/3}}{2\sqrt{3}\pi}, \\
&(3) \int_{-\infty}^{\infty} \phi(x+\sqrt{y})\phi(x-\sqrt{y}) d\Phi(x) = \frac{e^{-y}}{2\sqrt{3}\pi}, \\
&(4) \int_{-\infty}^{\infty} \Phi(x+\sqrt{y})\phi(x+\sqrt{y}) d\Phi(x) = \frac{e^{-y/4}}{2\sqrt{\pi}} \Phi\left(\sqrt{\frac{y}{6}}\right), \\
&(5) \int_{-\infty}^{\infty} \Phi(x+\sqrt{y})\phi(x-\sqrt{y}) d\Phi(x) = \frac{e^{-y/4}}{2\sqrt{\pi}} \Phi\left(\sqrt{\frac{3y}{2}}\right), \\
&(6) \int_{-\infty}^{\infty} \Phi(x-\sqrt{y})\phi(x+\sqrt{y}) d\Phi(x) = \frac{e^{-y/4}}{2\sqrt{\pi}} \Phi\left(-\sqrt{\frac{3y}{2}}\right),
\end{aligned}$$

$$(7)\ \int_{-\infty}^{\infty} \Phi(x-\sqrt{y})\phi(x-\sqrt{y})d\Phi(x) = \frac{e^{-y/4}}{2\sqrt{\pi}}\Phi\left(-\sqrt{\frac{y}{6}}\right),$$

$$(8)\ \int_{-\infty}^{\infty} x\Phi(x+\sqrt{y})\phi(x+\sqrt{y})d\Phi(x) = \frac{e^{-y/3}}{4\sqrt{3\pi}} - \frac{\sqrt{y}e^{-y/4}}{4\sqrt{\pi}}\Phi\left(\sqrt{\frac{y}{6}}\right),$$

$$(9)\ \int_{-\infty}^{\infty} x\Phi(x+\sqrt{y})\phi(x-\sqrt{y})d\Phi(x) = \frac{e^{-y}}{4\sqrt{3\pi}} + \frac{\sqrt{y}e^{-y/4}}{4\sqrt{\pi}}\Phi\left(\sqrt{\frac{3y}{2}}\right),$$

$$(10)\ \int_{-\infty}^{\infty} x\Phi(x-\sqrt{y})\phi(x+\sqrt{y})d\Phi(x) = \frac{e^{-y}}{4\sqrt{3\pi}} - \frac{\sqrt{y}e^{-y/4}}{4\sqrt{\pi}}\Phi\left(-\sqrt{\frac{3y}{2}}\right),$$

$$(11)\ \int_{-\infty}^{\infty} x\Phi(x-\sqrt{y})\phi(x-\sqrt{y})d\Phi(x) = \frac{e^{-y/3}}{4\sqrt{3\pi}} + \frac{\sqrt{y}e^{-y/4}}{4\sqrt{\pi}}\Phi\left(-\sqrt{\frac{y}{6}}\right),$$

$$(12)\ \Phi(x)-\Phi(0) = \frac{e^{-x^2/2}}{\sqrt{2\pi}}x + \frac{1}{2}F_3(x^2),$$

where $F_3(\cdot)$ denotes a chi-squared distribution function with 3 d.f. Then, a tedious calculation yields

$$G_3''(y) = -\frac{\sqrt{3}e^{-y/3}}{8\pi} - \frac{3e^{-y/4}}{4\sqrt{\pi y}}\left(\frac{1}{2}+\frac{1}{y}\right)F_3\left(\frac{y}{6}\right) \quad \text{(for MCA);}$$

$$G_3''(y) = -\frac{e^{-y/3}}{16\sqrt{3}\pi} - \frac{e^{-y/4}}{4\sqrt{\pi y}}\left(\frac{1}{2}+\frac{1}{y}\right)\left(\frac{1}{2}+\frac{1}{2}F_3\left(\frac{y}{6}\right)\right) \quad \text{(for MCB);}$$

$$G_3''(y) = -\frac{1}{8\sqrt{3}\pi}(e^{-y/3}+3e^{-y}) - \frac{e^{-y/4}}{4\sqrt{\pi y}}\left(\frac{1}{2}+\frac{1}{y}\right)\left(F_3\left(\frac{y}{6}\right)+F_3\left(\frac{3y}{2}\right)\right)$$
$$\text{(for MCC).}$$

Since $G_3''(y) < 0,\ y > 0$ for every method, the convexity of $-G_3(y)$ is proved.

(ii) We first note that

$$\begin{aligned}\Phi(x)-\Phi(0) &= \frac{1}{\sqrt{2\pi}}\int_0^{x^2}(\sqrt{t})'e^{-t/2}dt \\ &= \frac{e^{-x^2/2}}{\sqrt{2\pi}}\sum_{n=1}^{\infty}\frac{x^{2n-1}}{(2n-1)!!} + \frac{1}{\sqrt{2\pi}}\lim_{n\to\infty}\frac{2^{\frac{2n+1}{2}}\Gamma\left(\frac{2n+1}{2}\right)}{2(2n-1)!!}\lim_{n\to\infty}F_{2n+1}(x^2) \\ &= \frac{e^{-x^2/2}}{\sqrt{2\pi}}\sum_{n=1}^{\infty}\frac{x^{2n-1}}{(2n-1)!!},\end{aligned}$$

since

$$\lim_{n\to\infty}\frac{2^{\frac{2n+1}{2}}\Gamma\left(\frac{2n+1}{2}\right)}{2(2n-1)!!} = \sqrt{\frac{\pi}{2}} \quad \text{and} \quad \lim_{n\to\infty}F_{2n+1}(x^2) = 0.$$

So, when y (> 0) is small,

$$\begin{aligned}
&\Phi(x) - \Phi(x - \sqrt{y}) \\
&= \frac{e^{-x^2/2}}{\sqrt{2\pi}} \sum_{n=1}^{\infty} \frac{x^{2n-1}}{(2n-1)!!} - \frac{e^{-(x-\sqrt{y})^2/2}}{\sqrt{2\pi}} \sum_{n=1}^{\infty} \frac{(x-\sqrt{y})^{2n-1}}{(2n-1)!!} \\
&= \frac{e^{-x^2/2}}{\sqrt{2\pi}} \sum_{n=1}^{\infty} \frac{x^{2n-1}}{(2n-1)!!} \\
&\quad - \frac{e^{-x^2/2}}{\sqrt{2\pi}}(1+x\sqrt{y}+O(y))\left\{\sum_{n=1}^{\infty} \frac{x^{2n-1}}{(2n-1)!!} - \sum_{n=1}^{\infty} \frac{(2n-1)x^{2n-2}}{(2n-1)!!}\sqrt{y}+O(y)\right\} \\
&= \sqrt{y}\phi(x) + O(y).
\end{aligned}$$

Similarly, when y (> 0) is small,

$$\Phi(x+\sqrt{y}) - \Phi(x-\sqrt{y}) = 2\sqrt{y}\phi(x) + O(y).$$

Also, for small y (> 0) it holds that

$$\phi(x \pm \sqrt{y}) = \phi(x)(1 \mp x\sqrt{y} + O(y)).$$

After combining these results, a tedious calculation shows that

$$G_p''(y) = \frac{\sqrt{p}(p-1)(p-3)}{4(\sqrt{2\pi})^{p-1}} y^{\frac{p-5}{2}} + O(y^{\frac{p-4}{2}}) \quad \text{(for MCA)};$$

$$G_p''(y) = \frac{2^{p-3}(p-1)(p-3)}{(\sqrt{2\pi})^{p-1}\sqrt{p}} y^{\frac{p-5}{2}} + O(y^{\frac{p-4}{2}}) \quad \text{(for MCC)}.$$

So, when $p \geq 4$, one may conclude that $G_p''(y) > 0$ for small y (> 0). Hence, $-G_p(y)$ for MCA or MCC is not always convex when $p \geq 4$. ■

PROOF OF THEOREM 12.2.2. Let us take a constant t (> 0) instead of t_ν given by (12.10). Since $\nu S_i^2/\tau_i^2$ is distributed as the chi-squared distribution with $\nu = (p-1)(m-1)$ d.f., we have from (3.2) in Aoshima (2001) that

$$P_{\boldsymbol{\theta}}(\boldsymbol{\xi} \in R_{\boldsymbol{N}}) \geq E_{\boldsymbol{\theta}}\left\{G_p\left(\frac{t}{\nu}\sum_{i=1}^{k} c_i W_i^2\right)\right\}, \tag{12.21}$$

where W_i^2's are i.i.d. chi-squared random variables with ν d.f. and

$$c_i = \frac{|b_i|\tau_i/W_i}{\sum_{j=1}^{k}|b_j|\tau_j/W_j}.$$

In view of Lemma 12.4.2 and Remark 3, when $-G_p(y)$ is convex for $y > 0$, the expression (12.21) is bounded below by

$$\sum_{i=1}^{k} E_{\boldsymbol{\theta}}\left\{c_i G_p\left(\frac{t}{\nu}W_i^2\right)\right\}$$
$$= kE\left\{G_p\left(\frac{t}{\nu}W_i^2\right)\right\} - \sum_{i=1}^{k} E_{\boldsymbol{\theta}}\left\{\sum_{j\neq i} c_j G_p\left(\frac{t}{\nu}W_i^2\right)\right\}. \quad (12.22)$$

By using the technique given by Takada and Aoshima (1997, p. 358), we have for each i

$$E_{\boldsymbol{\theta}}\left\{\sum_{j\neq i} c_j G_p\left(\frac{t}{\nu}W_i^2\right)\right\} \leq E\left\{G_p\left(\frac{t}{\nu}\widetilde{W}_i^2\right)\right\}(1 - E_{\boldsymbol{\theta}}(c_i)), \quad (12.23)$$

where $\widetilde{W}_i^2$'s are i.i.d. chi–squared random variables with $\nu+1$ d.f. Substituting (12.23) into (12.22), we have

$$\sum_{i=1}^{k} E_{\boldsymbol{\theta}}\left\{c_i G_p\left(\frac{t}{\nu}W_i^2\right)\right\} \geq kE\left\{G_p\left(\frac{t}{\nu}W_i^2\right)\right\} - (k-1)E\left\{G_p\left(\frac{t}{\nu}\widetilde{W}_i^2\right)\right\}. \quad (12.24)$$

Then, part (i) is obtained.

In view of Lemma 12.4.2 and Remark 3, when $-G_p(y)$ is not always convex for $y > 0$, we note that the expression (12.21) is rewritten as

$$P_{\boldsymbol{\theta}}(\boldsymbol{\xi} \in R_{\boldsymbol{N}}) \geq E_{\boldsymbol{\theta}}\left\{G_p\left(t\tilde{X}_\nu\right)\right\}, \quad (12.25)$$

where $\tilde{X}_\nu = \nu^{-1}\sum_{i=1}^{k} c_i W_i^2$. Then, by using the technique similar to Theorem 2 in Aoshima and Takada (2004), the right-hand side in (12.25) is asymptotically expanded as

$$G_p(t) + \nu^{-1}\left\{G_p'(t)t(1-k) + G_p''(t)t^2\sum_{i=1}^{k} f_i^2\right\} + o(\nu^{-1}), \quad (12.26)$$

where $f_i = |b_i|\tau_i/\sum_{j=1}^{k}|b_j|\tau_j$. We formally put $t = a + \nu^{-1}b$ into (12.26) where a is the constant such that $G_p(a) = 1-\alpha$ for each method and b is a constant free from ν. Then, we have

$$\begin{aligned} &P_{\boldsymbol{\theta}}(\boldsymbol{\xi} \in R_{\boldsymbol{N}}) \\ &\geq 1-\alpha+\nu^{-1}\left\{G_p'(a)b + G_p'(a)a(1-k) + G_p''(a)a^2\sum_{i=1}^{k} f_i^2\right\} + o(\nu^{-1}) \\ &\geq 1-\alpha+\nu^{-1}\{G_p'(a)b + G_p'(a)a(1-k) + G_p''(a)a^2\} + o(\nu^{-1}), \quad (12.27) \end{aligned}$$

where the inequality (12.27) follows from the facts that $G''_p(a) < 0$ for a nominal level of α (see Table 12.1) and $\sum_{i=1}^k f_i^2 \le 1$. So, if the constant b is chosen as in (12.12), the two-stage procedure (12.9) possesses the asymptotic consistency in (12.13) as $\nu \to \infty$. Then, part (ii) is obtained. ■

PROOF OF THEOREM 12.2.3. We first recall from Remark 1 that both the design constant t_ν's defined by (12.11) and (12.12) have the same expression up to the order $O(\nu^{-1})$. Let $N_i^\star = [(t_\nu/d^2)|b_i|S_i\sum_{j=1}^k |b_j|S_j] + 1$. For $\boldsymbol{\theta}$ with (12.14), it follows that

$$mP_{\boldsymbol{\theta}}(N_i = m) = o(1) \quad \text{as} \quad d \to 0.$$

So, we have from (12.9) that

$$E_{\boldsymbol{\theta}}(N_i) = E_{\boldsymbol{\theta}}(N_i^\star) + o(1) \quad \text{as} \quad d \to 0. \tag{12.28}$$

Let

$$J_i = \frac{t_\nu}{d^2}|b_i|S_i\sum_{j=1}^k |b_j|S_j - \left[\frac{t_\nu}{d^2}|b_i|S_i\sum_{j=1}^k |b_j|S_j\right].$$

Then, we have as $d \to 0$,

$$\begin{aligned}
E_{\boldsymbol{\theta}}(N_i^\star) &= E_{\boldsymbol{\theta}}\left(\frac{t_\nu}{d^2}\sum_{j=1}^k |b_i||b_j|S_iS_j\right) + 1 - E_{\boldsymbol{\theta}}(J_i) \\
&= \frac{a}{d^2}\left\{1 + \frac{1}{\nu}\left(k - 1 - \frac{aG''_p(a)}{G'_p(a)}\right)\right\} \\
&\quad \times \left\{b_i^2\tau_i^2 + \sum_{j\neq i}|b_i||b_j|\tau_i\tau_j\left(1 - \frac{1}{2\nu}\right)\right\} + 1 - E_{\boldsymbol{\theta}}(J_i) + o(1) \\
&= n_i^\star + \frac{a}{2\nu d^2}\left\{\left(2k - 3 - \frac{2aG''_p(a)}{G'_p(a)}\right)\sum_{j=1}^k |b_i||b_j|\tau_i\tau_j + b_i^2\tau_i^2\right\} \\
&\quad +1 - E_{\boldsymbol{\theta}}(J_i) + o(1).
\end{aligned} \tag{12.29}$$

From Theorem 3 in Takada (2004), it follows that

$$E_{\boldsymbol{\theta}}(J_i) = 1/2 + o(1) \quad \text{as} \quad d \to 0. \tag{12.30}$$

By combining (12.29)–(12.30) with (12.28), the proof is completed. ■

References

1. Aoshima, M. (2001). Sample size determination for multiple comparisons with components of a linear function of mean vectors, *Communications in Statistics—Theory and Methods*, **30**, 1773–1788.

2. Aoshima, M. and Miyajima, H. (2001). Sample size reduction of two-stage procedure, *Kokyuroku,* **1224**, Research Institute for Mathematical Sciences, 61–72.

3. Aoshima, M. and Takada, Y. (2004). Asymptotic second-order efficiency for multivariate two-stage estimation of a linear function of normal mean vectors, *Sequential Analysis*, **23** (to appear).

4. Dunnett, C. W. (1955). A multiple comparison procedure for comparing several treatments with a control, *Journal of the American Statistical Association*, **50**, 1096–1121.

5. Hochberg, Y. and Tamhane, A. C. (1987). *Multiple Comparison Procedures*, John Wiley & Sons, New York.

6. Hsu, J. C. (1984). Constrained simultaneous confidence intervals for multiple comparisons with the best, *Annals of Statistics*, **12**, 1136–1144.

7. Hsu, J. C. (1996). *Multiple Comparisons: Theory and methods*, Chapman & Hall, New York.

8. Mukhopadhyay, N. and Duggan, W.T. (1997). Can a two-stage procedure enjoy second-order properties ?, *Sankhyā, Series A*, **59**, 435–448.

9. Stein, C. (1945). A two–sample test for a linear hypothesis whose power is independent of the variance, *Annals of Mathematical Statistics*, **16**, 243–258.

10. Takada, Y. (2004). Asymptotic second-order efficiency of a two-stage procedure for estimating a linear function of normal means, *Sequential Analysis*, **23**, 103–120.

11. Takada, Y. and Aoshima, M. (1997). Two-stage procedures for estimating a linear function of multinormal mean vectors, *Sequential Analysis*, **16**, 353–362.

12. Tukey, J. W. (1953). *The Problem of Multiple Comparisons*, Mimeographed Monograph.

PART IV
AGREEMENT ASSESSMENT

13

Measuring Agreement in Method Comparison Studies – A Review

Pankaj K. Choudhary[1] **and H. N. Nagaraja**[2]

[1]*Department of Mathematical Sciences, University of Texas at Dallas, Richardson, TX U.S.A.*
[2]*Department of Statistics, The Ohio State University, Columbus, OH U.S.A.*

Abstract: Assessment of agreement between two or more methods of measurement is of considerable importance in many areas. In particular, in medicine, new methods or devices that are cheaper, easier to use, or less invasive, are routinely developed. Agreement between a new method and a traditional reference or gold standard must be evaluated before the new one is put into practice. Various methodologies have been proposed for this purpose in recent years. We review the literature focussing on the assessment of agreement between two methods, and on the selection of the best when several methods are compared with a reference. A real data set is analyzed to illustrate the various approaches.

Keywords and phrases: Limits of agreement, gold standard, intersection-union tests, intraclass correlation, concordance correlation, selection of the best

13.1 Introduction and General Overview

When multiple methods are available for measuring a variable of interest, one is led to the task of some sort of comparison that depends on the objective of the study. General goals of such studies are [Lewis et al. (1991)]:

(a) *Comparison*: A new method has to be evaluated by comparison with an established standard, often called a *gold standard* or a *reference method*. Neither method may be accurate or precise. The goal is to learn the extent to which the measurements from the two methods agree and understand the nature of their differences. If the methods agree sufficiently well, we can use them interchangeably or use the new one, which is cheaper or more convenient, in place of the gold standard.

(b) *Calibration*: Compare an approximate method with a known accurate and precise method whose measurement error is negligible. The goal is to establish a mathematical relationship between their measurements so that the approximate method can be used as a predictor of the accurate and precise method (and hence of the true measurement).

(c) *Conversion*: Compare two approximate methods that measure two variables in different units that are surrogates for the same underlying quantity. The goal is to interpret the results of one in terms of the other.

The focus of this chapter is on surveying the recent, growing literature on the first topic. We concentrate only on the case when the measurements are continuous. Lin (2003), Lin et al. (2002) and Shoukri (1999) provide brief reviews of this area, but our scope is comprehensive and broader. The problem of assessing agreement in the categorical measurements case has been discussed extensively elsewhere. See the reviews by Kraemer, Periyakoil and Noda (2002), and Banerjee et al. (1999), and the books by Shoukri (2004) and Fleiss (1981). Also, see Cameron (1982) for an introduction to calibration problems, and to Lewis et al. (1991) for conversion problems.

Bland and Altman (1986) who present the *limits of agreement* approach and Lin (1989) who introduces the *concordance correlation coefficient* are the two classical references. The former, a favorite of medical researchers, has over 6100 citations in the Institute for Scientific Information database at the time of writing. When two methods are compared, the data consist of a random sample of paired measurements, (X_{1j}, X_{2j}), $j = 1, \ldots, n$, taken from a bivariate population (X_1, X_2), where X_1 and X_2 arise from the reference and the test method, respectively. The following model is often assumed:

$$X_{ij} = T_j + \beta_i + \epsilon_{ij};\ i = 1, 2,\ j = 1, \ldots, n;\ \text{ where,} \tag{13.1}$$

(a) T_j is the true unobservable measurement for the jth subject, distributed as $N(\mu_T, \sigma_T^2)$;

(b) β_i is the fixed bias of the ith method;

(c) ϵ_{ij} is the random error having $N(0, \sigma_{\epsilon_i}^2)$ distribution, $i = 1, 2$, and

(d) $(T_j, \epsilon_{1j}, \epsilon_{2j})$ are mutually independent for all j.

This model is known as the *Grubbs' model* in the literature when the objective of the experiment is to estimate the bias and precision of the methods [see e.g., Grubbs (1982) and Dunn and Roberts (1999)]. The quantity σ_T^2 is also known as the *between-subject variance* and $\sigma_{\epsilon_i}^2$ as the *within-subject variance* for, or the *measurement error variance* of the ith method. Let $E(X_i) = \mu_i = \mu_T + \beta_i$, $Var(X_i) = \sigma_i^2 = \sigma_T^2 + \sigma_{\epsilon_i}^2$, $i = 1, 2$, and $\rho = \sigma_T^2/(\sigma_1\sigma_2)$. Then

(X_1, X_2) is bivariate normal with means μ_1, μ_2, variances σ_1^2, σ_2^2, and correlation ρ. Notice that $\rho = (\rho_1\rho_2)^{1/2}$, where $\rho_i = \sigma_T^2/(\sigma_T^2 + \sigma_{\epsilon_i}^2)$, is the *reliability coefficient* of the ith method [see, e.g., Fleiss (1986)]. Now, define $D = X_2 - X_1$ and $D_j = X_{2j} - X_{1j}$; $j = 1, \ldots, n$. Thus, D is $N(\mu, \sigma^2)$, where $\mu = \mu_2 - \mu_1$ and $\sigma^2 = \sigma_1^2 + \sigma_2^2 - 2\rho\sigma_1\sigma_2 = \sigma_{\epsilon_1}^2 + \sigma_{\epsilon_2}^2$, and the D_j are a random sample from this population.

In the above setup, the two methods are in *perfect agreement* if all the paired measurements (X_{1j}, X_{2j}) lie on the 45° line through the origin. It can be characterized by any of the following equivalent conditions:

(A1) $\beta_1 = \beta_2$, $\sigma_{\epsilon_1}^2 = \sigma_{\epsilon_2}^2 = 0$;

(A2) $\mu_1 = \mu_2$, $\sigma_1^2 = \sigma_2^2$, $\rho = 1$;

(A3) $\mu = 0$, $\sigma^2 = 0$.

Typically, the criteria (A2) and (A3) are used in the literature. Although they are equivalent when there is perfect agreement, quantification of disagreements differ. In applications, point estimators are followed by confidence intervals on relevant parameters that are generally obtained by inverting a test of hypotheses of the form

$$\begin{aligned} &H: \text{The methods lack satisfactory agreement} \quad vs \\ &K: \text{The methods have satisfactory agreement.} \end{aligned} \tag{13.2}$$

The advantage of this formulation is that we look for evidence in the data to claim satisfactory agreement. This way the type-I error is actually the error of wrongly concluding satisfactory agreement. This formulation was first proposed by Lin (1992) and is now well established. However, there is no unanimity on what the best formulation of (H, K) is, in terms of parameters of the data. In the terminology of the US Food and Drug Administration, *satisfactory agreement* is also referred to as *substantial equivalence* [see, e.g., Lin, Whipple and Ho (1998)].

If insufficient agreement is inferred, it helps to pay attention to the nature and extent of disagreement. It may happen that X_1 and X_2 are highly correlated and have similar means, but $\sigma_{\epsilon_2}^2$ is smaller than $\sigma_{\epsilon_1}^2$. Then the new method is certainly worthy of adoption. Many times a simple linear calibration of the new method ($X_2' = a + bX_2$) may be enough for sufficient agreement between X_2' and X_1.

Let $\overline{X}_i$, S_i^2 and S_{12} be the usual unbiased estimators of $E(X_i)$, $Var(X_i)$ and $Cov(X_1, X_2)$. The components of variance in the model (13.1) are estimated as $\hat{\sigma}_T^2 = S_{12}$ and $\hat{\sigma}_{\epsilon_i}^2 = S_i^2 - S_{12}$. These are also known as *Grubbs' estimators* [see Grubbs (1982)]. Thus, the various parameter estimators are: $\hat{\mu}_i = \overline{X}_i$, $\hat{\sigma}_i^2 = S_i^2$, $\hat{\rho} = S_{12}/(S_1S_2)$, $\hat{\mu} = \overline{X}_2 - \overline{X}_1$ and $\hat{\sigma}^2 = S_1^2 + S_2^2 - 2S_{12}$.

A key issue is how much a method (say, the gold standard) agrees with itself, because it limits the amount of agreement that is possible between two methods. Popular terms for this phenomenon of agreement with itself include "reliability", "reproducibility" and "repeatability". But these terms have been used in other settings too [see, e.g., Lin (1989) and Bland and Altman (1999)]. The various measures of this agreement include the intraclass correlation (or the reliability coefficient) computed from one-way models [see, e.g., Fleiss (1986) and Dunn (1989, 1992)], intraclass correlation from two-way models [see e.g., Fleiss (1986)], within-subject variance [see Bland and Altman (1999)], and within-subject coefficient of variation [see Quan and Shih (1996) and the related correspondence]. See also the comments in Hawkins (2002), and Dunn and Roberts (1999). Generally, this issue is addressed in a separate reliability study and will not be discussed here.

This chapter is organized as follows. Sections 13.2 and 13.3 discuss various approaches for assessment of agreement between two methods. In Section 13.4 we provide an illustrative real example. In Section 13.5 we change our focus to the comparison of k (≥ 2) methods with a gold standard. The goal of this comparison is to select the best among k — the one that agrees most with the reference. Some concluding remarks are presented in Section 13.6.

In what follows, $\Phi(\cdot)$ and $\phi(\cdot)$ are the standard normal cdf and pdf, respectively, and $\mathbf{\Phi}_2(\cdot, \cdot; \rho)$ is the cdf of a bivariate standard normal distribution with correlation ρ. A χ^2-distribution will be denoted by χ^2_k, and a t-distribution will be denoted by t_k, where k is the degrees of freedom. An F distribution with degrees of freedom l and m will be denoted by $F(l, m)$. The notation H_i (K_i) is used for a null (alternative) hypothesis, with $i = 0, 1, \ldots$.

13.2 Early Approaches

The examples in Altman and Bland (1983) indicate that the correlation

$$\rho = \sigma_T^2 / \left((\sigma_T^2 + \sigma_{\epsilon_1}^2)(\sigma_T^2 + \sigma_{\epsilon_2}^2) \right)^{1/2},$$

arising from the model (13.1), was a widely used measure of "agreement" in the medical literature. Further, a statistically significant result of testing $H_0 : \rho = 0$ vs $K_0 : \rho \neq 0$ was often taken as evidence of agreement. But this test is generally useless because two methods designed to measure the same quantity will rarely be uncorrelated. Also, ρ is just a measure of strength of linear relationship, not of agreement. It is possible to have $\rho = 1$ even when $\mu_2 - \mu_1 = a$ ($\neq 0$) and $\sigma_2^2/\sigma_1^2 = b$ ($\neq 1$). Altman and Bland (1983) and Bland and Altman (1986, 1990, 1995a) draw attention to the following deficiencies of ρ:

(a) The value of ρ increases as $\sigma_T^2/\sigma_{\epsilon_i}^2$ increases. In practice, investigators try to assess the agreement of methods over a wide range of subjects resulting in large σ_T^2 and ρ values. This property makes ρ unsuitable as a measure of agreement, because how σ_T^2 and $\sigma_{\epsilon_i}^2$ compare is not related to the goal of assessing agreement.

(b) A consequence of (a) is that, on its own, the value of ρ does not tell us much about agreement. It is possible to have two data sets such that (i) the differences of the paired measurements in one set is identical with those in the other (and hence the two sets exhibit the same degree of agreement), but (ii) the estimate of ρ is quite low in one case and is very high in the other [see also Atkinson and Nevill (1997)].

These concerns also hold for other correlation type measures, namely concordance correlation and intraclass correlation, discussed later in this section.

Other popular, but inappropriate, early approaches include a paired-t test of $H_0 : \mu = 0$, and a test of H_0 : slope $= 1$, intercept $= 0$, when the test method is regressed on the reference method. Notice that the paired-t test only assesses whether the methods agree *on average*, not *for every subject.* Further, as Lin (1989) demonstrates with both graphs and real data, both these tests can be misleading. They may reject H_0 if the scatter around the 45° line is near zero (indicating good agreement), and may fail to reject it if the scatter is very high (indicating poor agreement). In addition, a correct regression approach must account for the fact that even the reference method measures with error. But again, since our goal is to quantify the disagreement between methods, and not how accurately one method can be predicted from the other, we do not discuss regression models with errors in variables or structural equations models here and refer the reader to Kelly (1985), Linnet (1993), Nix and Dunston (1991), and Fuller (1987).

13.2.1 The limits of agreement (LOA) approach

Bland-Altman Plot

The basic idea due to Bland and Altman (1986) is that if a large proportion (such as 95%) of the differences are sufficiently close to zero then the two methods have satisfactory agreement. The process of judging this agreement has two components: (a) 95% LOA, defined by $\hat{\mu} \pm 1.96\hat{\sigma}$, and (b) the plot of mean, $(X_1 + X_2)/2$, versus difference, D, with LOA superimposed. This plot is popularly known as the Bland-Altman plot. Statistical software SAS JMP produces such a plot for matched pair data.

The LOA estimate the set $(\mu - 1.96\sigma, \mu + 1.96\sigma)$. One could declare sufficient agreement if the differences within these limits are not practically (or clinically) important as determined by the investigator specified threshold δ_0 (> 0). Bland

and Altman recommend that this δ_0 be specified in advance and Bland and Altman (1999) observe that its choice "will depend on the use to which the result is put, and is a question of clinical judgement." See also Hawkins (2002).

The uncertainty in the estimation of LOA is accounted for by the approximate 95% confidence intervals (CIs) for the two limits. It is $(\hat{\mu} - 1.96\hat{\sigma}) \pm t_{n-1}(\alpha/2)1.71\hat{\sigma}/n^{1/2}$ for $\mu - 1.96\sigma$ (and similar for the upper limit), where $t_k(\alpha)$ is the upper αth percentile of a t_k distribution.

The Bland-Altman plot is an excellent supplement to the usual scatterplot of the data. It reveals interesting features of the data and also helps in diagnosing departures from the various model assumptions and suggesting remedies. Hawkins (2002) describes how the Bland-Altman plot can be used in conjunction with the standard regression diagnostics, and presents several real examples to illustrate the common model violations and discusses ways to handle them. He says, "the ideal plot resembles the ideal plot of residuals against the fitted values in a regression problem". Some common departures are:

(a) *LOA band not centered at zero or that is wide*: A plot not centered at zero indicates a bias between the methods. If the rest of the features of the plot are close to ideal, the new method can be recalibrated by adding a constant for good agreement with the old. However, if the points do not lie in a narrow band, it suggests that the variability of the differences is not small. This will result in wide LOA, and is a serious problem to resolve.

(b) *Linear trend*: The differences increase (or decrease) with increasing magnitude of the measurements in the range of measurement [see also Bland and Altman (1995b)]. Under the model in (13.1), this implies that the $\sigma^2_{\epsilon_i}$ are different. If $\sigma^2_{\epsilon_2} > \sigma^2_{\epsilon_1}$, a possible resolution is to recalibrate the new method by multiplying with a constant. Otherwise, the new is more precise than the old, and obviously is preferred. As another possible resolution, Bland and Altman (1986, 1999) suggest that a log transformation of the data may make the variabilities comparable. The analysis then proceeds with the log-scale differences, and the results are back-transformed to the ratio scale for interpretation.

(c) *Heteroscedasticity*: If the scatter of differences does not remain the same over the range of measurement, the model assumption that σ_i^2 is not related to μ_i is violated for one or both of the methods. A common violation corresponds to a "right opening megaphone" — an increase in scatter of differences as the magnitude increases. Often, a log transformation of the data corrects this problem. Sometimes a more sophisticated variance stabilizing transformation may be needed.

(d) *Outliers*: Vertical outliers flag the subjects for whom the measurements differ by an unusually large amount. It is recommended to assess the impact of the outliers on the results.

(e) *Non-linear curvature*: A non-linear recalibration may be needed to correct the problem.

Significance Tests

One can use the paired-t test for $H_0 : \mu_1 = \mu_2$, the Pitman-Morgan test for $H_0 : \sigma_1^2 = \sigma_2^2$, and the Bradley-Blackwood test for $H_0 : \mu_1 = \mu_2$, $\sigma_1^2 = \sigma_2^2$. The last two are described in Krummenauer (1999). Bartko (1994) shows how all of these can be implemented using the standard two-way ANOVA output. These tests (of point null hypotheses) are generally used only to supplement the graphical analysis, and rarely play a prominent role in the assessment of agreement [see also the discussion in Bland and Altman (1999)]. The hypotheses of whether the agreement is satisfactory or not is generally of the form (13.2). Bartko (1994) suggests adding an elliptical tolerance region of the bivariate distribution of the difference and the mean to the Bland-Altman plot, so that the relative magnitudes of the between-subject and the within-subject variations can be amplified.

The normality of differences can be assessed with the usual histogram and normal probability plot.

Confidence Regions

Beyond the plots, interest lies in quantifying disagreement by the estimation of the parameter region $(\mu - 1.96\sigma, \mu + 1.96\sigma)$, and its comparison with the threshold interval $(-\delta_0, \delta_0)$. For this comparison we must use a CI estimate of the region so that the uncertainty in the estimation is also accounted for. The usual LOA approach achieves this by constructing separate two-sided CIs for the two endpoints. However, Lin et al. (1998) argue that instead of the two-sided CIs, we need one-sided CIs, an upper confidence bound (UCB) for $\mu + 1.96\sigma$ and a lower confidence bound (LCB) for $\mu - 1.96\sigma$, because the interest lies in bounding the region $(\mu - 1.96\sigma, \mu + 1.96\sigma)$. Consequently, we can infer satisfactory agreement if

$$(\hat{\mu} - a_n\hat{\sigma}, \hat{\mu} + a_n\hat{\sigma}) \subset (-\delta_0, \delta_0),$$

where $a_n = 1.96 + 1.71n^{-1/2}t_{n-1}(\alpha)$. Based on this, Lin et al. (1998) derive a simple sample size formula for use in planning method comparison studies. They also indicate that the above rule can be thought of as a large-sample intersection-union test (IUT) [see Casella and Berger (2002, p. 380), for an introduction] of

$$H_1 : \text{Complement of } K_1 \quad vs. \quad K_1 : -\delta_0 < \mu - 1.96\sigma,\ \mu + 1.96\sigma < \delta_0, \qquad (13.3)$$

which is of the form (13.2). For these hypotheses, however, Liu and Chow (1997) have given an exact IUT in the context of assessment of individual bioequivalence. This test rejects H_1 if

$$(\hat{\mu} - b_n\hat{\sigma}, \hat{\mu} + b_n\hat{\sigma}) \subset (-\delta_0, \delta_0),$$

where $b_n = n^{-1/2}t_{n-1}(\alpha, n^{1/2}z(\frac{\pi_0}{2}))$, $z(\frac{\pi_0}{2})$ is the $(\frac{\pi_0}{2})$th upper percentile of a $N(0,1)$ distribution, and $t_k(\alpha, \Delta)$ is the upper αth percentile of a non-central t_k-distribution with non-centrality parameter Δ. As shown in Choudhary and Nagaraja (2004a), the interval $(\hat{\mu} - b_n\hat{\sigma}, \hat{\mu} + b_n\hat{\sigma})$ can also be interpreted as a large-sample two-sided tolerance interval with 0.95 content at confidence level $1 - \alpha$. See, e.g., David and Nagaraja (2003) or Guttman (1988) for an introduction to tolerance intervals. A comparison of these two tests of (13.3) and the associated sample size formulae will be of interest.

The strength of the LOA approach lies in its intuitive appeal and simplicity. Further, since the limits are based on difference, they are not affected by the between-subject variation in the data. This approach was recently generalized by Bland and Altman (1999) to accommodate replicate measurements on each subject from every method.

13.2.2 Intraclass correlation and related measures

Fleiss (1986, Ch. 1) gives an overview of the intraclass correlation coefficient (ICC) as a measure of agreement between k (≥ 2) methods. In our setting, the ICC is defined under the two-way mixed model (13.1) with additional assumptions that $i = 1, \ldots, k$ (≥ 2), $\sigma^2_{\epsilon_i} = \sigma^2_\epsilon$ for all i, and $\sum_{i=1}^k \beta_i = 0$. It serves as an index of agreement among k methods and is given by

$$\rho_I = \frac{\sigma^2_T}{\sigma^2_T + \sigma^2_\beta + \sigma^2_\epsilon}, \tag{13.4}$$

where $\sigma^2_\beta = \sum_{i=1}^k \beta_i^2/(k-1)$. This ICC is also known as the inter-method (or inter-rater) reliability. Note that $0 < \rho_I \leq 1$ and equals 1 only when there is perfect agreement among all the k methods.

The ANOVA table for this model is given in Table 13.1 [see also McGraw and Wong (1996)] where $\overline{X}_{i\cdot} = \sum_{j=1}^n X_{ij}/n$, $\overline{X}_{\cdot j} = \sum_{i=1}^k X_{ij}/k$, and $\overline{X}_{\cdot\cdot} = \sum_{i=1}^k \sum_{j=1}^n X_{ij}/(nk)$. The ICC is then estimated by

$$\hat{\rho}_I = \frac{SMS - EMS}{SMS + (k-1)EMS + (k/n)(IMS - EMS)}. \tag{13.5}$$

Its approximate $100(1-\alpha)\%$ LCB [McGraw and Wong (1996)] is

$$\frac{n(SMS - F_{n-1,\nu}(\alpha)\, EMS)}{F_{n-1,\nu}(\alpha)\,[k\, IMS + (kn - k - n)EMS] + n\, SMS},$$

Table 13.1: ANOVA Table for estimating the ICC ρ_I of (13.4)

Source	d.f.	SS	MS	E(MS)
Subjects	$n-1$	$k\sum_{j=1}^{n}(\overline{X}_{\cdot j}-\overline{X}_{\cdot\cdot})^2$	*SMS*	$k\sigma_T^2+\sigma_\epsilon^2$
Methods	$k-1$	$n\sum_{i=1}^{k}(\overline{X}_{i\cdot}-\overline{X}_{\cdot\cdot})^2$	*IMS*	$n\sigma_\beta^2+\sigma_\epsilon^2$
Error	$(n-1)(k-1)$	By Subtraction	*EMS*	σ_ϵ^2
Total	$nk-1$	$\sum_{i=1}^{k}\sum_{j=1}^{n}(X_{ij}-\overline{X}_{\cdot\cdot})^2$		

where

$$\nu=\frac{(a\,IMS+b\,EMS)^2}{(k-1)^{-1}(a\,IMS)^2+[(n-1)(k-1)]^{-1}(b\,EMS)^2},$$

$a=(k/n)[\hat{\rho}_I/(1-\hat{\rho}_I)]$, $b=1+(n-1)\,a$, and $F_{l,m}(\alpha)$ is the upper αth percentile of an $F(l,m)$ distribution.

This model can accommodate several methods, but the assumption of equal error variance is generally not reasonable. Further, ρ_I is non-negative but its estimates may be negative, and it is also sensitive to the between-subject variation. If ρ_I is low, it is unclear whether the lack of agreement is due to low between-subject variation and/or high error variation and/or location bias. See also Müller and Büttner (1994) for a critique of this measure.

Assuming X_1 and X_2 have the same means, St. Laurent (1998) has considered the model

$$X_{2j}=X_{1j}+\epsilon_j;\ j=1,\ldots,n,$$

and suggested the associated ICC as a measure of agreement. Here ϵ_j is the measurement error in the test method. It is assumed that the (i) X_{1j} and ϵ_j are independent and identically distributed (i.i.d.) with means μ_1 and zero, variances σ_1^2 and σ_ϵ^2, respectively, and (ii) X_{1j} and ϵ_j are mutually independent. The ICC is then given by

$$\rho_G=\frac{\sigma_1^2}{\sigma_1^2+\sigma_\epsilon^2}, \tag{13.6}$$

which is the squared correlation between X_1 and X_2. Clearly, $\rho_G>0$ and it equals 1 when $\sigma_\epsilon^2=0$. St. Laurent refers to $\rho_G^{1/2}$ as the *gold standard correlation*. When X_{1j} and ϵ_j are normally distributed, the maximum likelihood estimator (MLE) of ρ_G is

$$\hat{\rho}_G=\frac{(n-1)\hat{\sigma}_1^2}{(n-1)\hat{\sigma}_1^2+\sum_{i=1}^{n}D_i^2}$$

and the LCB of ρ_G is given by

$$\frac{F_{n,n-1}(\alpha)}{F_{n,n-1}(\alpha)+(\hat{\rho}_G^{-1}-1)(n-1)/n}.$$

St. Laurent (1998) also develops a large sample theory for the non-normal case. Recently, Harris, Burch, and St. Laurent (2001) have developed a family of estimators of ρ_G that includes the MLE and they indicate that at times other members of the family may be preferable to the MLE in terms of mean-squared error. Being an ICC, ρ_G also suffers from the drawback of being sensitive to the between-subject variation. In addition, the assumption of no bias between the methods may not be justified.

13.2.3 Concordance correlation approach

Lin (1989) proposed a concordance correlation coefficient (CCC) as an index of agreement and defined it as

$$\rho_C = 1 - \frac{E(X_1 - X_2)^2|_\rho}{E(X_1 - X_2)^2|_{\rho=0}} = \frac{2\rho\sigma_1\sigma_2}{(\mu_1 - \mu_2)^2 + \sigma_1^2 + \sigma_2^2}. \tag{13.7}$$

It represents the expected squared distance of a point (X_1, X_2) from the 45° line through the origin, scaled to lie between $[-1, 1]$. This distance, $\theta = E(X_1 - X_2)^2$, is also referred to as the mean squared deviation (MSD). Thus ρ_C measures how close the (paired) observations are to the 45° line. It can also be written as ρC_b, where

$$C_b = 2/(v + 1/v + u^2), v = \sigma_1/\sigma_2, \text{ and } u = (\mu_1 - \mu_2)/(\sigma_1\,\sigma_2)^{1/2}. \tag{13.8}$$

Thus the CCC has two components: (a) ρ, the correlation, which Lin calls the "precision" component, that measures how close the observations are to the best fit line, and (b) $C_b \in (0, 1]$, the "accuracy" component that measures how close the best fit line is to the 45° line. The CCC has the following properties:

(i) $|\rho_C| \leq |\rho| \leq 1$, (ii) $\rho_C = 0$ iff $\rho = 0$, (iii) $\rho_C = \rho$ iff $\sigma_1 = \sigma_2$, $\mu_1 = \mu_2$, and

(iv) $\rho_C = \pm 1$ iff $\rho = \pm 1$, $\sigma_1 = \sigma_2$, and $\mu_1 = \mu_2$.

The estimator of ρ_C, obtained by replacing the population moments with the sample moments, is

$$\hat{\rho}_C = \frac{2\,\hat{\rho}\hat{\sigma}_1\hat{\sigma}_2}{(\hat{\mu}_1 - \hat{\mu}_2)^2 + \hat{\sigma}_1^2 + \hat{\sigma}_2^2}. \tag{13.9}$$

Under bivariate normality for (X_1, X_2), Lin (1989) showed that $\hat{\rho}_C$ is asymptotically normal with mean ρ_C. He suggests the transformation $W = \tanh^{-1}(\hat{\rho}_C)$ for faster convergence; W is asymptotically normal with mean $\tanh^{-1}(\rho_C)$ and variance

$$\sigma_W^2 = \frac{1}{n-2}\left[\frac{\rho_C^2(1-\rho^2)}{\rho^2(1-\rho_C^2)} + \frac{2\rho_C^3(1-\rho_C)u^2}{\rho(1-\rho_C^2)^2} - \frac{\rho_C^4 u^4}{2\rho^2(1-\rho_C^2)^2}\right],$$

where u is given in (13.8). Another advantage of this transformation is that the asymptotic CI for ρ_C is constrained to lie within $[-1, 1]$ and an approximate level $(1-\alpha)$ LCB for ρ_C is

$$\tanh\left(\tanh^{-1}(\hat{\rho}_C) - z(\alpha)\hat{\sigma}_W\right),$$

where $\hat{\sigma}_W$ is obtained by replacing the population moments in σ_W^2 with the corresponding sample moments.

Based on the allowable losses in the "precision" and "accuracy" components described above, Lin (1992) first computes ρ_C^*, which represents the smallest acceptable value of ρ_C that the investigator is willing to consider as evidence of satisfactory agreement. Then he proposes to test

$$H_2 : \rho_C \leq \rho_C^* \;\; vs. \;\; K_2 : \rho_C > \rho_C^*,$$

using the LCB of ρ_C obtained as above. If it exceeds ρ_C^*, one rejects H_2 and infers satisfactory agreement. Lin also gave a sample size formula based on the above test of hypotheses.

In practice, however, an LCB for ρ_C is computed and compared with a cutoff such as 0.75. The practice of accepting satisfactory agreement if the LCB for CCC (or ICC) exceeds 0.75 [see, e.g., Lee, Koh and Ong (1989) and Atkinson and Nevill (1997)] may not be wise. When ρ is close to 1, a location/scale bias may not be reflected well in these measures, and hence they may lead to wrong conclusions. See Section 13.4 for an example.

The CCC formulation has motivated further research. Chinchilli et al. (1996) introduced a weighted version of ρ_C to handle repeated measurements data. King and Chinchilli (2001) have generalized ρ_C to incorporate distance functions other than the squared error, and constructed robust forms of ρ_C. They also demonstrated the relationship between ρ_C and the *kappa statistic*, a measure of agreement for nominal/ordinal measurements, and introduced further extensions [see also Robieson (1999)]. Barnhart, Haber and Song (2002) have constructed an overall CCC, a weighted average of all pairwise CCC's, for measuring agreement between k (> 2) methods. Additionally, Barnhart and Williamson (2001) suggested a generalized estimating equations approach to model ρ_C when covariates are present. Vonesh, Chinchilli and Pu (1996) have used CCC type of index in goodness-of-fit contexts – to evaluate the agreement between observed values and the values predicted by a model and for model selection.

Liao and Lewis (2000) allude to some deficiencies of CCC and suggest a slightly modified index. However, its performance tends to be similar. They also extend it to handle situations when parameters cannot be assumed to remain fixed over the entire range for measurement.

Lin (1989) pointed out that the CCC tends to produce results similar to the ICC. It was finally noted by Nickerson (1997) that, when $k = 2$, the estimate of the ICC ρ_I defined in (13.5), becomes

$$\hat{\rho}_I = \frac{2\,\hat{\rho}\hat{\sigma}_1\hat{\sigma}_2}{(\hat{\mu}_1 - \hat{\mu}_2)^2 + \hat{\sigma}_1^2 + \hat{\sigma}_2^2 + \hat{\sigma}^2/n}.$$

This differs from $\hat{\rho}_C$ in (13.9) in the denominator by $\hat{\sigma}^2/n$. She notes that this difference tends to be small resulting in close values for these coefficients.

There are two related issues with ICC, CCC and other such measures that have drawn much criticism in the literature [see, e.g., Müller and Büttner (1994), Bland and Altman (1995b), Atkinson and Nevill (1997,) and Lin and Chinchilli (1997)]. Firstly, these measures depend on the between-subject variation. Thus deficiencies of correlation that were listed on Page 218 continue to hold for CCC.

Secondly, although CCC is a convenient single index for agreement that lies in $[-1, 1]$ and combines components of a systematic bias, a difference in variabilities and a low correlation, there lies its major weakness. If a lack of satisfactory agreement is concluded using CCC, the reason is unclear. To deal with this issue Lin and Torbeck (1998) and Lin et al. (2002) suggest supplementing the CCC with CIs for its ρ ("precision") and C_b ("accuracy") components.

Recently, Lin et al. (2002) compared the power properties of the test (of agreement) based on CCC with those based on total deviation index [Lin (2000)] and the coverage probability, and found that the CCC based test has inferior power properties. These terms and tests are introduced next.

13.3 Recent Developments

13.3.1 Approaches based on percentiles and coverage probability

The LOA approach involved the following steps:

(a) Specify a threshold interval $(-\delta_0, \delta_0)$ such that the differences in this interval are practically equivalent to zero.

(b) Quantify the observed disagreement by estimating the range in which a pre-specified large proportion of differences (say π_0) are expected to lie.

(c) If the estimated range in (b) is contained in $(-\delta_0, \delta_0)$, declare satisfactory agreement.

In practice, the LOA approach takes $\pi_0 = 0.95$ and uses a CI estimate of $(\mu - 1.96\sigma, \mu + 1.96\sigma)$, the *centrally symmetric* region of 95% probability content. Often times δ_0 is not explicitly specified and the evaluation of agreement proceeds by assessing whether the estimated range contains any clinically important differences. We saw in Section 13.2.1 that this approach can be thought of as testing the hypotheses (13.3), which can also be expressed as

$$H_1 : \pi_c \leq \pi_0 \text{ vs. } K_1 : \pi_c > \pi_0, \tag{13.10}$$

where π_c is the proportion of centrally symmetric region contained in $(-\delta_0, \delta_0)$.

Lin (2000) and Lin et al. (2002) consider a slightly liberal variation on the above theme: instead of asking for π_0 proportion of *central* differences to lie in $(-\delta_0, \delta_0)$ for satisfactory agreement, they only ask for π_0 proportion of differences to lie in $(-\delta_0, \delta_0)$. There are now two theoretically equivalent ways to proceed.

Total Deviation Index (TDI)

For a specified π_0, consider $q(\pi_0)$, the π_0th percentile of $|D|$, as the measure of agreement and assess satisfactory agreement by testing

$$H_3 : q(\pi_0) \geq \delta_0 \text{ vs. } K_3 : q(\pi_0) < \delta_0. \tag{13.11}$$

Lin (2000) proposes this approach and calls $q(\pi_0)$ the *total deviation index*.

Since D is $N(\mu, \sigma^2)$ and $Pr\big(|D| < q(\pi_0)\big) = \pi_0$, $q(\pi_0)$ can be written as

$$q(\pi_0) = \sigma\big(\chi_1^2(1 - \pi_0, \mu^2/\sigma^2)\big)^{1/2}, \tag{13.12}$$

where $\chi_1^2(\alpha, \Delta)$ is the upper αth percentile of a non-central χ_1^2-distribution with non-centrality parameter Δ. Lin (2000) argues that the inference based on the estimate of $q(\pi_0)$ in (13.12) is intractable. So he approximates it by

$$q^*(\pi_0) = \big((\mu^2 + \sigma^2)\chi_1^2(1 - \pi_0, 0)\big)^{1/2} = (\mu^2 + \sigma^2)^{1/2} z((1 - \pi_0)/2), \tag{13.13}$$

and for assessing agreement, he modifies the hypotheses (13.11) to

$$H_3^* : q^*(\pi_0) \geq \delta_0 \text{ vs } K_3^* : q^*(\pi_0) < \delta_0. \tag{13.14}$$

He suggests estimating the MSD $\theta = (\mu^2 + \sigma^2) = E(D^2)$ by $\hat{\theta} = \sum_{i=1}^n D_i^2/(n-1)$ and performing inference using the large-sample $N(0, 1)$ distribution of $\big(\log(\hat{\theta}) - \log(\theta)\big)/\hat{\tau}$, where

$$\hat{\tau}^2 = 2(1 - \overline{D}^4/\hat{\theta}^2)/(n - 1) \tag{13.15}$$

is the estimated asymptotic variance of $\log(\hat{\theta})$. Thus the estimate of $q^*(\pi_0)$ and its approximate $100(1 - \alpha)\%$ UCB respectively become

$$\hat{q}^*(\pi_0) = \hat{\theta}^{1/2} z((1 - \pi_0)/2) \text{ and } \hat{q}^*(\pi_0) \exp\left[\tfrac{1}{2} z(\alpha)\hat{\tau}\right].$$

A test for (13.14) rejects H_3^* if this UCB is less than δ_0. Lin also gives a sample size formula associated with (13.11).

The approximation $q^*(\pi_0)$ for $q(\pi_0)$ will be good only when μ^2/σ^2 is small and Lin gives a range of values of π_0 and μ^2/σ^2 where this approximation can be considered reasonable. Further, the above test of (13.14) has asymptotic level α and is consistent. This leads to the undesirable property that in the limiting case where $n \to \infty$, with probability 1, there will be some regions where the agreement is satisfactory ($q(\pi_0) < \delta_0$), but the test will conclude otherwise (i.e., $q(\pi_0) \geq \delta_0$) and vice versa. Such regions depend on whether the approximation $q^*(\pi_0)$ is conservative (i.e., $q^*(\pi_0) > q(\pi_0)$) or liberal (i.e., $q^*(\pi_0) < q(\pi_0)$).

Coverage Probability Approach

This approach of Lin et al. (2002) takes the *coverage probability* (CP) $\pi = Pr(|D| < \delta_0)$ of the threshold interval $(-\delta_0, \delta_0)$ as the measure of agreement and tests

$$H_4 : \pi \leq \pi_0 \quad \text{vs.} \quad K_4 : \pi > \pi_0. \tag{13.16}$$

These hypotheses are equivalent to (13.11) for specified (δ_0, π_0).

Lin et al. (2002) estimate π as

$$\hat{\pi} = \Phi\big((\delta_0 - \hat{\mu})/\tilde{\sigma}\big) - \Phi\big((-\delta_0 - \mu)/\tilde{\sigma}\big),$$

where $\tilde{\sigma}^2 = (n-1)\hat{\sigma}^2/(n-3)$ and suggest performing inference through the large sample normality of $(\hat{\lambda} - \lambda)/\hat{\psi}$, where $\lambda = \log(\pi/(1-\pi))$, $\hat{\lambda} = \log(\hat{\pi}/(1-\hat{\pi}))$, and

$$\hat{\psi}^2 = \frac{1}{(n-3)\hat{\pi}^2(1-\hat{\pi})^2}\Big\{\Big[\phi\big((\delta_0 - \hat{\mu})/\tilde{\sigma}\big) - \phi\big((-\delta_0 - \hat{\mu})/\tilde{\sigma}\big)\Big]^2 + \frac{1}{2}\Big[\phi\big((\delta_0 - \hat{\mu})/\tilde{\sigma}\big)\big((\delta_0 - \hat{\mu})/\tilde{\sigma}\big) - \phi\big((-\delta_0 - \hat{\mu})/\tilde{\sigma}\big)\big((-\delta_0 - \hat{\mu})/\tilde{\sigma}\big)\Big]^2\Big\}$$

is the estimated asymptotic variance of $\hat{\lambda}$. Thus, in particular, the approximate $100(1-\alpha)\%$ LCB for π becomes

$$e^{\hat{\lambda}^-}/(1 + e^{\hat{\lambda}^-})$$

where $\hat{\lambda}^- = \hat{\lambda} - z(\alpha)\hat{\psi}$. When this LCB exceeds the cutoff π_0, H_4 of (13.16) is rejected.

Recent simulations by Choudhary and Nagaraja (2004a) reveal that this test is overly conservative for moderate sample sizes. When $30 \leq n \leq 50$ and $0.80 \leq \pi_0 \leq 0.95$, the empirical type-I error rate of this test is about 3% or less for a nominal 5% level. However, using the MLE of σ^2 in place of $\tilde{\sigma}^2$ greatly improves its performance.

Lin et al. (2002) provide a sample size formula for (13.16) and note that TDI and CP approaches yield similar powers.

Since the hypotheses (13.11) and (13.16) are equivalent, it is natural to expect that their tests should lead to the same conclusions when both are applied to the same data set. However, this is clearly not the case for the proposals given above. This issue is addressed in Choudhary and Nagaraja (2004a) where an exact level α test of (13.16) (or equivalently (13.11)) is given and a good, simple approximation to its critical value is presented.

In practice, it is easy to specify π_0 (generally, a number between 0.80 and 0.95) but we have seen in Section 13.2 that choosing δ_0 is relatively difficult and depends on the purpose. In this sense, it can be argued that the formulation (13.11) is better than (13.16) as the former produces an UCB, $\hat{q}^+(\pi_0)$, for $q(\pi_0)$ for a specified π_0. The interval $(-\hat{q}^+(\pi_0), \hat{q}^+(\pi_0))$ then can be used in the same way as the LOA without a specified δ_0. However, its value must be explicitly specified in advance for the formulation (13.16).

The hypotheses (13.16) are often used for the assessment of individual bioequivalence where X_1 and X_2 are respectively the measures of effectiveness of a reference drug and a test drug [see, e.g., Anderson and Hauck (1990) and Wang and Hwang (2001)]. There is also a connection between them and the ones used in statistical quality control. In *acceptance sampling* with two-sided specification limits, D refers to the quality characteristic of an item. If D falls in the specified interval (l, u), the item is conforming, and is non-conforming otherwise. The parameter of interest here is the lot quality as measured by $1 - \pi$ and the decision to accept or reject a lot of items is based on the test of

$$H_4' : 1 - \pi \leq 1 - \pi_0 \quad \text{vs.} \quad K_4' : 1 - \pi > 1 - \pi_0,$$

where $1-\pi_0$ is the specified *acceptable quality level.* Rejection of H_4' amounts to rejecting the lot. Notice that these hypotheses can be obtained by interchanging H_4 and K_4 of (13.16) while retaining the equality sign in the null. See Hamilton and Lesperance (1995) for a discussion of various tests of these hypotheses.

13.3.2 Approaches based on the intersection-union principle

All the formal approaches discussed so far use a single measure of agreement. Thus, when a lack of satisfactory agreement is inferred, we would not know the cause or extent of disagreement without additional investigation. Choudhary and Nagaraja (2004b) resolve this issue by giving two formulations of the hypotheses (13.2) that preserve the information on all causal indicators. The first tests

$$H_5 : \{|\mu| \geq \delta_\mu\} \cup \{\sigma \geq \delta_\sigma\} \quad \text{vs.} \quad K_5 : \{|\mu| < \delta_\mu\} \cap \{\sigma < \delta_\sigma\}, \tag{13.17}$$

where δ_μ and δ_σ reflect the extent of bias and variability in D, respectively, that the practitioner can tolerate and pre-specify. The second, more detailed formulation tests

$$\begin{aligned} &H_6 : \{|\mu_2 - \mu_1| \geq \delta_\mu\} \cup \{\sigma_2/\sigma_1 \leq \delta_1 \text{ or } \sigma_2/\sigma_1 \geq \delta_2\} \cup \{\rho \leq \delta_\rho\} \quad \text{vs.} \\ &K_6 : \{|\mu_2 - \mu_1| < \delta_\mu\} \cap \{\delta_1 < \sigma_2/\sigma_1 < \delta_2\} \cap \{\rho > \delta_\rho\}, \end{aligned} \tag{13.18}$$

where the pre-specified $\delta_\rho \in (0,1)$ is large, and $0 < \delta_1 < 1 < \delta_2$. Usually $\delta_1 = 1/\delta_2$ is taken so that they are symmetric about zero on the log-scale.

The hypotheses (13.17) and (13.18) are tested using the intersection-union principle [see Casella and Berger (2002, p. 380)], and the tests are inverted to give the CIs that quantify the extent of disagreement on individual indicators. The individual $100(1-\alpha)\%$ CIs for μ and σ associated with (13.17) are, respectively,

$$\{\min\big(0,\, \hat{\mu} - t_{n-1}(\alpha)n^{-1/2}\hat{\sigma}\big) \leq \mu \leq \max\big(0,\, \hat{\mu} + t_{n-1}(\alpha)n^{-1/2}\hat{\sigma}\big)\} \tag{13.19}$$

and

$$\{0 < \sigma \leq (n-1)^{1/2}\,\hat{\sigma}/\chi_{n-1}(1-\alpha)\}.$$

The $100(1-\alpha)\%$ CI for $\mu_2 - \mu_1$ associated with (13.18) is the same as (13.19) above. For σ_2/σ_1 this CI is given by

$$\{\min(1, \Delta^-) \leq \sigma_2/\sigma_1 \leq \max(1, \Delta^+)\},$$

where

$$\Delta^- = \frac{\hat{\sigma}_2}{\hat{\sigma}_1}\,\frac{\sqrt{1-t_1^2\hat{\rho}^2} - t_1\sqrt{1-\hat{\rho}^2}}{\sqrt{1-t_1^2}},\quad \Delta^+ = \frac{\hat{\sigma}_2}{\hat{\sigma}_1}\,\frac{\sqrt{1-t_1^2\hat{\rho}^2} + t_1\sqrt{1-\hat{\rho}^2}}{\sqrt{1-t_1^2}},$$

and $t_1 = \big(n-2+t_{n-2}^2(\alpha)\big)^{-1/2} t_{n-2}(\alpha)$. Finally, for ρ one could use the interval suggested by Fisher's z-transformation:

$$\{\rho \geq \tanh\big(\tanh^{-1}(\hat{\rho}) - z(\alpha)/(n-3)^{1/2}\big)\}.$$

A practical strategy for soliciting the various thresholds from the investigators is described in Choudhary and Nagaraja (2004c). They also illustrate how these CIs can be used in the assessment of agreement if the pre-specification of the thresholds is difficult. A simple sample size formula associated with the test of (13.17) is also given there.

Tests based on the IU principle tend to be conservative. Although the above overall IUT's have size α, it is attained in the limit as the variability diminishes to zero. But in practice, our major goal is to quantify the disagreement in a meaningful way. In this regard IU principle plays a natural role by producing CIs for individual disagreement indicators that are informative and easy to interpret.

Table 13.2 summarizes the basic features of the various approaches we have discussed thus far.

The hypotheses (13.17) and (13.18) resemble the ones used in bioequivalence studies [see, e.g., Berger and Hsu (1996)]. But for average bioequivalence, only the mean responses of the test and the reference drugs must be equivalent, and

Table 13.2: Summary of various approaches for assessing agreement

Approach	**Measure(s) of agreement**	**Remarks**
Limits of agreement	$(\mu - 1.96\sigma, \mu + 1.96\sigma)$	Easy to interpret; most popular
Concordance correlation	$\dfrac{2\rho\sigma_1\sigma_2}{(\mu_1 - \mu_2)^2 + \sigma_1^2 + \sigma_2^2}$	May be hard to interpret; sensitive to between-subject variation
ICC	$\rho_I = \sigma_T^2/(\sigma_T^2 + \sigma_\beta^2 + \sigma_\epsilon^2)$; $\rho_G = \sigma_1^2/(\sigma_1^2 + \sigma_\epsilon^2)$	$\rho_I \approx$ CCC; ρ_G requires a reference and assumes no bias; properties similar to CCC
Total deviation index(π_0)	$q^*(\pi_0)$ where $Pr(\|D\| \leq q^*(\pi_0)) \approx \pi_0$	Easy to interpret
Coverage probability (δ_0)	$Pr(\|D\| \leq \delta_0)$	Specifying δ_0 may be hard
IUT based on D	(μ, σ)	Identifies sources and extent of disagreement
IUT based on (X_1, X_2)	$(\mu_2 - \mu_1, \sigma_2/\sigma_1, \rho)$	Similar to IUT based on D, but more informative

for population bioequivalence, the marginal distributions of the responses must be equivalent. In contrast, here we also need the correlation to be close to 1.

A natural measure of agreement, already seen, is the MSD, $\theta = E(D^2) = (\mu^2+\sigma^2)$. The smaller the θ is, the better the agreement is between the methods. As noted earlier, Lin (2000) suggests estimating θ as $\hat{\theta} = \sum_{i=1}^n D_i^2/(n-1)$ and using the asymptotic normality of $\log(\hat{\theta})$ for inference. Hence, an approximate $100(1-\alpha)\%$ UCB for θ becomes $\hat{\theta}\exp\{z(\alpha)\hat{\tau}\}$ where $\hat{\tau}$ is given in (13.15). However, the practical utility of θ is limited by the fact that it is hard to interpret. Indeed, as Lin (1989, 2000) points out, the CCC and TDI approaches are attempts to translate this θ into more easily interpreted measures. But, θ (or equivalently $\log(\theta)$) has been the measure of choice in the selection of the instrument that agrees most with a reference. This is partly because the comparison of several methods in terms of this measure is more mathematically tractable. The problem of selection is the topic of Section 13.5.

13.4 An Example

We now illustrate the various approaches summarized in Table 13.2 using the plasma volume data from Bland and Altman (1999). The variable is measured as a percentage of expected values of normal individuals. Two sets of normal values, one due to Hurley (X_1) and the other due to Nadler (X_2) are being compared. We will take the Hurley method in the role of reference for the purpose of illustration. Figure 13.1 gives the scatter plot and the Bland-Altman plot. The solid line in the scatter plot represents the line of equality and the broken lines in the Bland-Altman plot represent the 95% LOA. We see that the two methods are highly correlated and the Nadler method consistently gives higher measurements. Most of the differences lie between 5% to 15%, are centered at around 10%, and increase as the magnitude of measurements increase.

Normal probability plots and formal tests indicate excellent normal fit for the differences and a reasonable bivariate normal fit for (X_1, X_2). Further, there is no evidence for heteroscedasticity or serious outliers in the mean versus difference plot. So we may assume that the model (13.1) holds along with normality. The positive trend observed in the mean versus difference plot is due to the higher variability of the Nadler method.

The measurements observed by the two methods fall in $(52.9, 133.2)$ and the differences $X_2 - X_1$ range from 2.5 to 17.40 with middle 50% in between 7.75 and 10.40. Since 7.75 and 10.40 respectively constitute about 10% and 13% of the measurement scale, these differences seem to be too high for the methods to have satisfactory agreement. This graphical analysis also indicates

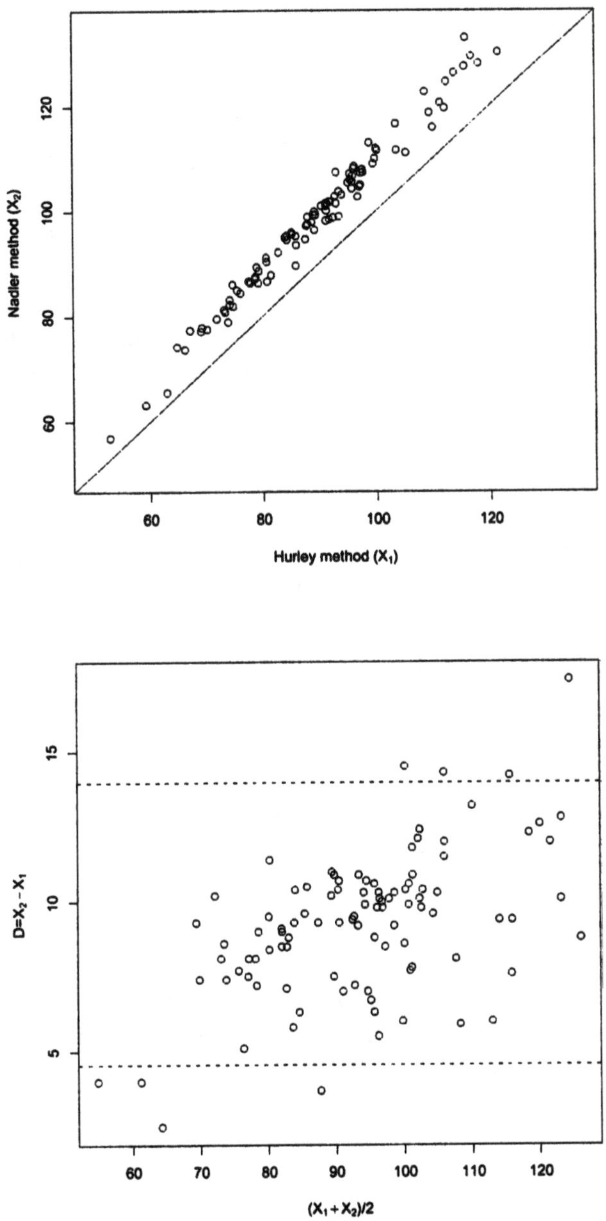

Figure 13.1: The scatter plot and the mean versus difference plot (or the Bland-Altman plot) of the plasma volume measurements

that the higher mean and variance of the Nadler method are the main sources of disagreement.

The various parameter estimates for these data are the following:

$$(\hat{\mu}, \hat{\sigma}) = (9.26, 2.40); \ (\hat{\mu}_1, \hat{\mu}_2, \hat{\sigma}_1, \hat{\sigma}_2, \hat{\rho}) = (89.24, 98.50, 13.89, 15.18, 0.99).$$

Table 13.3 presents the estimates of various measures of agreement summarized in Table 13.2 and their CIs when $(\delta_0, \pi_0) = (5, 0.95)$. All the approaches, except perhaps the CCC, confirm that there is substantial disagreement between the methods. The CCC may actually indicate satisfactory agreement since its LCB 0.78 is more than 0.75, a cutoff sometimes suggested for good agreement [see, e.g., Atkinson and Nevill (1997)]. The estimate of CCC is 0.82, which agrees with the estimate of ICC ρ_I (defined in (13.4), not shown in the table) up to two decimal places. This supports Nickerson's (1997) claim that these two quantities tend to be similar in practice. The LCB of ICC ρ_G is small.

From the LOA CIs, we infer that the interval in which the middle 95% of the differences are expected to lie can be as wide as (3.87, 14.66). The TDI approach says that the 95th percentile of the absolute differences can be as large as 19.69, and the CP approach says that the proportion of differences in the threshold interval $(-5, 5)$ can be as low as 2%. The CI for $\mu_2 - \mu_1$ from the two IUTs imply that μ_2 (the mean of Nadler method) is higher because the lower limit of the CI is zero, and this difference can be as high as 9.66. The difference based IUT also suggests that if the methods were considered equivalent, then the differences from $N(9.67, 2.74^2)$ must be considered acceptable. Note that the middle 95% of this distribution is between $9.67 \pm 1.96 \times 2.74 = (4.3, 15.0)$. From the IUT based on (X_1, X_2), we can also infer that σ_2 is higher than σ_1 and the ratio can be up to 1.12, and that the correlation ρ can only be as low as 0.99.

Our investigation thus far indicates that $\rho \approx 1$, $\mu_2 - \mu_1 \approx 9.50$ and $\sigma_2/\sigma_1 \approx 1.10$. Hence, a linear calibration of the Nadler method value (X_2) as $X_2' = aX_2 + b$ may make the methods agree well in the sense that $\mu_2' - \mu_1 \approx 0$ and $\sigma_2'/\sigma_1 \approx 1$ while their correlation remains close to one. Here $\mu_2' = E(X_2')$ and $\sigma_2' = SD(X_2')$. To find a and b, note that

$$\mu_2' - \mu_1 = a\mu_2 + b - \mu_1 \approx 9.50a + b - (1-a)\mu_1; \ \ \sigma_2'/\sigma_1 = a\sigma_2/\sigma_1 \approx 1.10a.$$

This suggests that $a = 1/1.10$ and since $\mu_1 \approx 90$, $b = (90(1.10-1)-9.50)/1.10 = -0.45 \approx -0.50$. All of the resulting measures of agreement for the recalibrated Nadler method $(X_2' = X_2/1.10 - 0.50)$ now show evidence of excellent agreement. This calibration, however, needs to be validated using an independent experiment.

Table 13.3: Estimates of various measures of agreement and their 95% CIs

Measure	Estimate	95% CI
95% LOA, $\mu \pm 1.96\sigma$	(4.55, 13.97)	$\mu - 1.96\sigma$: (3.87, ∞) $\mu + 1.96\sigma$: ($-\infty$, 14.66)
CCC, ρ_C	0.82	(0.78, 1.00)
ICC, ρ_G	0.68	(0.60, 1.00)
TDI, $q^*(0.95)$	18.84	(0, 19.69)
CP, $\pi(5)$	0.04	(0.02, 1.00)
IUT based on D, (μ, σ)	(9.26, 2.40)	μ: (0, 9.66) σ : (0, 2.73)
IUT based on (X_1, X_2), $(\mu_2 - \mu_1, \sigma_2/\sigma_1, \rho)$	(9.26,1.09, 0.99)	$\mu_2 - \mu_1$: (0, 9.66) σ_2/σ_1: (1.00, 1.12) ρ: (0.99, 1.00)

13.5 Selection Problems in Measuring Agreement

We now discuss a class of method comparison problems where $k\,(\geq 2)$ methods are compared with the gold standard with the goal of either

(a) selecting the best, i.e., the one that agrees most with the gold standard, or

(b) selecting the best when there is evidence that the best agrees sufficiently well with the gold standard.

In the literature, the problem (a) has been discussed in St. Laurent (1998), Hutson, Wilson and Geiser (1998), and Choudhary and Nagaraja (2004d), and the problem (b) in Choudhary and Nagaraja (2004c). We summarize them only for the $k = 2$ case.

Let G and W_i, $i = 1, 2$, represent measurements on a subject by the gold standard and the ith method. Also let D_i be the difference $W_i - G$. We assume that the vector $\mathbf{D} = (D_1, D_2)$ follows a bivariate normal (BVN) distribution with mean $\boldsymbol{\mu} = (\mu_1, \mu_2)$ and non-singular covariance matrix $\Sigma = (\sigma_{ij})_{2\times 2}$. Thus, the vector of squared differences $\mathbf{D}^{(2)} = (D_1^2, D_2^2)$ follows a continuous bivariate distribution with mean $\boldsymbol{\theta} = (\theta_1, \theta_2)$ and non-singular covariance matrix $\Gamma = (\gamma_{ij})_{2\times 2}$, where $\theta_i = \mu_i^2 + \sigma_{ii}$ and $\gamma_{ij} = 2\sigma_{ij}(\sigma_{ij} + 2\mu_i\mu_j)$, $i, j = 1, 2$. Finally, let $\boldsymbol{\lambda} = (\lambda_1, \lambda_2) = (\log(\theta_1), \log(\theta_2))$, and define

$$\Psi = (\psi_{ij})_{2\times 2} = \begin{pmatrix} \frac{\gamma_{11}}{\theta_1^2} & \frac{\gamma_{11}}{\theta_1^2} - \frac{\gamma_{12}}{\theta_1\theta_2} \\ \frac{\gamma_{11}}{\theta_1^2} - \frac{\gamma_{12}}{\theta_1\theta_2} & \frac{\gamma_{11}}{\theta_1^2} - 2\frac{\gamma_{12}}{\theta_1\theta_2} + \frac{\gamma_{22}}{\theta_2^2} \end{pmatrix}. \tag{13.20}$$

We take θ_i (or equivalently λ_i) as the measure of agreement between the ith instrument and the gold standard with preference for its smaller values. Thus the goal is to select the component of $\mathbf{D}^{(2)}$ having the smallest mean. In the literature on ranking and selection [see, e.g., Gupta and Panchapakesan (1979), for an excellent introduction], the problem of selecting the component having the smallest (or the largest) mean has been discussed in Mukhopadhyay and Chou (1984) for a k-variate normal population when all the correlations are non-negative. However in our case: (a) the multivariate normal assumption for $\mathbf{D}^{(2)}$ is generally not reasonable — we are assuming it for $\mathbf{D}$, and (b) the covariance matrix of $\mathbf{D}^{(2)}$ is not free of $\boldsymbol{\theta}$, the parameter of interest. So we cannot assume any structure for this matrix and hence the standard techniques of multiple comparisons with the best (MCB) [see Hsu (1996, Ch. 4) for an introduction] cannot be directly employed.

We assume each subject is measured only once by the three methods and suppose $\mathbf{D}_l = (D_{1l}, D_{2l})$, $l = 1, 2, \ldots$, is a sequence of i.i.d. observations on $\mathbf{D}$. Let $\hat{\boldsymbol{\mu}}_m = (\hat{\mu}_{1;m}, \hat{\mu}_{2;m})$ and $\hat{\Sigma}_m = (\hat{\sigma}_{ij;m})_{2\times 2}$ denote the usual unbiased estimators of $\boldsymbol{\mu}$ and Σ based on the first m observations on $\mathbf{D}$ (unbiasedness is not necessary for the validity of the large-sample result; it only requires that the estimators be consistent). The estimators $\hat{\boldsymbol{\theta}}_m$, $\hat{\Gamma}_m$, $\hat{\boldsymbol{\lambda}}_m$ and $\hat{\Psi}_m$ are then constructed by plugging-in $\hat{\boldsymbol{\mu}}_m$ and $\hat{\Sigma}_m$ for their population counterparts. We will omit the sample size from the notation of estimators when it is clear from the context.

13.5.1 Selection of the best

St. Laurent (1998) assumes a random effects model $W_i = G + \epsilon_i$ for the measurements, where $\epsilon_i, i = 1, 2$, are correlated random variables with zero means and distributed independently of G. This model assumes that the two instruments and the gold standard have the same means. Hence the equal agreement is equivalent to the equality of variances of ϵ_1 and ϵ_2. For the inference he uses a nonparametric bootstrap CI for the difference of the ICC's of (W_1, G) and (W_2, G). Recall from (13.6) that the ICC for (W_i, G) is $Var(G)/(Var(G) + Var(\epsilon_i))$. This approach is ad hoc and the equality of means assumption cannot always be justified. In addition, we have seen that the ICC is hard to interpret.

Hutson et al. (1998) consider a large sample $100(1-\alpha)\%$ CI for $\theta_1/(\theta_1+\theta_2)$ constructed from a sample of size n. They infer the first (second) instrument as the true best if the upper (lower) bound of this CI is less (greater) than 0.5, and remain indecisive if the interval contains 0.5. They estimate the mean vector and the covariance matrix of $\mathbf{D}^{(2)}$ with their sample counterparts. When bivariate normality for $\mathbf{D}$ is assumed, the large sample $100(1-\alpha)\%$ level CI for

$\theta_1/(\theta_1+\theta_2)$ becomes

$$\frac{\hat{\theta}_1}{(\hat{\theta}_1+\hat{\theta}_2)} \pm \frac{z(\alpha/2)}{n^{1/2}(\hat{\theta}_1+\hat{\theta}_2)^2}\left(\hat{\gamma}_{11}\hat{\theta}_2^2 - 2\hat{\gamma}_{12}\hat{\theta}_1\hat{\theta}_2 + \hat{\gamma}_{22}\hat{\theta}_1^2\right)^{1/2}.$$

Using Monte Carlo simulations, Choudhary (2002) shows that this CI has a substantial under-coverage. Instead, they recommend $\lambda_1 - \lambda_2$, where $\lambda_i = \log(\theta_i)$, as the metric for comparison since the asymptotic procedures based on the estimators of $\lambda_1 - \lambda_2$ tend to do well for samples with sizes as low as 15. For the purpose of selection, they suggest two approximate CI for $\lambda_1 - \lambda_2$ both of which have asymptotic level $1-\alpha$:

$$\left[\hat{\lambda}_1 - \hat{\lambda}_2 - n^{-1/2}t_{n-1}(\alpha/2)\hat{\psi}_{22}^{1/2},\ \hat{\lambda}_1 - \hat{\lambda}_2 + n^{-1/2}t_{n-1}(\alpha/2)\hat{\psi}_{22}^{1/2}\right],$$
$$\left[\min\{0, \hat{\lambda}_1 - \hat{\lambda}_2 - n^{-1/2}t_{n-1}(\alpha)\hat{\psi}_{22}^{1/2}\},\ \max\{0, \hat{\lambda}_1 - \hat{\lambda}_2 + n^{-1/2}t_{n-1}(\alpha)\hat{\psi}_{22}^{1/2}\}\right].$$

Here $\hat{\psi}_{22}$ is the estimate of ψ_{22} in (13.20). Using the first interval, one infers the first instrument to be the best if its upper (lower) bound is negative (positive), and is indecisive otherwise. The second interval is a constrained (to contain zero) MCB CI for $\lambda_1 - \lambda_2$ [see Hsu (1996, ch. 4)]. If the possibility that $\lambda_1 = \lambda_2$ is ruled out, one infers the first (second) instrument as the best if the upper (lower) bound of this interval equals zero. But when this procedure identifies one instrument to be the best, it does not give a negative upper bound on how much better it is when compared to the unselected. On the other hand, the unconstrained interval allows such an inference. However, the advantage of sacrificing this information is that the constrained interval identifies an instrument to be the best more frequently than the unconstrained one at the same asymptotic level. This sharper inference is desirable for us. Further, assuming $\lambda_1 \neq \lambda_2$ is reasonable from a practical viewpoint as it amounts to assuming that the two instruments do not agree *equally* with the gold standard.

In practice, single-stage CI procedures such as the above may fail to distinguish between two instruments. This difficulty can be avoided by using a two-stage procedure. However, for this, the investigator has to pre-specify a threshold δ (> 0) such that whenever $|\lambda_1 - \lambda_2| < \delta$, the two instruments are considered *practically equivalent*, and then the *correct* selection is not important. This δ is also known as the *indifference-zone* in the terminology of ranking and selection procedures [see, e.g., Gupta and Panchapakesan (1979)]. Consider the following two-stage procedure:

Stage 1: Select a random sample of size m, compute $\hat{\psi}_{22;m}$, and define

$$N_m = \max\left\{\left\lceil t_{m-1}^2(\alpha)\,\hat{\psi}_{22;m}\,\delta^{-2}\right\rceil, m\right\}, \tag{13.21}$$

where $\hat{\psi}_{22;m}$ is the estimate of ψ_{22} in (13.20).

Stage 2: Take $N_m - m$ additional i.i.d. observations if $N_m > m$ and compute the estimates $\hat{\lambda}_{i;N_m}, i = 1, 2$, using the complete two-stage sample. Then select the instrument that produces the smaller estimate as the best.

Choudhary and Nagaraja (2004d) show that when m is large and $|\lambda_1 - \lambda_2| \geq \delta$, the probability of correct selection with this procedure is approximately $1-\alpha$. When the differences are bivariate normal, $m = 15$ is a reasonable choice for the first-stage sample size.

13.5.2 Assessment of agreement and selection of the best

Above we focussed on the issue of selection of the best when two instruments are compared with a gold standard. However, for a practitioner, knowing the best instrument is unhelpful unless it also agrees sufficiently closely with the gold standard. Often times this information is not available in advance. Assuming $\lambda_1 \neq \lambda_2$, Choudhary and Nagaraja (2004c) address this problem by developing a two-stage procedure that first determines whether the best has satisfactory agreement with the gold standard through a test of

$$H_7 : \lambda_{[1]} \geq \lambda_0 \quad \text{vs} \quad K_7 : \lambda_{[1]} < \lambda_0$$

before proceeding to its selection. Here [1] is the unknown label of the true best among the two instruments and λ_0 is a user-specified cutoff such that $\{\lambda_i < \lambda_0\}$ is the region of satisfactory agreement.

Let δ be a threshold for practical equivalence of $\lambda_{[1]}$ with λ_0, and of λ_1 with λ_2, in the sense that whenever $|\lambda_{[1]} - \lambda_0| < \delta$ or $|\lambda_1 - \lambda_2| < \delta$ the distinction between the two quantities is not important from practical considerations. Finally, define

$$\hat{\lambda}_{(1)} = (\hat{\lambda}_1 - \hat{\lambda}_2) I(\hat{\lambda}_1 \leq \hat{\lambda}_2) + \hat{\lambda}_2, \; \hat{\psi}_{(11)} = \left(\frac{\hat{\gamma}_{11}}{\hat{\theta}_1^2} - \frac{\hat{\gamma}_{22}}{\hat{\theta}_2^2} \right) I(\hat{\lambda}_1 \leq \hat{\lambda}_2) + \frac{\hat{\gamma}_{22}}{\hat{\theta}_2^2},$$

where $I(A)$ is the indicator function of event A, and take

$$\hat{\nu} = \left(\hat{\psi}_{(11)} - \hat{\gamma}_{12}/(\hat{\theta}_1\hat{\theta}_2)\right) / \left(\hat{\psi}_{(11)}\hat{\psi}_{22}\right)^{1/2}.$$

For pre-specified α, β $(0 < \alpha < 1-\beta < 1)$, λ_0, δ (> 0) and m (≥ 2); Choudhary and Nagaraja (2004c) propose the following two-stage procedure:

Stage 1: Take a random sample of size m and compute the estimates $\hat{\boldsymbol{\mu}}_m$ and $\hat{\Sigma}_m$. Use them to compute $\hat{\psi}_{22;m}$, and $\hat{\psi}_{(11);m}$ and $\hat{\nu}_m$ defined above. Then solve the equation

$$\Phi_2 \left(\frac{L_m^{1/2} \delta}{\hat{\psi}_{(11);m}^{1/2}} - z(\alpha), \frac{L_m^{1/2} \delta}{\hat{\psi}_{22;m}^{1/2}} ; \hat{\nu}_m \right) = 1 - \beta$$

for L_m, and define $N_m = \max\{\lceil L_m \rceil, m\}$ as the second-stage sample size, where $\lceil L_m \rceil$ denotes the smallest integer $\geq L_m$.

Stage 2: Take $N_m - m$ additional i.i.d. observations if $N_m > m$. Compute $\hat{\lambda}_{(1);N_m}$ and $\hat{\psi}_{(11);N_m}$ using the entire sample. Reject H_7 when $\hat{\lambda}_{(1);N_m} + N_m^{-1/2}\, z(\alpha)\, \hat{\psi}^{1/2}_{(11);N_m} \leq \lambda_0$. Further, when H_7 is rejected, infer the instrument that produces $\hat{\lambda}_{(1);N_m}$ as the best.

This procedure has the property that, when m is large,

$$Pr(\text{reject } H_7, \text{ correct selection}) \approx 1 - \beta \text{ or more},$$

whenever $\lambda_0 - \lambda_{[1]} \geq \delta$ and $|\lambda_1 - \lambda_2| \geq \delta$. Using simulation studies to verify the small-sample properties of this procedure, Choudhary and Nagaraja (2004c) suggest $m = 15$ to be a reasonable choice for the first-stage sample size.

13.6 Concluding Remarks

For the assessment of agreement, we assumed that both the measurements are random. To handle the case when the reference measurements are fixed (nonrandom), some of the procedure presented here has been adapted by Lin et al. (2002). Further, the approaches discussed here are generally not robust to deviations from normality or outliers. When this assumption is a suspect, none of them are valid, but a simple nonparametric sign test can be used [see Bland and Altman (1999)]. Further, this sign test can be inverted to give a nonparametric UCB for $q(\pi_0)$.

Our discussion assumed the simple model (13.1) for the measurements. It can be easily extended to handle replicate measurements. More complicated models such as those including method-subject interactions, covariates or the effect of time on the repeated measurements may be called for as well. See Dunn and Roberts (1999), Bland and Altman (1999), and Chinchilli et al. (1996), for some such models. The CCC approach has been generalized in several directions. Similar extensions of the difference based approaches will be of interest.

Acknowledgments. The second author's research was supported by NIH (NCRR) grant # M01 RR00034 awarded to The Ohio State University.

References

1. Altman, D. G. and Bland, J. M. (1983). Measurement in medicine: The analysis of method comparison studies, *The Statistician*, **32**, 307–317.

2. Anderson, S. and Hauck, W. W. (1990). Consideration of individual bioequivalence, *Journal of Pharmacokinetics and Biopharmaceutics*, **18**, 259–274.

3. Atkinson, G. and Nevill, A. (1997). Comment on the use of concordance correlation to assess the agreement between two variables, *Biometrics*, **53**, 775–777.

4. Banerjee, M., Capozzoli, M., McSweeney, L., and Sinha, D. (1999). Beyond Kappa: A review of interrater agreement measures, *The Canadian Journal of Statistics*, **27**, 3–23.

5. Barnhart, H. X., Haber, M., and Song, J. L. (2002). Overall concordance correlation coefficient for evaluating agreement among multiple observers, *Biometrics*, **58**, 1020–1027.

6. Barnhart, H. X. and Williamson, J. M. (2001). Modeling concordance correlation via GEE to evaluate reproducibiltiy, *Biometrics*, **57**, 931–940.

7. Bartko, J. J. (1994). Measures of agreement: A single procedure, *Statistics in Medicine*, **13**, 737–745.

8. Berger, R. L. and Hsu, J. C. (1996). Bioequivalence trials, intersection-union tests and equivalence confidence sets, *Statistical Science*, **11**, 283–319.

9. Bland, J. M. and Altman, D. G. (1986). Statistical methods for assessing agreement between two methods of clinical measurement, *Lancet*, **i**, 307–310.

10. Bland, J. M. and Altman, D. G. (1990). A note on the use of the intraclass correlation coefficient in the evaluation of agreement between two methods of measurement, *Computers in Biology and Medicine*, **20**, 337–340.

11. Bland, J. M. and Altman, D. G. (1995a). Comparing two methods of clinical measurement: A personal history, *International Journal of Epidemiology*, **24**, S7–S14.

12. Bland, J. M. and Altman, D. G. (1995b). Comparing methods of measurement: Why plotting difference against standard method is misleading, *Lancet*, **346**, 1085–1087.

13. Bland, J. M. and Altman, D. G. (1999). Measuring agreement in method comparison studies, *Statistical Methods in Medical Research*, **8**, 135–160.

14. Cameron, J. M. (1982). Calibration, In *Encyclopedia of Statistical Sciences*, **1**, John Wiley & Sons, New York, pp. 346–351.

15. Casella, G. and Berger, R. (2002) *Statistical Inference*, 2nd edition, Duxbury Press, Pacific Grove, CA.

16. Chinchilli, V. M., Martel, J. K., Kumanyika, S., and Lloyd, T. (1996). A weighted concordance correlation coefficient for repeated measurement designs, *Biometrics*, **52**, 341–353.

17. Choudhary, P. K. (2002). Assessment of Agreement and Selection of the Best Instrument in Method Comparison Studies, *Ph.D. Dissertation*, The Ohio State University, Columbus, OH.

18. Choudhary, P. K. and Nagaraja, H. N. (2004a). Tests for assessment of agreement using probability criteria, *Submitted for publication.*

19. Choudhary, P. K. and Nagaraja, H. N. (2004b). Assessment of agreement using intersection-union principle, *Biometrical Journal* (to appear).

20. Choudhary, P. K. and Nagaraja, H. N. (2004c). A two-stage procedure for selection and assessment of agreement of the best with a gold standard, *Sequential Analysis* (to appear).

21. Choudhary, P. K. and Nagaraja, H. N. (2004d). Selecting the instrument closest to a gold standard, *Journal of Statistical Planning and Inference* (to appear).

22. David, H. A. and Nagaraja, H. N. (2003). *Order Statistics*, Third edition, John Wiley & Sons, New York.

23. Dunn, G. (1989). *Design and Analysis of Reliability Studies: The Statistical Evaluation of Measurement Errors*, Oxford University Press, New York.

24. Dunn, G. (1992). Design and analysis of reliability studies, *Statistical Methods in Medical Research*, **1**, 123–157.

25. Dunn, G. and Roberts, C. (1999). Modelling method comparison data, *Statistical Methods in Medical Research*, **8**, 161–179.

26. Fleiss, J. L. (1981). *Statistical Methods for Rates and Proportions*, John Wiley & Sons, New York.

27. Fleiss, J. L. (1986). *The Design and Analysis of Clinical Experiments*, John Wiley & Sons, New York.

28. Fuller, W. A. (1987). *Measurement Error Models*, John Wiley & Sons, New York.

29. Guttman, I. (1988). Statistical tolerance regions, *Encyclopedia of Statistical Sciences*, **9**, pp. 272–287, John Wiley & Sons, New York.

30. Grubbs, F. E. (1982). Grubbs' estimators, In *Encyclopedia of Statistical Sciences*, **2**, pp. 542–549, John Wiley & Sons, New York.

31. Gupta, S. S. and Panchapakesan, S. (1979). *Multiple Decision Procedures – Theory and Methodology of Selecting and Ranking Populations*, John Wiley, New York. Republished by SIAM, Philadelphia, 2002.

32. Hamilton, D. C. and Lesperance, M. L. (1995). A comparison of methods for univariate and multivariate acceptance sampling by variables, *Technometrics*, **37**, 329–339.

33. Harris, I. R., Burch, B. D. and St. Laurent, R. T. (2001). A blended estimator for measure of agreement with a gold standard, *Journal of Agricultural, Biological, and Environmental Statistics*, **6**, 326–339.

34. Hawkins, D. M. (2002). Diagnostics for conformity of paired quantitative measurements, *Statistics in Medicine*, **21**, 1913–1935.

35. Hsu, J. C. (1996). *Multiple Comparisons: Theory and Methods*, Chapman & Hall/CRC, Boca Raton, FL.

36. Hutson, A. D., Wilson, D. C., and Geiser, E. A. (1998). Measuring relative agreement: Echocardiographer versus computer, *Journal of Agricultural, Biological, and Environmental Statistics*, **3**, 163–174.

37. Kelly, G. E. (1985). Use of structural equations model in assessing the reliability of a new measurement technique, *Applied Statistics*, **34**, 258–263.

38. King, T. S. and Chinchilli, V. M. (2001). A generalized concordance correlation coefficient for continuous and categorical data, *Statistics in Medicine*, **20**, 2131 – 2147.

39. Kraemer, H. C., Periyakoil, V. S., and Noda, A. (2002). Kappa coefficients in medical research, *Statistics in Medicine*, **21**, 2109–2129.

40. Krummenauer, F. (1999). Intraindividual scale comparison in clinical diagnostic methods: A review of elementary methods, *Biometrical Journal*, **41**, 917–929.

41. Lee, J., Koh, D., and Ong, C. N. (1989). Statistical evaluation of agreement between two methods for measuring a quantitative variable, *Computers in Biology and Medicine*, **19**, 61–70.

42. Lewis, P. A., Jones, P. W., Polak, J. W., and Tillotson, H. T. (1991). The problem of conversion in method comparison studies, *Applied Statistics*, **40**, 105–112.

43. Liao, J. and Lewis, J. (2000). An agreement curve, Presented at the Joint Statistical Meetings, Indianapolis, IN.

44. Lin, L. I. (1989). A concordance correlation coefficient to evaluate reproducibility, *Biometrics*, **45**, 255–268. Corrections: 2000, **56**, 324–325.

45. Lin, L. I. (1992). Assay validation using the concordance correlation coefficient, *Biometrics*, **48**, 599–604.

46. Lin, L. I. (2000). Total deviation index for measuring individual agreement with applications in laboratory performance and bioequivalence, *Statistics in Medicine*, **19**, 255–270.

47. Lin, L. I. (2003). Measuring agreement. In *Encyclopedia of Biopharmaceutical Statistics*, 2nd edition, pp. 561–567, Marcel Dekker, New York.

48. Lin, L. I. and Chinchilli, V. (1997). Rejoinder to the letter to the editor from Atkinson and Nevill, *Biometrics*, **53**, 777–778.

49. Lin, L. I., Hedayat, A. S., Sinha, B. and Yang, M. (2002). Statistical methods in assessing agreement: Models, issues, and tools, *Journal of the American Statistical Association*, **97**, 257–270.

50. Lin, L. I. and Torbeck, L. D. (1998). Coefficient of accuracy and concordance correlation coefficient: New statistics for method comparison, *PDA Journal of Pharmaceutical Science and Technology*, **52**, 55–59.

51. Lin, S. C., Whipple, D. M., and Ho, C. S. (1998). Evaluation of statistical equivalence using limits of agreement and associated sample size calculation, *Communications in Statistics—Theory and Methods*, **27**, 1419–1432.

52. Linnet, K. (1993). Evaluation of regression procedures for method comparison studies, *Clinical Chemistry*, **39**, 424–432.

53. Liu, J.-P. and Chow, S.-C. (1997). A two one-sided tests procedure for assessment of individual bioequivalence, *Journal of Biopharmaceutical Statistics*, **7**, 49–61.

54. McGraw, K. O. and Wong, S. P. (1996). Forming inferences about some intraclass correlation coefficients, *Psychological Methods*, **1**, 30–46.

55. Mukhopadhyay, N. and Chou, W.-S. (1984). On selecting the best component of a multivariate normal population, *Sequential Analysis*, **3**, 1–22.

56. Müller, R. and Büttner, P. (1994). A critical discussion of intraclass correlation coefficients, *Statistics in Medicine*, **13**, 2465–2476.

57. Nickerson, C. A. (1997). Comment on "A concordance correlation coefficient to evaluate reproducibility", *Biometrics*, **53**, 1503–1507.

58. Nix, A. B. J. and Dunston, F. D. J. (1991). Maximum likelihood techniques applied to method comparison studies, *Statistics in Medicine*, **10**, 981–988.

59. Quan, H. and Shih, W. J. (1996). Assessing reproducibility by the within-subject coefficient of variation with random effects models, *Biometrics*, **52**, 1195–1203. Correspondence, *Biometrics*, **56**, 301–302.

60. Robieson, W. Z. (1999). On the weighted kappa and concordance correlation coefficient, *Ph.D. Dissertation*, University of Illinois at Chicago, IL.

61. Shoukri, M. M. (1999). Measurement of Agreement, In *Encyclopedia of Biostatistics*, **1**, pp. 103–117, John Wiley & Sons, New York.

62. Shoukri, M. M. (2004). *Measures of Interobserver Agreement*, Chapman & Hall/CRC, Boca Raton, FL.

63. St. Laurent, R. T. (1998). Evaluating agreement with a gold standard in method comparison studies, *Biometrics*, **54**, 537–545.

64. Vonesh, E. F., Chinchilli, V. P., and Pu, K. W. (1996). Goodness-of-fit in generalized nonlinear mixed-effects models, *Bometrics*, **52**, 572-587.

65. Wang, W. and Hwang, J. T. G. (2001). A nearly unbiased test for individual bioequivalence problems using probability criteria, *Journal of Statistical Planning and Inference*, **99**, 41–58.

14

Measures of Concordance for Assessing Agreement in Ratings and Rank Order Data

M. Raghavachari

Decision Sciences and Engineering Systems, Rensselaer Polytechnic Institute, Troy, NY U.S.A.

Abstract: Consider an array of ratings or rank orders provided by M judges or respondents on n subjects or products. We propose a new general measure of concordance (agreement) in this context. This measure when applied to rank order data reduces to Kendall's (1948) measure of concordance. A new measure for rank order data that depends on the average Kendall's measure of rank correlation is also proposed and is also a special case of our general measure. This particular rank order measure can be considered as an alternative to the well-known Friedman's statistic for the two-way analysis of variance. The general measure is also compared with another measure proposed by Lin (1989). Relations to the intraclass correlation coefficient of the ratings data are pointed out. Some distributional results are also presented. The proposed concordance measures provide the basis for testing the agreement between two or more methods, instruments or respondents in biometric research or market research surveys.

Keywords and phrases: Concordance coefficient, intraclass correlation, rank correlations, paired comparisons

14.1 Introduction

Consider M rankings of n subjects. Each ranking is a permutation of the integers $1, 2, \ldots, n$. A measure W of concordance among the M rankings was proposed by Kendall and Smith (1939) and Kendall (1948) and they provided a test of hypothesis to test for independence of the rankings. For $i = 1, 2, \ldots, n$, let R_{ip} denote the rank of the ith subject by judge p, $p = 1, 2, \ldots, M$, and $R_{i.}$ denote the sum of the ranks of the ith subject over all the M judges. The concordance measure W is defined as

$$W = \frac{12S}{M^2(n^3 - n)} \tag{14.1}$$

where

$$S = \sum_{i=1}^{n} \left[R_{i.} - \frac{M(n+1)}{2}\right]^2.$$

It measures the ratio of the variability of the n row sums $R_{i.}$ for the data to the variability of row sums when the M rankings are all identical. It is known that $0 \le W \le 1$. The measure W is linearly related to the average denoted by $\overline{r}_s$ of $M(M-1)/2$ Spearman correlation coefficients between pairs of rankings by the relation

$$\overline{r}_s = \frac{MW - 1}{M - 1}. \tag{14.2}$$

See Kendall (1948, p. 82) for details. The measure W is also functionally related to Friedman's statistic (1937) for two-way analysis of variance. Suppose the data are not rank ordered but represent, for example, the ratings of n subjects by M judges or n products by M respondents in a market research survey. There has been some research, primarily in the area of biometrics, on the development of measures of concordance to assess the agreement of the M judges or respondents or methods when the data represents ratings rather than ranks. See for example Lin et al. (2002) for the case $M = 2$, and the references cited therein. See also the review by Choudhary and Nagaraja (2005) in this volume.

The present paper studies the problem of concordance when the data set consists of interval scaled values. Here we propose a new general measure of concordance Q and investigate its properties. This measure is a modification of the intraclass correlation coefficient and is in the form of a quadratic form that depends on the correlation matrix. This measure can also be considered as a generalization of the Kendall measure of concordance W to interval scaled data. This is because Q reduces to Kendall's W for the rank order data. The measure Q for the case $M = 2$ is also similar but different from the concordance correlation coefficient proposed by Lin (1989).

Equation (14.2) shows the relationship of W with the Spearman correlations. While commenting upon the feasibility of constructing a concordance measure for the rank order data that depends on Kendall correlation coefficients, Kendall (1948, p. 82) states that "The case we are considering is one wherein ρ (Spearman correlation coefficient) is a more convenient coefficient than τ (Kendall's coefficient). There appears to be no simple method of expressing the average τ in terms of the sum of ranks". Earlier Kendall and Smith (1940) proposed a coefficient of agreement in the case of paired comparison data and related it for the rank order data to the average τ which is the average of $M(M-1)/2$ Kendall correlation coefficients between pairs of rankings. Following their work, Ehrenberg (1952) proposed the average τ as a measure of concordance for rank order data as an alternative to the earlier Kendall's measure of concordance W. While W ranges from 0 to 1, average τ proposed by Ehrenberg can be negative. Ehrenberg's measure is also linearly related to the measure of agreement proposed by Kendall and Smith (1940).

In this paper, we show that the average τ indeed can be expressed as a simple function of the rank order data. This leads to a new measure of concordance W_τ for the rank order data and is based on the Kendall's rank correlation coefficients. The measure W_τ is also related to the coefficient of agreement of Kendall and Smith (1940) and the average τ. We show that W, Q, and W_τ have the same structure and belong to a class of measures based on the appropriate correlation matrices and this fact unifies them according to a single principle. These measures are also shown to be functions of intraclass correlation coefficients applied to appropriate sets of data. Some discussion on the appropriate population measures of concordance for Q, W and W_τ is also presented. Distributions of the general measure Q for the independent and the equi-correlated cases are derived. Relations of Q with intraclass correlation coefficient r^* for the data are also pointed out.

Fisher (1921) derived the distribution of r^*. He did not give the complete derivation but stated the density function involving the terms of r^* only. In this paper we derive and state completely the pdf of r^*. This, in turn, provides what appears to be a new transformation of the intraclass correlation statistic that has an F distribution. Some future research work on the new measure Q and the new rank measure W_τ is also indicated.

14.2 A General Measure of Concordance

Consider an array of numbers in M rows and n columns. Let y_{ip} represent the element in the pth row and ith column, $p = 1, 2, \ldots, M$; $i = 1, 2, \ldots, n$. The element y_{ip} may denote, for example, a rating for the ith product or subject by respondent or judge p. It could also be a rank of product or subject i by respondent or judge p. The score matrix (y_{ip}) can be shown as

$$\begin{array}{cccc} y_{11} & y_{21} & \cdots & y_{n1} \\ y_{12} & y_{22} & \cdots & y_{n2} \\ \cdots & \cdots & \cdots & \cdots \\ y_{1M} & y_{2M} & \cdots & y_{nM} \end{array} \tag{14.3}$$

When the y_{ip}'s are rank orders i.e., for a given $p = 1, 2, \ldots, M$, $y_{1p}, y_{2p}, \cdots, y_{np}$ is a permutation of the integers $1, 2, \ldots, n$. Kendall proposed the measure of concordance W given by (14.1). First we define a measure of concordance for the general data given by (14.3). Let

$$y_{.p} = \sum_{i=1}^{n} y_{ip} \quad \text{and} \quad \overline{y}_{.p} = y_{.p}/n, p = 1, 2, \ldots, M.$$

Further, let $\overline{\overline{Y}} = \sum_{n=1}^{M} \overline{y}_{.p}/M$,

$$\lambda_p^2 = \frac{\sum_{i=1}^{n}(y_{ip} - \overline{y}_{.p})^2}{\sum_{i=1}^{n}\sum_{p=1}^{M}(y_{ip} - \overline{\overline{Y}})^2}, \quad p = 1, 2, \ldots, M, \tag{14.4}$$

and $\mu_p = \lambda_p/\sqrt{M}$, $p = 1, 2, \ldots, M$.

Let $\mathbf{R}$ denote the $M \times M$ correlation matrix (r_{pq}) where r_{pq} is the Pearson correlation coefficient between $\mathbf{y}'_p = (y_{1p}, y_{2p}, \ldots, y_{np})$ and $\mathbf{y}'_q = (y_{1q}, y_{2q}, \ldots, y_{nq})$. We propose a concordance measure Q defined by the quadratic form

$$Q = \boldsymbol{\mu}'\mathbf{R}\boldsymbol{\mu} \tag{14.5}$$

where $\boldsymbol{\mu}' = (\mu_1, \mu_2, \ldots, \mu_M)$. The rationale for the above expression arises from the consideration of the intraclass correlation coefficient of the data set (14.3). This measure is also motivated by the relationship between W in (14.1) and the average of the pairwise Spearman correlation coefficients of the rank order data.

Relation to Intraclass correlation coefficient r^*

The intraclass correlation coefficient arises in many fields; for example, in cluster sampling and in the determination of correlation among the members of a class with regard to a particular characteristic. We review briefly the definition, which forms the basis for further analysis.

Let $y_{i.}$ denote the total of the scores for the ith subject over the M judges, i.e., $y_{i.} = \sum_{p=1}^{M} y_{ip}$, $i = 1, 2, \ldots, n$. Let $\overline{Y} = \sum_{i=1}^{n} y_{i.}/n$ be the mean per subject and $\overline{\overline{Y}} = \overline{Y}/M$ be the mean per element. Let S_b^2 be the variance of the subject totals,

$$S_b^2 = \frac{\sum_{i=1}^{n}(y_{i.} - \overline{Y})^2}{M(n-1)}, \tag{14.6}$$

and S^2 be the overall variance

$$S^2 = \frac{\sum_{i=1}^{n}\sum_{p=1}^{M}(y_{ip} - \overline{\overline{Y}})^2}{(NM-1)}. \tag{14.7}$$

Then intraclass correlation coefficient r^* among the judges is defined as

$$r^* = \frac{(n-1)MS_b^2 - (NM-1)S^2}{(nM-1)(M-1)S^2}. \tag{14.8}$$

See Cochran (1977, p. 243) for details where the intraclass correlation coefficient is introduced in the context of cluster sampling. We show first that r^* is related to Q by the following relation.

Proposition 14.2.1 *The intraclass correlation can be expressed as*

$$r^* = \frac{M\boldsymbol{\mu}'\mathbf{R}\boldsymbol{\mu} - 1}{M-1} = \frac{MQ-1}{M-1},$$

where $\boldsymbol{\mu}' = (\mu_1, \mu_2, \ldots, \mu_M)$ *with* $\mu_p = \lambda_p/\sqrt{M}$, *and* $\mathbf{R}$ *is the* $M \times M$ *Pearson correlation matrix.*

PROOF. It follows from (14.6), (14.7) and (14.8) that

$$r^* = \frac{1}{M-1}\Big[\frac{\sum_{i=1}^{n}(y_{i\cdot} - \overline{Y})^2}{\sum_{i=1}^{n}\sum_{p=1}^{M}(y_{ip} - \overline{\overline{Y}})^2} - 1\Big]. \tag{14.9}$$

Note that

$$\sum_{i=1}^{n}(y_{i\cdot} - \overline{Y})^2 = \sum_{i=1}^{n}\sum_{p=1}^{M} y_{ip}^2 + \sum_{p=1}^{M}\sum_{q\neq p}^{M}\Big(\sum_{i=1}^{n} y_{ip}y_{iq}\Big) + n\overline{Y}^2 - 2\overline{Y}\sum_{i=1}^{n}\sum_{p=1}^{M} y_{ip}. \tag{14.10}$$

This follows from the two equations,

$$\begin{aligned}
\Big(\textstyle\sum_{p=1}^{M} y_{ip}\Big)^2 &= \sum_{p=1}^{M} y_{ip}^2 + \sum_{p=1}^{M}\sum_{q\neq p}^{M} y_{ip}y_{iq} \quad \text{and} \\
(y_{i\cdot} - \overline{Y})^2 &= \Big(\textstyle\sum_{p=1}^{M} y_{ip} - \overline{Y}\Big)^2 \\
&= \Big(\textstyle\sum_{p=1}^{M} y_{ip}\Big)^2 + \overline{Y}^2 - 2\overline{Y}\sum_{p=1}^{M} y_{ip}.
\end{aligned}$$

that hold for all $i = 1, 2, \ldots, n$. The right side of (14.10) can be written as

$$\begin{aligned}
&\sum_{i=1}^{n}\sum_{p=1}^{M} y_{ip}^2 + \sum_{p=1}^{M}\sum_{q\neq p}^{M}\Big\{\sum_{i=1}^{n} y_{ip}y_{iq} - \frac{1}{n}\Big(\sum_{i=1}^{n} y_{ip}\Big)\Big(\sum_{i=1}^{n} y_{iq}\Big)\Big\} \\
&\quad + \sum_{p=1}^{M}\sum_{q\neq p}^{M}\frac{1}{n}\Big(\sum_{i=1}^{n} y_{ip}\Big)\Big(\sum_{i=1}^{n} y_{iq}\Big) + n\overline{Y}^2 - 2\overline{Y}nM\overline{\overline{Y}}.
\end{aligned}$$

Note that $n\overline{Y}^2 - 2\overline{Y}nM\overline{\overline{Y}} = n\overline{Y}^2 - 2n\overline{Y}^2 = -n\overline{Y}^2$. Further,

$$\sum_{i=1}^{n}\sum_{p=1}^{M}(y_{ip} - \overline{\overline{Y}})^2 = \sum_{i=1}^{n}\sum_{p=1}^{M} y_{ip}^2 - nM\overline{\overline{Y}}^2.$$

After some algebra, we have

$$r^* = \frac{1}{M-1} \frac{\sum_{p=1}^{M} \sum_{q \neq p}^{M} \operatorname{cov}(\mathbf{y}_p, \mathbf{y}_q) + Z}{\sum_{i=1}^{n} \sum_{p=1}^{M} (y_{ip} - \overline{\overline{Y}})^2} \tag{14.11}$$

where

$$Z = \sum_{p=1}^{M} \sum_{q \neq p}^{M} \frac{y_{.p} y_{.q}}{n} - n\overline{Y}^2 + nM\overline{\overline{Y}}^2 .$$

The first term in Z is $\frac{1}{n}[(\sum_{p=1}^{M} y_{.p})^2 - (\sum_{p=1}^{M} y_{.p}^2)]$. Hence it follows that

$$Z = -\frac{1}{n}(\sum_{p=1}^{M} y_{.p}^2) + nM\overline{\overline{Y}}^2 = -n \sum_{p=1}^{M} (\overline{y}_{.p} - \overline{\overline{Y}})^2. \tag{14.12}$$

From the definition of λ_p in (14.4) and by (14.11) and (14.12), it follows that

$$r^* = \frac{1}{M-1} \left[\sum_{p=1}^{M} \sum_{q \neq p}^{M} r_{pq} \lambda_p \lambda_q - n \frac{\sum_{p=1}^{M} (\overline{y}_{.p} -)^2}{\sum_{i=1}^{n} \sum_{p=1}^{M} (y_{ip} - \overline{\overline{Y}})^2} \right]. \tag{14.13}$$

Since

$$\sum_{i=1}^{n} \sum_{p=1}^{M} (y_{ip} - \overline{y}_{.p})^2 = \sum_{i=1}^{n} \sum_{p=1}^{M} (y_{ip} -)^2 - n \sum_{p=1}^{M} (\overline{y}_{.p} - \overline{\overline{Y}})^2$$

and $r_{pp} = 1$, $p = 1, 2, \ldots, M$, (14.13) implies that

$$r^* = \frac{1}{M-1} \Big[\sum_{p=1}^{M} \sum_{q=1}^{M} r_{pq} \lambda_p \lambda_q - 1 \Big].$$

In matrix notation,

$$r^* = \frac{1}{M-1} (\boldsymbol{\lambda}' \mathbf{R} \boldsymbol{\lambda} - 1)$$

where $\boldsymbol{\lambda}' = (\lambda_1, \lambda_2, \ldots, \lambda_M)$. Since $\mu_p = \lambda_p / \sqrt{M}$, we have

$$r^* = \frac{1}{M-1} (M \boldsymbol{\mu}' \mathbf{R} \boldsymbol{\mu} - 1). \tag{14.14}$$

■

Remark 1. The intraclass correlation coefficient has traditionally been used as a correlation of a characteristic among the members of a class or family. It can also be used as a measure of agreement among these members. However, Q, while linearly related to r^*, is a better measure of agreement. It ranges from

0 to 1 and has the appealing structure of a quadratic form. Later we will show that the well-known Kendall's measure of concordance is a particular case of Q when applied to the rank order data.

Remark 2. The measure of concordance Q is related and yet different from the concordance correlation coefficient developed by Lin (1989, 1992). The measures developed by him and others are useful in assessing agreement of a new or generic process, methodology, and formulation in area of laboratory performance, and instrument or assay validation. See a review of such measures in Lin et al. (2002). Most of these studies are for comparing two samples of observations, and the measure proposed by Lin (1989) for this case takes the form

$$r_c = \frac{2rs_xs_y}{s_x^2 + s_y^2 + (\bar{y} - \bar{x})^2} \tag{14.15}$$

where the two samples are $x_1, x_2, \ldots, x_n$ and $y_1, y_2, \ldots, y_n$ and $\bar{y}, \bar{x}, s_x, s_y, r = s_{xy}/s_xs_y$ are the usual sample statistics. Applied to this case, our Q is similar to this and is verified to be

$$Q = \frac{s_x^2 + s_y^2 + 2s_{xy}}{2(s_x^2 + s_y^2 + \frac{1}{2}(\bar{y} - \bar{x})^2)}. \tag{14.16}$$

Our measure, however, is generalized to the case when there are more than two samples.

Proposition 14.2.2 *The concordance measure Q reduces to Kendall's measure of concordance W when the y_{ip}'s are rank order data.*

PROOF. It is well known that the Pearson correlation coefficient between two rank orders is the Spearman rank correlation coefficient . Hence $\mathbf{R}$ is the matrix of Spearman rank correlation coefficients and $\lambda_p = M^{-1/2}$ and $\mu_p = M^{-1}$. Then

$$\boldsymbol{\mu}'\mathbf{R}\boldsymbol{\mu} = \frac{1}{M^2}\mathbf{e}'\mathbf{R}\mathbf{e} \tag{14.17}$$

where $\mathbf{e}'$ is a $1 \times p$ row vector of 1's. This implies

$$\frac{\mathbf{e}'\mathbf{R}\mathbf{e} - M}{M(M-1)} = \frac{M^2(\boldsymbol{\mu}'\mathbf{R}\boldsymbol{\mu}) - M}{M(M-1)} = \frac{M\boldsymbol{\mu}'\mathbf{R}\boldsymbol{\mu} - 1}{M-1}. \tag{14.18}$$

However, $(\mathbf{e}'\mathbf{R}\mathbf{e} - M)/[M(M-1)]$ is clearly equal to $\bar{r}_s$, the average of $M(M-1)$ Spearman correlation coefficients. We have from (14.2)

$$\bar{r}_s = \frac{MW - 1}{M - 1}.$$

It follows that $W = \boldsymbol{\mu}'\mathbf{R}\boldsymbol{\mu}$, where $\mathbf{R}$ here is the matrix of Spearman correlation coefficients. ■

14.3 A New Measure of Concordance for Rank Order Data

Kendall's measure of concordance W is related to the Spearman correlation coefficients between the judges. We propose a measure of the concordance W_τ based on Kendall's τ correlation coefficients. We also show that W_τ is equivalent to the Q measure applied to a transformation of the rank order data.

For the pth ranking, $p = 1, 2, \ldots, M$, define for $i, j = 1, 2, \ldots, n$, $i \neq j$,

$$a_{ij}^{(p)} = \begin{cases} 1 & \text{if the rank of } i > \text{ rank of } j, \\ -1 & \text{if the rank of } i < \text{ rank of } j. \end{cases} \tag{14.19}$$

Let $T_{ij} = \sum_{p=1}^{M} a_{ij}^{(p)}$. Then we define the measure of concordance W_τ by

$$W_\tau = \frac{1}{n(n-1)M^2}\Big(\sum_{i=1}^{n}\sum_{j\neq i}^{n} T_{ij}^2\Big). \tag{14.20}$$

Proposition 14.3.1 *Let $\bar{\tau}$ denote the average of $M(M-1)$ Kendall 's τ correlation coefficients between pairs of the M judges. It can be expressed as*

$$\bar{\tau} = \frac{MW_\tau - 1}{M - 1}.$$

PROOF. Since $T_{ij} = \sum_{p=1}^{M} a_{ij}^{(p)}$, from (14.20) we have

$$\begin{aligned} n(n-1)M^2 W_\tau &= \sum_{i\neq j}\Big[\sum_{p=1}^{M} a_{ij}^{(p)}\Big]^2 \\ &= \sum_{i\neq j}\Big[\sum_{p=1}^{M}(a_{ij}^{(p)})^2 + \sum_{p\neq q}^{M} a_{ij}^{(p)}a_{ij}^{(q)}\Big] \\ &= \sum_{p=1}^{M}\sum_{i\neq j}(a_{ij}^{(p)})^2 + \sum_{p\neq q}^{M}\Big[\sum_{i\neq j} a_{ij}^{(p)}a_{ij}^{(q)}\Big]. \end{aligned} \tag{14.21}$$

Note that for each $p = 1, 2, \ldots, M$, $\sum_{i\neq j}(a_{ij}^{(p)})^2 = n(n-1)$ and $\sum_{i\neq j} a_{ij}^{(p)}a_{ij}^{(q)} = n(n-1)\tau_{pq}$ from (14.17), where τ_{pq} is Kendall's τ for the judges p and q. See also Daniels (1944). Expression (14.21) reduces, therefore, to

$$Mn(n-1) + n(n-1)\sum_{p\neq q}^{M}\tau_{pq}.$$

This implies that

$$MW_\tau = 1 + \bar{\tau}(M-1). \tag{14.22}$$

Hence

$$\bar{\tau} = \frac{MW_\tau - 1}{M-1}. \tag{14.23}$$

Thus, W_τ is a linear function of $\bar{\tau}$. Ehrenberg (1952) proposed $\bar{\tau}$ as a measure of concordance for the rank order data. Note that $\bar{\tau}$ was also proposed as a coefficient of agreement in paired data by Kendall and Smith (1940). They also noted the equivalence of their coefficient of agreement with $\bar{\tau}$. Note that $\bar{\tau}$ does not range from 0 to 1 and it can even be negative. We believe W_τ is a more appropriate concordance measure since it ranges from 0 to 1 and has also the representation $\boldsymbol{\mu}'\mathbf{R}\boldsymbol{\mu}$ as the following proposition shows. It is also related to the intraclass correlation coefficient of transformed data defined in (14.19).

■

Proposition 14.3.2 *The measure W_τ has the representation $\boldsymbol{\mu}'\mathbf{R}_\tau\boldsymbol{\mu}$ with $\mathbf{R}_\tau$ as the correlation matrix of the Kendall correlations and $\mu_p = M^{-1}$, $p = 1, 2, \ldots, M$.*

PROOF. Since the data is in the form of $a_{ij}^{(p)}$ with values $+1$ or -1, it is easily verified that $\mu_p = M^{-1}$, $p = 1, 2, \ldots, M$. This is because

$$\lambda_p^2 = \sum_{i \neq j} (a_{ij}^{(p)})^2 \Big/ \sum_{p=1}^{M} \sum_{i \neq j} (a_{ij}^{(p)})^2 = M^{-1}.$$

Then

$$\boldsymbol{\mu}'\mathbf{R}_\tau\boldsymbol{\mu} = \frac{1}{M^2}(M + \sum_{p \neq q}^{M} \tau_{pq}) = \frac{1}{M}[1 + \bar{\tau}(M-1)] = W_\tau \text{ by } (14.22).$$

■

Proposition 14.3.3 *Let τ^* be the intraclass correlation based on the a_{ij}'s. Then*

$$\tau^* = \frac{MW_\tau - 1}{M-1} = \bar{\tau}.$$

PROOF. From (14.8) it follows from the definition of the intra class correlation coefficient applied to the a_{ij}'s that

$$\begin{aligned}\tau^* &= \frac{1}{M-1}\Big[\frac{1}{n(n-1)M}(\sum_{i\neq j}^{n} T_{ij}^2) - 1\Big] \\ &= \frac{1}{M-1}\Big[\frac{1}{n(n-1)M^2}(M\sum_{i\neq j}^{n} T_{ij}^2) - 1\Big] \\ &= \frac{1}{M-1}[MW_\tau - 1] \\ &= \bar{\tau} \text{ by (14.23).}\end{aligned}$$

■

Proposition 14.3.4 *The intraclass correlation r_s^* for the $a_{ij}^{(p)}$ data based on the Spearman setup equals $\bar{r}_s$.*

PROOF. In this case,

$$a_{ij}^{(p)} = R_{ip} - R_{jp}$$

where R_{ip} is the rank of the ith subject in the pth ranking; $\mathbf{R}_s$ is the Spearman correlation matrix. It can be verified that for $p = 1, 2, \ldots, M$,

$$\begin{aligned}\sum_{i\neq j}^{n} a_{ij}^{(p)} &= 0, \\ \sum_{i\neq j}^{n} (a_{ij}^{(p)})^2 &= n^2(n^2-1)/6.\end{aligned}$$

Hence

$$\lambda_p^2 = \sum_{i\neq j}^{n} (a_{ij}^{(p)})^2 \Big/ \sum_{p=1}^{M}[\sum_{i\neq j}^{n}(a_{ij}^{(p)})^2] = M^{-1}.$$

The correlation matrix is given by $\mathbf{R}_s$, the matrix of Spearman correlation coefficients due to a well-known result, see Kendall (1948, p. 18) or Daniels (1944). Now

$$\boldsymbol{\mu}'\mathbf{R}_s\boldsymbol{\mu} = \frac{1}{M^2}(M + \sum_{p\neq q}^{M}(r_s)_{pq}) = \frac{1}{M}[1 + \bar{r}_s(M-1)] = W \text{ by (14.2).}$$

Since $\boldsymbol{\mu}'\mathbf{R}_s\boldsymbol{\mu} = W = (1 + r_s^*(M-1))/M$ by (14.14), we have $r_s^* = \bar{r}_s$. Since, by Proposition 14.2.2, Q reduces to W when the dataset is rank ordered, again by (14.14) we have that the intraclass correlation coefficient for the original rank order data is also the $\bar{r}_s$. ■

Summarizing, for the raw rank data or the $a_{ij}^{(p)}$ data based on the Spearman setup, the intraclass correlations are the same. This implies the concordance measures are also the same for both sets of data. However for $a_{ij}^{(p)}$ data based on Kendall's τ setup, the intraclass correlations differ for both sets of data. See also the numerical example in Section 14.4.

Population Analogues for Q, W and W_τ

Suppose that $Y_1, Y_2, \ldots, Y_M$ are M variables with means $\theta_1, \theta_2, \ldots, \theta_M$ and standard deviations $\sigma_1, \sigma_2, \ldots, \sigma_M$, respectively. The correlation matrix is given by the $M \times M$ matrix $\boldsymbol{\rho} = (\rho_{ij})$. Let

$$\bar{\theta} = \frac{1}{M}\sum_{p=1}^{M}\theta_p, \quad \lambda_p^0 = \frac{\sigma_p}{\left(\sum_{q=1}^{M}\sigma_q^2 + M\sum_{q=1}^{M}(\theta_q - \bar{\theta})^2\right)^{1/2}},$$

and $\mu_p^0 = \lambda_p^0/\sqrt{M}$. Then the population analogue for Q is

$$Q_0 = (\mu_1^0, \mu_2^0, \ldots, \mu_M^0)\boldsymbol{\rho}(\mu_1^0, \mu_2^0, \ldots, \mu_M^0)'.$$

The statistic Q can be considered as an estimate of Q_0 and the sample analogue of Q_0. The quantity $(MQ_0-1)/(M-1)$ can also be interpreted as the population analogue of the sample intraclass correlation coefficient.

The population analogues for W and W_τ for sample rank ordered data can also be defined. It is well known that population analogues for Spearman's rank correlation coefficient and Kendall's τ are the *grade* correlation ρ_G and Kendall's τ respectively. See Kendall (1948) for details. Thus we can suggest the population analogue for W as

$$W_0 = \frac{(M-1)\bar{\rho}_G + 1}{M}$$

where $\bar{\rho}_G$ is the average of the $[M(M-1)/2]$ population grade correlation coefficients. Similarly the analogue for W_τ is $\{(M-1)\bar{\tau} + 1\}/M$, where $\bar{\tau}$ is the average of the $[M(M-1)/2]$ population Kendall's correlation coefficients.

14.4 Example

Consider the following ranks awarded by $M = 4$ judges P–S while evaluating $n = 6$ subjects A–F [Kendall (1948), p. 80].

Judge	Subject					
	A	B	C	D	E	F
P	5	4	1	6	3	2
Q	2	3	1	5	6	4
R	4	1	6	3	2	5
S	4	3	2	5	1	6

Table 14.1 contains the corresponding $a_{ij}^{(p)}$ values for $i, j = 1, \ldots, 6$, and $p = 1, 2, 3, 4$.

Let S_{ij} denote the score for the pair of judges i and j, i.e., difference between the number of concordant and discordant pairs among the six subjects so that the Kendall rank correlation coefficient τ_{PQ} between the judges P and Q is given by $2S_{PQ}/n(n-1)$. Now,

$$S_{PQ} = 9 - 6 = 3; S_{PR} = 4 - 11 = -7; S_{PS} = 10 - 5 = 5;$$

$$S_{QR} = 4 - 11 = -7; S_{QS} = 8 - 7 = 1; S_{RS} = 9 - 6 = 3, \quad \text{and}$$

$$\begin{aligned} \bar{\tau} &= \frac{1}{3(30)} \{S_{PQ} + S_{PR} + S_{PS} + S_{QR} + S_{QS} + S_{RS}\} \\ &= \frac{1}{90}(3 - 7 + 5 - 7 + 1 + 3) = -\frac{1}{45}. \end{aligned}$$

Thus,

$$W_\tau = \frac{(M-1)\bar{\tau} + 1}{M} = \frac{7}{30}.$$

Also, from (14.20),

$$W_\tau = (\sum_{i \neq j} T_{ij}^2)/n(n-1)M^2 = 7/30.$$

Hence $\bar{\tau} = -1/45$ is also the intraclass correlation coeffcient for the a_{ij} data. To verify this, we use the formula [see Cochran (1977), p. 243]

$$r^* = \frac{\sum (y_i - \overline{Y})^2}{(M-1)\left[\sum_{i,j} y_{ij}^2 - \frac{n}{M}\overline{Y}^2\right]} - \frac{1}{M-1}.$$

Note that $\overline{Y} = 0$; $\sum_{i,j} y_{ij}^2 = 4(30) = 120$, and hence

$$\tau^* = \frac{112}{3(120)} - \frac{1}{3} = -\frac{1}{45}.$$

Table 14.1: Table of $a_{ij}^{(p)}$ values for the data (see (14.19)).

Judge	Pair of Subjects														
	AB	AC	AD	AE	AF	BC	BD	BE	BF	CD	CE	CF	DE	DF	EF
P	1	1	-1	1	1	1	-1	1	1	-1	-1	-1	1	1	1
Q	-1	1	-1	-1	-1	1	-1	-1	-1	-1	-1	-1	-1	1	1
R	1	-1	1	1	-1	-1	-1	-1	-1	1	1	1	1	-1	-1
S	1	1	-1	1	-1	1	-1	1	-1	-1	1	-1	1	-1	-1
T_{ij}	2	2	-2	2	-2	2	-4	0	-2	-2	0	-2	2	0	0
T_{ij}^2	4	4	4	4	4	4	16	0	4	4	0	4	4	0	0

Table 14.2: $a_{ij}^{(p)}$ Table for the data: Spearman case

Judge	Pair of Subjects														
	AB	AC	AD	AE	AF	BC	BD	BE	BF	CD	CE	CF	DE	DF	EF
P	1	4	-1	2	3	3	-2	1	2	-5	-2	-1	3	4	1
Q	-1	1	-3	-4	-2	2	-2	-3	-1	-4	-5	-3	-1	1	2
R	3	-2	1	2	-1	-5	-2	-1	-4	3	4	1	1	-2	-3
S	1	2	-1	3	-2	1	-2	2	-3	-3	1	-4	4	-1	-5
$y_{i.}$	4	5	-4	3	-2	1	-8	-1	-6	-9	-2	-7	7	2	- 5

Note: In both tables only half of the number of columns are shown. For the other pairs, the numbers will be the same with an opposite sign.

For the same example in Kendall (1948, p. 80), we can verify similar things based on the Spearman correlation coefficient applied to the original rank order data and corresponding a_{ij} data. We can also show that $-1/35$ is the intraclass correlation coefficient for the a_{ij} data in the Spearman context. Here $a_{ij}^{(p)} = R_{ip} - R_{jp}$ where R_{ip} is the rank of the ith subject in the pth ranking. Their values are given in Table 14.2. Since

$$\overline{Y} = 0; \sum_{i,j} Y_{ij}^2 = 2(420) = 840; \sum y_{i.}^2 = 768,$$

$$r_s^* = \frac{1}{3}\left[\frac{768}{840} - 1\right] = -\frac{1}{35}.$$

For the original rank order data

$$\overline{Y} = M(n+1)/2 = 4(7)/2 = 14,$$

and the $y_{i.}$'s are the totals for each subject, given below.

Subject	A	B	C	D	E	F
$y_{i.}$	15	11	10	19	12	17

Hence

$$\sum (y_{i.} - \overline{Y})^2 = 1 + 9 + 16 + 25 + 4 + 9 = 64, \sum_{i,j} y_{ij}^2 = 364$$

and

$$r_s^* = \frac{64}{3\left(364 - \frac{6}{4}(196)\right)} - \frac{1}{3} = -\frac{1}{35}.$$

Also,

$$W = \frac{12(64)}{210(16)} = \frac{8}{35}; \frac{4W-1}{3} = -\frac{1}{35} = r_s^*.$$

Thus the intraclass correlation coefficient $-1/35$ is the same whether we consider the rank order data or the transformed a_{ij} data in the Spearman case. However, it was shown that the intraclass correlation coefficient for the a_{ij} data defined for the Kendall case was $-1/45$.

14.5 Distribution of Q when the M Judges are Independent

We assume that $Y_{1p}, Y_{2p}, \ldots, Y_{np}$ are i.i.d. normal with mean θ and variance σ^2, that is, Y_{ip} is $N(\theta, \sigma^2)$, $p = 1, 2, \ldots, M$. Further $Y_{i1}, Y_{i2}, \ldots, Y_{iM}$ are i.i.d.

$N(\theta, \sigma^2)$, $i = 1, 2, \ldots, n$. In other words, there are M independent normal populations each with the same mean and variance and $Y_{1p}, Y_{2p}, \ldots, Y_{np}$ is a random sample from the pth population. We have

$$\begin{aligned}\sum_{i=1}^{n}\sum_{p=1}^{M}(y_{ip} - \overline{\overline{Y}})^2 &= \sum_{i=1}^{n}\sum_{p=1}^{M}(y_{ip} - \overline{Y}_{i.} + \overline{Y}_{i.} - \overline{\overline{Y}})^2 \\ &= \sum_{i=1}^{n}\sum_{p=1}^{M}(y_{ip} - \overline{Y}_{i.})^2 + M\sum_{i=1}^{n}(\overline{Y}_{i.} - \overline{\overline{Y}})^2 \\ &= \sum_{i=1}^{n}\sum_{p=1}^{M}(y_{ip} - \overline{Y}_{i.})^2 + \frac{1}{M}\sum_{i=1}^{n}(y_{i.} - \overline{Y})^2.\end{aligned}$$

This implies

$$1 = \frac{\sum_{i=1}^{n}(y_{i.} - \overline{Y})^2}{M\sum_{i=1}^{n}\sum_{p=1}^{M}(y_{ip} - \overline{\overline{Y}})^2} + \frac{\sum_{i=1}^{n}\sum_{p=1}^{M}(y_{ip} - \overline{Y}_{i.})^2}{\sum_{i=1}^{n}\sum_{p=1}^{M}(y_{ip} - \overline{Y})^2}.$$

The first term on the right is Q by Proposition 14.2.1 and (14.9). Hence

$$\frac{n(M-1)Q}{(n-1)(1-Q)} = \frac{M\sum_{i=1}^{n}(\overline{Y}_{i.} - \overline{\overline{Y}})^2/(n-1)}{\sum_{i=1}^{n}\sum_{p=1}^{M}(y_{ip} - \overline{Y}_{i.})^2/n(M-1)}.$$

This is the F statistic for the ANOVA one way layout model and hence is distributed as F distribution with $(n-1)$ and $n(M-1)$ degrees of freedom. It also follows from the relationship between an F and a Beta random variable, Q has a Beta distribution with parameters $(n-1)/2$ and $n(M-1)/2$.

Distribution of Q for the equicorrelated case

We have the following array of random variables

$$\begin{array}{cccc} Y_{11} & Y_{21} & \cdots & Y_{n1} \\ Y_{12} & Y_{22} & \cdots & Y_{n2} \\ \cdots & \cdots & \cdots & \cdots \\ Y_{1M} & Y_{2M} & \cdots & Y_{nM}. \end{array}$$

The vector $(Y_{i1}, Y_{i2}, \ldots, Y_{iM})$ has a multivariate normal distribution with $\mathrm{E}(Y_{ip}) = \mu$ and Variance$(Y_{ip}) = \sigma^2$, $p = 1, 2, \ldots, M$ and correlation coefficient between Y_{ip} and Y_{iq} is ρ , $p, q = 1, 2, \ldots, M; p \neq q$. Let $\Delta = \rho/(1-\rho)$. Define the concordance measure Q by (14.5) or equivalently

$$Q = \frac{\sum_{i=1}^{n}(y_{i.} - \overline{\overline{Y}})^2}{M\sum_{i=1}^{n}\sum_{p=1}^{M}(y_{ip} - \overline{\overline{Y}})^2}.$$

Define

$$G = \frac{n(M-1)}{(n-1)(1+\Delta M)} \frac{Q}{1-Q}.$$

Proposition 14.5.1 *The statistic G has F distribution with $(n-1)$ and $n(M-1)$ degrees of freedom.*

PROOF. Denote by $\mathbf{P}$ the correlation matrix. The eigenvalues of $\mathbf{P}$ are $1+(M-1)\rho, 1-\rho, 1-\rho, \ldots, 1-\rho$. Let C be an orthogonal matrix whose rows are the eigenvectors of $\mathbf{P}$ such that $C\mathbf{P}C'$ equals the diagonal matrix with diagonal elements $1+(M-1)\rho, 1-\rho, 1-\rho, \ldots, 1-\rho$. The eigenvector corresponding to the eigenvalue $1+(M-1)\rho$ is $M^{-1/2}(1,1,\ldots,1)$. Define $X_{ij} = \sum_{p=1}^{M} c_{jp} Y_{ip}$ where c_{jp} is the (j,p)th element of C. With this definition we can take

$$X_{i1} = \sum_{p=1}^{M} c_{1p} Y_{ip} = \frac{1}{\sqrt{M}} \sum_{p=1}^{M} Y_{ip} = \sqrt{M}\overline{Y}_{i.}$$

since $c_{1p} \equiv M^{-1/2}$, $p = 1,2,\ldots,M$. We also have $\sum_{p=1}^{M} c_{jp} = 0$ for $j = 2,3,\ldots,n$. This follows from the facts that the eigenvectors are orthogonal to each other and $c_{1p} = M^{-1/2}$, $p = 1,2,\ldots,M$. X_{ip} for $i = 2,3,\ldots,n$ are therefore i.i.d. $N(0,\sigma^2(1-\rho))$. They are also independent of X_{i1}, which is $N(\sqrt{M}\mu, \sigma^2(1+(M-1)\rho))$. Make an additional orthogonal transformation $(X_{11}, X_{21}, \ldots, X_{n1}) \to (Z_{11}, Z_{21}, \ldots, Z_{n1})$ with $Z_{11} = \sqrt{n}\overline{X}_{.1}$. The Z's are independent normal with common variance $\sigma^2(1+(M-1)\rho)$ and means $\mathrm{E}(Z_{11}) = \sqrt{Mn}\mu$ and $\mathrm{E}(Z_{i1}) = 0$ for $i > 1$. Put $Z_{ij} = X_{ij}$ for $j > 1$, and let

$$U = \sum_{i=2}^{n} Z_{i1}^2; V = \sum_{i=1}^{n}\sum_{p=2}^{M} Z_{ip}^2 = \sum_{i=1}^{n}\sum_{p=2}^{M} X_{ip}^2.$$

Consider the statistic

$$\begin{aligned} G &= \frac{U/[(1+(M-1)\rho)(n-1)]}{V/[n(M-1)(1-\rho)]} \\ &= \frac{U/(n-1)}{V/n(M-1)} \frac{1}{\{1+(M-1)\rho\}/(1-\rho)} = \frac{U/(n-1)}{V/n(M-1)} \frac{1}{(1+\Delta M)}, \end{aligned}$$

where $\Delta = \rho/(1-\rho)$. Thus G is an F statistic with $(n-1)$ and $n(M-1)$ degrees of freedom. It can also be verified that $U = M\sum_{i=1}^{n}(\overline{Y}_{i.} - \overline{\overline{Y}})^2$ and $V = \sum_{i=1}^{n}\sum_{j=1}^{M}(y_{ij} - \overline{Y}_{i.})^2$ so that

$$G = \frac{n(M-1)}{(n-1)(1+\Delta M)} \frac{Q}{1-Q}, \tag{14.24}$$

where $Q = \{1+(M-1)r^*\}/M$, and r^* is the intraclass correlation coefficient of the M judges.

■

14.6 Distribution of the Intraclass Correlation Coefficient

Fisher (1921) derived the distribution of intraclass correlation coefficient r^*. He did not give a complete derivation but stated the density function involving the terms of r^* only. In this section we derive completely the pdf of r^*. Since it is linearly related to Q defined in (14.14) by the relation $Q = [1 + (M-1)r^*]/M$, the distribution of r^* immediately follows from the distribution of G defined in (14.24) above where $\Delta = \rho/(1-\rho)$. It was shown in Proposition 14.5.1 that G has F distribution with $(n-1)$ and $n(M-1)$ degrees of freedom. It can be verified after some heavy algebra that the pdf of r^* is given by

$$c[1+(M-2)\rho-(M-1)\rho r]^{-(nM-1)/2}(1-r)^{(n(M-1)-2)/2}[1+(M-1)r]^{(n-3)/2}$$

for $r \geq -1/(M-1)$. The constant c is given by

$$c = \frac{\Gamma(\frac{nM-1}{2})}{\Gamma(\frac{n-1}{2})\Gamma(\frac{n(M-1)}{2})}\frac{(M-1)^{n(M-1)/2}}{M^{(nM-3)/2}}[1+\rho(M-1)]^{n(M-1)/2}(1-\rho)^{(n-1)/2}. \tag{14.25}$$

Since $Q = 1+(M-1)r^*$, the above complicated distribution can be transformed to an F distribution by the transformation (14.24). This transformation appears not to have been mentioned in the literature of intraclass correlation coefficient.

From the distribution of r^* or from the distribution of G defined in (14.24), it is easy to derive the pdf of Q. In particular for $M = 2$, it is verified that the pdf of Q is

$$c2^{(2n-3)/2}(1+\rho-2\rho q)^{-(2n-1)/2}q^{(n-3)/2}(1-q)^{(n-2)/2}, \quad 0 < q < 1,$$

where c is given by (14.25) with $M = 2$. The expression for Q when $M = 2$ is shown in (14.16). This distribution appears to be more tractable than the distribution of r_c proposed by Lin (1989). In particular for the independence case where $\rho = 0$, the distribution is a Beta distribution with parameters $(n-1)/2$ and $n/2$ – a result we derived in Section 14.5.

14.7 Summary and Discussion

This chapter develops a class of measures to assess the concordance or agreement of ratings data in the form of M rows and n columns. The rows could be judges or methods or, in general, entities that provide the assessment, and the columns could represent subjects or in general a sample of observations. The ratings data could be interval scaled or rank order data. The suggested general measure is a quadratic form that depends essentially on the sample variances and the correlation matrix, and is also a linear function of the intraclass correlation coefficients of the ratings data. When the distribution of judges is multivariate normal, the general measure has tractable and well-known distribution. For the rank order data, the proposed measure yields a new measure of concordance based on the Kendall rank correlation matrix and this can be considered as an alternative to the statistic proposed by Friedman (1937) for the two-way analysis of variance.

Here we have derived the distribution of Q under two special cases. The distribution of Q under the general setting, even when the judges are independent and in the case of multivariate normal with different means and standard deviations, appears to be complicated. Research is under way to study the measure for its general distribution for ratings and rank order data and the associated tests of significance. It would also be interesting to compare the two concordance measures for the rank order data and derive their distributions and asymptotic efficiencies. For the general data of ratings, the proposed measure could also be compared with the concordance correlation coefficient available for the case of biometric data with two assessors.

References

1. Choudhary, P. K. and Nagaraja, H. N. (2005). Measuring Agreement in Method Comparison Studies – A Review, In *Advances in Ranking and Selection, Multiple Comparisons, and Reliability* (Eds., N. Balakrishnan, N. Kannan and H. N. Nagaraja), pp. 215–244, Birkhäuser, Boston.

2. Cochran, W. G. (1977) *Sampling Techniques*, Third Edition, John Wiley & Sons, New York.

3. Daniels, H. E. (1944). The relation between measures of correlation in the universe of sample permutations, *Biometrika*, **33**, 129–135.

4. Ehrenberg, A. S. C. (1952). On Sampling from a Population of Rankers, *Biometrika* **39**, 82–87.

5. Fisher, R. A. (1921). On the "probable error" of a coefficient of correlation deduced from a small sample, *Metron* **1**, 3-32 (Also reprinted in *Collected Papers of R.A. Fisher*, **1** J.H. Bennett (Ed.), (1971) University of Adelaide).

6. Friedman, M. (1937). The use of ranks to avoid the assumption of normality implicit in the analysis of variance, *Journal of the American Statistical Association*, **32**, 675–701.

7. Kendall, M. G. (1948). *Rank Correlation Methods*, Charles Griffin, London.

8. Kendall, M. G. and Smith, B. B. (1939). The problem of m rankings, *Annals of Mathematical Statistics*, **10**, 275–287.

9. Kendall, M. G. and Smith, B. B. (1940). On the Method of Paired Comparisons, *Biometrika*, **31**, 324–425.

10. Lin, L. I. (1989). A concordance correlation coefficient to evaluate reproducibility, *Biometrics*, **45**, 255–268. Corrections: 2000, **56**, 324–325.

11. Lin, L. I. (1992). Assay validation using the concordance correlation coefficient, *Biometrics*, **48**, 599–604.

12. Lin, L. I., Hedayat, A. S., Sinha, B. and Yang, M. (2002). Statistical methods in assessing agreement: Models, issues, and tools, *Journal of the American Statistical Association*, **97**, 257–270.

PART V
Reliability

15

Cost-effective Analysis of Optimal Order-replacement Policies

T. Dohi[1], N. Kaio[2] and S. Osaki[3]

[1] *Department of Information Engineering, Hiroshima University, Japan*
[2] *Department of Economic Informatics, Hiroshima Shudo University, Japan*
[3] *Department of Information and Telecommunication Engineering, Nanzan University, Nagoya, Japan*

Abstract: In this chapter, we present a cost-effective analysis of several order-replacement policies for one-unit systems. More precisely, we treat two types of models, called the order-replacement model and the order-inspection one, and derive analytically the optimal ordering policies maximizing the so-called cost effectiveness, which is defined as the mean up time per unit mean operating cost, in both continuous and discrete time setting. Some examples are devoted to illustrate the optimal ordering policies numerically. A comprehensive reference list in this research area is also provided.

Keywords and phrases: Preventive maintenance, order-replacement, order-inspection, cost effectiveness, continuous time, discrete time, stochastic models

15.1 Introduction

In reliability and maintainability engineering, the potential failure of an operating unit (i.e., machine, equipment, part, etc.) may be predicted on the basis of any probabilistic scenario like physical deterioration and human error. Reliability engineers often face some replacement problems such as an age replacement and/or a block replacement. These have been discussed by many authors [Barlow and Proschan (1965, 1975)]. For these two replacement policies it is assumed that there exist an infinite number of spares on hand over an infinite time horizon, so that each spare can be provided immediately when it is desired. However, in practice, there may be a time lag between desire/order of spare and its supply/delivery, since spares are not always on hand. This type of delay is called the *lead time* and the maintenance policy with such a delay is called the *order-replacement policy* or simply *ordering policy*. Since the seminal works by Allen and D'Esopo (1968a,b) and Wiggins

(1967), many kinds of ordering policies have been discussed in the literature. Nakagawa and Osaki (1974), Nakagawa (1976), Osaki and Yamada (1976), Osaki (1977) treat several types of order-replacement models as extentions of the age replacement model. Dohi, Kaio and Osaki (1996a,b, 1998), Nakagawa and Osaki (1978), Sheu (1997a,b), Thomas and Osaki (1978a,b), Kaio and Osaki (1977, 1978a,b,c,d, 1981a, 1990b), Kalpakam and Shaul (1981), Park and Park (1986a,b), Sridharan (1991), Subramanian and Sridharan (1989) extend Osaki's model (1977) in terms of both lead time structure and cost component.

Kaio and Osaki (1979a, 1980a,b, 1990a) introduce the concept of discounting in the cost representation and derive the optimal order-replacement policies under the expected total discounted cost over an infinite time horizon. The discrete-time order-replacement models are considered in Kaio and Osaki (1979b) and Dohi et al. (2004). Osaki, Kaio and Yamada (1981), Park and Park (1986a), Kaio and Osaki (1980d, 1981b), Kapur, Garg and Bhalla (1991), Sheu and Liou (1992, 1993, 1994), Sheu, Liou and Tseng (1992), Sheu (1997a,b) analyze different order-replacement problems with minimal repair. Sung and Park (1986) and Sung and Kim (1987) apply the same technique to equipments with sensor and derive the other type of order-replacement policies. Kapur and Garg (1988) combine the order-replacement policy and the repair-cost limit replacement policy. Kaio, Dohi and Osaki (1998) formulate the determination problem of lead time for a fixed ordering time. Recently, Armstrong and Atkins (1996) solve the joint optimization problem of age replacement and the spare inventory. Csenki (1998, 1999) apply the well-known marginal cost analysis [Berg (1980)] and the asymptotic approach for the finite time horizon problem [Christer (1978)] to the order-replacement model, respectively. Kawai (1983a,b) extends the result in Mine and Kawai (1977) and prove the optimality of order-replacement policy in the framework of semi-Markov decision processes.

Dohi, Kaio and Osaki (1994, 1996a) consider the order-replacement policies for a two-unit standby redundant system, under so-called *cost effectiveness criterion* [Hunter (1963), Trott (1965) and Winlund (1965)] and the common long-run average cost, respectively. In the order-replacement problems based on age replacement, if the failure occurs, then it can be detected immediately. Kaio and Osaki (1978e, 1982) and Kaio, Dohi and Osaki (1992) assume the situation where the failure can be detected by a periodic inspection and call this problem *the order-inspection problem.* They derive the optimal order-inspection policies maximizing the cost effectiveness criteria under milder conditions. The technique of analysis for both the order-replacement problems and the order-inspection problems are applied to continuous review cyclic inventory models with emergency order [Dohi, Kaio and Osaki (1995a,b, 1997) and Dohi, Shibuya and Osaki (1997)]. The spare provisioning policy based on the block replacement is considered by Acharya, Nagabhushanam and Alam (1986). Kabir and his co-authors (1996a,b, 1997) use the simulation technique to determine both the optimal replacement timing and the spare provisioning policy. In the book chapters [Dohi, Kaio and Osaki (2003) and Kaio, Dohi and Osaki (2002)], monograph [Kapur, Garg and Kumar (1999)] and text books [Osaki (1985, 1992)], the reader will find surveys and tutorial articles on the order-replacement problems.

Here, we present a cost-effective analysis of several order-replacement policies for one-unit systems. More precisely, we treat two types of models, called the order-replacement model and the order-inspection one, and derive analytically the optimal ordering policies maximizing the cost effectiveness, in both continuous and discrete time setting. The order-replacement problem under cost effectiveness in continuous time is considered in Dohi, Kaio and Osaki (2001). The remaining parts on the order-inspection problem and the discrete-time analysis are all new results. The rest of this chapter is organized as follows. After describing notation and assumptions, we define cost effectiveness as a criterion of optimality to take into account the balance between the long-run average cost and the steady-state system availability. Next, we consider the order-replacement model and the order-inspection one in continuous time. The problems are at the moment formulated as two-dimensional optimization problems with the ordering time and inventory time limit. It can be shown that there exists a decomposition structure of two decision variables and that the two-dimensional problem can be essentially reduced to a simple one-dimensional one. We derive necessary and sufficient conditions for the existence of the optimal ordering policies analytically. The discrete time models are considered in a similar way. Some examples are devoted to compare the order-replacement policy with the order-inspection policy numerically.

15.2 Preliminaries

15.2.1 Notation (continuous time models)

X: failure time of unit (continuous random variable)

$F(t)$, $f(t)$, $1/\lambda$ (>0): c.d.f., p.d.f. and mean of X

$\overline{\psi}(\cdot) = 1 - \psi(\cdot)$ in general

$r(t) = f(t)/\overline{F}(t)$: failure rate of $F(t)$

c_1: cost per unit amount for expedited order

c_2: cost per unit amount for regular order

c_3: fixed inspection cost

k_i: inventory holding cost per unit time

k_s: shortage cost per unit time

k_m: monitoring cost per unit time

L_1, L_2: lead times for expedited and regular orders (continuous random variables)

$G_i(t), 1/\mu_j\ (> 0)$: c.d.f. and mean of L_i $(i = 1, 2)$

$R(t) = \{\int_0^\infty F(t + l_2)dG_2(l_2) - F(t)\}/\overline{F}(t)$: conditional expected failure rate function during the interval $(t, t + L_2]$

t_0: regular ordering time (decision variable)

t_1: inventory time limit (decision variable)

$C_j(t_0, t_1), V_j(t_0, t_1), T_j(t_0, t_1)$: long-run average cost, expected cost during one cycle and mean time length of one cycle in Model j $(= O, I)$, where Model O and Model I denote the order-replacement model and the order-inspection one, respectively.

$A_j(t_0, t_1), U_j(t_0, t_1), E_j(t_0, t_1)$: steady-state system availability, mean up time during one cycle and cost effectiveness in Model j $(= O, I)$

15.2.2 Notation (discrete time models)

N: failure time of unit (discrete random variable)

$P(n), p(n), 1/\lambda\ (> 0)$: c.d.f., p.m.f. and mean of N

$r(n) = p(n)/\overline{P}(n)$: discrete counterpart of the function $r(t)$

L_1, L_2: lead times for expedited and regular orders (discrete random variables)

$G_i(n)$, $g_i(n)$, $1/\mu_i\ (> 0)$: c.d.f., p.m.f. and mean of L_i $(i = 1, 2)$

$R(n) = \sum_{l_2=0}^{\infty}\{P(n + l_2) - P(n)\}g_2(l_2)/\overline{P}(n)$: discrete counterpart of the function $R(t)$

n_0: regular ordering time (decision variable)

n_1: inventory time limit (decision variable)

$C_j(n_0, n_1), V_j(n_0, n_1), T_j(n_0, n_1)$: long-run average cost, expected cost during one cycle and mean time length of one cycle in Model j $(= O, I)$

$A_j(n_0, n_1), U_j(n_0, n_1), E_j(n_0, n_1)$: steady-state system availability, mean up time during one cycle and cost effectiveness in Model j $(= O, I)$

15.2.3 Assumptions

(A-1) $c_1 + k_s/\mu_1 > c_2 + k_s/\mu_2$; the sum of ordering and shortage costs by an expedited order is strictly larger than that by a regular one.

(A-2) $k_i > k_m$; the monitoring cost per unit time is strictly less than the inventory holding cost.

15.2.4 Definition of cost effectiveness

In this chapter, we apply the *cost effectiveness* as a criterion of optimality. First, the cost effectiveness was introduced by Hunter (1963), Trott (1965) and Winlund (1965). For the continuous-time model, the cost effectiveness is defined by

$$\begin{aligned}
E_j(t_0, t_1) &= \frac{\lim_{t\to\infty} \mathrm{E}[\text{up time on } (0,t]]/t}{\lim_{t\to\infty} \mathrm{E}[\text{total cost on } (0,t]]/t} \\
&= \frac{\text{steady state system availability}}{\text{expected cost per unit time in the steady state}} \\
&= A_j(t_0, t_1)/C_j(t_0, t_1) \\
&= \frac{\mathrm{E}[\text{up time during one cycle}]}{\mathrm{E}[\text{total cost during one cycle}]} \\
&= U_j(t_0, t_1)/V_j(t_0, t_1) \qquad (15.1)
\end{aligned}$$

for Model j $(= O, I)$, where E denotes the mathematical expectation operator. That is, the cost effectiveness is the mean operative time per unit mean operating cost and is regarded as a unified criterion by taking account both the system availability and the economic justification. In a fashion similar to the continuous time model, we define the cost effectiveness in discrete time as

$$\begin{aligned}
E_j(n_0, n_1) &= A_j(n_0, n_1)/C_j(n_0, n_1) \\
&= U_j(n_0, n_1)/V_j(n_0, n_1) \qquad (15.2)
\end{aligned}$$

for Model j $(= O, I)$.

15.3 Order-Replacement Policies in Continuous Time

15.3.1 Model description

Consider an order-replacement problem for a one-unit system where each failed unit is scrapped and each spare is provided, after a lead time, by an order. The original unit begins operating at time $t = 0$, and the planning horizon

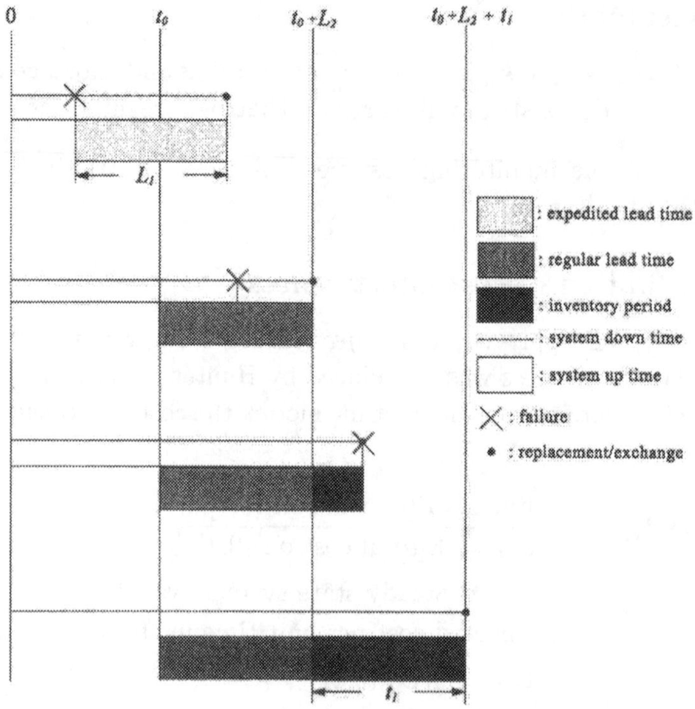

Figure 15.1: Configuration of order-replacement model in continuous time

is infinite. If the original unit does not fail up to a prespecified time $t_0 \in [0, \infty)$, the regular order for a spare is made at the time t_0 and after a lead time L_2 the spare is delivered. Then if the original unit has already failed by $t = t_0 + L_2$, the delivered spare takes over its operation immediately. But if the original unit is still working, the spare is put into the inventory and the original one is replaced/exchanged by the spare in the inventory when the original one fails/passes a prespecified time interval t_1 after the spare is delivered, whichever occurs first. It is assumed that the spare in the inventory does not fail or deteriorate. On the other hand, if the original unit fails before the time t_0, the expedited order is made immediately at the failure time and the spare takes over its operation just after it is delivered after a lead time L_1. In this situation, it should be noted that the regular order is not made. The same cycle repeats itself continually. The configuration of the basic model is illustrated in Figure 15.1.

Under this model, define the interval from one replacement/exchange to the following replacement/exchange as one cycle. Since the stochastic process under consideration is a renewal reward process and the points for replacement/exchange of a spare unit are the regenerative points, it is appropriate to consider the cyclic behavior of the stochastic process. The mean time length of one cycle is

$$\begin{aligned} T_O(t_0, t_1) &= \int_0^\infty \int_0^{t_0} (t + l_1) dF(t) dG_1(l_1) + \int_0^\infty \int_{t_0}^{t_0+l_2} (t_0 + l_2) dF(t) dG_2(l_2) \\ &\quad + \int_0^\infty \int_{t_0+l_2}^{t_0+l_2+t_1} t dF(t) dG_2(l_2) \\ &\quad + \int_0^\infty \int_{t_0+l_2+t_1}^{\infty} (t_0 + l_2 + t_1) dF(t) dG_2(l_2). \end{aligned} \tag{15.3}$$

The expected cost for one cycle is given by

$$\begin{aligned} V_O(t_0, t_1) &= k_s \Big\{ \int_0^\infty \int_0^{t_0} l_1 dF(t) dG_1(l_1) \\ &\quad + \int_0^\infty \int_{t_0}^{t_0+l_2} (t_0 + l_2 - t) dF(t) dG_2(l_2) \Big\} \\ &\quad + c_1 F(t_0) + c_2 \overline{F}(t_0) + k_i \int_0^\infty \int_{t_0+l_2}^{t_0+l_2+t_1} (t - t_0 - l_2) dF(t) dG_2(l_2) \\ &\quad + k_i \int_0^\infty \int_{t_0+l_2+t_1}^{\infty} t_1 dF(t) dG_2(l_2) + k_m \int_0^{t_0} \overline{F}(t) dt \\ &= k_s \Big\{ (1/\mu_1 - 1/\mu_2) F(t_0) + \int_0^\infty \int_{t_0}^{t_0+l_2} F(t) dt dG_2(l_2) \Big\} \\ &\quad + c_1 F(t_0) + c_2 \overline{F}(t_0) + k_i \int_0^\infty \int_{t_0+l_2}^{t_0+l_2+t_1} (t - t_0 - l_2) dF(t) dG_2(l_2) \\ &\quad + k_i \int_0^\infty \int_{t_0+l_2+t_1}^{\infty} t_1 dF(t) dG_2(l_2) + k_m \int_0^{t_0} \overline{F}(t) dt. \end{aligned} \tag{15.4}$$

The long-run average cost is, from the well-known renewal argument [see Ross (1970)],

$$\begin{aligned} C_O(t_0, t_1) &\equiv \lim_{t \to \infty} \frac{\text{E[the total cost on } (0, t]]}{t} \\ &= V_O(t_0, t_1) / T_O(t_0, t_1). \end{aligned} \tag{15.5}$$

On the other hand, since the mean up (operative) time during one cycle is given by

$$\begin{aligned} U_O(t_0, t_1) &= \int_0^\infty \int_0^{t_0+l_2+t_1} t dF(t) \\ &\quad + \int_0^\infty \int_{t_0+l_2+t_1}^{\infty} (t_0 + l_2 + t_1) dF(t) dG_2(l_2) \\ &= \int_0^\infty \int_0^{t_0+l_2+t_1} \overline{F}(t) dt dG_2(l_2), \end{aligned} \tag{15.6}$$

the steady-state system availability is obtained as

$$\begin{aligned} A_O(t_0, t_1) &\equiv \lim_{t\to\infty} \Pr\{\text{the system is operating at } t\} \\ &= U_O(t_0, t_1)/T_O(t_0, t_1). \end{aligned} \tag{15.7}$$

Then, from the definition in (15.1), the cost effectiveness is formulated as

$$E_O(t_0, t_1) = U_O(t_0, t_1)/V_O(t_0, t_1), \tag{15.8}$$

and the problem is to obtain the optimal schedule $(t_0{}^*, t_1{}^*) \in [0, \infty) \times [0, \infty)$ which maximizes the cost effectiveness, namely, it satisfies

$$E_O(t_0{}^*, t_1{}^*) = \max_{t_0, t_1} E_O(t_0, t_1). \tag{15.9}$$

Define the numerator of the partial derivative of $E_O(t_0, t_1)$ with respect to t_1 divided by $\int_0^\infty \overline{F}(t_0+l_2+t_1)dG_2(l_2)$, as $q_O(t_0, t_1)$. Then we obtain the following result, which represents a decomposition structure in the cost effectiveness between the inventory time limit for a spare (t_1) and the regular ordering time (t_0).

Lemma 15.3.1 *The function $q_O(t_0, t_1)$ is a univariate function of t_0, i.e., $q_O(t_0, t_1) = q_O(t_0)$.*

PROOF. The direct calculation yields

$$\begin{aligned} q_O(t_0, t_1) &= \{V_O(t_0, t_1)\}^2(\partial E_O(t_0, t_1)/\partial t_1)/\int_0^\infty \overline{F}(t_0 + l_2 + t_1)dG_2(l_2) \\ &= V_O(t_0, t_1) - k_i T_O(t_0, t_1) \\ &= k_s\Big\{(1/\mu_1 - 1/\mu_2)F(t_0) + \int_0^\infty \int_{t_0}^{t_0+l_2} F(t)dtdG_2(l_2)\Big\} \\ &\quad + c_1 F(t_0) + c_2\overline{F}(t_0) - k_i \int_0^\infty \int_0^{t_0+l_2} \overline{F}(t)dtdG_2(l_2) \\ &\quad + k_m \int_0^{t_0} \overline{F}(t)dt = q_O(t_0) \end{aligned} \tag{15.10}$$

and the result is trivial. ∎

Theorem 15.3.1 *For an arbitrary regular ordering time t_0 $(0 \le t_0 < \infty)$, if $q_O(t_0) \ge 0$, the optimal inventory time limit which maximizes $E_O(t_0, t_1)$ is $t_1{}^* \to \infty$, otherwise $t_1{}^* = 0$.*

PROOF. From Lemma 15.3.1, we have

$$\frac{\partial E_O(t_0, t_1)}{\partial t_1} = \frac{\int_0^\infty \overline{F}(t_0 + l_2 + t_1)dG_2(l_2)}{V_O(t_0, t_1)^2} q_O(t_0). \tag{15.11}$$

Thus, if $q_O(t_0) \geq 0$, then $E_O(t_0, t_1)$ is increasing in t_1 and $t_1{}^* \to \infty$, otherwise $t_1{}^* = 0$. ■

Theorem 15.3.1 implies that the two-dimensional maximization problem of the cost effectiveness, $\max_{(t_0,t_1)} E(t_0, t_1)$, can be reduced to one-dimensional problems in terms of t_0 when $t_1^* = 0$ and when $t_1^* \to \infty$. Hence, we focus our attention on these special cases.

15.3.2 Special case: $t_1{}^* \to \infty$

Consider the case of $t_1{}^* \to \infty$. Then the spare delivered by the regular order is put into the inventory until the original unit fails, if the original one is still operating (see Figure 15.1). In this case, the cost effectiveness is given by

$$E_O(t_0, \infty) = \lim_{t_1 \to \infty} E_O(t_0, t_1) = U_O(t_0, \infty)/V_O(t_0, \infty), \tag{15.12}$$

where $U_O(t_0, \infty) = 1/\lambda$ and

$$\begin{aligned} V_O(t_0, \infty) &= k_s\Big\{(1/\mu_1 - 1/\mu_2)F(t_0) + \int_0^\infty \int_{t_0}^{t_0+l_2} F(t)dtdG_2(l_2)\Big\} \\ &\quad + c_1 F(t_0) + c_2\overline{F}(t_0) + k_i \int_0^\infty \int_{t_0+l_2}^\infty \overline{F}(t)dtdG_2(l_2) \\ &\quad + k_m \int_0^{t_0} \overline{F}(t)dt. \end{aligned} \tag{15.13}$$

Define the numerator of the derivative of $E_O(t_0, \infty)$ with respect to t_0, divided by $\overline{F}(t_0)$, as $q_O^{(\infty)}(t_0)$,

$$\begin{aligned} q_O^{(\infty)}(t_0) &= -k_s\Big\{(1/\mu_1 - 1/\mu_2)r(t_0) + R(t_0)\Big\} - (c_1 - c_2)r(t_0) \\ &\quad + k_i\overline{R}(t_0) - k_m. \end{aligned} \tag{15.14}$$

The following lemma will be useful to characterize the optimal ordering policy.

Lemma 15.3.2 *The sign of the first derivative of the function $R(t_0)$ is the same as that of $r(t_0)$, i.e. $R(t_0)$ is increasing (decreasing) in t_0 if and only if the lifetime distribution $F(t_0)$ is IFR (DFR).*

See Dohi, Kaio and Osaki (1998) for the proof. From Lemma 15.3.2, we prove the following results.

Theorem 15.3.2 *(1) Suppose that $F(t)$ is strictly IFR (increasing failure rate) under (A-1) and (A-2).*

(i) *If* $q_O^{(\infty)}(0) > 0$ *and* $q_O^{(\infty)}(\infty) < 0$*, there exists a finite and non-zero optimal order-replacement time* t_0^* $(0 < t_0^* < \infty)$ *which satisfies* $E_O(t_0^*, \infty) = \max_{0<t_0<\infty} E_O(t_0, \infty)$ *and* $q_O^{(\infty)}(t_0^*) = 0$.

(ii) *If* $q_O^{(\infty)}(0) \leq 0$*, the optimal order-replacement time is* $t_0^* = 0$*. On the other hand, if* $q_O^{(\infty)}(\infty) \geq 0$*, the optimal order-replacement time is* $t_0^* \to \infty$.

(2) Suppose that $F(t)$ *is DFR (decreasing failure rate) under (A-1) and (A-2). If* $V_O(0, \infty) \geq V_O(\infty, \infty)$*, then the optimal order-replacement time is* $t_0^* \to \infty$*, otherwise,* $t_0^* = 0$.

PROOF. Differentiating $E_O(t_0, \infty)$ with respect to t_0 and setting it equal to zero implies that $q_O^{(\infty)}(t_0) = 0$. Further, with respect to t_0, we have

$$\frac{dq_O^{(\infty)}(t_0)}{dt_0} = -r'(t_0)\Big\{k_s(1/\mu_1 - 1/\mu_2) + c_1 - c_2\Big\} - (k_i + k_s)R'(t_0), \tag{15.15}$$

where in general $\psi'(t) = d\psi(t)/dt$. When the lifetime distribution is strictly IFR under the assumption (A-1), the function $q_O^{(\infty)}(t_0)$ is strictly decreasing. If $q_O^{(\infty)}(0) > 0$ and $q_O^{(\infty)}(\infty) < 0$, then there exists a finite and unique optimal ordering time $t_0{}^*$ $(0 < t_0{}^* < \infty)$ as a finite and unique solution of $q_O^{(\infty)}(t_0^*) = 0$ under (A-1) and (A-2), because

$$\begin{aligned} q_O^{(\infty)}(0) &= -r(0)\Big\{(c_1 + k_s/\mu_1) - (c_2 + k_s/\mu_2)\Big\} - (k_s + k_i)R(0) \\ &\quad + k_i - k_m, \end{aligned} \tag{15.16}$$

$$\begin{aligned} q_O^{(\infty)}(\infty) &= -r(0)\Big\{(c_1 + k_s/\mu_1) - (c_2 + k_s/\mu_2)\Big\} - (k_s + k_i)R(\infty) \\ &\quad + k_i - k_m. \end{aligned} \tag{15.17}$$

If $q_O^{(\infty)}(0) \leq 0$, then the function $E_O^{(\infty)}(t_0)$ is decreasing and the optimal ordering time is $t_0{}^* = 0$. If $q_O^{(\infty)}(\infty) \geq 0$, then the function $E_O^{(\infty)}(t_0)$ is increasing and the optimal ordering time is $t_0{}^* \to \infty$. On the other hand, if $F(t)$ is DFR under the assumptions (A-1) and (A-2), then the result is trivial. ■

Note in Theorem 15.3.2 that a sufficient condition that the cost effectiveness $E_O(t_0, \infty)$ is strictly concave (convex) in t_0 under strictly IFR (DFR) assumption, is given by the assumption (A-1). Also, if the assumption (A-2) does not hold, it can be shown that $q_O^{(\infty)}(0) < 0$ and $q_O^{(\infty)}(\infty) < 0$ and that the cost effectiveness is a decreasing function of t_0. In other words, if the monitoring cost is relatively more expensive than the inventory holding cost, the optimal order-replacement policy becomes trivial.

15.3.3 Special case: $t_1{}^* = 0$

Next, we consider the case of $t_1{}^* = 0$. In this model, the original unit is replaced/exchanged by the spare immediately when it is delivered by the regular order, irrespective of the state of the original one. Then, we have

$$E_O(t_0, 0) = U_O(t_0, 0)/V_O(t_0, 0), \tag{15.18}$$

where

$$U_O(t_0, 0) = \int_0^\infty \int_0^{t_0+l_2} \overline{F}(t)dG_2(l_2) \tag{15.19}$$

and

$$\begin{aligned} V_O(t_0, 0) &= k_s\Big\{(1/\mu_1 - 1/\mu_2)F(t_0) + \int_0^\infty \int_{t_0}^{t_0+l_2} F(t)dtdG_2(l_2)\Big\} \\ &\quad + c_1 F(t_0) + c_2\overline{F}(t_0) + k_m \int_0^{t_0} \overline{F}(t)dt. \end{aligned} \tag{15.20}$$

In a fashion similar to the case of $t_1 \to \infty$, define the numerator of the derivative of $E_O(t_0, 0)$ with respect to t_0, divided by $\overline{F}(t_0)$, as $q_O^{(0)}(t_0)$, namely,

$$\begin{aligned} q_O^{(0)}(t_0) &= \overline{R}(t_0)V_O(t_0, 0) - \Big\{\Big[k_s(1/\mu_1 - 1/\mu_2) + c_1 - c_2\Big]r(t_0) + k_s R(t_0) \\ &\quad + k_m\Big\}U_O(t_0, 0). \end{aligned} \tag{15.21}$$

We show the existence and uniqueness of the optimal ordering policy without the proof.

Theorem 15.3.3 *(1) Suppose that $F(t)$ is strictly IFR under (A-1).*

(i) *If $q_O^{(0)}(0) > 0$ and $q_O^{(0)}(\infty) < 0$, there exists a finite and non-zero optimal order-replacement time t_0^* $(0 < t_0^* < \infty)$ which satisfies $E_O(t_0^*, 0) = \max_{0<t_0<\infty} E_O(t_0, 0)$ and $q_O^{(0)}(t_0^*) = 0$.*

(ii) *If $q_O^{(0)}(0) \le 0$, the optimal order-replacement time is $t_0^* = 0$. On the other hand, if $q_O^{(0)}(\infty) \ge 0$, the optimal order-replacement time $t_0^* \to \infty$.*

(2) Suppose that $F(t)$ is DFR under (A-1). If

$$U_O(0, 0)V_O(\infty, 0) \ge U_O(\infty, 0)V_O(0, 0),$$

then the optimal order-replacement time is $t_0^* = 0$, otherwise, $t_0^* \to \infty$.

It should be noted in Theorem 15.3.3 that the assumption (A-2) is not always needed to guarantee the non-zero finite optimal order-replacement policy. This point is a unique feature for the case of $t_1^* = 0$, since the cost effectiveness in the case of $t_1^* = 0$ does not include the inventory holding cost.

15.4 Order-Inspection Policies in Continuous Time

15.4.1 Model description

Next, we consider the order-inspection model. The original unit begins operating at time $t = 0$, and the planning horizon is infinite. If the original unit does not fail up to a prespecified time $t_0 \in [0, \infty)$, the regular order for a spare is made at the time t_0 and after a lead time L_2 the spare is delivered. Then if the original unit has already failed as of $t = t_0 + L_2$, the delivered spare takes over its operation immediately. But if the original unit is still operating, the spare is put into the inventory and the original one is replaced/exchanged by the spare in the inventory when the original one fails/passes a prespecified time interval t_1 after the spare is delivered, whichever occurs first. On the other hand, even if the original unit fails before the time t_0, the failure can be detected at only the time t_0 and the expedited order is made immediately at the time t_0 instead of the regular order. The spare is delivered after a lead time L_1. The same cycle repeats itself again and again over an infinite time span. The configuration of this model called the order-inspection model is illustrated in Figure 15.2. Similar to the order-replacement model, define the interval from one replacement/exchange to the following replacement/exchange as one cycle.

The expected total cost for one cycle is

$$\begin{aligned} V_I(t_0, t_1) &= k_s\Big\{(1/\mu_1 - 1/\mu_2)F(t_0) + \int_0^\infty \int_0^{t_0+l_2} F(t)dtdG_2(l_2)\Big\} \\ &\quad + c_1 F(t_0) + c_2\overline{F}(t_0) + c_3 \\ &\quad + k_i \int_0^\infty \int_{t_0+l_2}^{t_0+l_2+t_1} \overline{F}(t)dtdG_2(l_2). \end{aligned} \tag{15.22}$$

On the other hand, the mean up (operative) time during one cycle is given by

$$U_I(t_0, t_1) = \int_0^\infty \int_0^{t_0+l_2+t_1} \overline{F}(t)dtdG_2(l_2). \tag{15.23}$$

Then, the cost effectiveness is formulated as

$$E_I(t_0, t_1) = U_I(t_0, t_1)/V_I(t_0, t_1), \tag{15.24}$$

and the problem is to obtain the optimal schedule $(t_0{}^*, t_1{}^*) \in [0, \infty) \times [0, \infty)$ which maximizes $E_I(t_0, t_1)$.

Define the numerator of the partial derivative of $E_I(t_0, t_1)$ with respect to t_1 divided by $\int_0^\infty \overline{F}(t_0 + l_2 + t_1)dG_2(l_2)$, as $q_I(t_0, t_1)$, i.e.,

$$\begin{aligned} q_I(t_0, t_1) &= \{V_I(t_0, t_1)\}^2\{\partial E_I(t_0, t_1)/\partial t_1\}/\int_0^\infty \overline{F}(t_0 + l_2 + t_1)dG_2(l_2) \\ &= V_I(t_0, t_1) - k_i U_I(t_0, t_1). \end{aligned} \tag{15.25}$$

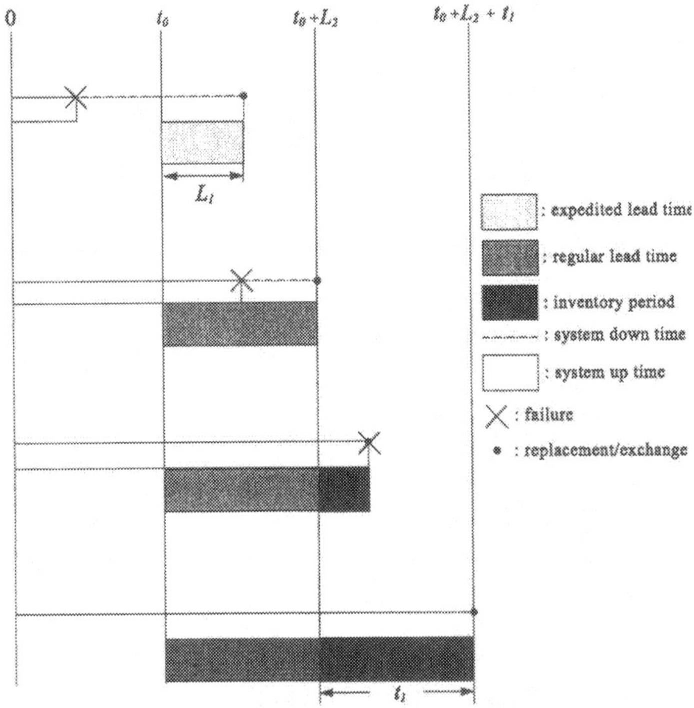

Figure 15.2: Configuration of order-inspection model in continuous time.

Lemma 15.4.1 *The function $q_I(t_0, t_1)$ is univariate in t_0, i.e.,*

$$\begin{aligned} q_I(t_0, t_1) &= k_s\Big\{(1/\mu_1 - 1/\mu_2)F(t_0) + \int_0^\infty \int_0^{t_0+l_2} F(t)dtdG_2(l_2)\Big\} \\ &\quad + c_1F(t_0) + c_2\overline{F}(t_0) + c_3 - k_i \int_0^\infty \int_0^{t_0+l_2} \overline{F}(t)dtdG_2(l_2) \\ &= q_I(t_0). \end{aligned} \tag{15.26}$$

Theorem 15.4.1 *For an arbitrary regular ordering time t_0 $(0 \le t_0 < \infty)$, if $q_I(t_0) \le 0$, the optimal inventory time limit which maximizes $E_I(t_0, t_1)$ is $t_1{}^* = 0$; otherwise $t_1{}^* \to \infty$.*

15.4.2 Special case: $t_1{}^* \to \infty$

Let us consider the case of $t_1{}^* \to \infty$. Then the spare delivered by the regular order is put into the inventory until the original unit fails, if the original one is still operating. In this case, the cost effectiveness is given by

$$E_I(t_0, \infty) = \lim_{t_1 \to \infty} E_I(t_0, t_1) = U_I(t_0, \infty)/V_I(t_0, \infty), \tag{15.27}$$

where $U_I(t_0, \infty) = 1/\lambda$ and

$$\begin{aligned} V_I(t_0, \infty) &= k_s\Big\{(1/\mu_1 - 1/\mu_2)F(t_0) + \int_0^\infty \int_0^{t_0+l_2} F(t)dtdG_2(l_2)\Big\} \\ &\quad + c_1F(t_0) + c_2\overline{F}(t_0) + c_3 \\ &\quad + k_i \int_0^\infty \int_{t_0+l_2}^\infty \overline{F}(t)dtdG_2(l_2). \end{aligned} \tag{15.28}$$

Define the numerator of the derivative of $E_I(t_0, \infty)$ with respect to t_0, divided by $\overline{F}(t_0)$, as $q_I^{(\infty)}(t_0)$,

$$\begin{aligned} q_I^{(\infty)}(t_0) &= -\Big(\frac{1}{\lambda}\Big)\Big\{\Big[k_s(1/\mu_1 - 1/\mu_2) + c_1 - c_2\Big]r(t_0) + k_s\Big[R(t_0) + \frac{F(t_0)}{\overline{F}(t_0)}\Big] \\ &\quad - k_i\overline{R}(t_0)\Big\}. \end{aligned} \tag{15.29}$$

Theorem 15.4.2 *Suppose that $F(t)$ is strictly IFR under (A-1).*

(i) *If $q_I^{(\infty)}(0) > 0$, there exists a finite and non-zero optimal ordering time t_0^* $(0 < t_0^* < \infty)$ which satisfies $E_I(t_0^*, \infty) = \max_{0<t_0<\infty} E_I(t_0, \infty)$ and $q_I^{(\infty)}(t_0^*) = 0$.*

(ii) *If $q_I^{(\infty)}(0) \le 0$, the optimal ordering time is $t_0^* = 0$.*

Theorem 15.4.3 *Under (A-1),*

(i) *if $dq_I^{(\infty)}(t_0)/dt_0 < 0$, then either (i) or (ii) in Theorem 15.4.2 holds,*

(ii) *if $dq_I^{(\infty)}(t_0)/dt_0 \ge 0$, then $t_0^* = 0$.*

15.4.3 Special case: $t_1{}^* = 0$

Next, we consider the case of $t_1{}^* = 0$. In this case, the original unit is replaced/exchanged by the spare immediately when it is delivered by the regular order, irrespective of the state of the original one. Then, we have

$$E_I(t_0, 0) = U_I(t_0, 0)/V_I(t_0, 0), \tag{15.30}$$

where

$$U_I(t_0, 0) = \int_0^\infty \int_0^{t_0+l_2} \overline{F}(t)dtdG_2(l_2) \tag{15.31}$$

and

$$\begin{aligned} V_I(t_0, 0) &= k_1\Big\{(1/\mu_1 - 1/\mu_2)F(t_0) + \int_0^\infty \int_0^{t_0+l_2} F(t)dtdG_2(l_2)\Big\} + c_1F(t_0) \\ &\quad + c_2\overline{F}(t_0) + c_3. \end{aligned} \tag{15.32}$$

Similar to the case of $t_1^* \to \infty$, define the numerator of the derivative of $E_I(t_0, 0)$ with respect to t_0, divided by $\overline{F}(t_0)$, as $q_I^{(0)}(t_0)$,

$$\begin{aligned} q_I^{(0)}(t_0) &= \overline{R}(t_0)V(t_0,0) - \Big\{\Big[k_1(1/\mu_1 - 1/\mu_2) + c_1 - c_2\Big]r(t_0) \\ &\quad + k_1\Big[R(t_0) + \frac{F(t_0)}{\overline{F}(t_0)}\Big]\Big\}U_I(t_0,0). \end{aligned} \tag{15.33}$$

Theorem 15.4.4 *Suppose that $F(t)$ is strictly IFR under (A-1).*

(i) *If $q_I^{(0)}(0) > 0$, there exists a finite and non-zero optimal ordering time t_0^* $(0 < t_0^* < \infty)$ which satisfies $E_I(t_0^*, 0) = \max_{0<t_0<\infty} E_I(t_0, 0)$ and $q_I^{(0)}(t_0^*) = 0$.*

(ii) *If $q_I^{(0)}(0) \le 0$, the optimal ordering time is $t_0^* = 0$.*

Theorem 15.4.5 *Under (A-1),*

(i) *if $dq_I^0(t_0)/dt_0 < 0$, then either (i) or (ii) in Theorem 15.4.4 holds,*

(ii) *if $dq_I^0(t_0)/dt_0 \ge 0$, then $t_0^* = 0$.*

15.5 Order-Replacement Policies in Discrete Time

15.5.1 Model description

Next, we consider similar models in discrete time setting. For a discrete time index $n = 0, 1, 2, \cdots$, consider an order-replacement problem for a one-unit system where each failed unit scrapped and each spare unit is provided after a lead time, in order. Let $P(n)$ be the unit failure time distribution with p.m.f. $p(n)$ and finite mean $1/\lambda$ (> 0). The original unit begins operating at time $n = 0$. If the original unit does not fail up to a prespecified time $n_0 \in [0, \infty)$, the regular order for a spare is made at that time and the spare is delivered after a lead time L_2 with p.m.f. $g_2(n)$ and finite mean $1/\mu_2$ (> 0). Then, if the original unit has already failed by time $n = n_0 + L_2$, the delivered spare takes over its operation from the delivery point. In this situation, if the original unit is still operating, the spare is put into the inventory and the original one is replaced/exchanged by the spare in the inventory when it fails/passes an allowable inventory period $n_1 \in [0, \infty)$ after the spare is delivered, whichever occurs first. It is assumed that the spare in the inventory does not fail or deteriorate with probability 1.

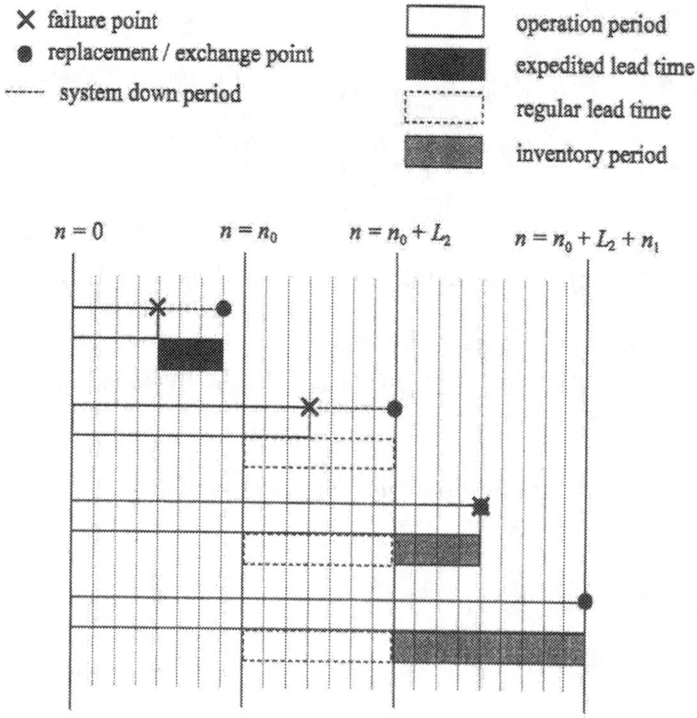

Figure 15.3: Configuration of order-replacement model in discrete time

On the other hand, if the original unit fails before the time n_0, an expedited order is made immediately at the failure time point and the spare takes over its operation just after it is delivered after a lead time L_1 with p.m.f. $g_1(n)$ and finite mean $1/\mu_1$ (> 0). In this situation, the regular order is not made. Define the time interval from one replacement or exchange of the unit to the following replacement or exchange as one cycle, where the same cycle repeats itself continually. Let k_s (> 0) and k_i (> 0) denote the shortage and inventory holding costs per unit time, respectively. Also, we define the fixed costs associated with expedited and regular orders by c_1 (> 0) and c_2 (> 0). Figure 15.3 depicts the possible behavior of the one-unit system under consideration.

We derive the expected cost per unit time in the steady state:

$$C_O(n_0, n_1) = V_O(n_0, n_1)/T_O(n_0, n_1), \tag{15.34}$$

where

$$V_O(n_0, n_1) = k_s\Big\{\sum_{l_1=0}^{\infty}\sum_{n=0}^{n_0-1} l_1 p(n) g_1(l_1) + \sum_{l_2=0}^{\infty}\sum_{n=n_0}^{n_0+l_2-1}(n_0 + l_2 - n)p(n)g_2(l_2)\Big\}$$

$$+k_i\Big\{\sum_{l_2=0}^{\infty}\sum_{n=n_0+l_2}^{n_0+l_2+n_1-1}(n-n_0-l_2)p(n)g_2(l_2)$$
$$+\sum_{l_2=0}^{\infty}\sum_{n=n_0+l_2+n_1}^{\infty}n_1p(n)g_2(l_2)\Big\}+c_1P(n_0-1)$$
$$+c_2\overline{P}(n_0-1)+k_m\sum_{n=0}^{n_0-1}\overline{P}(n) \tag{15.35}$$

and

$$\begin{aligned}T(n_0,n_1) &= \sum_{l_1=0}^{\infty}\sum_{n=0}^{n_0-1}(n+l_1)p(n)g_1(l_1)+\sum_{l_2=0}^{\infty}\sum_{n=n_0}^{n_0+l_2-1}(n_0+l_2)p(n)g_2(l_2)\\ &+\sum_{l_2=0}^{\infty}\sum_{n=n_0+l_2}^{n_0+l_2+n_1-1}np(n)g_2(l_2)\\ &+\sum_{l_2=0}^{\infty}\sum_{n=n_0+l_2+n_1}^{\infty}(n_0+l_2+n_1)p(n)g_2(l_2)\end{aligned} \tag{15.36}$$

are the expected total cost for one cycle and the mean time length of one cycle, respectively. On the other hand, the steady-state system availability is given by

$$A_O(n_1,n_0)=U_O(n_1,n_0)/T_O(n_1,n_0), \tag{15.37}$$

where

$$\begin{aligned}U_O(n_1,n_0) &= \sum_{l_1=0}^{\infty}\sum_{n=0}^{n_0-1}np(n)g_1(l_1)+\sum_{l_2=0}^{\infty}\sum_{n=n_0}^{n_0+l_2+n_1-1}np(n)g_2(l_2)\\ &+\sum_{l_2=0}^{\infty}\sum_{n=n_0+l_2+n_1}^{\infty}(n_0+l_2+n_1)p(n)g_2(l_2).\end{aligned} \tag{15.38}$$

Hence, from the definition in (15.2) the cost effectiveness is given by

$$E_O(n_0,n_1)=U_O(n_0,n_1)/V_O(n_0,n_1). \tag{15.39}$$

Define the numerator of the difference of $E_O(n_0,n_1)$ with respect to n_1 divided by $\sum_{l_2=0}^{\infty}\sum_{n=n_0+l_2+n_1+1}^{\infty}p(n)g_2(l_2)$, as $q_O(n_0,n_1)$, i.e.,

$$q_I(n_0,n_1)=\frac{V_O(n_0,n_1+1)V_O(n_0,n_1)\Big\{E_O(n_0,n_1+1)-E_O(n_0,n_1)\Big\}}{\sum_{l_2=0}^{\infty}\sum_{n=n_0+l_2+n_1+1}^{\infty}p(n)g_2(l_2)}. \tag{15.40}$$

Lemma 15.5.1 *The function $q_O(n_0, n_1)$ is univariate in n_0, i.e., $q_O(n_0, n_1) = q_O(n_0)$.*

PROOF. It is straightforward to see that

$$
\begin{aligned}
q_O(n_0, n_1) &= V_O(n_0, n_1 + 1) - k_i V_O(n_0, n_1) \\
&= k_s \Big\{ \sum_{l_1=0}^{\infty} \sum_{n=0}^{n_0-1} l_1 p(n) g_1(l_1) + \sum_{l_2=0}^{\infty} \sum_{n=n_0}^{n_0+l_2-1} (n_0 + l_2 - n) p(n) g_2(l_2) \Big\} \\
&\quad - k_i \Big\{ \sum_{l_1=0}^{\infty} \sum_{n=0}^{n_0-1} n p(n) g_1(l_1) + \sum_{l_2=0}^{\infty} \sum_{n=n_0}^{n_0-l_2-1} n p(n) g_2(l_2) \\
&\quad + \sum_{l_2=0}^{\infty} \sum_{n=n_0+l_2}^{\infty} (n_0 + l_2) p(n) g_2(l_2) \Big\} + k_m \sum_{n=0}^{n_0-1} \overline{P}(n) \\
&\quad + c_1 P(n_0 - 1) + c_2 \overline{P}(n_0 - 1). \qquad (15.41)
\end{aligned}
$$

■

Theorem 15.5.1 *For an arbitrary regular ordering time n_0 ($0 \leq n_0 < \infty$), if $q_O(t_0) \geq 0$, the optimal inventory time limit which maximizes $E_O(n_0, n_1)$ is $n_1{}^* \to \infty$, otherwise $n_1{}^* = 0$.*

PROOF. From Lemma 15.5.1, we have

$$
E_O(n_0, n_1 + 1) - E_O(n_0, n_1) = \frac{\sum_{l_2=0}^{\infty} \sum_{n=n_0+l_2+n_1+1}^{\infty} p(n) g_2(l_2)}{V_O(n_0, n_1 + 1) V_O(n_0, n_1)} q_O(n_0). \qquad (15.42)
$$

Thus, if $q_O(n_0) \geq 0$, then $E_O(n_0, n_1)$ is increasing in n_1 and $n_1{}^* \to \infty$, otherwise $n_1{}^* = 0$. ■

15.5.2 Special case: $n_1^* \to \infty$

When $n_1^* \to \infty$, the cost effectiveness is given by

$$
E_O(n_0, \infty) = U_O(n_0, \infty) / V_O(n_0, \infty), \qquad (15.43)
$$

where

$$
\begin{aligned}
V_O(n_0, \infty) &= k_s \Big\{ \sum_{l_1=0}^{\infty} \sum_{n=0}^{n_0-1} l_1 p(n) g_1(l_1) + \sum_{l_2=0}^{\infty} \sum_{n=n_0}^{n_0+l_2-1} (n_0 + l_2 - n) p(n) g_2(l_2) \Big\} \\
&\quad + k_i \sum_{l_2=0}^{\infty} \sum_{n=n_0+l_2}^{\infty} (n - n_0 - l_2) p(n) g_2(l_2) \\
&\quad + k_m \sum_{n=0}^{n_0-1} \overline{P}(n) + c_1 P(n_0 - 1) + c_2 \overline{P}(n_0 - 1) \qquad (15.44)
\end{aligned}
$$

and $U_O(n_0,\infty)=1/\lambda$. Taking the difference of $E_O(n_0,\infty)$ with respect to n_0, we define the function:

$$\begin{aligned} q_O^{(\infty)}(n_0) &= -k_s\Big\{(1/\mu_1-1/\mu_2)r(n_0)+R(n_0)\Big\} \\ &\quad -(c_1-c_2)r(n_0)+k_i\overline{R}(n_0)-k_m. \end{aligned} \tag{15.45}$$

Lemma 15.5.2 *The function $R(n)$ is decreasing (increasing) if the function $r(n)$ is decreasing (increasing).*

PROOF. It is seen that

$$\begin{aligned} R(n_0+1)-R(n_0) &= \sum_{l_2=0}^{\infty}\Big\{\frac{P(n_0+l_2+1)-P(n_0+1)}{\overline{P}(n_0+1)} \\ &\quad -\frac{P(n_0+l_2)-P(n_0)}{\overline{P}(n_0)}\Big\}g_2(l_2) \\ &= \sum_{l_2=0}^{\infty}\frac{g_2(l_2)}{\overline{P}(n_0)\overline{P}(n_0+l_2+1)}\Big\{r(n_0+l_2+1)-r(n_0+1)\Big\}. \end{aligned} \tag{15.46}$$

For an arbitrary l_2, if $r(n_0+l_2+1)\geq(\leq)r(n_0+1)$, then $R(n_0+1)\geq(\leq)R(n_0)$. ■

Strictly speaking, the function $r(n)=p(n)/\overline{P}(n)$ is not a failure rate of the discrete failure time distribution $P(n)$, because the failure rate should be defined as $p(n)/\overline{P}(n-1)$ [see Barlow and Proschan (1975)].

Using Lemma 15.5.2, we can characterize the optimal ordering policy n_0^* maximizing $E_O(n_0,\infty)$ as follows.

Theorem 15.5.2 *(1) Under the assumptions (A-1) and (A-2), suppose that the function $r(n)$ is a strictly increasing function of n.*

(i) *If $q_O^{(\infty)}(0)>0$ and $q_O^{(\infty)}(\infty)<0$, there exists (at least one, at most two) optimal ordering time n_0^* $(0<n_0^*<\infty)$ satisfying $q_O^{(\infty)}(n_0^*-1)>0$ and $q_O^{(\infty)}(n_0^*)\leq 0$.*

(ii) *If $q_O^{(\infty)}(0)\leq 0$, the optimal ordering time is $n_0^*=0$ with $E_O(0,\infty)=\{k_s\lambda\sum_{l_2=0}^{\infty}\sum_{n=0}^{l_2-1}(l_2-n)p(n)g_2(l_2)\}^{-1}$. If $q_O^{(\infty)}(\infty)\geq 0$ $n_0^*\to\infty$ with $E_O(\infty,\infty)=k_s/(\mu_1\lambda)$.*

(2) Under the assumptions (A-1) and (A-2), suppose that the function $r(n)$ is a decreasing function of n. Then, the optimal regular ordering time is $n_0^*=0$ or $n_0^*\to\infty$.

PROOF. Taking the difference of $E_O(n_0, \infty)$ with respect to n_0, we obtain

$$E_O(n_0+1,\infty) - E_O(n_0,\infty) = \frac{\overline{P}(n_0)}{V_O(n_0+1,\infty)V_O(n_0,\infty)} q_O^{(\infty)}(n_0). \quad (15.47)$$

Further difference yields

$$\begin{aligned} q_O^{(\infty)}(n_0+1) - q_O^{(\infty)}(n_0) &= -\Big\{r(n_0+1) - r(n_0)\Big\}\Big\{k_s(1/\mu_1 - 1/\mu_2) \\ &\quad + c_1 - c_2\Big\} - \Big\{R(n_0+1) - R(n_0)\Big\} \\ &\quad \times\Big\{k_s(1/\mu_1 - 1/\mu_2) + k_i\Big\}, \end{aligned} \quad (15.48)$$

where

$$\begin{aligned} q_O^{(\infty)}(0) &= -k_s\Big\{(1/\mu_1 - 1/\mu_2)r(0) + R(0)\Big\} \\ &\quad -(c_1 - c_2)r(0) + k_i\overline{R}(0) - k_m, \end{aligned} \quad (15.49)$$

$$\begin{aligned} q_O^{(\infty)}(\infty) &= -k_s\Big\{(1/\mu_1 - 1/\mu_2)r(\infty) + R(\infty)\Big\} \\ &\quad -(c_1 - c_2)r(\infty) + k_i\overline{R}(\infty) - k_m. \end{aligned} \quad (15.50)$$

Suppose that the function $r(n)$ is strictly increasing with respect to $n = 0, 1, \cdots$. Since $q_O^{(\infty)}(n_0) < 0$ for all n_0 under the assumptions (A-1) and (A-2), the function $C(n_0, \infty)$ is strictly concave in n_0. If $q_O^{(\infty)}(0) > 0$ and $q_O^{(\infty)}(\infty) < 0$, then the function $q_O^{(\infty)}(n_0)$ monotonically decreases with respect to n_0 and its sign changes at once from positive to negative. Hence, it is easily shown that there exists (at least one, at most two) optimal ordering time n_0^* ($0 < n_0^* < \infty$) satisfying $q_\infty(n_0^* - 1) > 0$ and $q_\infty(n_0^*) \le 0$. If $q_O^{(\infty)}(0) \le 0$ under the assumptions (A-1) and (A-2) with the strictly decreasing property, then the function $E_O(n_0, \infty)$ is a decreasing function of n_0, and the optimal ordering time is $n_0^* = 0$. Conversely, if $q_O^{(\infty)}(\infty) \ge 0$, then the function $E_O(n_0, \infty)$ is an increasing function of n_0, and the optimal ordering time is $n_0^* \to \infty$. The decreasing case of the function $r(n)$ is trivial. The proof is completed. ■

15.5.3 Special case: $n_1^* = 0$

On the other hand, when $n_1^* = 0$, we can obtain

$$\begin{aligned} V_O(n_0, 0) &= k_s\Big\{\sum_{l_1=0}^{\infty}\sum_{n=0}^{n_0-1} l_1 p(n) g_1(l_1) + \sum_{l_2=0}^{\infty}\sum_{n=n_0}^{n_0+l_2-1}(n_0 + l_2 - n)p(n)g_2(l_2)\Big\} \\ &\quad + c_1 P(n_0 - 1) + c_2\overline{P}(n_0 - 1) + k_m \sum_{n=0}^{n_0-1}\overline{P}(n) \end{aligned} \quad (15.51)$$

and

$$
\begin{aligned}
U_O(n_0, 0) &= \sum_{l_1=0}^{\infty}\sum_{n=0}^{n_0-1} np(n)g_1(l_1) + \sum_{l_2=0}^{\infty}\sum_{n=n_0}^{n_0+l_2-1} np(n)g_2(l_2) \\
&+ \sum_{l_2=0}^{\infty}\sum_{n=n_0+l_2}^{\infty} (n_0 + l_2)p(n)g_2(l_2). \qquad (15.52)
\end{aligned}
$$

Taking the difference of $E_O(n_0, 0)$ with respect to n_0, we have

$$
\begin{aligned}
q_O^{(0)}(n_0) &= -\Big\{k_s\Big[(1/\mu_1 - 1/\mu_2)r(n_0) + R(n_0)\Big] + (c_1 - c_2)r(n_0) \\
&+k_i\Big\}U_O(n_0, 0) + \overline{R}(n_0)V_O(n_0, 0). \qquad (15.53)
\end{aligned}
$$

Theorem 15.5.3 *(1) Under the assumption (A-1), suppose that the function $r(n)$ is a strictly increasing function of n.*

(i) *If $q_O^{(0)}(0) > 0$ and $q_O^{(0)}(\infty) < 0$, there exists (at least one, at most two) optimal ordering time n_0^* $(0 < n_0^* < \infty)$ satisfying $q_O^{(0)}(n_0^* - 1) > 0$ and $q_O^{(0)}(n_0^*) \leq 0$.*

(ii) *If $q_O^{(0)}(0) \leq 0$, the optimal ordering time is $n_0^* = 0$ with*

$$
E_O(0, 0) = \frac{\sum_{l_2=0}^{\infty}\sum_{n=0}^{l_2-1} np(n)g_2(l_2) + \sum_{l_2=0}^{\infty}\sum_{n=l_2}^{\infty} l_2p(n)g_2(l_2)}{k_s\sum_{l_2=0}^{\infty}\sum_{n=0}^{l_2-1}(l_2 - n)p(n)g_2(l_2)}, \qquad (15.54)
$$

otherwise $n_0^* \to \infty$ with $E_O(\infty, 0) = k_s/(\mu_1\lambda)$.

(2) Under the assumption (A-1), suppose that the function $r(n)$ is a decreasing function of n. Then, the optimal regular ordering time is $n_0^* = 0$ or $n_0^* \to \infty$.

15.6 Order-Inspection Policies in Discrete Time

15.6.1 Model description

Consider an order-inspection problem for a one-unit system where each failed unit is scrapped and each spare unit is provided after a lead time, placing order. The original unit begins operating at time $n = 0$. If it does not fail up to a prespecified time $n_0 \in [0, \infty)$, the regular order for a spare is made at that time and the spare is delivered after a lead time L_2. Then, if the original unit has already failed by time $n = n_0 + L_2$, the delivered spare takes

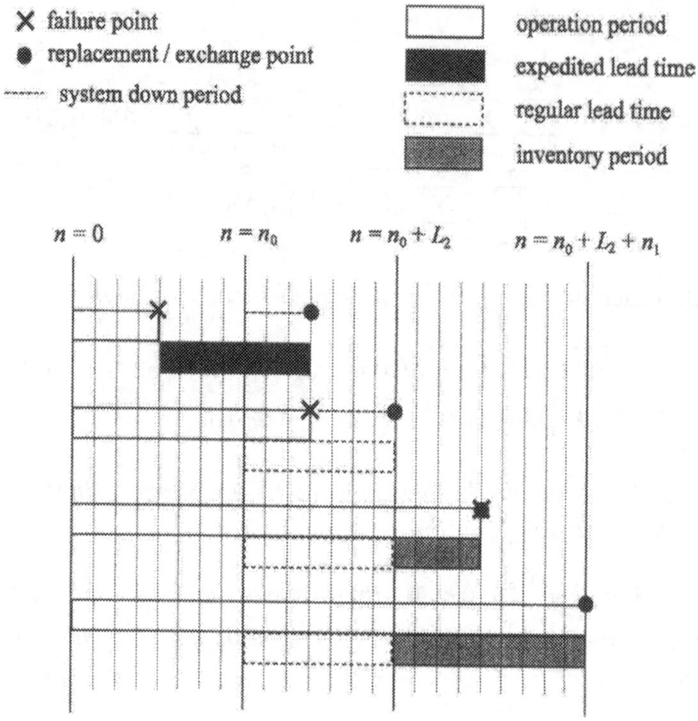

Figure 15.4: Configuration of order-inspection model in discrete time

over its operation from the delivery point. In this situation, if the original unit is still operating, the spare is put into the inventory and the original one is replaced/exchanged by the spare in the inventory when it fails/passes an inventory time limit $n_1 \in [0, \infty)$ after the spare is delivered, whichever occurs first. It is assumed that the spare in the inventory does not fail or deteriorate with probability 1.

On the other hand, if the original unit fails before the time n_0, an expedited order is made at the inspection time n_0. Then the spare takes over its operation just after it is delivered after a lead time L_1. In this case, the regular order is canceled. In Figure 15.4, the possible behavior of the one-unit system with order-inspection policy in discrete time setting is presented.

The cost effectiveness is formulated as

$$E_I(n_0, n_1) = U_I(n_0, n_1)/V_I(n_0, n_1), \tag{15.55}$$

where

$$V_I(n_0, n_1) \quad = \quad k_s\Big\{\sum_{l_1=0}^{\infty}\sum_{n=0}^{n_0-1}(n_0 + l_1 - n)p(n)g_1(l_1)$$

$$+\sum_{l_2=0}^{\infty}\sum_{n=n_0}^{n_0+l_2-1}(n_0+l_2-n)p(n)g_2(l_2)\Big\}$$

$$+k_i\Big\{\sum_{l_2=0}^{\infty}\sum_{n=n_0+l_2}^{n_0+l_2+n_1-1}(n-n_0-l_2)p(n)g_2(l_2)$$

$$+\sum_{l_2=0}^{\infty}\sum_{n=n_0+l_2+n_1}^{\infty}n_1p(n)g_2(l_2)\Big\}$$

$$+c_1P(n_0-1)+c_2\overline{P}(n_0-1)+c_3 \qquad (15.56)$$

and

$$U_I(n_0,n_1) = \sum_{l_1=0}^{\infty}\sum_{n=0}^{n_0-1}np(n)g_1(l_1)+\sum_{l_2=0}^{\infty}\sum_{n=n_0}^{n_0+l_2+n_1-1}np(n)g_2(l_2)$$

$$+\sum_{l_2=0}^{\infty}\sum_{n=n_0+l_2+n_1}^{\infty}(n_0+l_2+n_1)p(n)g_2(l_2). \qquad (15.57)$$

Define the numerator of the difference of $E_I(n_0,n_1)$ with respect to n_1 divided by $\sum_{l_2=0}^{\infty}\sum_{n=n_0+l_2+n_1+1}^{\infty}p(n)g_2(l_2)$, as $q_I(t_0,t_1)$.

Lemma 15.6.1 *The function $q_I(n_0,n_1)$ is univariate in n_0, i.e., $q_I(n_0,n_1)=V_I(n_0,n_1)-k_iU_I(n_0,n_1)=q_I(n_0)$.*

Theorem 15.6.1 *For an arbitrary regular ordering time n_0, if $q_I(n_0)\geq 0$, then the optimal inventory time limit is given by $n_1^*\to\infty$, otherwise $n_1^*=0$.*

15.6.2 Special case: $n_1^*\to\infty$

When $n_1^*\to\infty$, the cost effectiveness is given by

$$E_I(n_0,\infty)=V_I(n_0,\infty)/U_I(n_0,\infty),$$

where

$$V_I(n_0,\infty) = k_s\Big\{\sum_{l_1=0}^{\infty}\sum_{n=0}^{n_0-1}(n_0+l_1-l_2)p(n)g_1(l_1)$$

$$+\sum_{l_2=0}^{\infty}\sum_{n=n_0}^{n_0+l_2-1}(n_0+l_2-n)p(n)g_2(l_2)\Big\}$$

$$+k_i\sum_{l_2=0}^{\infty}\sum_{n=n_0+l_2}^{\infty}(n-n_0-l_2)p(n)g_2(l_2)+c_1P(n_0-1)$$

$$+c_2\overline{P}(n_0-1)+c_3 \qquad (15.58)$$

and $U_I(n_0,\infty) = 1/\lambda$. Taking the difference of $E_I(n_0,\infty)$ with respect to n_0, we define the function:

$$q_I^{(\infty)}(n_0) = -\Big\{k_s(1/\mu_1 - 1/\mu_2) + c_1 - c_2\Big\}r(n_0) \\ -k_i\Big\{R(n_0) + \frac{P(n_0)}{\overline{P}(n_0)} - k_i\overline{R}(n_0)\Big\}. \quad (15.59)$$

Theorem 15.6.2 *Suppose that $r(n)$ is increasing in n under (A-1).*

(i) *If $q_I^{(\infty)}(0) > 0$, there exists (at least one, at most two) optimal ordering time n_0^* $(0 < n_0^* < \infty)$ which satisfies $q_I^{(\infty)}(n_0^* - 1) > 0$ and $q_I^{(\infty)}(n_0^*) \leq 0$.*

(ii) *If $q_I^{(\infty)}(0) \leq 0$, the optimal ordering time is $n_0^* = 0$.*

Theorem 15.6.3 *Under (A-1),*

(i) *if $q_I^{(\infty)}(n_0+1) < q_I^{(\infty)}(n_0)$, then either (i) or (ii) in Theorem 15.6.2 holds,*

(ii) *if $q_I^{(\infty)}(n_0+1) \geq q_I^{(\infty)}(n_0)$, then $t_0^* = 0$.*

15.6.3 Special case: $n_1^* = 0$

On the other hand, when $n_1^* = 0$, we can obtain the cost effectiveness

$$E_I(n_0, 0) = V_I(n_0, 0)/U_I(n_0, 0),$$

where

$$V_I(n_0,0) = k_s\Big\{\sum_{l_1=0}^{\infty}\sum_{n=0}^{n_0-1}(n_0+l_1-n)p(n)g_1(l_1) \\ +\sum_{l_2=0}^{\infty}\sum_{n=n_0}^{n_0+l_2-1}(n_0+l_2-n)p(n)g_2(l_2)\Big\} \\ +c_1P(n_0-1) + c_2\overline{P}(n_0-1) + c_3 \quad (15.60)$$

and

$$U_I(n_0,0) = \sum_{l_1=0}^{\infty}\sum_{n=0}^{n_0-1}np(n)g_1(l_1) + \sum_{l_2=0}^{\infty}\sum_{n=n_0}^{n_0+l_2-1}np(n)g_2(l_2) \\ +\sum_{l_2=0}^{\infty}\sum_{n=n_0+l_2}^{\infty}(n_0+l_2)p(n)g_2(l_2). \quad (15.61)$$

Taking the difference of $E_I(n_0, 0)$ with respect to n_0, we get

$$q_I^{(0)}(n_0) = -\Big\{\Big[k_s(1/\mu_1 - 1/\mu_2) + c_1 - c_2\Big]r(n_0) + k_s\Big[R(n_0) + \frac{P(n_0)}{\overline{P}(n_0)}\Big]\Big\}U_I(n_0, 0) + \overline{R}(n_0)V_I(n_0, 0). \quad (15.62)$$

Theorem 15.6.4 *Suppose that $r(n)$ is strictly increasing in n under (A-1).*

(i) *If $q_I^{(0)}(0) > 0$, there exists (at least one, at most two) optimal ordering time n_0^* $(0 < n_0^* < \infty)$ which satisfies $q_I^{(0)}(n_0^* - 1) > 0$ and $q_I^{(0)}(n_0^*) \leq 0$.*

(ii) *If $q_I^{(0)}(0) \leq 0$, the optimal ordering time is $n_0^* = 0$.*

Theorem 15.6.5 *Under (A-1),*

(i) *if $q_I^{(0)}(n_0 + 1) < q_I^{(0)}(n_0)$, then either (i) or (ii) in Theorem 15.6.4 holds,*

(ii) *if $q_I^{(0)}(n_0 + 1) \geq q_I^{(0)}(n_0)$, then $n_0^* = 0$.*

15.7 Numerical Illustrations

15.7.1 Continuous time models

We present some examples that determine the optimal order-replacement and order-inspection policies. For the continuous time models, suppose that the lifetime distribution obeys the Weibull distribution:

$$F(x) = 1 - e^{-(x/\alpha)^\beta}, \quad x > 0, \quad (15.63)$$

where $1/\lambda = \alpha\Gamma(1 + 1/\beta)$ and $\Gamma(\cdot)$ denotes the standard gamma function. Also, it is assumed that the emergency lead time L_1 and the regular lead time L_2 are the exponentially distributed random variables having

$$G_1(x) = 1 - e^{-\mu_1 x} \quad (15.64)$$

and

$$G_2(x) = 1 - e^{-\mu_2 x}, \quad (15.65)$$

respectively. In Table 15.1 we present dependence of the scale parameter α of the Weibull distribution on the optimal ordering policies. As α monotonically increases, i.e., MTTF (mean time to failure) increases, the optimal ordering time becomes longer and its associated cost effectiveness increases. In this example, it can be seen that the case of $t_1^* = 0$ is better than that of $t_1^* \to \infty$ in terms of maximization of cost effectiveness. On the other hand, as the scale parameter increases, the cost-effectiveness of the order-inspection policy improves much faster than that of the order-replacement policy (the best policy is indicated by the underline). In Tables 15.2 and 15.3, we examine the optimal ordering policies for varying shape parameter β of the Weibull distribution and the inspection cost c_3, respectively. Similar to Table 15.1, the optimal policy is switched from the order-replacement policy (order-inspection policy) to the order-inspection policy (order-replacement policy) as β (c_3) increases in Tables 15.2 and 15.3, respectively. Also, in Table 15.3 the cost parameter c_3 is insensitive to the optimal policy except for the order-inspection policy with $t_1^* = 0$.

15.7.2 Discrete time models

Of our next interest is the investigation of sensitivity of model parameters on the optimal ordering policy in discrete time setting. Suppose that the (discrete) failure time obeys the following discrete Weibull distribution;

$$p(n) = (q)^{n^\beta} - (q)^{(n+1)^\beta}, \tag{15.66}$$

where $0 < q < 1$, $m > 0$ and $n = 1, 2, \ldots$. This interesting discrete distribution was introduced by Nakagawa and Osaki (1975). Later, Stein and Dattero (1984) defined a somewhat different discrete Weibull distribution. Also, it is assumed that the p.m.f. of the expedited and regular lead times are given by the following geometric distributions:

$$g_1(l) = p_1(1-p_1)^l, \tag{15.67}$$
$$g_2(l) = p_2(1-p_2)^l, \quad l = 1, 2, \cdots, \tag{15.68}$$

respectively, where p_j ($j = 1, 2$) $\in (0, 1)$. Table 15.4 presents the dependence of the parameter q of the discrete Weibull distribution on the optimal ordering policies. As q monotonically increases, the optimal ordering time and the corresponding cost effectiveness increase. In this example, it is observed that $n_1^* \to \infty$ is always better than $n_1^* = 0$ in terms of maximization of cost effectiveness. Also, in all the cases, it is optimal to perform the order-replacement policy. In Tables 15.5 and 15.6, we present the comparative results on the optimal ordering policies for varying shape parameter β and c_3, respectively. In Tables 15.5 and 15.6, it is obvious that the cost effectiveness does not always increase monotonically. Similar to the continuous case, the inspection cost c_3 tends to be insensitive in most cases.

Table 15.1: Dependence of failure parameter α on the optimal policy in continuous time: $k_i = 0.1$, $k_s = 0.05$, $k_m = 0.01$, $c_1 = 2$, $c_2 = 1$, $c_3 = 3$, $\mu_1 = 0.01$, $\mu_2 = 0.02$, $\beta = 2.0$

	Order-Inspection				Order-Replacement			
α	t_0^*	$E_I(t_0^*, 0)$	t_0^*	$E_I(t_0^*, \infty)$	t_0^*	$E_O(t_0^*, 0)$	t_0^*	$E_O(t_0^*, \infty)$
1000	309.2	56.2	547.8	29.8	164.5	69.6	1584.2	42.7
1200	359.2	64.3	674.9	30.8	205.2	73.2	2290.3	47.0
1400	406.5	72.1	802.1	31.5	246.3	75.8	3124.6	50.8
1600	451.8	79.5	929.4	32.1	287.6	78.0	4084.2	54.2
1800	495.2	86.7	1056.6	32.6	329.0	79.9	5217.8	57.0
2000	537.1	93.6	1183.9	32.9	370.4	81.4	6834.5	59.6
2200	577.7	100.2	1311.2	33.3	412.0	82.8	7672.9	61.9
2400	617.0	106.7	1438.5	33.5	453.6	83.9	9724.5	63.9

Table 15.2: Dependence of failure parameter β on the optimal policy in continuous time: $k_i = 0.1$, $k_s = 0.05$, $k_m = 0.01$,. $c_1 = 2$, $c_2 = 1$, $c_3 = 3$, $\mu_1 = 0.01$, $\mu_2 = 0.02$, $\alpha = 1600$

	Order-Inspection				Order-Replacement			
β	t_0^*	$E_I(t_0^*, 0)$	t_0^*	$E_I(t_0^*, \infty)$	t_0^*	$E_O(t_0^*, 0)$	t_0^*	$E_O(t_0^*, \infty)$
1.25	329.3	47.1	686.7	24.8	278.2	62.5	23080	55.4
1.50	370.3	58.1	786.5	27.2	245.4	69.2	13883	54.6
1.75	411.5	68.9	865.5	29.6	261.0	74.3	6921	54.3
2.00	451.8	79.5	929.4	32.1	287.6	78.0	4084	54.2
2.50	527.3	99.6	1026.1	37.0	348.7	83.1	2565	54.3
3.00	595.3	118.0	1095.6	41.8	410.1	86.2	2073	54.9
3.50	655.9	134.6	1148.0	46.5	467.8	88.2	1844	55.6
4.00	709.9	149.6	1188.9	51.0	521.1	89.6	1718	56.5

Table 15.3: Dependence of cost parameter c_3 on the optimal policy in continuous time: $k_i = 0.1$, $k_s = 0.05$, $k_m = 0.01$, $c_1 = 2$, $c_2 = 1$, $\mu_1 = 0.01$, $\mu_2 = 0.02$, $\alpha = 1600$, $\beta = 2.0$

	Order-Inspection				Order-Replacement			
c_3	t_0^*	$E_I(t_0^*, 0)$	t_0^*	$E_I(t_0^*, \infty)$	t_0^*	$E_O(t_0^*, 0)$	t_0^*	$E_O(t_0^*, \infty)$
0.5	307.2	150.1	929.4	34.0	287.6	78.0	4084.2	54.2
1.0	345.0	124.8	929.4	33.6	287.6	78.0	4084.2	54.2
1.5	376.8	108.0	929.4	33.2	287.6	78.0	4084.2	54.2
2.0	404.6	95.9	929.4	32.8	287.6	78.0	4084.2	54.2
2.5	429.4	86.8	929.4	32.4	287.6	78.0	4084.2	54.2
3.0	451.8	79.5	929.4	32.1	287.6	78.0	4084.2	54.2
3.5	472.4	73.6	929.4	31.7	287.6	78.0	4084.2	54.2
4.0	491.4	68.7	929.4	31.4	287.6	78.0	4084.2	54.2

Table 15.4: Dependence of failure parameter q on the optimal policy in discrete time: $k_i = 0.1$, $k_s = 0.05$, $k_m = 0.01$, $c_1 = 2$, $c_2 = 1$, $c_3 = 3$, $p_1 = 0.1$, $p_2 = 0.2$, $\beta = 2.0$

	Order-Inspection				Order-Replacement			
q	n_0^*	$E_I(n_0^*, 0)$	n_0^*	$E_I(n_0^*, \infty)$	n_0^*	$E_O(n_0^*, 0)$	n_0^*	$E_O(n_0^*, \infty)$
0.992	9	1.57	1	2.14	5	4.32	0	6.73
0.993	10	1.68	1	2.29	5	4.59	1	7.11
0.994	11	1.80	1	2.46	6	4.90	1	7.57
0.995	12	1.97	1	2.69	6	5.31	1	8.11
0.996	13	2.18	2	2.99	7	5.85	1	8.78
0.997	15	2.50	3	3.41	9	6.64	2	9.69
0.998	19	3.00	4	4.09	12	7.90	4	11.01
0.999	26	4.09	9	5.51	18	10.61	9	13.39

Table 15.5: Dependence of failure parameter β on the optimal policy in discrete time: $k_i = 0.1$, $k_s = 0.05$, $k_m = 0.01$, $c_1 = 2$, $c_2 = 1$, $c_3 = 3$, $p_1 = 0.1$, $p_2 = 0.2$, $q = 0.999$

	Order-Inspection				Order-Replacement			
β	n_0^*	$E_I(n_0^*, 0)$	n_0^*	$E_I(n_0^*, \infty)$	n_0^*	$E_O(n_0^*, 0)$	n_0^*	$E_O(n_0^*, \infty)$
1.25	103	7.10	53	0.04	112	17.29	96	0.07
1.50	68	8.61	38	0.01	86	22.42	103	0.02
1.75	41	5.93	18	0.10	36	15.14	24	0.22
2.00	26	4.09	9	5.51	18	10.61	9	13.39
2.50	12	2.33	3	3.03	7	6.53	3	9.05
3.00	7	1.57	2	1.95	4	4.65	1	6.34
3.50	5	1.16	1	1.40	3	3.57	0	4.69
4.00	4	0.92	0	1.08	2	2.84	0	3.66

Table 15.6: Dependence of cost parameter c_3 on the optimal policy in discrete time: $k_i = 0.1$, $k_s = 0.05$, $k_m = 0.01$, $c_1 = 2$, $c_2 = 1$, $p_1 = 0.1$, $p_2 = 0.2$, $q = 0.999$, $\beta = 2.0$

	Order-Inspection				Order-Replacement			
c_3	n_0^*	$E_I(n_0^*, 0)$	n_0^*	$E_I(n_0^*, \infty)$	n_0^*	$E_O(n_0^*, 0)$	n_0^*	$E_O(n_0^*, \infty)$
0.05	15	10.33	9	13.46	18	10.61	9	13.39
0.10	15	10.02	9	13.13	18	10.61	9	13.39
0.50	17	8.14	9	11.03	18	10.61	9	13.39
1.00	20	6.69	9	9.19	18	10.61	9	13.39
1.50	22	5.73	9	7.87	18	10.61	9	13.39
2.00	23	5.04	9	6.89	18	10.61	9	13.39
2.50	25	4.51	9	6.12	18	10.61	9	13.39
3.00	26	4.09	9	5.51	18	10.61	9	13.39

Acknowledgement. The authors would like to thank Professor N. Balakrishnan, Department of Mathematics and Statistics, McMaster University, Canada, for his invitation to contribute to this special volume. The authors were partially supported by the Ministry of Education, Science, Sports and Culture, Grant-in-Aid for Exploratory Research; Grant No. 15651076 (2003-2005), the Research Program 2004 under the Institute for Advanced Studies of the Hiroshima Shudo University and the Nanzan University Pache Research Subsidy for 2004.

References

1. Acharya, D., Nagabhushanam, G. and Alam, S. S. (1986). Jointly optimal block-replacement and spare provisioning policy, *IEEE Transactions on Reliability*, **R-35**, 447–451.

2. Armstrong, M. J. and Atkins, D. R. (1996). Joint optimization of maintenance and inventory policies for a simple system, *IIE Transactions*, **28**, 415–424.

3. Allen, S. G. and D'Esopo, D. A. (1968a). An ordering policy for repairable stock items, *Operations Research*, **16**, 669–674.

4. Allen, S. G. and D'Esopo, D. A. (1968b). An ordering policy for stock items when delivery can expedited, *Operations Research*, **16**, 880–883.

5. Barlow, R. E. and Proschan, F. (1965). *Mathematical Theory of Reliability*, John Wiley & Sons, New York.

6. Barlow, R. E. and Proschan, F. (1975). *Statistical Theory of Reliability and Life Testing: Probability Models*, Holt, Rinehart and Winston, New York.

7. Berg, M. (1980). A marginal cost analysis for preventive replacement policies, *European Journal of Operational Research*, **4**, 136–142.

8. Christer, A. H. (1978). Refined asymptotic costs for renewal reward processes, *Journal of the Operational Research Society*, **29**, 577–588.

9. Csenki, A. (1998). Marginal cost analysis of some basic ordering policies for a spare unit and extensions, *International Journal of Reliability, Quality and Safety Engineering*, **5**, 293–312.

10. Csenki, A. (1999). Refined asymptotic analysis of two basic order-replacement models for a spare unit, *IMA Journal of Mathematics Applied in Business & Industry*, **9**, 177–199.

11. Dohi, T., Kaio, N. and Osaki, S. (1994). Optimal ordering policies for a two-unit standby redundant system under cost effectiveness criterion, *Proceedings of 1994 Pacific Conference on Manufacturing*, 1047–1054.

12. Dohi, T., Kaio, N. and Osaki, S. (1995a). Monitored replenishment rules for a continuous review cyclic inventory system, *IMA Journal of Mathematics Applied in Business and Industry*, **6**, 223–237.

13. Dohi, T., Kaio, N. and Osaki, S. (1995b). Continuous review cyclic inventory models with emergency order, *Journal of Operations Reserach Society of Japan*, **38**, 212–229.

14. Dohi, T., Kaio, N. and Osaki, S. (1996a). Optimal planned maintenance with salvage cost for a two-unit standby redundant system, *Microelectronics and Reliability*, **36**, 1581–1588.

15. Dohi, T., Kaio, N. and Osaki, S. (1996b). Optimal ordering policies with time-dependent delay structure, *Journal of Quality in Maintenance Engineering*, **26**, 50–62.

16. Dohi, T., Kaio, N. and Osaki, S. (1997). Spare parts inventory control policies under cost effectiveness criterion, *Proceedings of 14th International Conference on Production Research*, 1036–1039.

17. Dohi, T., Shibuya, T. and Osaki, S. (1997). Models for 1-out-of-Q systems with stochastic lead times and expedited ordering options for spares inventory, *European Journal of Operational Research*, **103**, 255–272.

18. Dohi, T., Kaio, N. and Osaki, S. (1998). On the optimal ordering policies in maintenance theory – survey and applications, *Applied Stochastic Models and Data Analysis*, **14**, 309–321.

19. Dohi, T., Kaio, N. and Osaki, S. (2001). The cost-effective optimal order-replacement policies, *Proceedings of 10th International Symposium on Applied Stochastic Models and Data Analysis*, 381–386.

20. Dohi, T., Kaio, N. and Osaki, S. (2003). Preventive maintenance models: replacement, repair, ordering and inspection, In *Handbook of Reliability* (Ed., H. Pham), pp. 349–366, Springer-Verlag, London.

21. Dohi, T., Iwamoto, K., Kaio, N. and Osaki, S. (2004). A generalized discrete-time order-replacement model, *IMA Journal of Management Mathematics* (to appear).

22. Hunter, L. C. (1963). Optimum checking procedures, In *Statistical Theory of Reliability* (Ed. M. Zelen), The University of Wisconsin Press, Madison, Wisconsin.

23. Kabir, A. B. M. Z. and Al-Olayan, A. S. (1996a). A stocking policy for spare part provisioning undser age based preventive replacement, *European Journal of Operational Research*, **90**, 171–181.

24. Kabir, A. B. M. Z. and Farrash, S. H. A. (1996b). Simulation of an integrated age replacement and spare provisioning policy using SLAM, *Reliability Engineering and System Safety*, **52**, 129–138.

25. Kabir, A. B. M. Z. and Farrash, S. H. A. (1997). A fixed interval ordering policy for joint optimization of age replacement and spare part provisioning, *International Journal of Systems Science*, **28**, 1299–1309.

26. Kaio, N. and Osaki, S. (1977). Ordering policies with two types of lead times, *Microelectronics and Reliability*, **16**, 225–229.

27. Kaio, N. and Osaki, S. (1978a). Optimum ordering policies with lead time for an operating unit in preventive maintenance, *IEEE Transactions on Reliability*, **R-27**, 270–271.

28. Kaio, N. and Osaki, S. (1978b). Optimum ordering policies when order costs depend on time, *R.A.I.R.O. Recherche Operationnelle – Operations Research*, **12**, 93–99.

29. Kaio, N. and Osaki, S. (1978c). Optimum ordering policies with two kinds of lead times and non-linear ordering costs, *International Journal of Systems Science*, **9**, 265–272.

30. Kaio, N. and Osaki, S. (1978d). Optimum planned maintenance with salvage cost, *International Journal of Production Research*, **16**, 249–257.

31. Kaio, N. and Osaki, S. (1978e). Optimum inspection-ordering policies with salvage cost, *Microelectronics and Reliability*, **18**, 253–257.

32. Kaio, N. and Osaki, S. (1979a). Optimum ordering policies with discounting when order costs depend on time, *International Journal of Systems Science*, **10**, 539–555.

33. Kaio, N. and Osaki, S. (1979b). Discrete-time ordering policies, *IEEE Transactions on Reliability*, **R-28**, 405–406.

34. Kaio, N. and Osaki, S. (1980a). Extended optimum ordering policies with discounting, *International Journal of Systems Science*, **11**, 149–157.

35. Kaio, N. and Osaki, S. (1980b). Optimum planned maintenance with discounting, *International Journal of Production Research*, **18**, 515–523.

36. Kaio, N. and Osaki, S. (1980d). Discrete time ordering policies with minimal repair, *R.A.I.R.O. Recherche Operationnelle – Operations Research*, **14**, 257–263.

37. Kaio, N. and Osaki, S. (1981a). Optimum planned maintenance policies with lead time, *IEEE Transactions on Reliability*, **R-30**, 79.

38. Kaio, N. and Osaki, S. (1981b). Extended discrete time ordering policies with minimal repair, *Proceedings of International Conference on Policy Analysis and Information Systems*, 665–672.

39. Kaio, N. and Osaki, S. (1982). A note on inspection-ordering policies, *Proceedings of Asia-Pacific Conference on Operational Research*, 341–348.

40. Kaio, N. and Osaki, S. (1990a). Optimal ordering policies with two types of randomized lead times, *Computers & Mathematics with Applications*, **19**, 43–52.

41. Kaio, N. and Osaki, S. (1990b). Optimum ordering policies with nonlinear running and salvage costs, *Microelectronics and Reliability*, **30**, 785–793.

42. Kaio, N., Dohi, T. and Osaki, S. (1992). Optimal maintenance policies with lead times and repair, *International Journal of Systems Science*, **23**, 1299–1308.

43. Kaio, N., Dohi, T. and Osaki, S. (1998). A note on determination of lead time in ordering policy, *Proceedings of the 4th China-Japan International Symposium on Industrial Management*, 320–324.

44. N. Kaio, Dohi, T. and Osaki, S. (2002). Classical replacment models, In *Stochastic Models in Reliability and Maintenance* (Ed., S. Osaki), pp. 65–87, Springer-Verlag, Berlin.

45. Kalpakam, S. and Shaul H. M. A. (1981). Optimum ordering policies with random lead times, *Microelectronics and Reliability*, **21**, 737–741.

46. Kapur, P. K. and Garg, R. B. (1988). Ordering policies based on repair cost limit criterion, *Opsearch*, **25**, 245–258.

47. Kapur, P. K., Garg, R. B. and Bhalla, V. K. (1991). Generalised ordering policies under age dependent minimal failures with random lead time, *IAPQR Transactions*, **16**, 33–44.

48. Kapur, P. K., Garg, R. B. and Kumar, S. (1999). *Contributions to Hardware and Software Reliability*, World Scientific Publishers, Singapore.

49. Kawai, H. (1983a). An optimal ordering and replacement policy of a Markovian degradation system under complete observation: Part I, *Journal of Operations Research Society of Japan*, **26**, 279–292.

50. Kawai, H. (1983b). An optimal ordering and replacement policy of a Markovian degradation system under incomplete observation: Part II, *Journal of Operations Research Society of Japan*, **26**, 293–308.

51. Mine, H. and Kawai, H. (1977). Optimal ordering and replacement for a 1-unit system, *IEEE Transactions on Reliability*, **R-26**, 273–276.

52. Nakagawa, T. (1976). On scheduling the delivery of spare units, *IEEE Transactions on Reliability*, **R-25**, 35–37.

53. Nakagawa, T. and Osaki, S. (1974). Optimum replacement policies with delay, *Journal of Applied Probability*, **11**, 102–110.

54. Nakagawa, T. and Osaki, S. (1975). The discrete Weibull distribution, *IEEE Transactions on Reliability*, **R-24**, 300–301.

55. Nakagawa, T. and Osaki, S. (1978). Optimum ordering policies with lead time for an operating unit, *R.A.I.R.O. Recherche Operationnelle – Operations Research*, **12**, 383–393.

56. Osaki, S. (1977). An ordering policy with lead time, *International Journal of Systems Science*, **8**, 1091–1095.

57. Osaki, S. and Yamada, S. (1976). Age replacement with lead time, *IEEE Transactions on Reliability*, **R-25**, 344–345.

58. Osaki, S., Kaio, N. and Yamada, S. (1981). A summary of ordering policies, *IEEE Transactions on Reliability*, **R-30**, 272–277.

59. Osaki, S. (1985). *Stochastic System Reliability Modeling*, World Scientific Publishers, Singapore.

60. Osaki, S. (1992). *Applied Stochastic System Modeling*, Springer-Verlag, Berlin.

61. Park, K. S. and Park, Y. T. (1986a). Ordering policies under minimal repair, *IEEE Transactions on Reliability*, **R-35**, 82–84.

62. Park, Y. T. and Park, K. S. (1986b). Generalized spare ordering policies with random lead time, *European Journal of Operational Research*, **23**, 320–330.

63. Ross, S. M. (1970). *Applied Probability Models with Optimization Applications*, Holden-Day, San Francisco.

64. Sheu, S. and Liou, C. (1992). An ordering policy with age-dependent minimal repair and age-dependent random repair costs, *Microelectronics and Reliability*, **32**, 1105–1113.

65. Sheu, S. and Liou, C. (1993). Optimum ordering policies with age-dependent minimal repair and two types of lead times, *International Journal of Systems Science*, **24**, 253–264.

66. Sheu, S. and Liou, C. (1994). Generalized ordering policies with general random minimal repair costs and random lead times, *R.A.I.R.O. Recherche Operationnelle – Operations Research*, **28**, 97–118.

67. Sheu, S., Liou, C. and Tseng, B. (1992). Optimal ordering policies and optimal number of minimal repairs before replacement, *Microelectronics and Reliability*, **32**, 995–1002.

68. Sheu, S. (1997a). An optimal ordering policy of a system subject to shocks, *International Journal of Systems Science*, **28**, 241–247.

69. Sheu, S. (1997b). A generalized model for jointly determining the optimal ordering point and the optimal number of minimal repairs before replacement, *International Journal of Systems Science*, **28**, 759–766.

70. Sridharan, V. (1991). A note on ordering policies with cost observation and lead times, *International Journal of Production Economics*, **32**, 91–97.

71. Stein, W. E. and Dattero, R. (1984). A new discrete Weibull distribution, *IEEE Transactions on Reliability*, **R-33**, 196–197.

72. Subramanian, R. and Sridharan, V. (1989). Optimum ordering policies with random lead times and salvage cost, *Microelectronics and Reliability*, **29**, 523–527.

73. Sung, C. S. and Park, Y. K. (1986). Optimum ordering policies for an equipment with a sensing device, *Microelectronics and Reliability*, **26**, 87–98.

74. Sung, C. S. and Kim, C. H. (1987). Optimum ordering policy under random lead time for equipment with sensor, *IEEE Transactions on Reliability*, **R-36**, 467–469.

75. Thomas, L. C. and Osaki, S. (1978a). An optimal ordering policy for a spare unit with lead time, *European Journal of Operational Research*, **2**, 409–419.

76. Thomas, L. C. and Osaki, S. (1978b). A note on ordering policy, *IEEE Transactions on Reliability*, **R-27**, 380–381.

77. Trott, E. P. (1965). Maintainability and reliability cost effectiveness program (MARCEP), *Proceedings of 4th Reliability and Maintainability Conference*, 219–228.

78. Wiggins, A. D. (1967). A minimum cost model of spare parts inventory control, *Technometrics*, **9**, 661–665.

79. Winlund, E. S. (1965). Cost-effectiveness management, *Proceedings of 4th Reliability and Maintainability Conference*, 87–96.

16

Estimating Reliabilities Following Purely Sequential Sampling from Exponential Populations

Nitis Mukhopadhyay[1] **and Greg Cicconetti**[2]

[1]*Department of Statistics, University of Connecticut, Storrs, CT U.S.A.*
[2]*Department of Mathematical Sciences, Muhlenberg College, Allentown, PA U.S.A.*

Abstract: This problem stems from already implemented sequential minimum risk point estimation methodologies for a scale parameter or for comparing two scale parameters in exponential distributions. We move on to address the estimation problem for the reliability parameter in a one-sample scenario. We investigate both invariance (of fixed-sample-size MLE) -based estimators as well as those depending upon bias-corrected forms using already gathered sequential data and the stopped sample size. While we derive important guidance from Mukhopadhyay et al. (1997), the present investigation is significantly broader in scope. The merits of the proposed estimators are critically examined through simulation. These ideas are subsequently extended to the two-sample problem where our goal is to estimate how many times more likely it is for an observation from one population to "survive" beyond time point t, compared with another population. Twelve estimators are developed and compared. Again, performances of both biased and bias-corrected estimators based on sample means from already gathered sequential data and the stopped sample sizes are investigated via simulation. Some real data examples are included.

Keywords and phrases: Exponential distribution, mean, reliability, MLE, sequential sampling, bias correction, one-sample problem, two-sample problem, Taylored estimators, estimator templates, simulations, real data examples

16.1 Introduction

Survival times with a constant failure rate are typically modeled by an exponential distribution. Of particular interest to us is the probability that an observation survives beyond time t. Let us denote the *probability density func-*

tion (p.d.f.) for an exponential random variable with mean $\lambda_1 > 0$ as

$$f(x; \lambda_1) = \lambda_1^{-1} \exp(-x/\lambda_1), x > 0. \tag{16.1}$$

In a one-sample problem, let X represent the life span of a randomly chosen manufactured product, for example, a particular kind of activation mechanism of an emergency sprinkler system. Assuming the exponential distribution (16.1) with an unknown mean survival time $\lambda_1(> 0)$, we may wish to estimate the *reliability parameter*, namely

$$\theta_X(t) = P_{\lambda_1}(X > t) \equiv \exp(-t/\lambda_1), \tag{16.2}$$

where $t\ (> 0)$ is some fixed number in time-scale. The distributional assumption leads to the so-called *memoryless property*. Many products, however, tend to exhibit "problems" near the beginning of their life or they start to deteriorate beyond a certain point, implying a non-constant failure rate. The notion of a constant failure rate is most useful when the failure of a system is perhaps due to some random shock, for example, a sudden surge of electricity may cause a computer chip to fail instantaneously. There are many models for survival times more sophisticated than (16.1), but in this chapter, we stay with exponential models for the sake of simplicity.

For overviews on reliability theory, we refer to Lomnicki (1973), Barlow and Proschan (1975), and Ansell and Phillips (1989). The edited volume of Balakrishnan and Basu (1995) presented the whole wide spectrum of statistical research involving exponential distributions. A couple of articles addressed topics that were in spirit similar to our present investigation, including Tong (1977), Brown (1977), Beg and Singh (1979), and Beg (1980). Unfortunately, these articles were rather few and far between. One may look at this investigation as a rebirth of the possible union of reliability estimation and sequential analysis in the light of the seminal papers of Epstein and Sobel (1953, 1954, 1955) and Sobel (1956). From what one will see here, it should be clear that marrying the notions of reliability estimation and sequential analysis still remains a novel idea worthy of revival and advancement. It is our hope that this report will energize analogous investigations in the future for models that are more sophisticated than (16.1).

If we have two comparable products with different constant rates of failure $\lambda_1\ (> 0)$ and $\lambda_2\ (> 0)$, it is often of interest to compare the reliability parameters beyond some time point $t(> 0)$. Let us suppose that X and Y are the lifetimes, each modeled as an exponential random variable, with respective mean survival times λ_1 and λ_2. Now, we consider the parameter,

$$\tau_{X,Y}(t) \equiv \theta_X(t)/\theta_Y(t) = P_{\lambda_1}(X > t)/P_{\lambda_2}(Y > t). \tag{16.3}$$

Interpretation of $\tau_{X,Y}$ is very similar to that of a *risk-ratio*. As an example, suppose we know that $P_{\lambda_1}(X > t) = 0.5$ and $P_{\lambda_2}(Y > t) = 0.25$. Then, it is natural to say that the chances of X lasting beyond t is twice as likely as Y lasting beyond t.

Section 16.2 begins with a review of some of the results available concerning sequential estimation of the mean, λ_1. This is an essential starting point for the tasks at hand and we discuss four sequential estimators for λ_1 in Section 16.2.1. Section 16.2.2 addresses estimation of the reliability parameter, $\theta_X(t)$, and simulations are used to explore eight different estimators of $\theta_X(t)$ in Section 16.2.2. A small partition of these findings was presented at the MMR'2000-Bordeaux conference as evidenced by the printed abstract [Mukhopadhyay and Cicconetti (2000)].

Section 16.3 is structured similarly as we turn our attention to the two-sample problem: After sequentially estimating the difference of the means, $\lambda_1 - \lambda_2$, we subsequently address twelve estimators for $\tau_{X,Y}(t)$ from (16.3).

We first focus on existing sequential procedures for estimating λ_1 or $\lambda_1 - \lambda_2$. In a one-sample problem, we work with the available dataset $(M, X_1, ..., X_M)$ where M is a randomly stopped sample size associated with the purely sequential sampling scheme of Starr and Woodroofe (1972). In other words, starting with the dataset $(M, X_1, ..., X_M)$ that was exclusively designed and implemented to solve one specific estimation problem for λ_1, we propose and compare a class of estimators for λ_1 and $\theta_X(t)$. This is the subject matter of Section 16.2. In a two-sample problem, we work with the available dataset $(M, N, X_1, ..., X_M, Y_1, ..., Y_N)$ where M, N are the randomly stopped sample sizes associated with the purely sequential sampling scheme of Mukhopadhyay and Purkayastha (1994). In other words, starting with the dataset $(M, N, X_1, \ldots, X_M, Y_1, ..., Y_N)$, we propose and compare a class of estimators for $\lambda_1 - \lambda_2$ and $\tau_{X,Y}(t)$ which is the subject matter of Section 16.3. The derivations of some of these estimators and evaluating their performances are non-trivial whether one deals with a one- or two-sample problem. The two-sample scenario is admittedly the more involved of the two. An Appendix provides proofs of some of the important results contained in Section 16.3.

We end this investigation by demonstrating performances of these estimators with the help of a well-known dataset from Proschan (1963) concerning the failure times of air conditioners.

16.2 A One-sample Problem: Sequential Estimation of the Mean

For completeness, we quickly summarize some of the initial results from Cicconetti's (2002) doctoral dissertation. These were concerned with the one-sample problem of sequential estimation of λ_1. The estimators discussed herein serve as building blocks to come up with the estimators of the reliability parameter $\theta_X(t)$.

Let us suppose that the observations $X_1, X_2, \ldots$ originate from the distribution (16.1). Having observed a random sample of size m, we start with point estimation for λ_1 via the maximum likelihood estimator (MLE), $\overline{X}_m$, under the weighted squared error plus observational cost as our loss function:

$$L_m\left(\overline{X}_m, \lambda_1\right) = A\left(\overline{X}_m - \lambda_1\right)^2 + cm. \tag{16.4}$$

Here, A (> 0) and c (> 0) are known to the experimenter where c is the cost incurred for recording each observation. In light of an alternative estimator, T, say, for λ_1 we would consider the loss function $L_m(T_m, \lambda_1)$, simply replacing $\overline{X}_m$ with T_m in (16.4). Starr and Woodroofe (1972) investigated the problem of finding a *minimum risk point estimator* of λ_1 and worked with the risk associated with (16.4) which is given by

$$R_m(c) \equiv E_{\lambda_1}\left[L_m\left(\overline{X}_m, \lambda_1\right)\right] = A\lambda_1^2 m^{-1} + m. \tag{16.5}$$

One may 'treat' m as a continuous variable and find m^*, the smallest fixed-sample-size that minimizes the risk function (16.5), given by

$$m^* = (A/c)^{1/2}\lambda_1. \tag{16.6}$$

This expression of m^* motivated Starr and Woodroofe's (1972) to propose a purely sequential stopping rule $\left(M, \overline{X}_M\right)$ where

$$M \equiv M(c) = \inf\left\{m\,(\geq k) : m \geq (A/c)^{1/2}\,\overline{X}_m\right\}. \tag{16.7}$$

That is, a pilot sample of size k (≥ 1) is chosen first. The need for any additional observation is evaluated prior to its collection while sampling terminates with the first observation causing m $(\geq k)$ to exceed (or equal) $(A/c)^{1/2}\,\overline{X}_m$. Starr and Woodroofe (1972) proved that the associated *regret*, defined by Robbins (1959) as

$$\omega(c) = E_{\lambda_1}[L_M] - R_{m^*}(c), \tag{16.8}$$

had the order $O(c)$ if and only if $k \geq 2$. Woodroofe (1977) returned to provide the following finer *second-order* result:

$$\omega(c) = 3c + o(c) \text{ as } c \to 0 \text{ if } k \geq 3. \tag{16.9}$$

Therefore, the difference between the *sequential risk*, $E_{\lambda_1}[L_M]$, and the *fixed-sample-size optimal risk*, $R_{m^*}(c)$, amounts to the cost of approximately three observations. Woodroofe also offered the following second-order expansion for the average sample size:

$$E_\lambda(M) = m^* - 0.254965 + o(1) \text{ if } k \geq 2. \tag{16.10}$$

Let us mention in passing that the limiting operation "$c \to 0$" is viewed as a simple mathematical device that makes the sample size "large". To review the relevant literature, one may refer to Mukhopadhyay (1988, 1995) and Ghosh, Mukhopadhyay and Sen (1997). For completeness, let us mention that in a fixed-width confidence interval estimation problem for λ_1, the associated second-order results were provided by Mukhopadhyay and Datta (1996).

Upon termination of the purely sequential procedure (16.7), the customary estimator for λ_1 has been the sample mean, $\overline{X}_M$. This estimator is intuitively pleasing since it mimics the fixed-sample-size MLE for λ_1. However, Mukhopadhyay and Chattopadhyay (1991) and Mukhopadhyay (1988, 1995) argued that $\overline{X}_M$ was a biased estimator for λ_1, even asymptotically, and in fact they obtained the following expression for the bias:

$$\text{Bias}_c \equiv E_{\lambda_1}[\overline{X}_M] - \lambda_1 = -\frac{\lambda_1}{m^*} + o\left(\frac{1}{m^*}\right) \quad \text{if } k \geq 2. \tag{16.11}$$

In light of (16.11), Mukhopadhyay and Chattopadhyay (1991) came up with a bias-corrected estimator, namely $M(M-1)^{-1}\overline{X}_M$, for λ_1.

On the other hand, as the purely sequential sampling procedure (16.7) terminates, one genuinely expects M to be in a close proximity to m^*. This idea prompted Mukhopadhyay and Duggan (2000) to consider $(c/A)^{1/2} M$ as an alternative estimator of λ_1. It should be noted that this latter estimator is based solely on the stopping sample size (plus known design constants c and A). This estimator does not involve $X_1, ..., X_M$ at all! It can be shown that $(c/A)^{1/2} M$ is also a biased estimator of λ_1. We are prompted, then, to consider the bias-corrected version of the Mukhopadhyay and Duggan estimator, namely $(c/A)^{1/2}(M + 0.254965)$. Let us denote

$$\begin{aligned} T_{1M} &= \overline{X}_M, T_{2M} = M(M-1)^{-1}\overline{X}_M, T_{3M} = (c/A)^{1/2} M, \\ &\quad \text{and } T_{4M} = (c/A)^{1/2}(M + 0.254965). \end{aligned} \tag{16.12}$$

It should be noted that these four estimators, T_{jM}, $j = 1, 2, 3, 4$, are each *asymptotically risk efficient* for estimating λ_1 in the sense of Robbins (1959). That is, one has:

$$\eta_i \equiv \eta_i(c) = E_{\lambda_1}[L_M(T_{iM}, \lambda_1)] / R_{m^*}(c) \to 1 \text{ as } c \to 0, \ i = 1, 2, 3, 4. \tag{16.13}$$

16.2.1 Comparing estimators of a population mean

We look to compare the performances among the estimators $T_{iM}, i = 1, 2, 3, 4$ for moderate values of m^* and fixed $\lambda_1 = 5, A = 1, k = 5, 10$ and $m^* = 50, 100$. In each configuration, we used 5,000 independent simulation runs to obtain the average estimated sample size $\overline{M}$, its estimated standard error $s(\overline{M})$, as well as, the average $\overline{T}_i$ to estimate λ_1 and its estimated standard error $s(\overline{T}_i)$, for

$i = 1, 2, 3, 4$. The entity $\overline{M}$ estimates $E_{\lambda_1}(M)$. We used the same simulation runs to obtain the average estimated risk $\overline{R}_i$ and its estimated standard error $s(\overline{R}_i)$, $i = 1, 2, 3, 4$. We treat $\overline{R}_i$ as a simulated estimate of $E[L_M(T_{iM}, \lambda_1)]$. We also provide the estimated values $\overline{\eta}_i$, $\overline{\omega}_i$ respectively for η_i and ω_i, $i = 1, 2, 3, 4$. We refrain from supplying their estimated standard errors since one notes the following obvious relationships:

$$s^2(\overline{\eta}_i) = s^2(\overline{R}_i)/R^2_{m^*}(c) \text{ and } s^2(\overline{\omega}_i) = s^2(\overline{R}_i).$$

It is true that we seem rather ignorant about λ_1 before data collection begins, but as soon as we record the pilot sample of size k, we start "building" useful information. A larger pilot sample starts off with more information about λ_1, and so the performance of an estimator is expected to improve when $k = 10$ instead of $k = 5$. This feature is validated when we consider the overall message obtained from the Table 16.1.

The bias-corrected form of the estimator T_{1M}, namely T_{2M}, decisively performs better than T_{1M}, which is reflected by the fact that $\overline{T}_2$ is much closer to the fixed value $\lambda_1 = 5$, whereas the estimated standard errors $s(\overline{T}_1), s(\overline{T}_2)$ stay comparable. The bias-corrected estimator T_{2M} also performs better than T_{1M} in relation to their associated risks, $\overline{R}_1$, $\overline{R}_2$, as well as, the associated risk efficiency and regret measures $\overline{\eta}_1, \overline{\eta}_2$ and $\overline{\omega}_1, \overline{\omega}_2$.

A similar feature is also noticed when the new bias-corrected estimator T_{4M} is compared with the estimator T_{3M}. What is truly remarkable about T_{4M}, which depends on the observed data *only* through the stopping variable M, is that it performs head-to-head with T_{2M}. In other words, for all practical purposes, the *Fisher-Information* about the unknown parameter λ_1 provided by the data $(M, \overline{X}_M)$, which is sufficient for λ_1, appears to be nearly equivalent to the amount of Fisher-Information retained by a portion of the data through the stopping variable M alone.

It will be very interesting to critically examine the process that lets the stopping variable M to eventually soak in the Fisher-Information from the sample mean $\overline{X}_M$.

16.2.2 Estimation of a reliability parameter following purely sequential procedure (16.7)

Consider our observations $X_1, X_2, ...$ originating from the distribution (16.1). Given a time point t, we are interested in estimating the reliability parameter, $\theta_X(t)$. Having been initially charged with the estimation of the mean λ_1, assume that the purely sequential algorithm (16.7) has been employed first and we have in hand the dataset $(M, \mathbf{X}_M)$ where $\mathbf{X}_M \equiv (X_1, X_2, ..., X_M)$.

Had our dataset been based on some fixed number of observations m, the MLE of $\theta_X(t)$ would be $\exp(-t/\overline{X}_m)$. While $\exp(-t/\overline{X}_m)$ is not an unbiased

Table 16.1: Comparing the estimators of λ_1

Case (i): $m^* = 50$, $c = 0.010$, $R_{m^*}(c) = 1.000$, $\lambda_1 = 5$	
$k = 5$	$k = 10$
$\overline{m} = 49.5822, s(\overline{m}) = 0.09858$	$\overline{m} = 49.5906, s(\overline{m}) = 0.09737$
$\overline{T}_1 = 4.90046, \overline{T}_2 = 4.98419$	$\overline{T}_1 = 4.87221, \overline{T}_2 = 4.98718$
$\overline{T}_3 = 4.97408, \overline{T}_4 = 4.99958$	$\overline{T}_3 = 4.94560, \overline{T}_4 = 4.97110$
$s(\overline{T}_1) = 0.01047, s(\overline{T}_2) = 0.01049$	$s(\overline{T}_1) = 0.01039, s(\overline{T}_2) = 0.01026$
$s(\overline{T}_3) = 0.01050, s(\overline{T}_4) = 0.01050$	$s(\overline{T}_3) = 0.01042, s(\overline{T}_4) = 0.01042$
$\overline{R}_1 = 1.05543, \overline{R}_2 = 1.04632$	$\overline{R}_1 = 1.05031, \overline{R}_2 = 1.02324$
$\overline{R}_3 = 1.04927, \overline{R}_4 = 1.04860$	$\overline{R}_3 = 1.04069, \overline{R}_4 = 1.03856$
$s(\overline{R}_1) = 0.01464, s(\overline{R}_2) = 0.01326$	$s(\overline{R}_1) = 0.01191, s(\overline{R}_2) = 0.01182$
$s(\overline{R}_3) = 0.01446, s(\overline{R}_4) = 0.01437$	$s(\overline{R}_3) = 0.01157, s(\overline{R}_4) = 0.01150$
$\overline{\eta}_1 = 1.05543, \overline{\eta}_2 = 1.04632$	$\overline{\eta}_1 = 1.05031, \overline{\eta}_2 = 1.02324$
$\overline{\eta}_3 = 1.04927, \overline{\eta}_4 = 1.04860$	$\overline{\eta}_3 = 1.04069, \overline{\eta}_4 = 1.03856$
$\overline{\omega}_1 = 0.05543, \overline{\omega}_2 = 0.04632$	$\overline{\omega}_1 = 0.05031, \overline{\omega}_2 = 0.02324$
$\overline{\omega}_3 = 0.04927, \overline{\omega}_4 = 0.04860$	$\overline{\omega}_3 = 0.04069, \overline{\omega}_4 = 0.03856$
Case (ii): $m^* = 100$, $c = 0.003$, $R_{m^*}(c) = 0.500$, $\lambda_1 = 5$	
$k = 5$	$k = 10$
$\overline{m} = 99.6026, s(\overline{m}) = 0.12464$	$\overline{m} = 99.8124, s(\overline{m}) = 0.12258$
$\overline{T}_1 = 4.94339, \overline{T}_2 = 4.99353$	$\overline{T}_1 = 4.95378, \overline{T}_2 = 5.00392$
$\overline{T}_3 = 4.98013, \overline{T}_4 = 4.99288$	$\overline{T}_3 = 4.99062, \overline{T}_4 = 5.00337$
$s(\overline{T}_1) = 0.00717, s(\overline{T}_2) = 0.00717$	$s(\overline{T}_1) = 0.00721, s(\overline{T}_2) = 0.00721$
$s(\overline{T}_3) = 0.00718, s(\overline{T}_4) = 0.00718$	$s(\overline{T}_3) = 0.00723, s(\overline{T}_4) = 0.00723$
$\overline{R}_1 = 0.50948, \overline{R}_2 = 0.50630$	$\overline{R}_1 = 0.51187, \overline{R}_2 = 0.50973$
$\overline{R}_3 = 0.50737, \overline{R}_4 = 0.50703$	$\overline{R}_3 = 0.51080, \overline{R}_4 = 0.51072$
$s(\overline{R}_1) = 0.00529, s(\overline{R}_2) = 0.00525$	$s(\overline{R}_1) = 0.00542, s(\overline{R}_2) = 0.00537$
$s(\overline{R}_3) = 0.00525, s(\overline{R}_4) = 0.00525$	$s(\overline{R}_3) = 0.00539, s(\overline{R}_4) = 0.00539$
$\overline{\eta}_1 = 1.01897, \overline{\eta}_2 = 1.01261$	$\overline{\eta}_1 = 1.02374, \overline{\eta}_2 = 1.01946$
$\overline{\eta}_3 = 1.01474, \overline{\eta}_4 = 1.01405$	$\overline{\eta}_3 = 1.02160, \overline{\eta}_4 = 1.02145$
$\overline{\omega}_1 = 0.00948, \overline{\omega}_2 = 0.00630$	$\overline{\omega}_1 = 0.01187, \overline{\omega}_2 = 0.0097$
$\overline{\omega}_3 = 0.00737, \overline{\omega}_4 = 0.00703$	$\overline{\omega}_3 = 0.01080, \overline{\omega}_4 = 0.01072$

estimator for $\theta_X(t)$, it is consistent and asymptotically normal with asymptotic variance given in terms of the inverse of the Fisher-Information.

In terms of applications, it is our hope that such attractive asymptotic properties will begin to show at reasonable sample sizes. Indeed these are the fixed-sample-size properties that may motivate a practitioner to work with $\exp\left(-t/\overline{X}_M\right)$. Given the analogy between the fixed-sample-size MLE $\overline{X}_m$ and $T_{1M}(\equiv \overline{X}_M)$, we may intuitively expect

$$\widehat{\theta}_{1X}(t) \equiv \widehat{\theta}_{1X}\left(t, \overline{X}_M\right) = \exp\left(-t/T_{1M}\right) \tag{16.14}$$

to perform well by "appealing" to the invariance property of an MLE [Zehna (1966)]. We remind readers that as $c \to 0$, the bias in $\overline{X}_M$ tends to zero as seen from (16.11). Mukhopadhyay, Padmanaghan and Solanky (1997) were thus inclined to seek an approximately unbiased estimator for $\theta_X(t)$. They utilized the Taylor expansion of $\widehat{\theta}_X(t)$ in a successful bid to find an approximately unbiased estimator for $\theta_X(t)$ which is precisely stated below without a proof.

Theorem 16.2.1 *Given some fixed* $t\,(>0)$, *the estimator*

$$\widehat{\theta}^*_{1X}(t) \equiv \widehat{\theta}^*_{1X}\left(t, \overline{X}_M\right) = \left\{1 + \left(\frac{1}{2}t - \overline{X}_M - 1\right) t\overline{X}_M^{-2} M^{-1}\right\} \exp\left(-t/\overline{X}_M\right) \tag{16.15}$$

is approximately unbiased for θ, *that is,* $E\left[\widehat{\theta}^*_{1X}(t)\right] = \theta_X(t) + o\left(\sqrt{c}\right)$ *as* $c \to 0$.

Incidentally, we should add that this expression corrects an error found in the 1997 paper of Mukhopadhyay, Padmanaghan and Solanky. We shall expand this idea in Section 16.3 by considering a bivariate Taylor expansion for the two-sample problem, and we shall outline a derivation in the Appendix.

Also, we draw attention of our readers to a new phrase that we have coined. Incidentally, bias-corrected estimators and other variants obtained through applications of *univariate* (or *bivariate*) Taylor expansions are consciously referred to as *univariate* (or *bivariate*) *Taylored* estimators in the sequel.

Estimator templates

To facilitate our discussion here (and later when we consider the two-sample problem), we introduce a notion of *estimator templates.* Let us be reminded that it was the sample average $\overline{X}_M$ that motivated the form of both $\widehat{\theta}_{1X}(t)$ in (16.14) and $\widehat{\theta}^*_{1X}(t)$ in (16.15). Clearly, we are interested in exploring the gain from employing (16.15) over (16.14). However, notwithstanding the fact that (16.15) was Taylored to be an estimator of $\theta_X(t)$ based on $\overline{X}_M$, the performances of T_{2M}, T_{3M}, T_{4M} in simulations from Section 16.2 beg the question: What would happen if we replaced $T_{1M} \equiv \overline{X}_M$ with these other estimators? We may view

$$\widehat{\theta}_X(t, Z) = \exp(-t/Z), \tag{16.16}$$

$$\widehat{\theta}^*_X(t, Z) = \left\{1 + \left(\frac{1}{2}t - Z - 1\right) tZ^{-2}M^{-1}\right\} \exp(-t/Z), \tag{16.17}$$

as *estimator templates* and consider the performance of these templates under a generic estimator, Z, for λ_1. We are particularly interested in the performance of these templates when we replace Z with T_{iM}, $i = 1, 2, 3, 4$. To distinguish which estimator of λ_1 is used in a template, we shall add a subscript i. We shall refer to $\widehat{\theta}_{iX}$, $i = 1, 2, 3, 4$ as invariance-based estimators since the estimators from this template are motivated by the fixed-sample-size invariance property of an MLE. Taylored estimators $\widehat{\theta}^*_{iX}$, $i = 1, 2, 3, 4$, take their name because the template is obtained via the Taylor expansion of $\exp\left(-t/\overline{X}_M\right)$. Next, we summarize the results from our simulations.

Simulation results

Simulations were used to mimic sampling from an exponential distribution with a scale parameter $\lambda_1 = 5$. For each combination of $m^* = 50, 75, 100, 150$ and $k = 5, 10$, we performed 5000 repetitions of the one-sample sequential sampling scheme (16.7). We collected the average performances of the univariate estimators: M, T_{1M}, T_{2M}, T_{3M}, T_{4M}. Reliability estimates based on T_{iM}, $i = 1, 2, 3, 4$, and bias-corrected reliability estimates developed from the univariate Taylor expansion (Theorem 16.2.1) were computed. The values of t used in our simulations were chosen so that $P_{\lambda_1}(X > t) = 0.875, 0.75, 0.625, 0.5, 0.25$. In addition, we also recorded the standard errors of these estimators.

We naturally investigated the performances of our four estimators of λ_1, namely T_{iM}, $i = 1, 2, 3, 4$. These simulations were shown to correspond with our previous study and one may refer to the comments on the performances of T_{iM}, $i = 1, 2, 3, 4$, given in Section 16.2.1. We include summary statistics as a part of Tables 16.2–16.3 when $m^* = 100$, $\lambda_1 = 5$, $A = 1$ with $k = 5, 10$ respectively.

We find that the reliability estimators performed more consistently for larger values of k and m^*. The most dramatic improvement was seen when comparing the results in the cases corresponding with $m^* = 50$ and $m^* = 75$. The effect of the pilot sample size k dwindles as m^* increases. Among eight estimators for $\theta_X(t)$ under consideration, arising from a combination of the invariance-based estimator (16.16) and Taylored estimator (16.17) templates with the T_{iM}'s. Our simulations supported the following empirical findings:

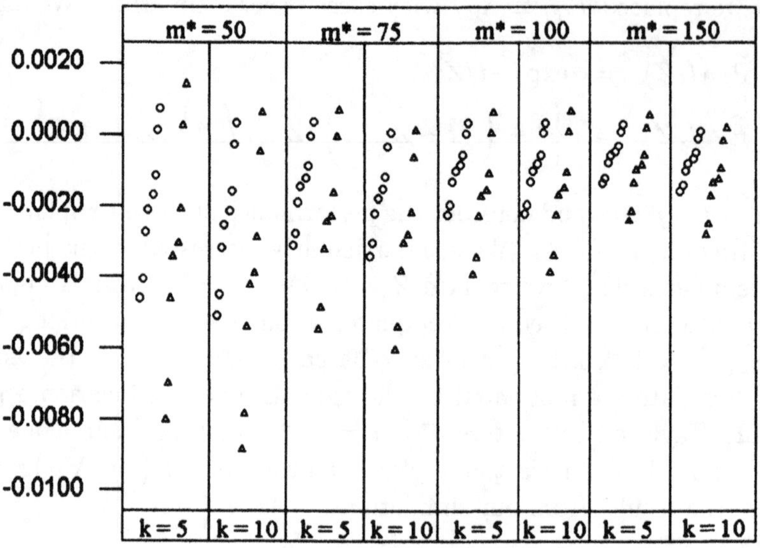

Figure 16.1: Performance of template estimators for the one-sample problem. Triangles: Invariance-Based $\widehat{\theta}_{iX}$, $i = 1, 2, 3, 4$; Circles: Taylored $\widehat{\theta}^*_{iX}$, $i = 1, 2, 3, 4$, corresponding to estimators based on t for which $\theta_X(t) = 0.875$ and 0.75, respectively

$$
\begin{aligned}
E_{\lambda_1}[\widehat{\theta}_{1X}(t)] &< E_{\lambda_1}[\widehat{\theta}_{2X}(t)] < E_{\lambda_1}[\widehat{\theta}_{3X}(t)] < E_{\lambda_1}[\widehat{\theta}_{4X}(t)] < E_{\lambda_1}[\widehat{\theta}^*_{1X}(t)] \\
&< E_{\lambda_1}[\widehat{\theta}^*_{2X}(t)] < E_{\lambda_1}[\widehat{\theta}^*_{3X}(t)] < E_{\lambda_1}[\widehat{\theta}^*_{4X}(t)],
\end{aligned}
$$

for $\theta = 0.875, 0.75$.

These findings are illustrated in Figure 16.1. In this figure, we explore the quantities $(\widehat{\theta}_{iX}(t) - \theta_X(t))$ and $(\widehat{\theta}^*_{iX}(t) - \theta_X(t))$, $i = 1, 2, 3, 4$, and establish a benchmark for the proposed estimators as follows: Those falling closest to zero perform the "best". Circles used in Figure 16.1 correspond to the estimators based on t which yield $\theta_X(t) = 0.875$, whereas triangles correspond to the estimators based on t such that $\theta_X(t) = 0.75$. For each value of m^* and k, these symbols highlight the average performance of an estimator. The first four circles (triangles), correspond to estimators $\widehat{\theta}_{iX}(t)$, $i = 1, 2, 3, 4$, respectively and the last four circles (triangles) correspond to estimators $\widehat{\theta}^*_{iX}(t)$, $i = 1, 2, 3, 4$, respectively when $t = 0.875$ (0.75).

For the two particular values of θ, we have noticed some interesting features:

(i) Estimators for θ incorporating bias-corrected estimators of λ tend to be larger than their uncorrected counterparts;

Table 16.2: Summary for estimation of the mean $\lambda_1 = 5$, $m^* = 100$, $\lambda_1 = 5$, $k = 5$, $A = 1$

	Mean	Risk Eff. η	Regret ω	se(est)	se(risk)
T_1	4.94914	1.00451	0.00225	0.00700	0.36306
T_2	4.95928	1.00055	0.00027	0.00706	0.36020
T_3	4.98606	1.00150	0.00075	0.00709	0.36104
T_4	4.99880	1.00112	0.00056	0.00709	0.36085
M	99.7212	-	-	0.14179	-

Table 16.3: Comparison of invariance-based estimators and Taylored estimators: $m^* = 100$, $\lambda_1 = 5$, $k = 5$, $A = 1$

	t_1	t_2	t_3	t_4	t_5
λ_1-Estimator	Invariance-Based Estimator Template (16.16)				
T_1	0.87263	0.74582	0.61963	0.49411	0.24543
T_2	0.87288	0.74629	0.62025	0.49484	0.24614
T_3	0.87352	0.74746	0.62185	0.49671	0.24799
T_4	0.87383	0.74803	0.62261	0.49760	0.24887
λ_1-Estimator	Taylored Estimator Template (16.17)				
T_1	0.87402	0.74819	0.62255	0.49712	0.24715
T_2	0.87427	0.74865	0.62317	0.49784	0.24786
T_3	0.87490	0.74982	0.62476	0.49972	0.24973
T_4	0.87520	0.75038	0.62552	0.50061	0.25061
Max se(.) $\times 10^4$	1.77	3.24	4.36	5.08	4.95
Min se(.) $\times 10^4$	1.70	3.12	4.24	4.98	4.93

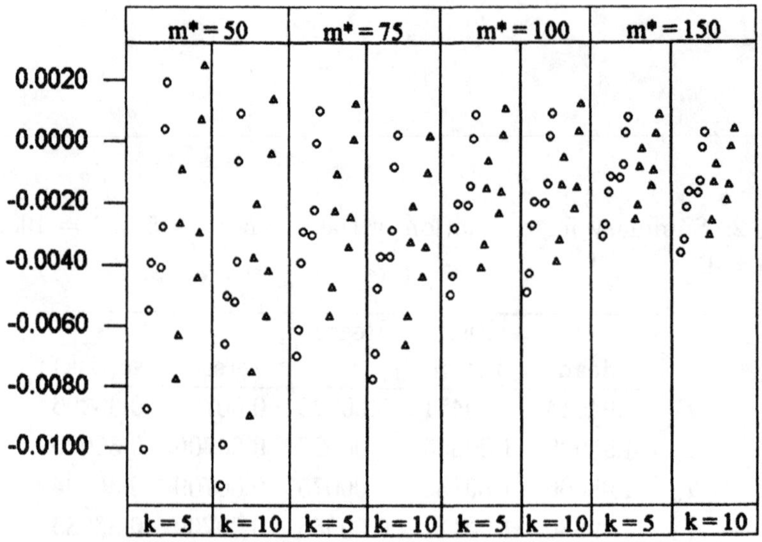

Figure 16.2: Performance of template estimators for the one-sample problem. Triangles: Invariance-Based $\widehat{\theta}_{iX}$, $i = 1,2,3,4$; Circles: Taylored $\widehat{\theta}^*_{iX}$, $i = 1,2,3,4$, corresponding to estimators based on t for which $\theta_X(t) = 0.25$ and 0.625, respectively

(ii) Estimators for θ defined via stopping variable alone tend to be larger than the sample mean based counterparts; and

(iii) Invariance-based estimators tend to be smaller than Taylored estimators.

Changes occurred when we considered the results for $\theta_X(t) = 0.625, 0.5$, and 0.25. Figure 16.2 is constructed analogously. Here, the thicker curve refers to $\theta_X(t) = 0.25$ and the thinner curve corresponds to $\theta_X(t) = 0.625$.

For these values of $\theta_X(t)$, the performances of $\widehat{\theta}^*_{1X}(t)$, $\widehat{\theta}^*_{2X}(t)$ dropped slightly. That is, the bias-corrected estimators took hits, supporting the following idea:

$$\begin{aligned} E_{\lambda_1}[\widehat{\theta}_{1X}(t)] &< E_{\lambda_1}[\widehat{\theta}_{2X}(t)] < E_{\lambda_1}[\widehat{\theta}^*_{1X}(t)] < E_{\lambda_1}[\widehat{\theta}^*_{2X}(t)] < E_{\lambda_1}[\widehat{\theta}_{3X}(t)] \\ &< E_{\lambda_1}[\widehat{\theta}_{4X}(t)] < E_{\lambda_1}[\widehat{\theta}^*_{3X}(t)] < E_{\lambda_1}[\widehat{\theta}^*_{4X}(t)]. \end{aligned}$$

Hence, under these values of $\theta_X(t)$, we note that "sample size"- based estimators for $\theta_X(t)$ dominate "sample mean"-based estimators. We again notice $\widehat{\theta}_{iX}(t) < \widehat{\theta}^*_{iX}(t)$.

Overall, the proposed estimators are fairly similar in performance with discrepancies occurring only in the third decimal place or later. However, regardless of the true values of $\theta_{3X}(t)$ and t, $\widehat{\theta}_{3X}(t)$ appears to perform the "best",

Table 16.4: Summary for estimation of the mean $\lambda_1 = 5, m^* = 100$, $\lambda_1 = 5$, $k = 10$, $A = 1$

	Mean	Risk Eff. η	Regret ω	se(est)	se(risk)
T_1	4.95386	1.02521	0.01260	0.00722	0.37764
T_2	4.96399	1.02135	0.01067	0.00721	0.37474
T_3	4.99038	1.02263	0.01131	0.00723	0.37666
T_4	5.00312	1.02247	0.01124	0.00723	0.37652
M	99.8076	-	-	0.14470	-

Table 16.5: Comparison of invariance-based estimators and Taylored estimators: $m^* = 100$, $\lambda_1 = 5$, $k = 10$, $A = 1$

	t_1	t_2	t_3	t_4	t_5
λ_1-Estimator	Invariance-Based Estimator Template (16.16)				
T_1	0.87269	0.74595	0.61981	0.49434	0.24572
T_2	0.87295	0.74641	0.62043	0.49507	0.24643
T_3	0.87358	0.74757	0.62201	0.49691	0.24825
T_4	0.87388	0.74814	0.62277	0.49781	0.24913
λ_1-Estimator	Taylored Estimator Template (16.17)				
T_1	0.87408	0.74831	0.62273	0.49735	0.24743
T_2	0.87433	0.74877	0.62335	0.49808	0.24815
T_3	0.87495	0.74992	0.62491	0.49992	0.24998
T_4	0.87526	0.75048	0.62567	0.50081	0.25087
Max se $\times 10^4$	1.81	3.31	4.45	5.19	5.05
Min se $\times 10^4$	1.73	3.19	4.33	5.08	5.02

always falling closest to the zero benchmark. Recall that this estimator incorporated the "sample size"-based estimator T_{3M} for λ_1 within the bias-corrected expression for $\theta_X(t)$. We note that there is no fixed-sample-size analog of this estimator. One is encouraged to visually examine the results of these simulations when $m^* = 100, \lambda_1 = 5, A = 1$ and $k = 5, 10$. These may be found in Tables 16.3 and 16.5 respectively.

We conclude by noting that the standard errors for these reliability estimators are comparable for a fixed value of t and larger values of t yielded larger standard errors. We also noted that as t decreased, the estimators tended to under-estimate $\theta_X(t)$ to some degree. That is, estimators for $\theta_X(t)$ are closer to their targets for larger values of t.

16.3 A Two-sample Problem: Sequential Estimation of the Difference of Means

We consider independent observations $X_1, X_2, \ldots$ and $Y_1, Y_2, \ldots$ originating from distribution (16.1) with respective means λ_1, λ_2. We say that the X's and Y's originate from population 1 and population 2 respectively. While Section 16.2.2 was concerned with estimating $\theta_X(t) = P_{\lambda_1}(X > t)$ following sequential estimation of λ_1, now we wish to find estimators for the ratio

$$\tau_{X,Y}(t) = \theta_X(t)/\theta_Y(t), t > 0,$$

following sequential estimation of $\lambda_1 - \lambda_2 (\equiv \delta$, say).

Imagine that we have data, $X_1, X_2, \ldots, X_m$ and $Y_1, Y_2, \ldots, Y_n$, where m, n are fixed sample sizes. We measure the loss in estimating δ by $\overline{X}_m - \overline{Y}_n (\equiv W_{m,n}$, say) with

$$L_{m,n} \equiv L_{m,n}(W_{m,n}; \delta) = A(W_{m,n} - \delta)^2 + c(m+n), \tag{16.18}$$

where A, c are known positive numbers as before. The risk associated with (16.18) amounts to

$$R_{m,n}(c) \equiv E_{\lambda_1,\lambda_2}[L_{m,n}(W_{m,n}; \delta)] = A(\lambda_1 m^{-1} + \lambda_2 n^{-1}) + c(m+n). \tag{16.19}$$

The risk (16.19) is minimized when

$$m \equiv m^* = (A/c)^{1/2}\lambda_1, n \equiv n^* = (A/c)^{1/2}\lambda_2, \widetilde{n}^* \equiv m^* + n^* = (A/c)^{1/2}(\lambda_1 + \lambda_2), \tag{16.20}$$

with the fixed-sample-size minimum risk given by

$$R_{m^*,n^*}(c) = 2c(m^* + n^*) = 2c\widetilde{n}^*; \tag{16.21}$$

the magnitudes of m^*, n^* remains unknown, but the expressions of m^*, n^* would motivate a purely sequential stopping rule $(M, N, W_{M,N})$ where

$$\begin{aligned} M \equiv M(c) &= \inf\left\{m\,(\geq k) : m \geq (A/c)^{1/2}\,\overline{X}_m\right\}, \\ N \equiv N(c) &= \inf\left\{n\,(\geq k) : n \geq (A/c)^{1/2}\,\overline{Y}_n\right\}. \end{aligned} \quad (16.22)$$

That is, a pilot sample of size k (≥ 1) is chosen first from both populations. The need for any additional observation from X or Y is evaluated prior to its collection. Sampling terminates with the first observation causing m to exceed (or equal) $(A/c)^{1/2}\,\overline{X}_m$ *as well as* n to exceed (or equal) $(A/c)^{1/2}\,\overline{Y}_n$. These two sampling schemes are performed independently of one another so that M and N are independent random variables. Finally, we estimate the difference $\lambda_1 - \lambda_2$ with $\overline{X}_M - \overline{Y}_N$. The two stopping rules in (16.22) can be run independently because $m^*(n^*)$ involves the only unknown parameter $\lambda_1(\lambda_2)$ so that any allocation scheme along the lines of Robbins et al. (1967) is deemed irrelevant.

This kind of purely sequential estimation procedure was investigated by Mukhopadhyay and Chattopadhyay (1991) in the exponential case and by Mukhopadhyay and Purkayastha (1994) when the parent population distributions are assumed unknown.

We define $\widetilde{N} = M + N$ to be the total number of observations required for point estimation of δ. Under the loss function (16.18), the associated risk function for estimating δ is given by:

$$\begin{aligned} E_{\lambda_1,\lambda_2}[L_{M,N}] &= E_{\lambda_1,\lambda_2}\left[A(W_{M,N} - \delta)^2 + c(M+N)\right] \\ &= E_{\lambda_1}\left[A\left(\overline{X}_M - \lambda_1\right)^2 + cM\right] + E_{\lambda_2}\left[A\left(\overline{Y}_N - \lambda_2\right)^2 + cN\right] \\ &\quad -2AE_{\lambda_1,\lambda_2}\left[\left(\overline{X}_M - \lambda_1\right)\left(\overline{Y}_N - \lambda_2\right)\right]. \end{aligned} \quad (16.23)$$

Now, the associated *regret* function can be expressed as:

$$\omega(c) = E_{\lambda_1,\lambda_2}[L_{M,N}] - R_{m^*,n^*}(c). \quad (16.24)$$

Then, we combine (16.21), (16.23) with (16.9) and (16.11) to rewrite $\omega(c)$. For $k \geq 2$, we have the following *second-order* expansion as $c \to 0$:

$$\omega(c) = 2\{3c + o(c)\} - 2A\left[\lambda_1\lambda_2(m^*n^*)^{-1}\right] + o(c) = 4c + o(c). \quad (16.25)$$

For the average sample sizes, however, we can express the following second-order results. With $k \geq 2$, as $c \to 0$:

$$\begin{aligned} E_{\lambda_1}(M) &= m^* - 0.254965 + o(1), \quad E_{\lambda_2}(N) = n^* - 0.254965 + o(1), \\ E_{\lambda_1,\lambda_2}\left(\widetilde{N}\right) &= \widetilde{n}^* - 0.509930 + o(1). \end{aligned} \quad (16.26)$$

Now, we return to seek estimators for $\tau_{X,Y}(t)$.

Recall the notion of an *estimator template*, introduced in Section 16.2.2. We proposed (16.14)–(16.15) involving the sample mean, $T_{1M} = \overline{X}_M$, that led to *templates* for finding further estimators of $\theta_X(t)$. In short, we replaced T_{1M} with other estimators for λ_1, namely T_{2M}, T_{3M}, T_{4M} from (16.12), in the expressions given by (16.14)–(16.15). Indeed, we found empirical evidence to suggest that some of these ad-hoc estimators of $\theta_X(t)$ performed better than those given by (16.14)–(16.15). We shall continue on that path in our quest for finding a reasonable collection of estimators of $\tau_{X,Y}(t)$.

In the two-sample situation, we offer three estimator templates for $\tau_{X,Y}(t)$. Should Z be an estimator for a population parameter, the addition of the subscript M(or N) would alert us to the fact that the estimator Z_M(or Z_N) estimates a parameter from population 1(or 2) based on M(or N) observations from that population. This notation allows us to easily keep track of stopping times and estimators under consideration. Three templates are now summarized:

1. **Invariance (of MLE) Based Estimators:** We consider mimicking the fixed-sample-size MLE for $\tau_{X,Y}(t)$ since we have reasons to believe, based on the results presented for the one-sample problem, that this route will continue to provide success in this setting. Should Z_M, Z_N be estimators of λ_1, λ_2 respectively, then the invariance-based estimator of $\tau_{X,Y}(t)$ is the ratio of our univariate invariance-based estimators (16.14), that is,

$$\widehat{\tau}_{X,Y}(t) \equiv \widehat{\tau}_{X,Y}(t, Z_M, Z_N) = \widehat{\theta}_X(t, Z_M)/\widehat{\theta}_Y(t, Z_N) = \exp\left(-t(Z_M^{-1} - Z_N^{-1})\right). \tag{16.27}$$

2. **Univariate Taylored Estimators:** It seems natural to consider the ratio of our univariate Taylored estimator from (16.15). The univariate Taylored estimators from a one-sample problem were found to be in close proximity of invariance-based estimators, and indeed outperformed the invariance-based estimators in the one-sample problem. So, it is intuitive to consider the ratio of univariate Taylored estimators in the present setting. Should Z_M, Z_N be estimators of λ_1, λ_2 respectively, this template is simply the ratio of estimators of the form (16.15), that is,

$$\begin{aligned}\widehat{\tau}^*_{X,Y}(t) &\equiv \widehat{\tau}^*_{X,Y}(t, Z_M, Z_N) = \widehat{\theta}^*_X(t, Z_M)/\widehat{\theta}^*_Y(t, Z_N)\\ &= \{a(t, M, Z_M)\}^{-1} a(t, N, Z_N) \exp\left(-t(Z_M^{-1} - Z_N^{-1})\right)\end{aligned} \tag{16.28}$$

where $a(t, r, Z_r) = 1 + \left(\frac{1}{2}t - Z_r - 1\right) t Z_M^{-2} r^{-1}$, $r = M, N$.

3. **Bivariate Taylored Estimators:**

$$\begin{aligned}\widehat{\tau}^{**}_{X,Y}(t) &\equiv \widehat{\tau}^{**}_{X,Y}(t, Z_M, Z_N)\\ &= \left(1 + \left[\frac{-t(Z_M - \frac{1}{2}t(t-2Z_M))}{Z_M^2 M} + \frac{t(Z_N + \frac{1}{2}t(t+2Z_N))}{Z_N^2 N} + \frac{2t^2}{Z_M Z_N M N}\right]\right)^{-1}\\ &\quad \times \exp\left(-t(Z_M^{-1} - Z_N^{-1})\right).\end{aligned} \tag{16.29}$$

The bivariate Taylored estimator (16.29) is developed in the Appendix, but for now let us motivate it. Given the results of Mukhopadhyay, Padmanaghan and Solanky (1997), we may intuitively propose to consider an estimator of θ_1/θ_2 as follows:

$$\widehat{\theta_X(t)/\theta_Y(t)} = \exp\left(-t\left(\overline{X}_M^{-1} - \overline{Y}_N^{-1}\right)\right). \tag{16.30}$$

But, even the fixed-sample-size version of the estimator from (16.30) is not unbiased for the ratio of parameters. Hence, we pursue a bivariate Taylor expansion of this estimator and then find the associated bias-corrected form.

Theorem 16.3.1 *For* $k \geq 2$*, the bias-corrected version of the estimator from (16.30) based on the bivariate Taylor expansion, is given by*

$$\begin{aligned}
&\widehat{\tau}^{**}_{X,Y}\left(\overline{X}_M, \overline{Y}_N\right) \\
&= \left(1 + \left[\frac{-t(\overline{X}_M - \frac{1}{2}t(t-2\overline{X}_M))}{\overline{X}_M^2 M} + \frac{t(\overline{Y}_N + \frac{1}{2}t(t+2\overline{Y}_N))}{\overline{Y}_N^2 N} + \frac{2t^2}{\overline{X}_M\overline{Y}_N MN}\right]\right)^{-1} \\
&\quad \times \exp\left(-t(\overline{X}_M^{-1} - \overline{Y}_N^{-1})\right).
\end{aligned} \tag{16.31}$$

In light of Theorem 16.3.1, our third template shall take the form (16.29) with Z_M and Z_N replacing $\overline{X}_M$, and $\overline{Y}_S$, respectively.

16.3.1 A simulation study

In this simulation study, we have chosen to follow closely the precedence set in the case of the one-sample situation discussed in Section 16.2.2. We have conducted simulations of 1000 runs on each combination of the following parameters:

$$m^* = n^* = 50, 75, 100, 150;\ k = 5, 10;$$
$$\lambda_1 = \lambda_2 = 5;\ \text{and}\ A = 1.$$

More specifically, for each of the eight combinations above, we simulate independent purely sequential sampling strategies from two exponential distributions with mean $\lambda_1 = \lambda_2 = 5$ along the line of (16.22). Once the sampling operation was completed for a fixed set of parameter configuration, we subsequently investigated the performances of estimators of $\theta_X(t)$, $\theta_Y(t)$ and $\tau_{X,Y}(t)$. The values of t used in our simulations were again chosen so that

$$P_{\lambda_1}(X > t) = P_{\lambda_2}(Y > t) = 0.875,\ 0.75,\ 0.625,\ 0.5,\ 0.25.$$

We have recorded the following entities from each run:

T_{iM}, T_{iN}, $i = 1, 2, 3, 4$, allow us to compute the invariance-based estimators (3.10), namely

$$\widehat{\tau}_{iX,Y}(t) = \exp\left(-t(T_{iM}^{-1} - T_{iN}^{-1})\right),\ i = 1,\ 2,\ 3,\ 4;$$

$\widehat{\theta}_{iX}, \widehat{\theta}_{iY}, i = 1, 2, 3, 4$, allow us to compute the univariate Taylored estimators (16.28), namely

$$\widehat{\tau}^{*}_{iX,Y}(t) = \widehat{\theta}_{iX}(t) / \widehat{\theta}_{iY}(t), \quad i = 1, 2, 3, 4.$$

Together with bivariate Taylored estimators, namely $\widehat{\tau}^{**}_{iX,Y}(t)$, $i = 1, 2, 3, 4$, we perform fairly exhaustive comparisons of these twelve estimators.

Our interest, however, lies in the long-run performance of these estimators. This said, we concentrate on the average performance of these competing estimators in the light of each of the eight configurations over 1000 runs. Additionally, we also obtain the corresponding estimated standard errors for all aforementioned entities. One-sample analyses on the stopping sample sizes and average performance of the estimators T_{iM}, T_{iN} have strengthened our conviction favoring the kinds of comments we made earlier in Section 16.2.2.

Obviously $\tau_{X,Y}(t) = \tau^{-1}_{Y,X}(t)$, so that we think of $\tau_{X,Y}(t)$ as a means for comparing population 1 with population 2, but then $\tau_{Y,X}(t)$ should reverse the perspective. Since we have run simulations assuming equal means for populations 1 and 2, the true value of $\tau_{X,Y}(t)$ is identically 1 under all configurations. Hence, we shall consider these estimators to be doing their "job" well if we find them to be close to 1. That is, we need only consider the magnitude of an estimator's deviation from 1, although we have noted when such estimator has under-(or over-)estimated the target. Tables 16.6–16.7 summarize some of the findings. These tables provide the deviations of estimators from the value 1.

Let us comment first on features of the simulation study that were common to all the twelve estimators of $\tau_{X,Y}(t)$. The role of the pilot sample size k reflects our comments made earlier and confirms our intuition. Generally speaking, larger pilot sample size leads to better performances of estimators. This feature was also noted in Section 16.2.1 when we compared the performances of T_{iM}, $i = 1, 2, 3, 4$. This sentiment continued to hold true in Section 16.2.2 when discussing the eight estimators of $\theta_X(t)$. When it comes to estimating $\tau_{X,Y}(t)$, we again note that estimators based on pilot samples of size $k = 10$ perform better than pilots based on $k = 5$ observations.

We fixed $m^* = n^*$ but the common value plays its usual role as well. Recall that these values represent the optimal fixed-sample-sizes needed to estimate λ_1, λ_2. From our findings discussed in Section 16.2, we know that the estimators T_i, $i = 1, 2, 3, 4$, converge in probability to 5 as $c \to 0$ (or equivalently, as the optimal fixed-sample-sizes tend to infinity). Naturally, since the templates (16.27), (16.28), and (16.29) incorporated variants of these *first-order risk-efficient estimators*, performances of the templates are expected to improve as sample sizes became larger. These sentiments are expressed well by what one sees in Tables 16.6–16.7. Figures 16.3–16.6 have been constructed in a fashion similar in spirit to those in Figures 16.1–16.2. Since we are now discussing the performance of twelve estimators, we have allocated a separate

Table 16.6: Performance of templates and λ-estimators seen through the average values $(\widehat{\tau} - 1) \times 10^3$; $m^* = n^* = 50$, $\lambda_1 = \lambda_2 = 5$, $k = 5$

	t_1	t_2	t_3	t_4	t_5
	Invariance-Based Estimator Template (16.27)				
λ-Estimator					
T_1	0.692	2.989	7.839	17.115	77.905
T_2	0.671	2.878	7.521	16.395	69.336
T_3	0.628	2.727	7.162	15.649	69.336
T_4	0.614	2.663	6.986	15.245	67.113
	Univariate Taylored Estimator Template (16.28)				
λ-Estimator					
T_1	0.622	2.685	7.135	15.959	80.745
T_2	0.605	2.589	6.849	15.247	74.845
T_3	0.563	2.448	6.515	14.537	70.905
T_4	0.551	2.391	6.355	14.158	68.426
	Bivariate Taylored Estimator Template (16.29)				
λ-Estimator					
T_1	0.295	−2.516	−12.319	−38.549	−234.793
T_2	0.281	−2.493	−12.183	−38.143	−233.071
T_3	0.247	−2.463	−11.934	−37.339	−229.207
T_4	0.239	−2.438	−11.800	−36.933	−227.308

Table 16.7: Performance of templates and λ-estimators seen through the average values $(\widehat{\tau} - 1) \times 10^3$; $m^* = n^* = 50$, $\lambda_1 = \lambda_2 = 5$, $k = 10$

	t_1	t_2	t_3	t_4	t_5
	Invariance of MLE-Based Estimator Template (16.27)				
λ-Estimator					
T_1	0.099	1.569	5.160	12.289	54.492
T_2	0.097	1.516	4.984	11.870	52.600
T_3	0.116	1.530	4.852	11.720	51.556
T_4	0.110	1.489	4.827	11.431	50.292
	Univariate Taylored Estimator Template (16.28)				
λ-Estimator					
T_1	0.078	1.385	4.671	11.392	53.919
T_2	0.076	1.338	4.511	10.998	51.997
T_3	0.096	1.357	4.489	10.862	50.909
T_4	0.092	1.320	4.374	10.590	49.625
	Bivariate Taylored Estimator Template (16.29)				
λ-Estimator					
T_1	−0.242	−2.951	−12.650	−38.927	−236.708
T_2	−0.239	−2.919	−12.513	−38.524	−234.950
T_3	−0.213	−2.833	−12.222	−37.704	−231.068
T_4	−0.213	−2.803	−12.085	−37.292	−229.145

figure for each value of t that we shall discuss. These figures also help to describe the behavior of these estimators when sample sizes increase. As we consider these figures, let us be reminded that it is the large deviation from the value 1 that will mark a "poor" estimator. Let us not make the mistake of saying that some estimators appear to be overestimating or underestimating $\tau_{X,Y}(t)$. That is, we would simply prefer those estimators with deviations near zero. Notice that each figure consists of eight slides where each slide corresponds to a particular combination of $n^* = 50, 75, 100, 150$ and $k = 5, 10$. Inside each slide within a figure, one finds twelve symbols. The first set of four symbols (circles) corresponds to the invariance-based estimators (16.27), the second set of four symbols (triangles) corresponds to the univariate Taylored estimators (16.28), while the final set of four symbols (squares) corresponds to the bivariate Taylored estimators (16.29). Within a set of four symbols, we retain the basic structure of Figures 16.1–16.2, namely that the ith symbol in a set corresponds to the template when T_{iR}, T_{iS} are used, $i = 1, 2, 3, 4$. We do emphasize that each symbol represents the average performance of that particular estimator over a course of 1000 simulations.

The value of t plays an important role in how accurate the estimators are. In particular, each estimator tends to perform better when t is relatively small. Figures 16.3–16.4 provide a visual summary of performances of all twelve estimators when t is chosen to yield $\theta_X(t_1) = 0.875$ and $\theta_X(t_2) = 0.75$ respectively. In Figure 16.3, we note that all the estimators are performing admirably and it is quite difficult to decide which estimator is the best. Figures 16.4–16.6 each share a common feature: As n^* increased, the estimators tended to do much better, that is, the range of the estimators shrank as m^*, n^* increased. Since this feature is absent in Figure 16.3, we are led to believe that asymptotics start affecting performances in a positive way fairly quickly for this particular choice of t. In further support of this claim, we note that as t increased, the axis length of the figures also increased. That is, for larger values of t (or equivalently, smaller values of $\theta_X(t)$), the range of the vertical axis in these figures became larger.

We mention that Figures 16.3–16.4 provided the best support in favor of the bivariate Taylored estimators. For chosen values of t, there are configurations that find these estimators performing better than their invariance-based and univariate Taylored peers. In particular, we note in Figure 16.3 that the bivariate Taylored estimators performed best in all but two cases. On the other hand, in Figure 16.4, these estimators performed best in half of the cases, and comparably in the remaining four.

Figures 16.5–16.6 correspond to t values chosen so that $\theta_X(t) = 0.5$ and $\theta_X(t) = 0.25$, respectively. From these figures we realize some limitations of the bivariate Taylored estimators. In each slide within these figures, we find that the invariance-based estimators (16.27) and univariate Taylored estimators

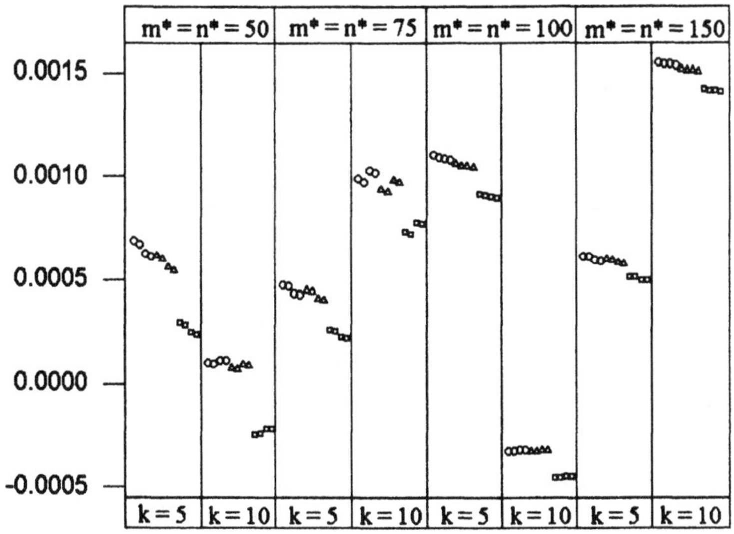

Figure 16.3: Performance of template estimators for the two sample problem t is chosen such that $\theta_X(t) = \theta_Y(t) = 0.875$; Triangles: Invariance-based estimators, $i = 1, 2, 3, 4$; Circles: Univariate Taylored estimators, $i = 1, 2, 3, 4$; Squares: Bivariate Taylored estimators, $i = 1, 2, 3, 4$

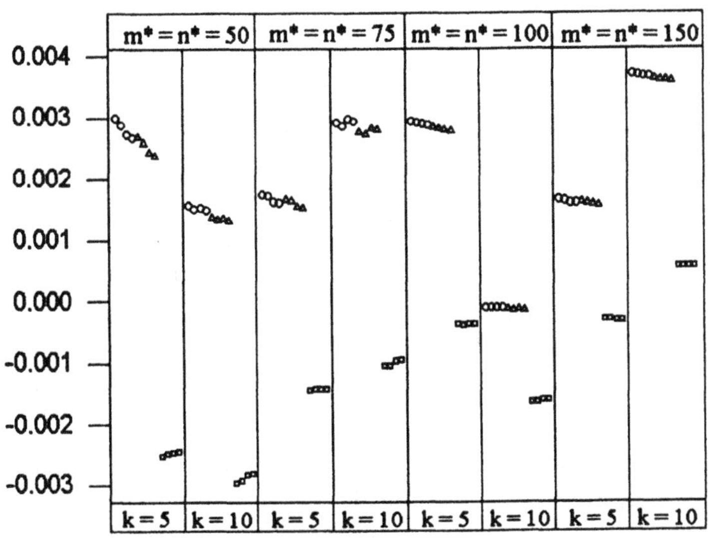

Figure 16.4: Performance of template estimators for the two sample problem t is chosen such that $\theta_X(t) = \theta_Y(t) = 0.75$; Triangles: Invariance-based estimators, $i = 1, 2, 3, 4$; Circles: Univariate Taylored estimators, $i = 1, 2, 3, 4$; Squares: Bivariate Taylored estimators, $i = 1, 2, 3, 4$

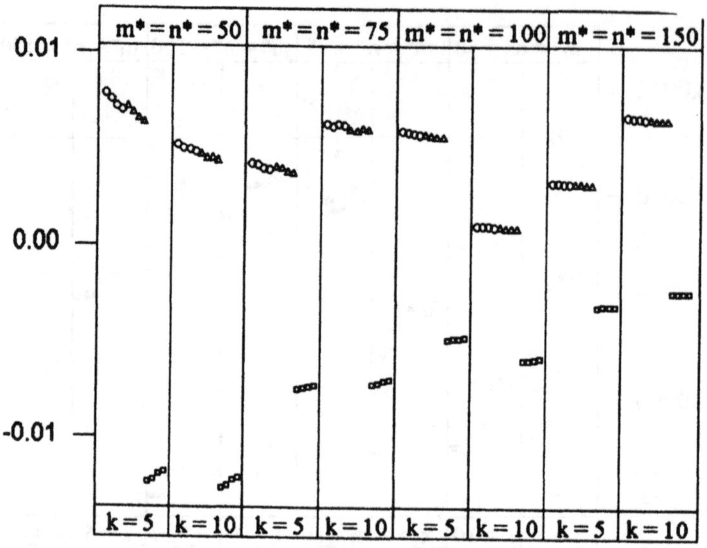

Figure 16.5: Performance of template estimators for the two sample problem t is chosen such that $\theta_X(t) = \theta_Y(t) = 0.50$; Triangles: Invariance-based estimators, $i = 1, 2, 3, 4$; Circles: Univariate Taylored estimators, $i = 1, 2, 3, 4$; Squares: Bivariate Taylored estimators, $i = 1, 2, 3, 4$

(16.28) perform in a superior fashion. However, we do take solace in the fact that as m^*, n^* increase, the discrepancy among these estimators became less pronounced.

From these figures, we are tempted to offer some general guidelines concerning the behavior of the estimators based on the particular mean-estimator T_{iR} and T_{iS} used in the template. We find that the following sentiments are supported by these figures:

$$\left|\widehat{\tau}_{4X,Y}(t) - 1\right| < \left|\widehat{\tau}_{3X,Y}(t) - 1\right| < \left|\widehat{\tau}_{2X,Y}(t) - 1\right| < \left|\widehat{\tau}_{1X,Y}(t) - 1\right|$$
$$\left|\widehat{\tau}^*_{4X,Y}(t) - 1\right| < \left|\widehat{\tau}^*_{3X,Y}(t) - 1\right| < \left|\widehat{\tau}^*_{2X,Y}(t) - 1\right| < \left|\widehat{\tau}^*_{1X,Y}(t) - 1\right|$$
$$\left|\widehat{\tau}^{**}_{4X,Y}(t) - 1\right| < \left|\widehat{\tau}^{**}_{3X,Y}(t) - 1\right| < \left|\widehat{\tau}^{**}_{2X,Y}(t) - 1\right| < \left|\widehat{\tau}^{**}_{1X,Y}(t) - 1\right|$$

Let us remark that these broad range of comments simply reflect what is observed in the majority of slides found in Figures 16.3–16.6. This in turn leads one to believe that regardless of a template, the mean-estimator (T_4) seems to provide the best performance in estimating $\tau_{X,Y}(t)$. From these figures we may also suggest that in general, the univariate Taylored estimators offer the overall best possible template. With a fixed mean-estimator T_i, its performance is uniformly superior to the invariance-based estimators. Its merit is best highlighted in the case of larger values of t when the bivariate Taylored estimators

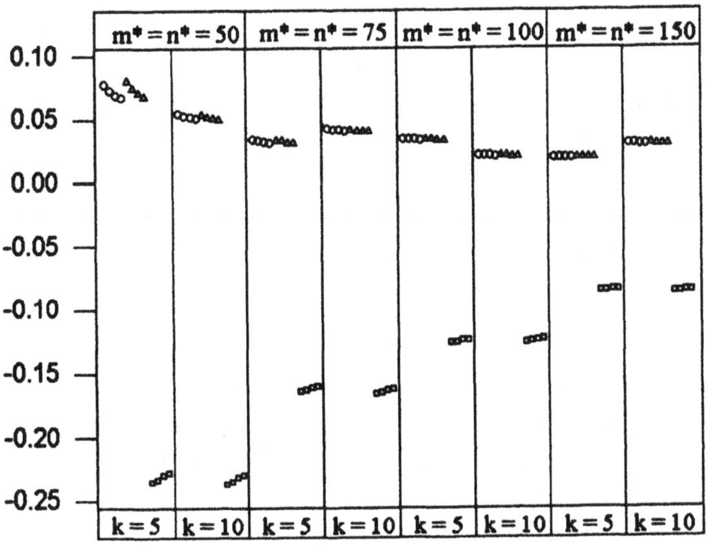

Figure 16.6: Performance of template estimators for the two sample problem t is chosen such that $\theta_X(t) = \theta_Y(t) = 0.25$; Triangles: Invariance-based estimators, $i = 1,2,3,4$; Circles: Univariate Taylored estimators, $i = 1,2,3,4$; Squares: Bivariate Taylored estimators, $i = 1,2,3,4$

do not perform as well as the other templates. In our final section, we mention some examples and apply these techniques to a historic survival dataset.

16.4 Some Examples and Data Analyses

Example 16.4.1: Time to Relapse in Drug Rehabilitation Programs

Cicconetti (2002) considered a dataset of Hosmer and Lemeshow (1998) which considered patients from drug rehabilitation programs. The response variable of interest was 'time to relapse'. A two-sample scenario was envisioned by considering patients from two treatment centers. In such a situation, it seemed canonical to estimate λ_1, λ_2 sequentially and it also seemed prudent to continually update databases as patients returned. Hosmer and Lemeshow (1998) dataset prompted us to recreate the scenario. After sequentially sampling from a permutation of the data set, Cicconetti (2002) estimated the means, λ_1, λ_2 using T_i, $i = 1,2,3,4$, of Section 16.2. Recall that an exponential distribution is inherently right-skewed and the mean often provides a distorted view of a population. That is, we have an inflated view of 'central tendency'. By considering $\theta(t)$, better decisions might be possible. For example,. Cicconetti

(2002) showed that for both locations $\theta(t) \approx 0.52$ for $t = 100$ days; that is, the proportion of patients making it beyond 100 days without a relapse is nearly 52%. This information might be used to suggest installing mandatory follow-up meetings at both 1 and 2 months after release. Comparisons of estimators for $\tau_{X,Y}(t)$ were similar and close to the ones for this example except in the case of the bivariate Taylored estimators. These provided estimates that were grossly different from the others. Should all estimators of $\tau_{X,Y}(t)$ be sizably different from one another, an investigation into the practices at both sites would hopefully reveal reasons for inequity.

Example 16.4.2: Proschan's Air Conditioner Data

Proschan (1963) provides records of the duration of time between successive failures of air-conditioning systems of each member of a fleet of 13 Boeing 720 jet airplanes. We came across this dataset in Olkin, Gleser and Derman's (1978) text that is riddled with great historical datasets. We shall use this particular dataset to demonstrate the techniques used here. In this dataset the airplanes have increasing identification codes. Let us assume that in the manufacturing stage of older planes, certain 'bugs' might have been encountered, and these were addressed while building the newer planes. Hence, we consider the first six planes (IDs: 7907, 7908, 7909, 7910, 7911, 7912) to represent population 1 and the last seven planes (IDs: 7913, 7914, 7915, 7916, 7917, 8044, 8045) to represent population 2 because it seems fair to believe that planes built at a later date should possibly have more reliability. We might expect that this difference in reliability would be captured when we consider estimators of $\theta_X(t)$, $\theta_Y(t)$, and $\tau_{X,Y}(t)$. In this demonstration, malfunction times have been permuted.

Let us forget for the moment that we are working from a historical dataset and pretend that we are retrospectively soliciting this information. Perhaps we wish to examine the merit of the implemented changes retrospectively. In the absence of a database, one would find the data collection process to be very time-consuming and potentially costly. It conceivably may involve trips to repair facilities to scour log-books. For this example, we envision a value of $A = 100$ and cost, $c = 200$. In addition, we chose a pilot sample of size $k = 30$. Probability plots based on the pilot samples indicated that an exponential distribution would be reasonable.

Table 16.8 provides univariate summary statistics. We fix two values for t: 40 hours and 60 hours. The observations we make here are in line with those made concerning our simulation study of Section 16.2. We find each estimator T_i, $i = 1, 2, 3, 4$ to behave fairly similarly. This feature is also echoed in the performance of $\widehat{\theta}_X(40, T_i)$ and $\widehat{\theta}_X(60, T_i)$. Recall that our previous simulation studies gave us reasons to look toward $\widehat{\theta}_X(40, T_3) = 0.634595$ and $\widehat{\theta}_X(60, T_3) = 0.6925823$ among eight contending estimators. This information might suggest a prudent mandatory maintenance check at, say, every 40 or 50 hours for the older and newer planes respectively.

Table 16.8: Estimators of $\lambda_1, \lambda_2, \theta(t)$ for Proschan's (1963) air-conditioner dataset

Sample Size	Sample I $M = 56$	Sample II $N = 77$		Sample I	Sample II
Mean	T_1 : 77.55357	107.07792	Bias-Corrected	T_2 : 77.57143	107.09091
Sample Size Based	T_3 : 79.19596	108.89444	Bias-Corrected	T_4 : 79.55653	109.25502
	MLE-Based	MLE-Based		Taylor-Based	Taylor-Based
$\widehat{\theta}(40, T_1)$	0.5970392	0.6882799	$\widehat{\theta}^*(40, T_1)$	0.6012200	0.6910375
$\widehat{\theta}(40, T_2)$	0.5971101	0.6883111	$\widehat{\theta}^*(40, T_2)$	0.6012907	0.6910686
$\widehat{\theta}(40, T_3)$	0.6034595	0.6925823	$\widehat{\theta}^*(40, T_3)$	0.6076251	0.6953206
$\widehat{\theta}(40, T_4)$	0.6048425	0.6934224	$\widehat{\theta}^*(40, T_4)$	0.6090045	0.6961568
$\widehat{\theta}(60, T_1)$	0.4613221	0.5710151	$\widehat{\theta}^*(60, T_1)$	0.4653470	0.5740612
$\widehat{\theta}(60, T_2)$	0.4614043	0.5710539	$\widehat{\theta}^*(60, T_2)$	0.4654295	0.5741000
$\widehat{\theta}(60, T_3)$	0.4687834	0.5763776	$\widehat{\theta}^*(60, T_3)$	0.4728379	0.5794196
$\widehat{\theta}(60, T_4)$	0.4703959	0.5774267	$\widehat{\theta}^*(60, T_4)$	0.4744564	0.5804677

Results displayed in Table 16.9 also reflect the findings of our prior simulation studies. For both $t = 40$ and $t = 60$, we observed some grouping among first eight estimators. The majority of estimators of $\tau_{X,Y}(40)$ fell around 0.87. One may interpret the reciprocal and say the newer planes are about 1.15 times more likely to last longer than 40 hours. However, when we came to the bivariate Taylored estimators, we recall observing some disheartening performance through simulations. For the two particular choices of t here, we see the bivariate Taylored estimators return ratios that are much larger (smaller) than its competitors when $t = 40$ (60). The estimators based on the bivariate Taylor expansion have proven to be a rather unstable group and more investigations may be warranted.

Appendix: Proof of Theorem 16.3.1

Suppose that we have a function of two variables,

$$f(x,y) = \exp\left(-t(x^{-1} - y^{-1})\right), \ x > 0, y > 0. \tag{A.1}$$

Let us simply write $\overline{X} \equiv \overline{X}_M$ and $\overline{Y} \equiv \overline{Y}_N$. Then, using bivariate Taylor expansion [Thomas and Finney (1990)], we can rewrite (A.1) as follows:

$$\begin{aligned} f\left(\overline{X}, \overline{Y}\right) &= \exp\left(\tfrac{t(\lambda_1-\lambda_2)}{\lambda_1\lambda_2}\right)\left\{1 + \left[\tfrac{t(\overline{X}-\lambda_1)}{\lambda_1^2} - \tfrac{t(\overline{Y}-\lambda_2)}{\lambda_2^2}\right]\right\} \\ &\quad + \tfrac{1}{2}\exp\left(\tfrac{t(U-V)}{UV}\right)\left[\left(\overline{X}-\lambda_1\right)^2 \tfrac{t(t-2U)}{U^4} + \left(\overline{Y}-\lambda_2\right)^2 \tfrac{t(t+2V)}{V^4}\right] \\ &\quad - \left(\overline{X}-\lambda_1\right)\left(\overline{Y}-\lambda_2\right)\tfrac{t^2}{U^2V^2}\exp\left(\tfrac{t(U-V)}{UV}\right), \end{aligned} \tag{A.2}$$

Table 16.9: Estimators of $\tau_{X,Y}(t)$ for Proschan's (1963) air-conditioner dataset

Estimator	$t = 40$ hrs.	$t = 60$ hrs.
$\widehat{\tau}_{X,Y}(Z_R, Z_S);\ Z = T_1$	0.8674367	0.8078982
$\widehat{\tau}_{X,Y}(Z_R, Z_S);\ Z = T_2$	0.8675004	0.8079872
$\widehat{\tau}_{X,Y}(Z_R, Z_S);\ Z = T_3$	0.8713182	0.8133269
$\widehat{\tau}_{X,Y}(Z_R, Z_S);\ Z = T_4$	0.872257	0.8146418
$\widehat{\tau}^{*}_{X,Y}(Z_R, Z_S);\ Z = T_1$	0.8700251	0.8106226
$\widehat{\tau}^{*}_{X,Y}(Z_R, Z_S);\ Z = T_2$	0.8700883	0.8107116
$\widehat{\tau}^{*}_{X,Y}(Z_R, Z_S);\ Z = T_3$	0.8738776	0.8160544
$\widehat{\tau}^{*}_{X,Y}(Z_R, Z_S);\ Z = T_4$	0.8748094	0.8173692
$\widehat{\tau}^{**}_{X,Y}(Z_R, Z_S);\ Z = T_1$	0.9105683	0.7735883
$\widehat{\tau}^{**}_{X,Y}(Z_R, Z_S);\ Z = T_2$	0.9106254	0.7737010
$\widehat{\tau}^{**}_{X,Y}(Z_R, Z_S);\ Z = T_3$	0.9152088	0.7841508
$\widehat{\tau}^{**}_{X,Y}(Z_R, Z_S);\ Z = T_4$	0.9162191	0.7863754

where (U, V) is a random vector, with $U \equiv U_{M,N}, V \equiv V_{M,N}$, that lies on the line segment joining the points (λ_1, λ_2) and $(\overline{X}, \overline{Y})$. In what follows, we suppress "as $c \to 0$" for convenience.

Now, from Mukhopadhyay (1988) or (16.11), we have

$$E_{\lambda_1}\left(\overline{X} - \lambda_1\right) = -\lambda_1 m^{*-1} + o\left(m^{*-1}\right).$$

A similar result for $E_{\lambda_2}\left(\overline{Y} - \lambda_2\right)$, together with the fact that $U \xrightarrow{P} \lambda_1, V \xrightarrow{P} \lambda_2$ leads to the following asymptotic behaviors:

$$\begin{aligned}
&\tfrac{t(t-2U)}{U^4} \exp\left(\tfrac{t(U-V)}{UV}\right) \xrightarrow{P} \tfrac{t(t-2\lambda_1)}{\lambda_1^4} \exp\left(\tfrac{t(\lambda_1-\lambda_2)}{\lambda_1\lambda_2}\right);\\
&\tfrac{t(t+2V)}{V^4} \exp\left(\tfrac{t(U-V)}{UV}\right) \xrightarrow{P} \tfrac{t(t+2\lambda_2)}{\lambda_2^4} \exp\left(\tfrac{t(\lambda_1-\lambda_2)}{\lambda_1\lambda_2}\right);\\
&\tfrac{t^2}{U^2V^2} \exp\left(\tfrac{t(U-V)}{UV}\right) \xrightarrow{P} \tfrac{t^2}{\lambda_1^2\lambda_2^2} \exp\left(\tfrac{t(\lambda_1-\lambda_2)}{\lambda_1\lambda_2}\right).
\end{aligned} \tag{A.3}$$

Next, Anscombe's (1952) random central limit theorem allows us to claim:

$$m^*\left(\overline{X} - \lambda_1\right)^2 \xrightarrow{\pounds} \lambda_1^2\chi_1^2 \text{ and } n^*\left(\overline{Y} - \lambda_2\right)^2 \xrightarrow{\pounds} \lambda_2^2\chi_1^2. \tag{A.4}$$

Moreover, we know that

$$E\left[\left(\overline{X} - \lambda_1\right)\left(\overline{Y} - \lambda_2\right)\right] = \lambda_1\lambda_2(m^*n^*)^{-1} + o(c), \tag{A.5}$$

which was used earlier in (16.23). Cicconetti (2002) proved the following result regarding uniform integrability of the crucial terms involved in (A.2). For completeness, we state Cicconetti's result without reproducing its proof.

Lemma A.1 *Recall that* $\overline{X} \equiv \overline{X}_M$ *and* $\overline{Y} \equiv \overline{Y}_N$. *Let us denote*

$$I_{1c} = \left(\overline{X} - \lambda_1\right)^2 \tfrac{t(t-2U)}{U^4} \exp\left(\tfrac{t(U-V)}{UV}\right), I_{2c} = \left(\overline{Y} - \lambda_2\right)^2 \tfrac{t(t+2V)}{V^4} \exp\left(\tfrac{t(U-V)}{UV}\right),$$
$$I_{3c} = 2\left(\overline{X} - \lambda_1\right)\left(\overline{Y} - \lambda_2\right) \tfrac{t^2}{U^2V^2} \exp\left(\tfrac{t(U-V)}{UV}\right).$$

where (U, V) *is a random vector, with* $U \equiv U_{M,N}, V \equiv V_{M,N}$, *that lies on the line segment joining the points* (λ_1, λ_2) *and* $\left(\overline{X}, \overline{Y}\right)$. *Then,* I_{jc} *is uniformly integrable for* $0 < c \le c_0$, $j = 1, 2, 3$.

Now, we can combine (A.3)–(A.5) with Lemma A.1 and (A.2) in order to express $f\left(\overline{X}, \overline{Y}\right)$ up to the order $o(c^{1/2})$, and hence we are able to write

$$\begin{aligned} f\left(\overline{X}, \overline{Y}\right) &= \tau_{X,Y}(t)\left[1 + \left\{\frac{t}{\lambda_1^2}\left[-\lambda_1 m^{*-1}\right] - \frac{t}{\lambda_2^2}\left[-\lambda_2 n^{*-1}\right]\right\}\right] \\ &+ \frac{1}{2}\tau_{X,Y}(t)\left[\frac{\lambda_1^2}{m^*}\frac{t\,(t-2\lambda)}{\lambda_1^4} + \frac{\lambda_2^2}{n^*}\frac{t\,(t+2\lambda)}{\lambda_2^4} - 2\frac{t^2}{\lambda_1^2\lambda_2^2}\left(\lambda_1\lambda_2 m^{*-1} n^{*-1}\right)\right] \\ &+ o(c^{1/2}). \end{aligned} \tag{A.6}$$

The expression of $\widehat{\tau}_{X,Y}(t)$ is now obtained by dividing $f\left(\overline{X}, \overline{Y}\right)$ with the expression that is given in the right-hand side of (A.6), disregarding the term $o(c^{1/2})$, and then replacing $\lambda_1, \lambda_2, m^*, n^*$ with $\overline{X}, \overline{Y}, M, N$ respectively. ■

References

1. Anscombe, F.J. (1952). Large sample theory of sequential estimation, *Proc. Camb. Phil. Soc.*, **48**, 600–607.

2. Ansell, J. and Phillips, M. (1989). Practical problems in the statistical analysis of reliability data, *Applied Statistics*, **38**, 205–247.

3. Balakrishnan, N. and Basu, A. P. (1995). *The Exponential Distribution*, Gordon and Breach, Amsterdam.

4. Barlow, R. and Proschan, F. (1975). *Statistical Theory of Reliability and Life Testing*, Holt, Rinehart and Winston, New York.

5. Beg, M and Singh, N. (1979). Estimation of $P(Y < X)$ for the Pareto distribution, *IEEE Transactions on Reliability*, **28**, 411–414.

6. Beg, M. (1980). Estimation of $P(Y < X)$ for exponential families, *IEEE Transactions on Reliability*, **29**, 158–160.

7. Brown, R. (1977). Evaluation of $P(X > Y)$ when both X and Y are from 3-parameter Weibull Distributions, *IEEE Transactions on Reliability,* **22,** 100–105.

8. Cicconetti, G. (2002). Contributions to Sequential Analysis, *Ph.D. Dissertation*, Department of Statistics, University of Connecticut, Storrs.

9. Epstein, B. and Sobel, M. (1953). Life testing, *Journal of the American Statistical Association*, **48**, 486–502.

10. Epstein, B. and Sobel, M. (1954). Some theorems relevant to life testing from an exponential distribution, *Annals of Mathematical Statistics*, **25**, 373–381.

11. Epstein, B. and Sobel, M. (1955). Sequential procedures in life testing from an exponential distribution, *Annals of Mathematical Statistics* **26**, 82–93.

12. Ghosh, M., Mukhopadhyay, N. and Sen, P.K. (1997). *Sequential Estimation*, John Wiley & Sons, New York.

13. Hosmer, D. W. and Lemeshow, S. (1998). *Applied Survival Analysis: Regression Modeling of Time to Event Data,* John Wiley & Sons, New York.

14. Lomnicki, Z. (1973). Some aspects of the statistical approach to reliability, *Journal of the Royal Stastical Society, Series A*, **136**, 395–420.

15. Mukhopadhyay, N. (1988). Sequential estimation problems for negative exponential populations, *Communications in Statistics—Theory and Methods*, **17**, 2471–2506.

16. Mukhopadhyay, N. (1995). Two-stage and multi-stage estimation. In *The Exponential Distribution* (Eds., N. Balakrishnan and A.P. Basu), pp. 429-452, Gordon and Breach, Amsterdam.

17. Mukhopadhyay, N. and Chattopadhyay, S. (1991). Sequential methodologies for comparing exponential mean survival times, *Sequential Analysis*, **10**, 139–148.

18. Mukhopadhyay, N. and Cicconetti, G. (2000). Estimation of the reliability function after sequential experimentation, *Second International Conference on Mathematical Methods in Reliability, MMR'2000, Abstract Book,* **2** (Eds., M. Nikulin and N. Limnios), pp. 788–791. Université Victor Segalen, Bordeaux.

19. Mukhopadhyay, N. and Datta, S. (1996). On sequential fixed-width confidence intervals for the mean and second-order expansions of the associated coverage probabilities, *Annals of the Institute of Statistical Mathematics*, **48**, 497–507.

20. Mukhopadhyay, N. and Duggan, W. T. (2000). New results on two-stage estimation of the mean of an exponential distribution. In *Perspectives in Statist. Sci.* (Eds., A. K. Basu, J. K. Ghosh, P. K. Sen and B. K. Sinha), pp. 219–231, Oxford University Press, New Delhi.

21. Mukhopadhyay, N., Padmanabhan, A. R. and Solanky, T. K. S. (1997). On estimation of the reliability after sequentially estimating the mean: the exponential case, *Metrika*, **45**, 235–252.

22. Mukhopadhyay, N. and Purkayastha, S.(1994). On sequential estimation of the difference of means, *Statistics & Decisions*, **12**, 41–52.

23. Olkin, I., Gleser, L. and Derman, C. (1978). *Probability Models and Applications*, Macmillan Publishing, New York.

24. Proschan, F. (1963). Theoretical explanation of observed decrease failure rate, *Technometrics*, **5**, 375–384.

25. Robbins, H. (1959). Sequential estimation of the mean of a normal population, In *Probability and Statistics, H. Cramer Volume* (Ed., U. Grenander), pp. 235–245, Almquist Wiksell, Uppsala.

26. Robbins, H., Simons, G. and Starr, N. (1967). A sequential analog of the Behrens-Fisher problem, *Annals of Mathematical Statistics*, **38**, 1384–1391.

27. Sobel, M. (1956). Statistical techniques for reducing experiment time in reliability, *Bell. System Tech. J.*, **36**, 179–202.

28. Starr, N. and Woodroofe, M. (1972). Further remarks on sequential estimation: the exponential case, *Annals of Mathematical Statistics*, **43**, 1147–1154.

29. Thomas, G. and Finney, R. (1990). *Calculus and Analytic Geometry*, 7th edition, Addison-Wesley, New York.

30. Tong, H. (1977). On estimation of $P(Y < X)$ for exponential families, *IEEE Transactions on Reliability*, **26**, 54–57.

31. Woodroofe, M. (1977). Second-order approximations for sequential point and interval estimation, *Annals of Statistics*, **5**, 984–995.

32. Zehna, P..(1966). Invariance of maximum likelihood estimators, *Annals of Mathematical Statistics*, **37**, 3, 744.

17

Empirical Bayes Estimation of Mean Lifetime for an Exponential Distribution: Unequal Sample Sizes Case

Tachen Liang

Department of Mathematics, Wayne State University, Detroit, MI U.S.A.

Abstract: In this chapter, we study the empirical Bayes estimation of the mean lifetime θ in an exponential distribution with unequal sample sizes. It is assumed that θ is in the interval $[a, b]$ where $0 < a < b < \infty$. We investigate a method for constructing an empirical Bayes estimator $\varphi^*_{n+1,n}$ under unequal sample sizes situation. The asymptotic optimality of $\varphi^*_{n+1,n}$ is studied. We have proved that $\varphi^*_{n+1,n}$ is asymptotically optimal, and its regret converges to zero at a rate $O\left((\ln n)^{3M-1}/n\right)$ when both a and b are known, or at a rate $O\left((\ln n)^{3M-1}(\ln\ln n)^2/n\right)$ when both a and b are unknown, where M is an upper bound of sample sizes.

Keywords and phrases: Asymptotically optimal, empirical Bayes, rate of convergence, regret, unequal sample sizes

17.1 Introduction

In the empirical Bayes context as introduced by Robbins (1956, 1964), one considers a sequence of statistical problems, called the component problems, which have the same generic structure. In the ith component problem, let X_i be a random variable having a probability density $f_i(x|\theta_i)$, $x \in \mathcal{X}_i$, $\theta_i \in \Omega$, where $\mathcal{X}_i$ is the sample space of X_i. X_i may be viewed as a sufficient statistic for the parameter θ_i based on a sample of size $m(i)$. One is interested in the nature of state of the parameter θ_i. Let $\mathcal{A}$ be an action space, and $L \geq 0$ be a loss function defined on $\Omega \times \mathcal{A}$. A decision procedure δ is a mapping from $\mathcal{X}_i$ into $\mathcal{A}$ such that $\delta(x_i)$ is an action about the parameter θ_i when $X_i = x_i$ is observed.

It is assumed that θ_i is a realization of a random variable Θ_i having a prior distribution G. Thus, the Bayes risk associated with the decision procedure δ is:

$$R_i(G,\delta) = \int_\Omega \int_{\mathcal{X}_i} L(\theta, \delta(x)) f_i(x|\theta)\, dx dG(\theta).$$

Let C be the class of all decision procedures. Thus, $R_i(G) = \min_{\delta \in C} R_i(G,\delta)$ is the minimum Bayes risk. A decision procedure δ_G such that $R_i(G,\delta_G) = R_i(G)$ is called a Bayes procedure with respect to the prior distribution G. When G is unknown, it is not possible to implement the Bayes procedure δ_G.

In the standard empirical Bayes framework, it is assumed that the sample sizes $m(1), m(2)...,$ are all equal and $X_1, X_2...,$ are marginally mutually independent and identically distributed. At the present stage, say stage $n+1$, we let $X_1, ..., X_n$ denote the n past data and X_{n+1} stand for the present random observation. Let θ_{n+1} be a realized value of the current random variable Θ_{n+1}. One shall take a decision about the nature of state of the present parameter θ_{n+1} based on the present observation $X_{n+1} = x$ and the n past data $\underset{\sim}{X}(n) = (X_1, ..., X_n)$. An empirical Bayes procedure $\delta_{n+1,n}(x, \underset{\sim}{X}(n)) \equiv \delta_{n+1,n}(x)$ is then considered to be an action about the parameter θ_{n+1}. The Bayes risk of the empirical Bayes procedure $\delta_{n+1,n}$ for the $(n+1)$-st component decision problem is:

$$R_{n+1}(G, \delta_{n+1,n}) = \int_\Omega \int_{\mathcal{X}_{n+1}} E_n\left[L(\theta, \delta_{n+1,n}(x))\right] f_{n+1}(x|\theta)\, dx dG(\theta),$$

where the expectation E_n is taken with respect to $\underset{\sim}{X}(n)$. Since $R_{n+1}(G)$ is the minimum Bayes risk for the $(n+1)$-st component problem, $R_{n+1}(G, \delta_{n+1,n}) - R_{n+1}(G) \geq 0$ for all n. A sequence of empirical Bayes procedures $\{\delta_{n+1,n}\}_{n=1}^{\infty}$ is said to be asymptotically optimal if $\lim_{n\to\infty} [R_{n+1}(G, \delta_{n+1,n}) - R_{n+1}(G)] = 0$.

In the practical applications, however, components with equal sample sizes are rare. For example, we may often have samples of different sizes from different component problems. Thus, it is essential to generalize the standard empirical Bayes approach to unequal sample sizes situation. In the literature, certain work has been done on component problems with unequal sample sizes. For example, see O'Bryan (1976, 1979), O'Bryan and Susarla (1976a,b), Stijnen and van Houwelingen (1990), van Houwelingen and Stijnen (1993), and Datta (2000). As pointed out by several authors, [for example, see O'Bryan (1976)], generalizations of the standard empirical Bayes approach beyond identical components is not easy.

In this paper, we study the empirical Bayes estimation of the mean lifetime for an exponential distribution with unequal sample sizes components. The paper is organized as follows. We introduce the concerned empirical Bayes estimation problem in Section 17.2. In Section 17.3, we study a method for constructing empirical Bayes estimators, and an empirical Bayes estimator $\varphi^*_{n+1,n}$ is proposed. The associated asymptotic optimality of $\varphi^*_{n+1,n}$ is investigated in Section 17.4. Assuming that the mean lifetime is in the interval $[a,b]$,where $0 < a < b < \infty$, we have proved that $\varphi^*_{n+1,n}$ is asymptotically optimal, and its regret converges to zero at a rate $O\left((\ln n)^{3M-1}/n\right)$ when both a and b are known, or at a rate $O\left((\ln n)^{3M-1}(\ln\ln n)^2/n\right)$ when both a and b are unknown, where M is a known, positive integer.

17.2 The Empirical Bayes Estimation Problem

First, we introduce the empirical Bayes framework for the concerned estimation problem involving unequal sample sizes components. At stage i, let $X_{i1},...,X_{i,m(i)}$ be a sample of size $m(i)$ obtained from an exponential distribution with mean lifetime θ_i, where θ_i is a realization of a positive random variable Θ_i. It is assumed that $\left(X_{i1},...,X_{i,m(i)},\Theta_i\right)$, $i = 1,2,...$ are mutually independent, and $\Theta_1, \Theta_2, ...$ are identically distributed, having the common, unknown prior distribution G. Let $Y_i = \sum_{j=1}^{m(i)} X_{ij}$. Then Y_i is a sufficient statistic for the parameter θ_i. Given θ_i, Y_i follows a gamma distribution with a probability density

$$f\left(y|m(i),\theta_i\right) = \frac{y^{m(i)-1}}{\Gamma(m(i))\theta_i^{m(i)}}\exp\left(-y/\theta_i\right). \tag{17.1}$$

Let $u\left(y|m(i)\right) = \frac{y^{m(i)-1}}{\Gamma(m(i))}$, $c\left(\theta_i|m(i)\right) = \frac{1}{\theta_i^{m(i)}}$.

At the present stage, say stage $n+1$, Y_{n+1} is the present random observation and $\underset{\sim}{Y}(n) = (Y_1,...,Y_n)$ is the n past data. We shall estimate the parameter θ_{n+1} using the squared error loss. For convenience, in the following, we use $m \equiv m(n+1)$ to denote the sample size of the present stage. Suppose that $\int_0^\infty \theta^2 dG(\theta) < \infty$. If G were known, under the squared error loss, the Bayes estimator for θ_{n+1} given $Y_{n+1} = y$ is the posterior mean

$$\varphi_{n+1,G}(y|m) = E\left[\Theta_{n+1}|Y_{n+1} = y\right] = \frac{\psi_G(y|m)}{f_G(y|m)}, \tag{17.2}$$

where

$$\begin{aligned} f_G(y|m) &= \int_0^\infty f(y|m,\theta)\,dG(\theta) = u(y|m)\int c(\theta|m)\exp(-y/\theta)\,dG(\theta) \\ &= u(y|m)\,\alpha_G(y|m), \end{aligned} \tag{17.3}$$

$$\alpha_G(y|m) = \int c(\theta|m) \exp(-y/\theta)\, dG(\theta) = \int \frac{1}{\theta^m} \exp(-y/\theta)\, dG(\theta), \quad (17.4)$$

$$\psi_G(y|m) = \int \theta f(y|m,\theta)\, dG(\theta) = u(y|m) \int \frac{\theta}{\theta^m} \exp(-y/\theta)\, dG(\theta). \quad (17.5)$$

The minimum Bayes risk for the $(n+1)$-st component estimation problem is:

$$R_{n+1}(G) = R_{n+1}(G, \varphi_{n+1,G}) = E_{(Y_{n+1},\Theta_{n+1})} \left[\varphi_{n+1,G}(Y_{n+1}|m) - \Theta_{n+1}\right]^2. \quad (17.6)$$

The expectation $E_{(Y_{n+1},\Theta_{n+1})}$ is taken with respect to (Y_{n+1}, Θ_{n+1}).

Note that the Bayes estimator $\varphi_{n+1,G}(y|m)$ is a function of the prior distribution G. When G is unknown, it is not possible to implement the Bayes estimator $\varphi_{n+1,G}(y|m)$ for practical applications. Following the empirical Bayes idea of Robbins (1956,1964), we may incorporate information from the past data $\underset{\sim}{Y}(n)$ to construct a "good" estimator, say $\varphi_{n+1,n}(y, \underset{\sim}{Y}(n)) = \varphi_{n+1,n}(y)$, for the present parameter θ_{n+1}, where y is an observed value of Y_{n+1}. The Bayes risk of $\varphi_{n+1,n}$ is:

$$R_{n+1}(G, \varphi_{n+1,n}) = E_n E_{(Y_{n+1},\Theta_{n+1})} \left[\varphi_{n+1,n}(Y_{n+1}|m) - \Theta_{n+1}\right]^2. \quad (17.7)$$

Since $\varphi_{n+1,G}(y|m)$ is the Bayes estimator for the $(n+1)$-st component problem, $R_{n+1}(G, \varphi_{n+1,n}) - R_{n+1}(G, \varphi_{n+1,G}) \geq 0$ for all n. A sequence of empirical Bayes estimators $\{\varphi_{n+1,n}\}_{n=1}^{\infty}$ is said to be asymptotically optimal, relative to the prior distribution G, at a rate $O(\varepsilon_n)$ if $R_{n+1}(G, \varphi_{n+1,n}) - R_{n+1}(G, \varphi_{n+1,G}) = O(\varepsilon_n)$, where $\{\varepsilon_n\}$ is a sequence of positive, decreasing numbers such that $\lim_{n\to\infty} \varepsilon_n = 0$.

In the following, we seek a way to construct empirical Bayes estimators possessing the desired asymptotic optimality. Throughout the chapter, we assure that Θ_i, $i = 1, 2, \ldots$, are bounded random variables, such that $0 < a \leq \Theta_i \leq b < \infty$, where the values of a and b may be known or unknown. We also assume the sample sizes $m(j)$, $j = 1, 2, \ldots$, satisfying $1 \leq m(j) \leq M < \infty$ for some finite, known integer M.

17.3 Construction of Empirical Bayes Estimators

17.3.1 Kernel function

Let J_v denote the Bessel function of the first kind of order v. Define

$$K(t) = \frac{1}{\sqrt{t}} J_1\left(2\sqrt{t}\right) I(t) \quad (17.8)$$

where $I(t) = 1$ if $t > 0$, and 0 otherwise. Then,

$$K^{(j)}(t) = (-1)^j \frac{1}{\sqrt{t^{j+1}}} J_{j+1}\left(2\sqrt{t}\right) I(t), \; j = 1, 2, ... \tag{17.9}$$

The kernel K has been used by Pensky and Singh (1999) for empirical Bayes estimation of reliability characteristics in gamma family distributions and by Liang (2002) for an empirical Bayes testing problem in a positive exponential family. The kernel K has the properties that [see Gradshteyn and Ryzhik (1994)],

$$K^{(j)}(0) = \frac{(-1)^j}{(j+1)!}, \int_0^\infty \left[K^{(j)}(t)\right]^2 dt = \frac{1}{(j!)^2 (2j+1)}, \tag{17.10}$$

$$\int_0^\infty K^{(j)}(t) \exp\left(-\frac{t}{z}\right) dt = \sum_{i=0}^{j} \frac{(-1)^{j-i}}{(j-i)! z^i} - \frac{1}{z^j} e^{-z}, z > 0. \tag{17.11}$$

Define double arrays $\{(a_{kj}); k = 0, 1, 2, ...; j = 0, 1, ..., k\}$ as follows:

$$a_{kk} = 1, k = 0, 1, ...; a_{10} = 1, \text{and for } k \geq 2 \text{ and each } l = 0, 1, ..., k-1, \\ \sum_{j=l}^{k} \frac{(-1)^{j-l}}{(j-l)!} a_{kj} = 0. \tag{17.12}$$

Note that for each k, the values of a_{kl}, $l = k-1, k-2, ...$ can be obtained recursively. Solving (17.12), we obtain: $a_{kl} = \frac{1}{(k-l)!}, l = 0, 1, .., k;\ k = 0, 1, 2, ...$

17.3.2 The proposed empirical Bayes estimators

Under the assumption that $0 < a \leq \Theta_i \leq b < \infty$,we have: $a \leq \varphi_{n+1,G}(y|m) \leq b$. This property should be used when empirical Bayes estimators are constructed. Empirical Bayes estimators will be proposed according to whether the values of a and b are known or unknown. When both a and b are known, let $h \equiv h(n) = \frac{a}{\ln n}$, $a_n = a$, $b_n = b$, $c_n = \frac{\ln n}{b}$. When both a and b are unknown, let $h = \frac{2}{(\ln n)(\ln \ln n)^{1/(2M-1)}}$, $a_n = \frac{1}{\sqrt{\ln n}}$, $b_n = (\ln \ln n)^{1/(M+2)}$ and $c_n = (\ln n)(\ln \ln n)^{1/(M+2)}$.

For each $Y_{n+1} = y$, $j = 1, ..., n$, define

$$\alpha_{n+1}(y, j, m(j) - m) \\ = \begin{cases} \frac{(Y_j - y)^{m(j)-m} u(y|m) I(Y_j - y)}{u(Y_j|m(j)) \Gamma(m(j)-m)} & \text{if } m(j) - m \geq 1, \\ \sum_{i=0}^{m-m(j)} a_{m-m(j),i} \frac{I(Y_j - y) u(y|m)}{h^{m-m(j)+1} u(Y_j|m(j))} K^{(i)}\left(\frac{Y_j - y}{h}\right) & \text{if } m(j) - m \leq 0. \end{cases} \tag{17.13}$$

$$f_{n+1}(y) = \frac{1}{n} \sum_{j=1}^{n} \alpha_{n+1}(y, j, m(j) - m). \tag{17.14}$$

$$\beta_{n+1}(y,j,m(j)-m) = \begin{cases} \frac{(Y_j-y)^{m(j)-m}u(y|m)I(Y_j-y)}{u(Y_j|m(j))\Gamma(m(j)+1-m)} \text{ if } m(j)-m \geq 0, \\ \sum_{i=0}^{m-m(j)-1} a_{m-m(j)-1,i}\frac{u(y|m)I(Y_j-y)}{h^{m-m(j)}u(Y_j|m(j))}K^{(i)}\left(\frac{Y_j-y}{h}\right), \text{ otherwise.} \end{cases} \tag{17.15}$$

$$\psi_{n+1}(y) = \frac{1}{n}\sum_{j=1}^{n}\beta_{n+1}(y,j,m(j)-m). \tag{17.16}$$

From Lemmas A.2–A.3, we have:

$$|E_n f_{n+1}(y) - f_G(y|m)| \leq \frac{e^{-a/h}f_G(y|m)}{n}\sum_{j=1}^{n}\left(\frac{b}{h}\right)^{m-m(j)} I(m-m(j)+1)\sum_{i=0}^{m-m(j)}\left(\frac{h}{b}\right)^{i}, \tag{17.17}$$

$$Var(f_{n+1}(y)) \leq \frac{f_G(y|m)}{n^2}\sum_{j=1}^{n}\Big[\frac{\Gamma(m(j))}{\Gamma(m)\Gamma(m(j)-m)}\frac{1}{a}I(m(j)-m) + \frac{[m-m(j)+1]^2}{h^{2(m-m(j))+1}}(yb)^{m-m(j)}I(m-m(j)+1)\Big], \tag{17.18}$$

$$|E_n\psi_{n+1}(y) - \psi_G(y|m)| \leq \frac{e^{-a/h}f_G(y|m)}{n}\sum_{j=1}^{n}\left(\frac{b}{h}\right)^{m-m(j)} I(m-m(j))\sum_{i=0}^{m-m(j)-1}\left(\frac{h}{b}\right)^{i}, \tag{17.19}$$

$$Var(\psi_{n+1}(y)) \leq \frac{f_G(y|m)}{n^2}\sum_{j=1}^{n}\Big[\frac{\Gamma(m(j))b}{\Gamma(m(j)-m+1)\Gamma(m)}I(m(j)-m+1) + \frac{(m-m(j))^2}{h^{2(m-m(j))-1}}(yb)^{m-m(j)}I(m-m(j))\Big]. \tag{17.20}$$

Thus, $f_{n+1}(y)$ and $\psi_{n+1}(y)$ are consistent estimators of $f_G(y|m)$ and $\psi_G(y|m)$, respectively.

Now, we propose an empirical Bayes estimator as follows: For each observed $Y_{n+1}=y$, define

$$\varphi^*_{n+1,n}(y) = \begin{cases} \left(\frac{\psi_{n+1}(y)}{f_{n+1}(y)} \vee a_n\right) \wedge b_n & \text{if } 0<y\leq c_n, \\ \varphi^*_{n+1,n}(c_n) & \text{if } y>c_n. \end{cases} \tag{17.21}$$

The Bayes risk of the empirical Bayes estimator $\varphi^*_{n+1,n}$ is:

$$R_{n+1}\left(G,\varphi^*_{n+1,n}\right) = E_n E_{(Y_{n+1},\Theta_{n+1})}\left[\varphi^*_{n+1,n}\left(Y_{n+1}\right) - \Theta_{n+1}\right]^2. \tag{17.22}$$

17.4 Rate of Asymptotic Optimality

In this section, the analysis is made on the case that n is sufficiently large such that $\frac{h}{b} \leq \frac{1}{2}$, and when both a and b are unknown, $0 < a_n \leq a$ and $b_n \geq b > 0$. The empirical Bayes estimator $\varphi^*_{n+1,n}$ possesses the following asymptotic optimality.

Theorem 17.4.1 *Suppose the prior distribution G is such that $G(a) = 0$ and $G(b) = 1$ for some $0 < a < b < \infty$. Also the sample sizes $m(j)$ are such that $1 \leq m(j) \leq M$ for all j for some known integer M. Then the empirical Bayes estimator $\varphi^*_{n+1,n}$ is asymptotically optimal; and*

(a) When both a and b are known,
$R_{n+1}\left(G,\varphi^*_{n+1,n}\right) - R_{n+1}\left(G,\varphi_{n+1,G}\right) = O\left(\frac{(\ln n)^{3M-1}}{n}\right)$;

(b) When both a and b are unknown,
$R_{n+1}\left(G,\varphi^*_{n+1,n}\right) - R_{n+1}\left(G,\varphi_{n+1,G}\right) = O\left(\frac{(\ln n)^{3M-1}(\ln\ln n)^2}{n}\right)$.

PROOF. From (17.22) and (17.6), the regret of $\varphi^*_{n+1,n}$ is

$$\begin{aligned} &R_{n+1}\left(G,\varphi^*_{n+1,n}\right) - R_{n+1}\left(G,\varphi_{n+1,G}\right) \\ &= E_n E_{Y_{n+1}}\left[\varphi^*_{n+1,n}\left(Y_{n+1}\right) - \varphi_{n+1,G}\left(Y_{n+1}|m\right)\right]^2 \\ &= \int_0^\infty E_n\left[\varphi^*_{n+1,n}(y) - \varphi_{n+1,G}\left(y|m\right)\right]^2 f_G\left(y|m\right)dy. \end{aligned} \tag{17.23}$$

From Lemma A.4 and by the fact that $a \leq \varphi_G(y|m) = \frac{\psi_G(y|m)}{f_G(y|m)} \leq b \leq b_n$, it follows that

$$E_n\left[\varphi^*_{n+1,n}(y) - \varphi_{n+1,G}\left(y|m\right)\right]^2 = E_n\left[\left(\frac{\psi_{n+1}(y)}{f_{n+1}(y)} \vee a_n\right) \wedge b_n - \frac{\psi_G(y|m)}{f_G(y|m)}\right]^2$$

$$\leq \frac{2}{f_G^2(y|m)}\left\{E_n\left[\psi_{n+1}(y) - \psi_G\left(y|m\right)\right]^2 + (2b_n)^2 E_n\left[f_{n+1}(y) - f_G(y|m)\right]^2\right\}.$$

Substituting the preceding inequality into (17.23) for $0 < y \le c_n$, we obtain

$$\begin{aligned}
& R_{n+1}\left(G,\varphi^{*}_{n+1,n}\right) - R_{n+1}\left(G,\varphi_{n+1,G}\right) \\
& \le \int_0^{c_n} \frac{2}{f_G\left(y|m\right)} E_n\left[\psi_{n+1}(y) - \psi_G\left(y|m\right)\right]^2 dy \\
& \qquad + \int_0^{c_n} \frac{8b_n^2}{f_G\left(y|m\right)} E_n\left[f_{n+1}(y) - f_G\left(y|m\right)\right]^2 dy \\
& \qquad + \int_{c_n}^{\infty} E_n\left[\varphi^{*}_{n+1,n}(y) - \varphi_{n+1,G}(y)\right]^2 dy \qquad (17.24) \\
& \equiv A\left(n\right) + B\left(n\right) + C\left(n\right).
\end{aligned}$$

From (17.19)–(17.20),

$$\begin{aligned}
A\left(n\right) &= \int_0^{c_n} \frac{2}{f_G\left(y|m\right)} \left[Var\left(\psi_{n+1}\left(y\right)\right) + \left|E_n\psi_{n+1}\left(y\right) - \psi_G\left(y|m\right)\right|^2\right] dy \\
&\le \int_0^{c_n} \frac{2}{n^2} \sum_{j=1}^{n} \left[\frac{\Gamma\left(m\left(j\right)\right) bI\left(m\left(j\right) - m + 1\right)}{\Gamma\left(m\left(j\right) - m + 1\right)\Gamma\left(m\right)}\right. \\
&\quad \left. + \frac{\left(m - m\left(j\right)\right)^2}{h^{2(m-m(j))-1}} \left(yb\right)^{m-m(j)} I\left(m - m\left(j\right)\right)\right] dy \\
&\quad + \int_0^{c_n} \frac{e^{-2a/h}h^2 f_G\left(y|m\right)}{n^2} \\
&\qquad \times \left[\sum_{j=1}^{n} \left(\frac{b}{h}\right)^{m-m(j)} I\left(m - m\left(j\right)\right) \sum_{i=0}^{m-m(j)-1} \left(\frac{h}{b}\right)^{i}\right]^2 dy \\
&\le \frac{2}{n^2} \sum_{j=1}^{n} \left[\Gamma\left(M\right) bc_n I\left(m\left(j\right) - m + 1\right)\right. \\
&\qquad \left. + \frac{M^2 b^{m-m(j)}}{h^{2(m-m(j))-1}} c_n^{m-m(j)+1} I\left(m - m\left(j\right)\right)\right] \\
&\quad + \frac{e^{-2a/h}h^2}{n^2} \left[\sum_{j=1}^{n} \left(\frac{b}{h}\right)^{m-m(j)} I\left(m - m\left(j\right)\right) \sum_{i=0}^{m-m(j)-1} \left(\frac{h}{b}\right)^{i}\right]^2 \\
&= O\left(\frac{c_n^M}{nh^{2M-1}}\right). \qquad (17.25)
\end{aligned}$$

From (17.17)-(17.18),

$$
\begin{aligned}
B(n) &= \int_0^{c_n} \frac{8b_n^2}{f_G(y|m)} \left[Var\left(f_{n+1}(y)\right) + \left|E_n f_{n+1}(y) - f_G(y|m)\right|^2 \right] dy \\
&\leq \int_0^{c_n} \frac{8b_n^2}{n^2} \sum_{j=1}^{n} \left[\frac{\Gamma(M)}{a} I\left(m(j) - m\right) \right. \\
&\qquad \left. + \frac{M^2}{h^{2M-1}} (yb)^{m-m(j)} I\left(m - m(j) + 1\right) \right] dy \\
&\quad + \int_0^{c_n} \frac{8b_n^2 e^{-2a/h} f_G(y|m)}{n^2} \\
&\qquad \times \left[\sum_{j=1}^{n} \left(\frac{b}{h}\right)^{m-m(j)} I\left(m - m(j) + 1\right) \sum_{i=0}^{m-m(j)} \left(\frac{h}{b}\right)^i \right]^2 dy \\
&\leq \frac{8\Gamma(M)\, b_n^2 c_n}{na} + \frac{8b_n^2}{n^2} \times \frac{M^2}{h^{2M-1}} \sum_{j=1}^{n} c_n^{m-m(j)+1} b^{m-m(j)} I\left(m - m(j) + 1\right) \\
&\quad + \frac{8b_n^2 e^{-2a/h}}{n^2} \left[\sum_{j=1}^{n} \left(\frac{a}{h}\right)^{m-m(j)} I\left(m - m(j) + 1\right) \sum_{i=0}^{m-m(j)} \left(\frac{h}{b}\right)^i \right]^2 \\
&= O\left(\frac{b_n^2 c_n^M}{nh^{2M-1}} \right). \qquad (17.26)
\end{aligned}
$$

By Lemma A.1

$$
\begin{aligned}
C(n) &= \int_{c_n}^{\infty} E_n \left[\varphi_{n+1,n}^*(y) - \varphi_{n+1,G}(y|m) \right]^2 f_G(y|m)\, dy \\
&\leq b_n^2 \int_{c_n}^{\infty} f_G(y|m)\, dy \leq b_n^2 \int_{c_n}^{\infty} \frac{y^{m-1}}{\Gamma(m)} \times \frac{1}{a^m} e^{-y/b} dy \\
&= \frac{b_n^2}{\Gamma(m) a^m} \left[\frac{c_n^{m-1} e^{-c_n/b}}{b} + \int_{c_n}^{\infty} \frac{e^{-y/b}}{b} (m-1)\, y^{m-2} dy \right] \\
&\approx \frac{b_n^2 c_n^{m-1} e^{-c_n/b}}{b a^m \Gamma(m)} = O\left(b_n^2 c_n^{M-1} e^{-c_n/b} \right). \qquad (17.27)
\end{aligned}
$$

Combining (17.24)-(17.27) yields that

$$
R_{n+1}\left(G, \varphi_{n+1,n}^*\right) - R_{n+1}\left(G, \varphi_{n+1,G}\right) = O\left(\frac{c_n^M b_n^2}{nh^{2M-1}} \right) + O\left(b_n^2 c_n^{M-1} e^{-c_n/b} \right).
$$

When a and b are known, $h = \frac{a}{\ln n}$, $c_n = \frac{\ln n}{b}$, $b_n = b$. Hence,

$$R_{n+1}\left(G, \varphi^*_{n+1,n}\right) - R_{n+1}\left(G, \varphi_{n+1,G}\right) = O\left(\frac{(\ln n)^{3M-1}}{n}\right).$$

When a and b are unknown, $h = \frac{2}{(\ln n)(\ln\ln n)^{1/(2M-1)}}$, $c_n = (\ln n)(\ln\ln n)^{1/(M+2)}$, $b_n = (\ln\ln n)^{1/(M+2)}$. Then,

$$R_{n+1}\left(G, \varphi^*_{n+1,n}\right) - R_{n+1}\left(G, \varphi_{n+1,G}\right) = O\left(\frac{(\ln n)^{3M-1}(\ln\ln n)^2}{n}\right).$$

Thus, the proof of Theorem 17.4.1 is complete. ■

Appendix

We recall that the prior distribution G satisfies Assumption A:

Assumption A: For some $0 < a < b < \infty$, $G(a) = 0$ and $G(b) = 1$.

Lemma A.1 *Under Assumption A,* $f_G(y|m) \leq \frac{y^{m-1}}{\Gamma(m)a^m} \exp(-y/b)$.

PROOF: A straightforward computation will lead to the result immediately. ■

Lemma A.2 Under Assumption A, the following hold.

(a)

$$\begin{aligned}&\left|E_n\alpha_{n+1}(y, j, m(j) - m) - f_G(y|m)\right| \\ &\quad\leq \frac{e^{-a/h} f_G(y|m)}{h^{m-m(j)}} b^{m-m(j)} \sum_{i=0}^{m-m(j)} \left(\frac{h}{b}\right)^i I(m - m(j) + 1),\end{aligned}$$

$$\begin{aligned}&\left|E_n f_{n+1}(y) - f_G(y|m)\right| \\ &\quad\leq \frac{e^{-a/h} f_G(y|m)}{n} \sum_{j=1}^{n} \frac{b^{m-m(j)}}{h^{m-m(j)}} I(m - m(j) + 1) \sum_{i=0}^{m-m(j)} \left(\frac{h}{b}\right)^i.\end{aligned}$$

(b)

$$\begin{aligned}&\left|E_n\beta_{n+1}(y, j, m(j) - m) - \psi_G(y|m)\right| \\ &\quad\leq \frac{e^{-a/h} f_G(y|m)\, b^{m-m(j)}}{h^{m-m(j)-1}} I(m - m(j)) \sum_{i=0}^{m-m(j)-1} \left(\frac{h}{b}\right)^i,\end{aligned}$$

$$
\begin{aligned}
&|E_n\psi_{n+1}(y) - \psi_G(y|m)| \\
&\le \frac{e^{-a/h} f_G(y|m)}{n} \sum_{j=1}^{n} \frac{b^{m-m(j)}}{h^{m-m(j)-1}} I(m-m(j)) \sum_{i=0}^{m-m(j)-1} \left(\frac{h}{b}\right)^i .
\end{aligned}
$$

PROOF. (a) Let $l = |m(j) - m|$. If $m(j) - m \ge 1$, then $l = m(j) - m$, $m(j) - l = m$.

$$
\begin{aligned}
E_n\alpha_{n+1}(y, j, m(j) - m) &= E_n\left[\frac{(Y_j - y)^{l-1} u(y|m) I(Y_j - y)}{u(Y_j|m(j)) \Gamma(l)}\right] \\
&= \int_{t=y}^{\infty} \frac{(t-y)^{l-1} u(y|m)}{\Gamma(l)} \alpha_G(t|m(j))\, dt \\
&= \int_{s=0}^{\infty} \frac{s^{l-1} u(y|m)}{\Gamma(l)} \left[\int c(\theta|m(j)) \exp(-(s+y)/\theta)\, dG(\theta)\right] ds \\
&= \int_{\theta=0}^{\infty} u(y|m) c(\theta|m(j)) \exp(-y/\theta) \left[\int_{s=0}^{\infty} \frac{s^{l-1}}{\Gamma(l)} \exp(-s/\theta)\, ds\right] dG(\theta) \\
&= \int u(y|m) c(\theta|m(j))\theta^l dG(\theta) = \int u(y|m) c(\theta|m)\, dG(\theta) = f_G(y|m),
\end{aligned} \tag{17.28}
$$

since $c(\theta|m(j)) \times \theta^l = \frac{1}{\theta^{m(j)-l}} = \frac{1}{\theta^m} = c(\theta|m)$.

If $m(j) - m \le 0$, then $l = m - m(j)$, and

$$
\begin{aligned}
E_n\alpha_{n+1}(y, j, m(j) - m) &= E_n\left[\sum_{i=0}^{l} a_{l,i} \tfrac{I(Y_j - y) u(y|m)}{h^{l+1} u(Y_j|m(j))} K^{(i)}\left(\tfrac{Y_j - y}{h}\right)\right] \\
&= \sum_{i=0}^{l} a_{l,i} E_n\left[\tfrac{I(Y_j - y) u(y|m)}{h^{l+1} u(Y_j|m(j))} K^{(i)}\left(\tfrac{Y_j - y}{h}\right)\right],
\end{aligned} \tag{17.29}
$$

where

$$
\begin{aligned}
&E_n\left[\frac{I(Y_j - y) u(y|m)}{h^{l+1} u(Y_j|m(j))} K^{(i)}\left(\frac{Y_j - y}{h}\right)\right] \\
&= \int_{t=y}^{\infty} \frac{u(y|m)}{h^{l+1}} K^{(i)}\left(\frac{t-y}{h}\right) \alpha_G(t|m(j))\, dt \\
&= \int_{s=0}^{\infty} \frac{u(y|m)}{h^{l+1}} K^{(i)}(s) \left[\int_\theta c(\theta|m(j)) \exp(-(y+hs)/\theta)\, dG(\theta)\right] h ds \\
&= \int_\theta \frac{u(y|m)}{h^l} c(\theta|m(j)) \exp(-y/\theta) \left[\int_{s=0}^{\infty} K^{(i)}(s) \exp\left(-\frac{hs}{\theta}\right) ds\right] dG(\theta) \\
&= \frac{1}{h^l} \int_\theta u(y|m) c(\theta|m(j)) \exp(-y/\theta) \\
&\qquad \times \left[\sum_{k=0}^{i} \frac{(-1)^{i-k}}{(i-k)!} \left(\frac{h}{\theta}\right)^k - \left(\frac{h}{\theta}\right)^i \exp\left(-\frac{\theta}{h}\right)\right] dG(\theta).
\end{aligned} \tag{17.30}
$$

Substituting (17.30) into (17.29) and noting that $\sum_{i=k}^{l} \frac{(-1)^{i-k}}{(i-k)!} a_{li} = 0$ for each $k = 0, 1, ..., l-1$, and $= 1$ for $k = l$, it follows that

$$
\begin{aligned}
& E_n \left[\alpha_{n+1} \left(y, j, m\left(j\right) - m\right)\right] \\
& = \sum_{i=0}^{l} \frac{a_{li}}{h^l} \int_\theta u\left(y|m\right) c\left(\theta|m\left(j\right)\right) e^{-y/\theta} \\
& \qquad \times \left[\sum_{k=0}^{i} \frac{(-1)^{i-k}}{(i-k)!} \left(\frac{h}{\theta}\right)^k - \left(\frac{h}{\theta}\right)^i e^{-\frac{\theta}{h}}\right] dG\left(\theta\right) \\
& = \frac{1}{h^l} \int_\theta u\left(y|m\right) c\left(\theta|m\left(j\right)\right) e^{-y/\theta} \\
& \qquad \times \left[\sum_{i=0}^{l} a_{li} \sum_{k=0}^{i} \frac{(-1)^{i-k}}{(i-k)!} \left(\frac{h}{\theta}\right)^k - \sum_{i=0}^{l} a_{li} \left(\frac{h}{\theta}\right)^i e^{-\theta/h}\right] dG\left(\theta\right) \\
& = \frac{1}{h^l} \int_\theta u\left(y|m\right) c\left(\theta|m\left(j\right)\right) e^{-y/\theta} \\
& \qquad \times \left[\sum_{k=0}^{l} \left[\sum_{i=k}^{l} a_{li} \frac{(-1)^{i-k}}{(i-k)!}\right] \left(\frac{h}{\theta}\right)^k - \sum_{i=0}^{l} a_{li} \left(\frac{h}{\theta}\right)^i e^{-\theta/h}\right] dG\left(\theta\right) \\
& = \frac{1}{h^l} \int_\theta u\left(y|m\right) c\left(\theta|m\left(j\right)\right) e^{-y/\theta} \left[\left(\frac{h}{\theta}\right)^l - \sum_{i=0}^{l} a_{li} \left(\frac{h}{\theta}\right)^i e^{-\theta/h}\right] dG\left(\theta\right) \\
& = \int u\left(y|m\right) \frac{c\left(\theta|m\left(j\right)\right)}{\theta^l} e^{-y/\theta} dG\left(\theta\right) \\
& \qquad - \sum_{i=0}^{l} a_{li} \int_\theta u\left(y|m\right) \frac{c\left(\theta|m\left(j\right)\right)}{\theta^i} e^{-y/\theta} \times \frac{1}{h^{l-i}} e^{-\theta/h} dG\left(\theta\right) \\
& = \int u\left(y|m\right) c\left(\theta|m\right) e^{-y/\theta} dG\left(\theta\right) - \sum_{i=0}^{l} \frac{a_{li}}{h^{l-i}} \int_\theta \frac{u\left(y|m\right)}{\theta^{m(j)+i}} e^{-y/\theta} e^{-\theta/h} dG\left(\theta\right) \\
& = f_G\left(y|m\right) - \sum_{i=0}^{l} a_{li} \int \frac{u\left(y|m\right)}{\theta^{m(j)+i} h^{l-i}} e^{-y/\theta} e^{-\theta/h} dG\left(\theta\right). \qquad (17.31)
\end{aligned}
$$

Here,

$$\begin{aligned} 0 &< \int \frac{u(y|m)}{\theta^{m(j)+i}} \times \frac{e^{-y/\theta} e^{-\theta/h}}{h^{l-i}} dG(\theta) \\ &\leq \frac{e^{-a/h}}{h^{l-i}} \int \frac{u(y|m)}{\theta^m} \theta^{m-m(j)-i} e^{-y/\theta} dG(\theta) \\ &\leq \frac{e^{-a/h} b^{m-m(j)-i}}{h^{l-i}} \int \frac{u(y|\theta)}{\theta^m} e^{-y/\theta} dG(\theta) \\ &= \frac{e^{-a/h} b^{m-m(j)-i}}{h^{l-i}} f_G(y|m). \end{aligned} \tag{17.32}$$

Since $a_{li} = \frac{1}{(l-i)!}$, thus,

$$\begin{aligned} 0 &< \sum_{i=0}^{l} a_{li} \int \frac{u(y|m) e^{-y/\theta} e^{-\theta/h}}{\theta^{m(j)+i} h^{l-i}} dG(\theta) \leq \sum_{i=0}^{l} \frac{e^{-a/h} b^{m-m(j)-i}}{h^{l-i}} f_G(y|m) \\ &= \frac{e^{-a/h} f_G(y|m)}{h^l} b^{m-m(j)} \sum_{i=0}^{l} \left(\frac{h}{b}\right)^i. \end{aligned} \tag{17.33}$$

Combining (17.28)–(17.33), we obtain

$$\begin{aligned} &|E_n \alpha_{n+1}(y, j, m(j) - m) - f_G(y|m)| \\ &\leq \frac{e^{-a/h} f_G(y|m) b^{m-m(j)}}{h^l} I(m - m(j) + 1) \sum_{i=0}^{m-m(j)} \left(\frac{h}{b}\right)^i. \end{aligned}$$

Next,

$$\begin{aligned} |E_n f_{n+1}(y) - f_G(y|m)| &= \left| \frac{1}{n} \sum_{j=1}^{n} E_n \alpha_{n+1}(y, j, m(j) - m) - f_G(y|m) \right| \\ &\leq \frac{1}{n} \sum_{j=1}^{n} |E_n \alpha_{n+1}(y, j, m(j) - m) - f_G(y|m)| \\ &\leq \frac{e^{-a/h} f_G(y|m)}{nh^l} \sum_{j=1}^{n} b^{m-m(j)} I(m - m(j) + 1) \sum_{i=0}^{m-m(j)} \left(\frac{h}{b}\right)^i. \end{aligned}$$

Hence, the proof of part (a) is complete.

Part (b) can be proved in a similar discussion. The detail is thus omitted. ■

Lemma A.3 *Under Assumption A, the following hold.*

(a)

$$
\begin{aligned}
&Var\left(\alpha_{n+1}\left(y, j, m\left(j\right)-m\right)\right) \\
&\quad \leq \frac{\Gamma\left(m\left(j\right)\right)}{\Gamma\left(m\right)\Gamma\left(m\left(j\right)-m\right)a} f_G\left(y|m\right) I\left(m\left(j\right)-m\right) \\
&\qquad + \frac{\left[m-m\left(j\right)+1\right]^2}{h^{2\left(m-m\left(j\right)\right)+1}}\left(yb\right)^{m-m\left(j\right)} f_G\left(y|m\right) I\left(m-m\left(j\right)+1\right).
\end{aligned}
$$

$$
\begin{aligned}
&Var\left(f_{n+1}\left(y\right)\right) \\
&\quad \leq \frac{f_G\left(y|m\right)}{n^2}\sum_{j=1}^{n}\frac{\Gamma\left(m\left(j\right)\right)}{\Gamma\left(m\right)\Gamma\left(m\left(j\right)-m\right)a} I\left(m\left(j\right)-m\right) \\
&\qquad + \frac{f_G\left(y|m\right)}{n^2}\sum_{j=1}^{n}\frac{\left[\left(m-m\left(j\right)\right)+1\right]^2}{h^{2\left(m-m\left(j\right)\right)+1}}\left(yb\right)^{m-m\left(j\right)} I\left(m-m\left(j\right)+1\right).
\end{aligned}
$$

(b)

$$
\begin{aligned}
&Var\left(\beta_{n+1}(y, j, m\left(j\right)-m)\right) \\
&\quad \leq \frac{\Gamma\left(m\left(j\right)\right) b}{\Gamma\left(m\left(j\right)-m+1\right)\Gamma\left(m\right)} f_G\left(y|m\right) I\left(m\left(j\right)-m+1\right) \\
&\qquad + \frac{\left[m-m\left(j\right)\right]^2}{h^{2\left(m-m\left(j\right)\right)-1}}\left(yb\right)^{m-m\left(j\right)} f_G\left(y|m\right) I\left(m-m\left(j\right)\right).
\end{aligned}
$$

$$
\begin{aligned}
&Var\left(\psi_{n+1}(y)\right) \\
&\quad \leq \frac{f_G\left(y|m\right)}{n^2}\sum_{j=1}^{n}\frac{\Gamma\left(m\left(j\right)\right) b}{\Gamma\left(m\left(j\right)-m+1\right)\Gamma\left(m\right)} I\left(m\left(j\right)-m+1\right) \\
&\qquad + \frac{f_G\left(y|m\right)}{n^2}\sum_{j=1}^{n}\frac{\left(m-m\left(j\right)\right)^2}{h^{2\left(m-m\left(j\right)\right)-1}}\left(yb\right)^{m-m\left(j\right)} I\left(m-m\left(j\right)\right).
\end{aligned}
$$

PROOF. (a) Let $l = |m(j) - m|$. As $m(j) - m \geq 1$, $l = m(j) - m$, and

$$
\begin{aligned}
Var\left(\alpha_{n+1}(y, j, m(j) - m)\right) &= Var\left(\frac{(Y_j - y)^{l-1} u(y|m) I(Y_j - y)}{u(Y_j|m(j)) \Gamma(l)}\right) \\
&\leq E_n\left[\frac{(Y_j - y)^{l-1} u(y|m) I(Y_j - y)}{u(Y_j|m(j)) \Gamma(l)}\right]^2 \\
&= \int_{t=y}^{\infty} \frac{u^2(y|m)(t-y)^{2(l-1)}}{u(t|m(j)) \Gamma^2(l)} \left[\int_\theta c(\theta|m(j)) e^{-t/\theta} dG(\theta)\right] dt \\
&= u(y|m) \int_{s=0}^{\infty} \frac{u(y|m) s^{2(l-1)}}{u(y+s|m(j)) \Gamma^2(l)} \left[\int c(\theta|m(j)) e^{-(y+s)/\theta} dG(\theta)\right] ds \\
&= u(y|m) \int_\theta c(\theta|m(j)) e^{-y/\theta} \frac{\Gamma(m(j))}{\Gamma(m)\Gamma^2(l)} \left[\int_{s=0}^{\infty} \frac{y^{m-1} s^{2(l-1)}}{(y+s)^{m(j)-1}} e^{-s/\theta} ds\right] dG(\theta) \\
&\leq u(y|m) \int c(\theta|m(j)) e^{-y/\theta} \frac{\Gamma(m(j))}{\Gamma(m)\Gamma^2(l)} \Gamma(l-1) \theta^{l-1} dG(\theta) \\
&= \frac{\Gamma(m(j)) \Gamma(l-1)}{\Gamma(m)\Gamma^2(l)} \int \frac{u(y|m)}{\theta^{m+1}} e^{-y/\theta} dG(\theta) \qquad (17.34) \\
&\leq \frac{\Gamma(m(j))}{\Gamma(m)\Gamma(l) a} \int \frac{u(y|m)}{\theta^m} e^{-y/\theta} dG(\theta) \\
&= \frac{\Gamma(m(j))}{\Gamma(m)\Gamma(m(j) - m) a} f_G(y|m).
\end{aligned}
$$

In (17.34), the second inequality is obtained based on the fact that

$$
\begin{aligned}
\int_{s=0}^{\infty} \frac{y^{m-1} s^{2(l-1)}}{(y+s)^{m(j)-1}} e^{-s/\theta} ds &= \int_{s=0}^{\infty} \frac{y^{m-1} s^{2(l-1)}}{(y+s)^{m-1} (y+s)^l} e^{-s/\theta} ds \\
\leq \int_{s=0}^{\infty} \frac{y^{m-1} s^{2(l-1)}}{y^{m-1} s^l} e^{-s/\theta} ds &= \int_{s=0}^{\infty} s^{l-2} e^{-s/\theta} ds = \Gamma(l-1) \theta^{l-1}.
\end{aligned}
$$

As $m(j) - m \leq 0$, $l = m - m(j)$, and

$$
\begin{aligned}
& Var\left(\alpha_{n+1}(y, j, m(j) - m)\right) \\
&= Var\left(\sum_{i=0}^{l} a_{li} \frac{I(Y_j - y) u(y|m)}{h^{l+1} u(Y_j|m(j))} K^{(i)}\left(\frac{Y_j - y}{h}\right)\right) \\
&\leq E_n\left[\sum_{i=0}^{l} a_{li} \frac{u(y|m) I(Y_j - y)}{h^{l+1} u(Y_j|m(j))} K^{(i)}\left(\frac{Y_j - y}{h}\right)\right]^2 \\
&\leq \frac{(l+1)}{h^{2(l+1)}} \sum_{i=0}^{l} E_n\left[a_{li} \frac{u(y|m) I(Y_j - y)}{u(Y_j|m(j))} K^{(i)}\left(\frac{Y_j - y}{h}\right)\right]^2. \qquad (17.35)
\end{aligned}
$$

Since $a_{li} = \frac{1}{(l-i)!}$, in (17.35),

$$\begin{aligned}
&E_n\left[a_{li}\frac{u(y|m)\,I(Y_j-y)}{u(Y_j|m(j))}K^{(i)}\left(\frac{Y_j-y}{h}\right)\right]^2 \\
\le\;& \int_{t=y}^{\infty}\frac{u^2(y|m)}{u(t|m(j))}\left[K^{(i)}\left(\frac{t-y}{h}\right)\right]^2\alpha_G(t|m(j))\,dt \\
=\;& \int_{s=0}^{\infty}\frac{u^2(y|m)}{u(y+sh|m(j))}\left[K^{(i)}(s)\right]^2\int_\theta c(\theta|m(j))\,e^{-(y+sh)/\theta}dG(\theta)\,hds \\
=\;& hu(y|m)\int c(\theta|m(j))\,e^{-y/\theta} \\
&\times\left[\int_{s=0}^{\infty}\frac{u(y|m)}{u(y+sh|m(j))}\left[K^{(i)}(s)\right]^2e^{-sh/\theta}ds\right]dG(\theta) \\
\le\;& hu(y|m)\int c(\theta|m(j))\,e^{-y/\theta}\cdot y^{m-m(j)}dG(\theta) \\
=\;& hy^{m-m(j)}\int\frac{u(y|m)}{\theta^m}\theta^{m-m(j)}e^{-y/\theta}dG(\theta)\le hy^{m-m(j)}b^{m-m(j)}f_G(y|m).
\end{aligned} \tag{17.36}$$

In (17.36), the second inequality is obtained due to the facts that $m-m(j)\ge 0$ and

$$\begin{aligned}
&\int_{s=0}^{\infty}\frac{u(y|m)}{u(y+sh|m(j))}\left[K^{(i)}(s)\right]^2e^{-hs/\theta}ds \\
&\quad=\int_{s=0}^{\infty}\frac{y^{m-1}}{(y+sh)^{m(j)-1}}\left[K^{(i)}(s)\right]^2ds \\
&\quad\le y^{m-m(j)}\int_{s=0}^{\infty}\left[K^{(i)}(s)\right]^2ds\le y^{m-m(j)},
\end{aligned}$$

since $\int_{s=0}^{\infty}\left[K^{(i)}(s)\right]^2ds=\frac{1}{(i!)^2(2i+1)}\le 1$. Substituting (17.36) into (17.35), we obtain

$$\begin{aligned}
Var(\alpha_{n+1}(y,j,m(j)-m)) &\le \frac{l+1}{h^{2(l-1)}}\sum_{i=0}^{l}hy^{m-m(j)}b^{m-m(j)}f_G(y|m) \\
&\le \frac{(l+1)^2}{h^{2l+1}}y^{m-m(j)}b^{m-m(j)}f_G(y|m) \\
&= \frac{[m-m(j)+1]^2}{h^{2(m-m(j))+1}}y^{m-m(j)}b^{m-m(j)}f_G(y|m).
\end{aligned} \tag{17.37}$$

From (17.34) and (17.37), we have

$$
\begin{aligned}
&Var\left(\alpha_{n+1}\left(y, j, m\left(j\right)-m\right)\right) \\
&\quad\leq \frac{\Gamma\left(m\left(j\right)\right)}{\Gamma\left(m\right)\Gamma\left(m\left(j\right)-m\right)a} f_G\left(y|m\right) I\left(m\left(j\right)-m\right) \\
&\quad\quad+\frac{\left[m-m\left(j\right)+1\right]^2}{h^{2(m-m(j))+1}} y^{m-m(j)} b^{m-m(j)} f_G\left(y|m\right) I\left(m-m\left(j\right)+1\right).
\end{aligned}
$$

Thus, the proof of part (a) is complete.

Part (b) can be obtained in a similar discussion. The detail is omitted. ■

The following lemma is cited from Singh and Wei (1992).

Lemma A.4 *Let Y, Z be random variables and y, $z \neq 0$, $C > 0$ be real numbers. Then, for $0 < \lambda \leq 2$, the following hold:*

$$
E\left[\left|\frac{Y}{Z}-\frac{y}{z}\right| \wedge C\right]^{\lambda} \leq \frac{2}{|z|^{\lambda}}\left\{E\left[|Y-y|^{\lambda}\right]+\left(\left|\frac{y}{z}\right|+C\right)^{\lambda} E\left[|Z-z|^{\lambda}\right]\right\}.
$$

References

1. Datta, S. (2000). Empirical Bayes estimation with non-identical components, *Nonparametric Statistics*, **12**, 709–725.

2. Gradshteyn, I. S. and Ryzhik, I. M. (1994). *Tables of Integrals, Series and Products*, 5th edition, Academic Press, New York.

3. Liang, T. (2002). An improved empirical Bayes test for positive exponential families, *Statistica Neerlandica*, **56**, 346–361.

4. O'Bryan, T. E. (1976). Some empirical Bayes results in the case of component problems with varying sample sizes for discrete exponential families. *Annals of Statistics*, **4**, 1290–1293.

5. O'Bryan, T. E. (1979). Rates of convergence in a modified empirical Bayes problem involving Poisson distributions, *Communications in Statistics–Theory and Methods*, **8**, 167–174.

6. O'Bryan, T. E. and Susarla, V. (1976a). Empirical Bayes estimation with non-identical components: Continuous case, *Australian Journal of Statistics*, **19,** 115–125.

7. O'Bryan, T. E. and Susarla, V. (1976b). Rates in the empirical Bayes estimation problem with non–identical components. Case of normal distributions, *Annals of the Institute of Statistical Mathematics*, **28**, 389–397.

8. Pensky, M. and Singh, R. S. (1999). Empirical Bayes estimation of reliability characteristics for an exponential family, *Canadian Journal of Statistics*, **27**, 127–136.

9. Robbins, H. (1956). An empirical Bayes approach to statistics, *Proceedings of the Third Berkeley Symposium on Mathematical Statistics and Probability*, **1**, 157–163, University of California Press, Berkeley, CA.

10. Robbins, H. (1964). The empirical Bayes approach to statistical decision problems, *Annals of Mathematical Statistics*, **35**, 1–20.

11. Singh, R. S. and Wei, L. (1992). Empirical Bayes with rates and best rates of convergence in $u(x)c(\theta)\exp(-x/\theta)$-family: estimation case, *Annals of the Institute of Statistical Mathematics*, **44**, 435–449.

12. Stijnen, T. and van Houwelingen, J. C. (1990). Empirical Bayes methods in clinical trials meta-analysis, *Biometrical Journal*, **32**, 335–346.

13. Van Houwelingen, J. C. and Stijnen, T. (1993). Monotone empirical Bayes estimators based on more informative samples, *Journal of the American Statistical Association*, **88**, 1438–1443.

PART VI

Biostatistics

18

Bayesian Analysis of Mixtures of Improper Survival Distributions

Kaushik Patra, Dipak K. Dey and Samiran Ghosh

Department of Statistics, University of Connecticut, Storrs, Connecticut, U.S.A.

Abstract. A mixture model is proposed to model the failure data where each component has a surviving fraction i.e., admits a positive probability of cure. The proposed model is fitted to a real data set describing recidivism of prisoners i.e., return to prison for any crime. The proposed model is computationally appealing, and the Markov Chain Monte Carlo (MCMC) methods are developed to sample from the posterior distribution of the parameters. Classification of a prisoner to different groups is also done based on the proposed model.

Keywords and phrases: Improper distribution, Markov Chain Monte Carlo (MCMC) method, mixture distribution, surviving fraction

18.1 Introduction

A generalization of the usual survival models in order to allow for possible heterogeneity is useful in medical, criminology and other contexts. In particular, the distribution of the failure times, F, is allowed to be improper, i.e., to have $F(\infty) < 1$, so that only a proportion $\alpha = F(\infty)$ of the individuals under study are assumed eventually to fail. The complementary proportion $1 - \alpha$ is considered to be "immune" or possibly "cured", and never to fail. Most users of this method have modelled the distribution of observed failure or censored times by

$$F(t) = \alpha F_0(t) \tag{18.1}$$

where $F_0(t)$ is a proper distribution function describing the failure of "susceptible" individuals, i.e., those who will eventually fail. A parametric form is chosen for $F_0(.)$ and the parameters, together with α, are estimated from the sample. Ghitany and Maller (1992) considered a parametric approach, using the exponential distribution for F_0. It is also customary to use log normal or Weibull distributions to model F_0.

In general we cannot identify the immune individuals, if any, in a sample. We can only anticipate their presence if a relatively large number of failure times are censored at large values. If (18.1) is correct, we would, in fact, expect that a proportion of approximately $1-\alpha$ of the observations with the largest failure times will be censored. In addition, the failure times of non-immunes may be censored after being lost to follow-up. A plot of the Kaplan and Meier (1958) empirical distribution function $\widehat{F}_n(t)$ from such a sample of size n, say, will tend to level off near the value α, provided the observation time is sufficiently long. Maller and Zhou (1992) have used the estimator

$$\widehat{\alpha}_n = \widehat{F}_n(T_n), \tag{18.2}$$

where T_n is the maximum observed failure or the censored time. They investigated the asymptotic properties of the distribution of this estimator. Later, Maller and Zhou (1994) investigated the presence of the "immune" proportion via an asymptotic test and illustrated using a criminological data set.

In this chapter, we extend their modelling approach. Our objective is to extract information about the "cured" fraction when we have a mixture population; in particular, we consider a mixture model in which each improper component may have its own distinct fraction of immunes. The methodology is used to analyze the criminological data presented in Maller and Zhou (1994), although it can be applied to other fields. In their original work, Maller and Zhou considered two populations of prisoners separately; we treat them as one sample as coming from an overall population, where the subpopulations correspond to different groups but group membership labels are not available. Hence, the notion of a mixture model comes into consideration. Each of the mixture components can have its own proportion of not failing where we define failure in a formal way in Section 18.2. After fitting the model, we can classify each individual to one of the groups in order to investigate the extent of misclassification. Thus, the data set gives us a benchmark to validate the proposed model since we can actually compare the classification through our model with the truth. Of course, in practice, we would implement this methodology in situations where we do not know anything about the individual components of the mixture model.

The format of the chapter is as follows. In Section 18.2, we propose a two-component model for the recidivism data. A detailed description of the data is given in Section 18.3. Section 18.4 describes the likelihood and the full conditionals for the parameters of the model proposed in Section 18.2. In Section 18.5, the results from fitting the model are outlined with relevant Bayesian inferences.

Section 18.6 proposes another competing model for the same data with associated results in Section 18.7. Finally, we conclude in Section 18.8 with description of possible future work.

18.2 A Two-component Model

In this section, we develop methods to handle mixture models where at least one component arises from an improper distribution. The notion of being "improper" can be introduced with the idea described earlier. We illustrate our methodology with two components, with extension to more than two components being straightforward. Let the survival functions of group 1, group 2 and overall population be denoted by $S_1(t), S_2(t)$ and $S_T(t)$, respectively. Then

$$S_T(t) = pS_1(t) + (1-p)\,S_2(t), \tag{18.3}$$

where $0 < p < 1$. Now even though this form is widely reported in the literature, the crucial difference between this model and others is that both of the individual components have their distinct surviving fraction, i.e., $S_j(\infty) = 1 - \alpha_j$, $0 < \alpha_j < 1$, $j = 1, 2$. The presence of the surviving fraction can be established from the context of the problem. Our approach is fully Bayesian with an intuitive introduction of the surviving fraction. We define failure as "recidivism", i.e., a return to prison for any offense. Suppose we have two groups of prisoners convicted for committing serious offenses, one with No-Prior conviction (referred to as group 1) and the other with Prior conviction (referred to as group 2). Then the survival function is given by (18.3), where $p = \Pr$ [A prisoner belongs to the No-Prior conviction group]. Evidently, each of the subpopulations will have a surviving fraction (a fraction of the population which never returns to prison). We anticipate that the surviving fraction for group 1 will be higher than that for group 2. Let us derive the survival function incorporating the surviving fraction for each of the subpopulation. Suppose $\phi = 1$ if recidivism occurs, and 0 otherwise. The unconditional cumulative distribution function corresponding to the first subpopulation is then given by

$$\begin{aligned} F_1(t) &= \Pr(T \le t) \\ &= \Pr(T \le t|\phi = 1)\Pr(\phi = 1) + \Pr(T \le t|\phi = 0)\Pr(\phi = 0) \\ &= \Pr(T \le t|\phi = 1)\Pr(\phi = 1) \\ &= \alpha_1 F_{10}(t), \end{aligned} \tag{18.4}$$

where $\alpha_1 = \Pr(\text{recidivism}|\text{ group 1})$. Hence $S_1(t) = 1 - \alpha_1 F_{10}(t)$. Defining $\alpha_2 = \Pr(\text{recidivism}|\text{ group 2})$, we have similarly that $S_2(t) = 1 - \alpha_2 F_{20}(t)$.

Note that $F_{10}(.)$ and $F_{20}(.)$ are proper cdfs whereas $F_1(.)$ and $F_2(.)$ are not. Hence the survival function corresponding to the overall population, $S(t)$, is not a proper survival function, but rather a mixture of two improper survival functions. We adopt a Weibull distribution to model each of $F_{10}(.)$ and $F_{20}(.)$ i.e., $F_{i0}(t) = 1 - \exp(-b_i t^{a_i})$ for $i = 1, 2$, as in Maller and Zhou (1994) where separate models were fit for the two different groups. Now, (18.3) reduces to

$$\begin{aligned} S_T(t) = \quad & p\{1 - \alpha_1(1 - \exp(-b_1 t^{a_1}))\} + q\{1 - \alpha_2(1 - \exp(-b_2 t^{a_2}))\} \\ = \quad & 1 - p\alpha_1(1 - \exp(-b_1 t^{a_1})) - q\alpha_2(1 - \exp(-b_2 t^{a_2})). \end{aligned} \tag{18.5}$$

Note that, $S_T(\infty) = 1 - (p\alpha_1 + q\alpha_2)$. In practice, we will be able to obtain Kaplan-Meier estimates for the overall population. Usually we will have no idea about the values of the individual Weibull parameters and the surviving fractions. However, in the context of the current problem, we will assume that $\alpha_1 \le \alpha_2$, as remarked before. This kind of prior knowledge will provide us a guideline to develop an appropriate model.

18.3 Data

The criminological data presented in Maller and Zhou (1994) are drawn from a large data set described by Broadhurst et al. (1988) that consists of two groups of prisoners convicted of serious sexual offenses, one with no prior records (No-Prior group) and the other with prior convictions (Prior group) and released, following imprisonment for serious offense, after June 30, 1975. "Failure" for these individuals is "recidivism" (a return to prison for any offense), and we observe their times to return, possibly censored by the necessity to cease observation at a predetermined time (June 30, 1987). The data has been reported by failure times (in years). There are 296 prisoners in the No-Prior group and 121 in the Prior group. For our analysis, we suppose a total sample of 417 prisoners coming from an overall population consisting of two groups. However, for future benchmarking, in Figure 18.1, for the two groups individually and for the joint sample, we plot the Kaplan-Meier estimates of the cumulative distribution function treating the upper tails as described below.

Assume that n possibly censored failure times $t_1, \cdots, t_n$ are observations on the minima of random variables with failure distribution F of the form (18.1) and random variables with a censoring distribution G. Censoring indicator δ_i, taking the value 0 or 1 according to whether or not t_i is censored, $i = 1, .., n$, are also observed. We construct $\widehat{F}_n(t)$, the Kaplan-Meier estimate of $F(t)$, by

$$\widehat{F}_n(t) = 1 - \prod_{i:t_{(i)} \le t} \left(\frac{n - i - \delta_i + 1}{n - i + 1} \right), \tag{18.6}$$

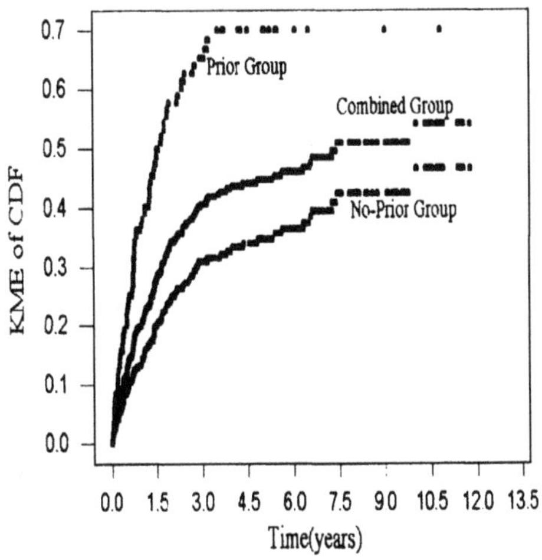

Figure 18.1: Kaplan-Meier estimate of CDF for individual and combined group

where $t_{(1)} \leq \cdots \leq t_{(n)}$ are ordered failure times (with uncensored observations indexed before censored observations in the case of ties), and $\delta_{(i)}$ is the censoring indicator associated with $t_{(i)}$. Let $T_n = \max_{1 \leq i \leq n} t_{(i)}$ be the largest failure time. If T_n is uncensored, then (18.6) implies $\widehat{F}_n(T_n) = 1$; but if T_n is censored, then $\widehat{F}_n(T_n) < 1$. We do not redefine, $\widehat{F}_n(T_n) = 1$ in this case, as is sometimes advocated, because $\widehat{F}_n(T_n) < 1$ may be indicative of immunes in the population. In fact, Maller and Zhou (1992) and Sposto, Sather and Baker (1992) have suggested the nonparametric estimator $\widehat{\alpha}_n = \widehat{F}_n(T_n)$ as an estimator of α in (18.1). The step function $\widehat{F}_n(t)$ does not jump at the censored observations, so it is constant on the interval (T_n^*, T_n), where T_n^* is the largest uncensored failure time. Thus we also have $\widehat{\alpha}_n = \widehat{F}_n(T_n^*)$. From Figure 18.1, we see that the cumulative distribution function levels off near 0.47 for the No-Prior group and 0.70 for the Prior group. Again, for our purposes, we can only presume to have the estimate of the overall surviving fraction which is approximately 0.55.

18.4 The Likelihood and Full Conditionals

18.4.1 Model 1

Let us denote the density and the survival function corresponding to overall population by $f_T(.)$ and $S_T(.)$ respectively. The likelihood function can be written as

$$L \propto \prod_{i=1}^{n} f_T(y_i)^{1-\delta_i} S_T(y_i)^{\delta_i} = \left(\prod_{i:\delta_i=0} f_T(t_i) \right) \left(\prod_{i:\delta_i=1} S_T(C_i) \right)$$

where y_i is the observed time and δ_i is the censoring indicator, i.e., $Y_i = T_i$ if $T_i < C_i, = C_i$ if $T_i \geqslant C_i$ and $\delta_i = I(Y_i = C_i)$, $i = 1, .., n$. The conditional survival function $S_T(y)$ given z is $S_T(y) = S_1(y)^z S_2(y)^{1-z}$ where z is the latent random variable which takes value 1 or 0 according to whether the observation corresponds to the first or second component of the mixture distribution with survival functions $S_1(y)$ and $S_2(y)$ defined earlier. Consequently the conditional density $f_T(y)$ given z is $f_T(y) = f_1(y)^z f_2(y)^{1-z}$ where $f_i(y) = \alpha_i a_i b_i y^{a_i-1} \exp(-b_i y^{a_i})$ for $i = 1, 2$. We fit the model using a Bayesian approach adopting weak but proper priors on the parameters. The prior distribution for p is taken as $Beta(d_1, d_2) \propto p^{d_1-1}(1-p)^{d_2-1}$. For the scale and the shape parameters of the individual Weibull distributions, a diffuse $Gamma(m_1, m_2)$ prior distribution is used, where we assume that $m_1 = m_2 = 0.01$. Assuming $\alpha_1 \leq \alpha_2$, we specify $\pi(\alpha_1, \alpha_2) = \pi(\alpha_1|\alpha_2)\pi(\alpha_2)$ where $\pi(\alpha_1|\alpha_2) \equiv Uniform(0, \alpha_2)$ and $\pi(\alpha_2) \equiv Uniform(0, 1)$. Let $[\mathbf{x}]$ denote the distribution of the vector $\mathbf{x}$. Then the joint posterior of the model parameters and the latent random variables are given as

$$[\mathbf{z}, p, \mathbf{a}, \mathbf{b}, \alpha] \propto [\mathbf{t} \mid \mathbf{z}, \mathbf{a}, \mathbf{b}, \alpha] [\mathbf{z} \mid p] [p] [\mathbf{a}] [\mathbf{b}] [\alpha], \tag{18.7}$$

where $[\mathbf{t} \mid \mathbf{z}, \mathbf{a}, \mathbf{b}, \alpha]$ is the likelihood. In the sequel, we refer to (18.7) as Model 1. Note that we can write

$$\prod_{i:\delta_i=0} f_T(t_i) = \left[\prod_{i:\delta_i=0, z_i=1} f_1(t_i) \right] \left[\prod_{i:\delta_i=0, z_i=0} f_2(t_i) \right]$$

and

$$\prod_{i:\delta_i=1} S_T(C_i) = \left[\prod_{i:\delta_i=1, z_i=1} S_1(C_i) \right] \left[\prod_{i:\delta_i=1, z_i=0} S_2(C_i) \right].$$

Also $[\mathbf{z} \mid p] = \prod_{i=1}^{n} \left[p^{z_i} (1-p)^{1-z_i}\right]$. The full conditional of $\mathbf{z}$ is proportional to $[\mathbf{t} \mid \mathbf{z}, \mathbf{a}, \mathbf{b}, \alpha] [\mathbf{z} \mid p]$ and the full conditional of p is exactly $Beta(\sum z_i + d_1, n - \sum_{i=1}^{n} z_i + d_2)$. The full conditional of the shape parameter a_1 is proportional to

$$\left[\prod_{i:\delta_i=0,z_i=1} f_1(t_i)\right] \left[\prod_{i:\delta_i=1,z_i=1} S_1(C_i)\right] [a_1]$$

and that for a_2 is proportional to

$$\left[\prod_{i:\delta_i=0,z_i=0} f_2(t_i)\right] \left[\prod_{i:\delta_i=1,z_i=0} S_2(C_i)\right] [a_2].$$

We get similarly expressions for b_1 and b_2 with $[a_i]$ replaced by $[b_i]$, $i = 1, 2$. The full conditionals for b_1 and b_2 will be $Gamma\left(n + v, \sum y_i^{a_j} + u\right)$ for $j = 1, 2$.

The full conditionals for α_1 and α_2 are proportional to

$$\left[\prod_{i:\delta_i=0,z_i=1} f_1(t_i)\right] \left[\prod_{i:\delta_i=1,z_i=1} S_1(C_i)\right] \times \frac{1}{\alpha_2} 1(0 \leq \alpha_1 \leq \alpha_2),$$

$$\left[\prod_{i:\delta_i=0,z_i=0} f_2(t_i)\right] \left[\prod_{i:\delta_i=1,z_i=0} S_2(C_i)\right] 1(\alpha_1 \leq \alpha_2 \leq 1),$$

respectively. Finally, we transform α_1 and α_2 on the real line by taking logit transformation and hence the full conditional of $\alpha_1^* = \log\left(\frac{\alpha_1}{1-\alpha_1}\right)$ is proportional to

$$\left[\prod_{i:\delta_i=0,z_i=1} f_1(t_i)\right] \left[\prod_{i:\delta_i=1,z_i=1} S_1(C_i)\right] \times \frac{1}{\alpha_2} \times \frac{\exp(\alpha_1^*)}{(1 + \exp(\alpha_1^*))^2}$$

and that of $\alpha_2^* = \log\left(\frac{\alpha_2}{1-\alpha_2}\right)$ is proportional to

$$\left[\prod_{i:\delta_i=0,z_i=0} f_2(t_i)\right] \left[\prod_{i:\delta_i=1,z_i=0} S_2(C_i)\right] \times \frac{\exp(\alpha_2^*)}{(1 + \exp(\alpha_2^*))^2}.$$

We use Gibbs sampling to obtain the posterior distributions of the model parameters. Since the full conditionals of a_1, a_2, α_1 and α_2 are non-standard, we use a Metropolis-Hastings step within each stage of Gibbs sampling. The proposal density for a_1 and a_2 are taken to be $Gamma(n + v, b_j \sum_{i:y_i>1} \ln y_i + u)$ for $j = 1, 2$. For logit(α_1) and logit(α_2), we use the normal distribution as the proposal density with mean zero and standard deviation 1.

Table 18.1: Posterior Estimates of Model 1 parameters

Parameters	Posterior Mean	Posterior Median	Posterior Std. Dev	95% credible interval
α_1	0.465	0.465	0.002	(0.460, 0.469)
α_2	0.701	0.701	0.002	(0.696, 0.704)
p	0.709	0.709	0.023	(0.665, 0.751)
a_1	0.986	0.986	0.057	(0.875, 1.090)
b_1	0.329	0.329	0.019	(0.291, 0.366)
a_2	1.166	1.156	0.110	(0.975, 1.407)
b_2	0.760	0.749	0.072	(0.641, 0.931)

18.5 Results From Fitting Model 1

Table 18.1 gives the posterior summary for the model parameters under Model 1 and Figure 18.2 gives the posterior distribution for the model parameters. The posterior distribution of α_1 and α_2 are quite symmetric and highly centred around the empirical estimates obtained from the Kaplan-Meier plot from Figure 18.1. The posterior mode of p is close to 0.7 which is consistent with the sample estimate. The estimated posterior predictive cumulative distribution functions for the No-Prior and Prior groups are superimposed on the Kaplan-Meier estimators and are given in Figure 18.3a and Figure 18.3b, respectively. The overall estimated cumulative distribution function is plotted in Figure 18.3c. From the graphs, Model 1 appears to give a reasonably good fit.

Now let us consider the issue of classifying each individual into one of the prisoner groups based on the estimates. Note that the posterior distribution of Z given $T = t$ will be a Bernoulli distribution with success probability

$$p_z(t) = E_{p,\theta_1,\theta_2|T=t}\left[\frac{pf_1(t|\theta_1)^{1-\delta} S_1(t|\theta_1)^{\delta}}{pf_1(t|\theta_1)^{1-\delta} S_1(t|\theta_1)^{\delta} + (1-p) f_2(t|\theta_2)^{1-\delta} S_2(t|\theta_2)^{\delta}}\right]$$

where θ_1 and θ_2 are the model parameters for first and second components, respectively. Using the posterior sample of size M, we obtain a Monte Carlo estimate of $p_z(t)$ as
$\widehat{p_z}(t) =$

$$\frac{1}{M}\sum_{k=1}^{M}\frac{p^{(k)} f_1\left(t|\theta_1^{(k)}\right)^{1-\delta} S_1\left(t|\theta_1^{(k)}\right)^{\delta}}{p^{(k)} f_1\left(t|\theta_1^{(k)}\right)^{1-\delta} S_1\left(t|\theta_1^{(k)}\right)^{\delta} + \left(1-p^{(k)}\right) f_2\left(t|\theta_2^{(k)}\right)^{1-\delta} S_2\left(t|\theta_2^{(k)}\right)^{\delta}}$$

Table 18.2: Posterior Estimates of Model 2 parameters

Parameters	Posterior Mean	Posterior Median	Posterior Std. Dev	95% Credible Interval
α_1	0.465	0.465	0.0002	$(0.464, 0.466)$
α_3	0.700	0.700	0.0002	$(0.699, 0.710)$
p_1	0.316	0.301	0.162	$(0.037, 0.659)$
p_2	0.420	0.430	0.145	$(0.107, 0.672)$
p_3	0.266	0.267	0.068	$(0.130, 0.402)$
a_1	0.658	0.655	0.539	$(0.568, 0.758)$
b_1	0.609	0.605	0.057	$(0.518, 0.719)$
a_2	1.034	1.032	0.053	$(0.907, 1.133)$
b_2	0.278	0.277	0.018	$(0.247, 0.314)$
a_3	1.230	1.229	0.088	$(1.066, 1.407)$
b_3	0.758	0.757	0.052	$(0.658, 0.864)$

where $p^{(k)}, \theta_1^{(k)}$ and $\theta_2^{(k)}$, $(k = 1, .., M)$ are the kth posterior sample values from the MCMC output. If $\widehat{p_z}\left(t_{(i)}\right) > 0.5$ we classify individual i to group 1, otherwise to group 2. Since we know the actual classification, we can calculate the proportion of misclassification. For Model 1, we have 28.8% misclassification, i.e., we would classify roughly 71.2% correctly without knowing anything about the underlying heterogeneity.

18.6 A Three-component Model: Model 2

The prisoners from group 1, i.e., those with no prior conviction record are expected to be more heterogeneous compared to the group with prior conviction. This leads us to propose a competing model where we anticipate that the group 1 is a mixture of two other improper Weibull distributions, i.e., the overall model is

$$f_T(y) = p_1 f_1(y) + p_2 f_2(y) + p_3 f_3(y)$$

with corresponding survival function

$$S_T(y) = p_1 S_1(y) + p_2 S_2(y) + p_3 S_3(y),$$

where $p_1 + p_2 + p_3 = 1$. Without going into detail, this model can be built in the same spirit as the two-component model. Each of the components will have a surviving fraction, say, α_1, α_2 and α_3 with the constraint $\alpha_1 \leq \alpha_2 \leq \alpha_3$.

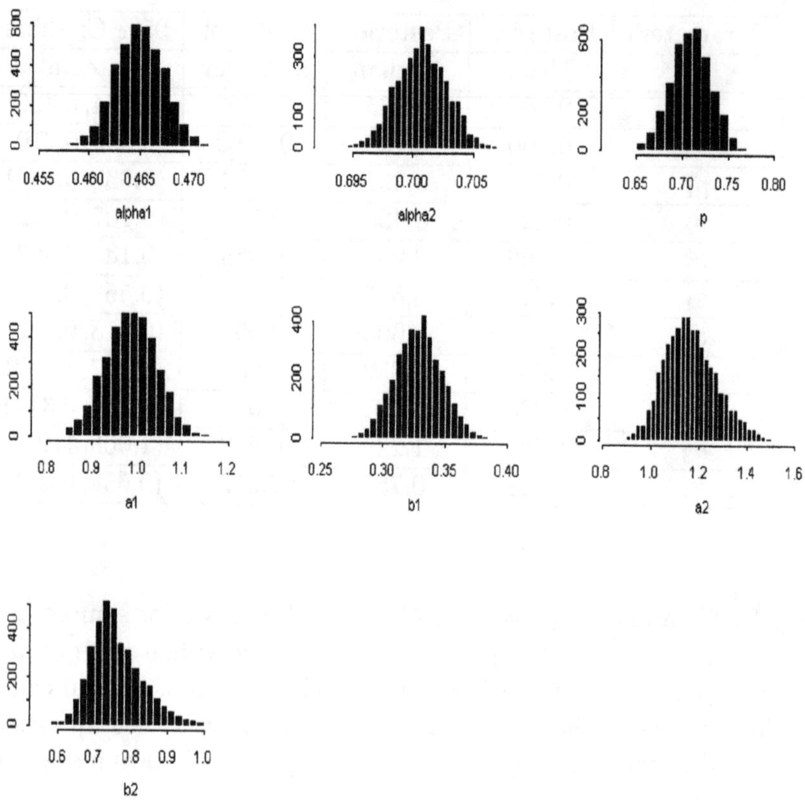

Figure 18.2: Posterior distribution of Model 1 parameters

Again, we write

$$\prod_{i:\delta_i=0} f_T(t_i) = \left[\prod_{\substack{i:\delta_i=0\\ \mathbf{z}_i=(1,0,0)^T}} f_1(t_i)\right]\left[\prod_{\substack{i:\delta_i=0\\ \mathbf{z}_i=(0,1,0)^T}} f_2(t_i)\right]\left[\prod_{\substack{i:\delta_i=0\\ \mathbf{z}_i=(0,0,1)^T}} f_3(t_i)\right]$$

and

$$\prod_{i:\delta_i=0} S_T(C_i) = \left[\prod_{\substack{i:\delta_i=0\\ \mathbf{z}_i=(1,0,0)^T}} S_1(C_i)\right]\left[\prod_{\substack{i:\delta_i=0\\ \mathbf{z}_i=(0,1,0)^T}} S_2(C_i)\right]\left[\prod_{\substack{i:\delta_i=0\\ \mathbf{z}_i=(0,0,1)^T}} S_3(C_i)\right]$$

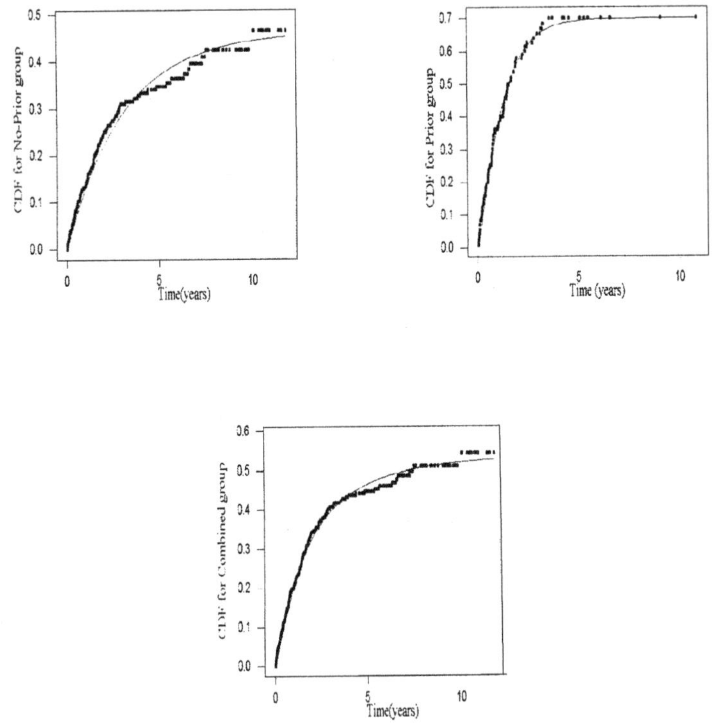

Figure 18.3: Estimated CDF for model 1. Top left (Figure 18.3a) for No-Prior group. Top right (Figure 18.3b) for Prior group. Bottom (Figure 18.3c) for combined group

where $\mathbf{z}_i = (z_{1i}, z_{2i}, z_{3i})^T \sim Multinomial\,(p_1, p_2, p_3)$ and as before, $f_i\,(y) = \alpha_i a_i b_i y^{a_i - 1} \exp\,(-b_i y^{a_i})$, $i = 1, 2, 3$. In fact, for the two conceptual components of group 1, we adopt distinct survival functions but assume a common surviving fraction where $\alpha_1 = \alpha_2 \leq \alpha_3$. We also fit a model with different surviving fractions for group 1 components but virtually they are not distinct, hence omit the detail. The prior distributions of the model parameters are chosen in the same way as in the case of the two-component model, i.e., the Weibull parameters follow a Gamma distribution and the joint prior of α_1 and α_3 is the same as the previous one.

18.7 Results From Fitting Model 2

It is obvious that Model 2 is a direct extension of Model 1, and thus can be fitted by the same approach as described in Section 18.4. Figure 18.4 gives the posterior distribution of the model parameters and Table 18.2 gives the corresponding estimates. Figure 18.5 gives the estimate of the overall cdf for model 2. Using the same argument as before we get the posterior distribution of z given $T = t$ as a $Trinomial\left(1, p_z^1(t), p_z^2(t), p_z^3(t)\right)$, where

$$p_z^i(t) = E_{\mathbf{p},\theta_1,\theta_2,\theta_3|T=t}\left(\frac{p_i f_i(t|\theta_i)^{1-\delta} S_i(t|\theta_i)^{\delta}}{\sum_{j=1}^{3} p_j f_j(t|\theta_j)^{1-\delta} S_j(t|\theta_j)^{\delta}}\right), i = 1, 2, 3.$$

The classification rule would be as follows: an individual will be classified to group 1, if $\widehat{p}_z^1(t) + \widehat{p}_z^2(t) > \widehat{p}_z^3(t)$. The misclassification percentage in this case is 29.0%. Model 2 is no better than Model 1 with respect to prediction but this is likely due to heavy censoring in the No-Prior group as evident from the histogram in Figure 18.6. The dark region of the histogram corresponds to the censored observations. Censoring in this group can occur for a variety of reasons unknown to us. Thus in the interest of parsimony, we would choose Model 1.

18.8 Conclusion

In this paper, we developed a model to fit a mixture distribution where all components contain surviving fractions. Bayesian analysis effectively extracted information about the unknown immune fractions. Extension to incorporate covariates into the model, where they are available, is straightforward.

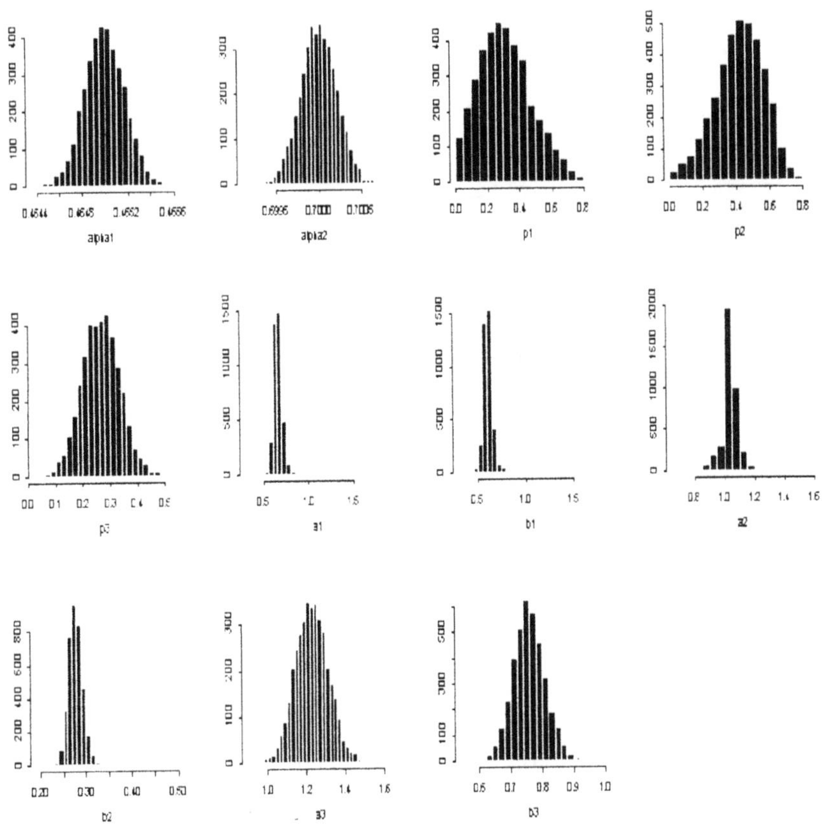

Figure 18.4: Posterior distribution of Model 2 parameters

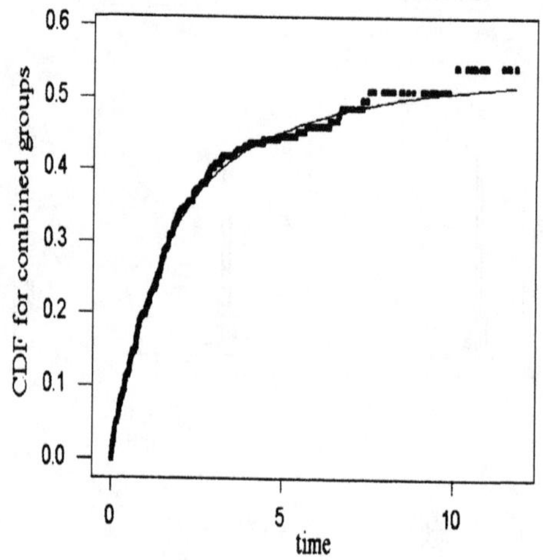

Figure 18.5: Estimated CDF for combined sample from model 2

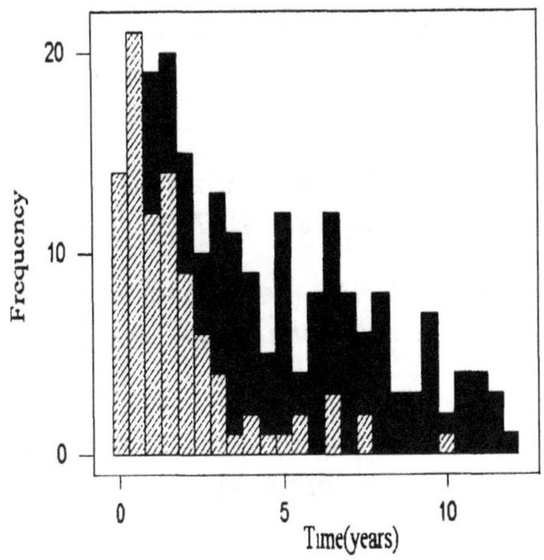

Figure 18.6: Histogram of censored and uncensored observations for No-Prior group

References

1. Broadhurst, R. G., Maller, R. A., Maller, M. G. and Duffecy, J. (1988). Aboriginal and nonaboriginal recidivism in Western Australia: a failure rate analysis, *Journal of Research in Crime Delinquency*, **25**, 83–107.

2. Ghitany, M. E. and Maller, R. A.(1992). Asymptotic results for exponential mixture models with long-term survivors, *Statistics*, **23**, 321–336.

3. Ghitany, M. E., Maller, R. A. and Zhou, S. (1995). Exponential mixture models with long-term survivors and covariates. *Journal of Multivariate Analysis*, **49**, 218–241.

4. Kaplan, E. L. and Meier, P. (1958). Nonparametric estimation from incomplete observations, *Journal of the American Statistical Association*, **53**, 457–481.

5. Maller, R. A. and Zhou, S. (1992). Estimating the proportion of immunes in a censored sample, *Biometrika*, **79**, 731–739.

6. Maller, R. A. and Zhou, S. (1994). Testing for sufficient follow-up and outliers in survival data, *Journal of the American Statistical Association*, **89**, 1499–1506.

7. Sposto, R. and Sather, H. and Baker, S. (1992). A comparison of tests of the difference in the proportion of patients who are cured, *Biometrics*, **48**, 87–99.

19

Multivariate Survival Analysis with PVF Frailty Models

Madhuja Mallick and Nalini Ravishanker

Department of Statistics, University of Connecticut, Storrs, CT U.S.A.

Abstract: In this chapter, we describe inference for multivariate lifetimes data using a conditional proportional hazards model with a power variance family (PVF) frailty distribution and a Weibull baseline hazard. The likelihood function is derived via the joint density of tilted positive stable random variables. Inference is carried out in the Bayesian framework, using Markov chain Monte Carlo techniques. We illustrate our approach on data involving recurrent infections due to insertion of a catheter in patients on portable dialysis machines.

Keywords and phrases: Multivariate survival model, tilted positive stable density, proportional hazards model, Weibull baseline hazard

19.1 Introduction

The proportional hazards shared frailty regression model for multivariate survival times has the form

$$h(t_{ij}|X_i, \tilde{z}_{ij}) = h_0(t_{ij}) \exp\left[\tilde{\beta}' \tilde{z}_{ij}\right] X_i, \tag{19.1}$$

where for the jth ($j = 1, \cdots, m$) subject in the ith ($i = 1, \cdots, n$) group, t_{ij} is the observed value of the random survival time T_{ij}, $\tilde{z}_{ij}$ is a fixed, possibly time dependent covariate vector of dimension p, and $\tilde{\beta}$ is the p-dimensional vector of regression parameters. In (19.1), the dependence is generated by a frailty random variable X_i, which is assumed to follow a power variance family (PVF) frailty distribution [Hougaard (2000)], while

$$h_0(t_{ij}) = \lambda \gamma t_{ij}^{\gamma - 1}, \ \lambda > 0, \ \gamma > 0$$

denotes the Weibull baseline hazard function. Alternate parametric frailty specifications used in the recent literature include the gamma model [Clayton and

Cuzick (1985)], the log-normal model [Gustafson (1997)], and the positive stable model [Qiou, Ravishanker and Dey (1999)]. Under the infinite variance positive stable frailty specification, the proportional hazards model holds unconditionally as well as conditionally [Hougaard (1986a,b) and Oakes (1994)], and also allows for a much higher degree of heterogeneity among the common covariates than would be possible under the finite variance gamma frailty. Inference for positive stable frailty models has been described in the literature by Hougaard (1986a,b, 2000), Manatunga and Oakes (1999), and Oakes (1994). Qiou, Ravishanker and Dey (1999) extended Buckle's (1995) idea for computing the density of a four-parameter stable distribution to frailty modeling, and described fully Bayesian inference for all parameters in the conditional proportional hazard model with a positive stable frailty distribution and a correlated piecewise exponential prior process as baseline hazard.

The PVF distribution, which was discussed in the context of frailty modeling by Hougaard (1986a, 1995) and Hougaard, Harvald and Holm (1992) includes the positive stable and gamma frailties as special cases and thereby offers a flexible framework for modeling. Although the PVF distribution is conceptually simple, inference for the PVF frailty model is complicated due to the lack of a closed form expression for the density function of the PVF random variable. Hougaard (2000) discussed the usefulness of inference for the PVF frailty which incorporates intermediate dependence, but mentioned that all existing estimation procedures for this frailty model are complicated. In this article, we describe simultaneous inference for multivariate survival times with a PVF shared frailty distribution. For this, we derive the density function of the PVF variable by viewing this density as a tilted positive stable density Hougaard (1986a), and we employ the Bayesian approach using Markov chain Monte Carlo methods for carrying out inference.

The format of this chapter is as follows. In Section 19.2, we review some general properties of the PVF distributions, with special attention to their role in modeling frailty. In Section 19.3, we derive the density function of the PVF frailty random variable by viewing it as a tilted positive stable density, thereby producing a computational vehicle for its evaluation. In Section 19.4, we present the likelihood for the multivariate survival model and derive the joint posterior distribution of the parameters. Section 19.5 describes the complete conditional distributions and the implementation of Bayesian inference. In Section 19.6, we present an illustration of this approach for data on multiple recurrence times (in days) to infection from the time of insertion of a catheter for patients on portable dialysis machines [McGilchrist and Aisbett (1991)]. We conclude with a few remarks in Section 19.7. The sampling algorithms are described in the Appendix.

19.2 Properties of the PVF Frailty

In the three-parameter power variance family, denoted by $PVF(\alpha, \delta, \theta)$, the variance of the random variable is a power function of its mean. The parameter space of the PVF distribution consists of $\alpha \leq 1, \delta > 0$, and $\theta \geq 0$ for $\alpha > 0$ and $\theta > 0$ for $\alpha \leq 0$. Under this parametrization, $PVF(\alpha, \delta, \theta)$ includes as special cases the gamma (for $\alpha = 0$), the positive stable (for $\theta = 0$, $\delta = \alpha$), and the inverse Gaussian (for $\alpha = 1/2$) distributions, among others. For frailty modeling in this article, we focus on the parameter region $\alpha \geq 0$, for which the distribution is concentrated on $\mathcal{R}^+$. When $\alpha < 0$, the distribution is concentrated on $\mathcal{R}^+ \cup \{0\}$, implying that some groups have zero risk; this, of course, is impossible in the context of lifetimes because of its implication of a person's immortality.

Let $X \sim PVF(\alpha, \delta, \theta)$, where $\alpha \leq 1$, $\delta > 0$, and $\theta \geq 0$; then

$$E(X) = \delta\theta^{\alpha-1} \text{ and } Var(X) = \delta\,(1-\alpha)\,\theta^{\alpha-2}. \tag{19.2}$$

When $\theta > 0$, all (positive) moments of the distribution exist. Let $H(t) = \int_0^t h_0\,(u)\,du$ denote the cumulative baseline hazard function; for the Weibull hazard, $H(t) = \lambda t^\gamma$. Assuming a single covariate Z taking on the value 0 or the value 1, the ratio of unconditional hazards is

$$\frac{h_\nu(t)}{h_0(t)} = \frac{-\frac{d}{dt}\log S_\nu\,(t)}{-\frac{d}{dt}\log S_0\,(t)} = \frac{\{\theta + \nu H(t)\}^{\alpha-1}\nu}{\{\theta + H(t)\}^{\alpha-1}},$$

which is a function of t; here, $\nu = \exp(\beta)$, β being the regression coefficient corresponding to Z. Unlike the positive stable, the unconditional hazards $h_\nu(t)$ and $h_0(t)$ for the PVF family are not related via proportionality.

Measures of dependence enable us to compute the association among t_{ij}'s when the model is continuous. When $\alpha > 0$, Kendall's τ is given by the expression

$$\tau = (1-\alpha) - 2\theta + \left(4\theta^2/\alpha\right)\exp\left(2\theta/\alpha\right)E_{(1/\alpha)-1}\left(2\theta/\alpha\right), \tag{19.3}$$

where $E_m\,(x) = \int_1^\infty t^{-m}e^{-xt}dt$ is the generalized exponential integral. Hougaard (2000) and Hougaard, Harvald and Holm (1992) pointed out that although it is difficult to identify both parameters of the PVF frailty, it should be possible to determine the degree of dependence with reasonable precision, using say, Kendall's τ. The median concordance is

$$\varphi = 4\exp\left[-\{2\,(\theta/\alpha + \log 2)^{1/\alpha} - (\theta/\alpha)^{1/\alpha}\}^\alpha + \theta/\alpha\right] - 1. \tag{19.4}$$

The cross ratio function is another useful measure of local dependence and has the form

$$\begin{aligned}\chi(t) &= \frac{S(t_1,t_2)\frac{\partial^2 S(t_1,t_2)}{\partial t_1 \partial t_2}}{\{\frac{\partial}{\partial t_1}S(t_1,t_2)\}\{\frac{\partial}{\partial t_2}S(t_1,t_2)\}} \\ &= 1+(1-\alpha)\{\theta+H_1(t_1)+H_2(t_2)\}^{-\alpha}/\delta, \end{aligned} \tag{19.5}$$

where $H_i(t_i)=\lambda t_i^{\gamma}, i=1,2$ is the integrated hazard function.

19.3 PVF as a Tilted Positive Stable

When $\alpha \geq 0$, the power variance family is obtained as the exponential family generated from the positive stable distributions [Hougaard (1986) and Jorgensen (1987)]. Specifically, suppose W follows a positive stable distribution, $P(\alpha,\alpha,0)$, where $\alpha \in (0,1)$. For $\delta > 0$, the distribution of $X = \left(\frac{\delta}{\alpha}\right)^{1/\alpha} W$ is $P(\alpha,\delta,0)$. Since the Laplace transform of the positive stable distribution is

$$L(s) = E\exp(-sW) = \exp(-s^{\alpha}), \; s \geq 0,$$

it follows that the Laplace transform of X is

$$L(s) = E\exp[-s\left(\frac{\delta}{\alpha}\right)^{1/\alpha} W] = \exp\left[-s^{\alpha}\left(\frac{\delta}{\alpha}\right)\right].$$

For fixed α, the exponential dispersion model generated by the positive stable distribution is denoted by $P(\alpha,\delta,\theta)$, and the PVF random variable X has density [Jorgensen (1987)]

$$\frac{f_{\alpha}(x|\zeta)\exp(-\theta x)}{\zeta L_{\alpha}(\zeta\theta)}, \text{ where } \zeta = \left(\frac{\delta}{\alpha}\right)^{1/\alpha},$$

where $f_{\alpha}(.)$ and $L_{\alpha}(.)$ are, respectively, the positive stable density and Laplace transformation. This density is characterized by three parameters and its Laplace transform is

$$L(s) = \exp\left[-\frac{\delta\{(\theta+s)^{\alpha}-\theta^{\alpha}\}}{\alpha}\right]. \tag{19.6}$$

For fixed α and δ, this corresponds to a one-parameter exponential family with canonical parameter $-\theta$, and is the tilted positive stable distribution. For $\theta = 0$ and $\delta = \alpha$, (19.6) is the Laplace transform of a positive stable variable, and when $\alpha \to 0$, it is the Laplace transform of a gamma variable. For $0 < \alpha < 1$, (corresponding to lifetimes data), Hougaard (1986a) expressed the unimodal PVF density as

$$f(x) = -\frac{1}{\pi x}\left[\sum_{k=1}^{\infty}\frac{k\alpha!}{k!}\left(-\frac{x^{-\alpha}\delta}{\alpha}\right)^k \sin(\alpha k\pi)\right]\exp(-\theta x)\exp\left(\frac{\delta\theta^{\alpha}}{\alpha}\right). \quad (19.7)$$

Since the PVF density has the form of exponential dispersion of a positive stable density, parameter estimation with the PVF frailty has the same difficulty as with the positive stable frailty; the complication is due to the absence of a closed form expression for the density of a stable random variable. The density in (19.7) involves an infinite sum, and its use in a likelihood based framework is therefore unwieldy. Hougaard, Harvald and Holm (1992) proposed an estimation of PVF frailty using the first two stages of Hougaard's three-stage estimation procedure but this technique conflicts with the probability mechanism behind the data. It would be also difficult to extend the marginal likelihood approach described by Lam and Kuk (1997) for the PVF frailty model with constant baseline hazard ($h_0(t) = 1$) to a model with more general baseline hazards, such as the parametric Weibull baseline hazard. Here, we implement fully Bayesian inference for a PVF frailty model with Weibull baseline hazard; this permits simultaneous likelihood based inference for all the parameters (baseline hazard parameters, regression coefficients and frailty parameters). We first give the derivation of a suitable form for the PVF density.

The support of the positive stable distribution is $\mathcal{R}^+$, the distribution is totally skewed to the right, $\alpha \in (0,1]$, and the scale parameter is held fixed at 1 for identifiability purposes. Buckle (1995) provided an expression for the joint density of n iid observations from a four-parameter stable distribution by utilizing a bivariate density function $f(w_i, y_i|\alpha)$ whose marginal density with respect to y_i is exactly the stable density. Specifically, let $f(w_i, y_i|\alpha)$ be a bivariate function which projects $(-\infty, 0) \times (-1/2, l_\alpha) \cup (0,\infty) \times (l_\alpha, 1/2)$ to $(0,\infty)$:

$$f(w_i, y_i|\alpha) = \frac{\alpha}{|\alpha-1|}\exp\left[-\left|\frac{w_i}{\tau_\alpha(y_i)}\right|^{\alpha/(\alpha-1)}\right]\left|\frac{w_i}{\tau_\alpha(y_i)}\right|^{\alpha/(\alpha-1)}\frac{1}{w_i}, \quad (19.8)$$

where

$$\tau_\alpha(y_i) = \frac{\sin(\pi\alpha y_i + \psi_\alpha)}{\cos\pi y_i}\left[\frac{\cos\pi y_i}{\cos\{\pi(\alpha-1)y_i + \psi_\alpha\}}\right]^{(\alpha-1)/\alpha}$$

with $w_i \in (-\infty,\infty)$, $y_i \in (-1/2, 1/2)$, $\psi_\alpha = \min(\alpha, 2-\alpha)\pi/2$ and $l_\alpha = -\psi_\alpha/\pi\alpha$. Then

$$f(w_i|\alpha) = \frac{\alpha|w_i|^{1/(\alpha-1)}}{|\alpha-1|}\int_{-1/2}^{1/2}\exp\left[-\left|\frac{w_i}{\tau_\alpha(y_i)}\right|^{\alpha/(\alpha-1)}\right]\left|\frac{1}{\tau_\alpha(y_i)}\right|^{\alpha/(\alpha-1)} dy_i. \quad (19.9)$$

Following Jorgensen (1987) and Hougaard (1986a), the density for the PVF family when $\alpha \in (0,1]$ becomes

$$\begin{aligned} f(x_i|\alpha,\delta,\theta) &= \frac{\alpha|x_i|^{1/(\alpha-1)}\left(\frac{\alpha}{\delta}\right)^{1/(\alpha-1)}\exp(-\theta x_i+\frac{\delta\theta^\alpha}{\alpha})}{|\alpha-1|} \\ &\times \int_{-1/2}^{1/2}\exp\left[-\left|\frac{x_i}{\tau_\alpha(y_i)}\right|^{\alpha/(\alpha-1)}\left(\frac{\alpha}{\delta}\right)^{1/(\alpha-1)}\right] \\ &\times \left|\frac{1}{\tau_\alpha(y_i)}\right|^{\alpha/(\alpha-1)} dy_i, \end{aligned} \tag{19.10}$$

where X is $\left(\frac{\delta}{\alpha}\right)^{1/\alpha}W$. For $\delta=\theta^{1-\alpha}$, the mean is equal to 1 in (19.2). Hougaard (2000) discussed a reparametrization $\delta=\eta^{1-\alpha}$ and $\theta=\eta$; the resulting distribution has mean 1 and variance $(1-\alpha)/\eta$. After reparametrization of δ and θ in terms of η, (in order to yield mean 1), the density (19.10) becomes

$$\begin{aligned} f(x_i|\alpha,\eta) &= \frac{\alpha^{\alpha/(\alpha-1)}|x_i|^{1/(\alpha-1)}\eta\exp(-\eta x_i+\frac{\eta}{\alpha})}{|\alpha-1|} \\ &\times \int_{-1/2}^{1/2}\exp\left[-\left|\frac{x_i}{\tau_\alpha(y_i)}\right|^{\alpha/(\alpha-1)}\alpha^{\frac{1}{\alpha-1}}\eta\right]\left|\frac{1}{\tau_\alpha(y_i)}\right|^{\alpha/(\alpha-1)} dy_i. \end{aligned} \tag{19.11}$$

The next section incorporates (19.11) into the likelihood formulation for multivariate times to events.

19.4 Likelihood and Prior Specification

Let δ_{ij} denote the indicator variable which is 1 if the jth subject is in the ith group fails, and is 0 otherwise. Then, t_{ij} denotes the failure time of the jth subject in the ith group if $\delta_{ij}=1$; otherwise, t_{ij} is the censoring time. The triplet $(t_{ij},\delta_{ij},\widetilde{z}_{ij})$ is observed for the jth subject in the ith group. Let all such triplets be denoted by $\widetilde{Z}$, and let $\widetilde{x}=(x_1,\ldots,x_n)$ denote the frailty. We refer to $(\widetilde{x},\widetilde{Z})$ as the *complete data,* and $\widetilde{x}$ is treated as an unknown parameter vector in the Bayesian formulation.

The complete data likelihood is

$$\begin{aligned} L\left(\lambda,\gamma,\widetilde{\beta},\eta,\alpha|\widetilde{Z},\widetilde{x}\right) &= \Pi_{i=1}^n\Pi_{j=1}^m\left\{\lambda\gamma t_{ij}^{\gamma-1}\exp\left[\widetilde{\beta}'\widetilde{z}_{ij}\right]x_i\right\}^{\delta_{ij}} \\ &\times \exp\left\{-\lambda t_{ij}^\gamma\exp(\widetilde{\beta}'\widetilde{z}_{ij})x_i\right\}. \end{aligned} \tag{19.12}$$

The observed data likelihood $L\left(\lambda, \gamma, \widetilde{\beta}, \eta, \alpha | \widetilde{Z}\right)$, which depends only on the observed data $\widetilde{Z}$ is obtained by integrating out the x_is from (19.12) using the PVF density in (19.11) as

$$\begin{aligned} L\left(\lambda, \gamma, \widetilde{\beta}, \eta, \alpha | \widetilde{Z}\right) &= \Pi_{i=1}^{n} \int \Pi_{j=1}^{m} \left\{ \lambda \gamma t_{ij}^{\gamma-1} \exp(\widetilde{\beta}' \widetilde{z}_{ij}) x_i \right\}^{\delta_{ij}} \\ &\quad \times \exp\left\{ -\lambda t_{ij}^{\gamma} \exp\left(\widetilde{\beta}' \widetilde{z}_{ij}\right) x_i \right\} \\ &\quad \times \frac{\alpha^{\alpha/(\alpha-1)} |x_i|^{1/(\alpha-1)} \eta \exp(-\eta x_i + \frac{\eta}{\alpha})}{|\alpha - 1|} \\ &\quad \times \int_{-1/2}^{1/2} \exp\left[- \left| \frac{x_i}{\tau_\alpha(y_i)} \right|^{\alpha/(\alpha-1)} \alpha^{\frac{1}{\alpha-1}} \eta \right] \\ &\quad \times \left| \frac{1}{\tau_\alpha(y_i)} \right|^{\alpha/(\alpha-1)} dy_i \, dx_i, \end{aligned} \tag{19.13}$$

where y_i denotes an auxiliary variable [see (19.8)]. The complete data likelihood corresponds to a conditional model given the frailty X_i while the observed data likelihood corresponds to a marginal model in which the frailty is integrated out.

The prior specification for the model parameters follows. We choose a Gamma(ρ, ρ) prior for λ, a Gamma(κ, κ) prior for γ, a Normal$(\widetilde{e}, D)$ prior for the p-dimensional vector $\widetilde{\beta}$, where D is a $p \times p$ matrix, a Uniform(0,1) prior for α, and a Gamma(c, c) prior for η. The posterior density based on the observed data likelihood is proportional to the product of the likelihood and the joint prior (assuming independence of all parameters in the model), i.e.,

$$p\left(\lambda, \gamma, \widetilde{\beta}, \eta, \alpha | \widetilde{Z}\right) \propto L\left(\lambda, \gamma, \widetilde{\beta}, \eta, \alpha | \widetilde{Z}\right) p(\lambda) p(\gamma) p\left(\widetilde{\beta}\right) p(\eta) p(\alpha). \tag{19.14}$$

Notice that this posterior density involves a double integration with respect to $\widetilde{x} = (x_1, \ldots, x_n)$ and $\widetilde{y} = (y_1, \ldots, y_n)$. However, we replace integration by Monte Carlo simulation and thereby compute the posterior density, and run the Gibbs sampler to generate samples from $p\left(\lambda, \gamma, \widetilde{\beta}, \eta, \alpha | \widetilde{Z}\right)$. Given initial values for $\lambda, \gamma, \widetilde{\beta}, \eta, \alpha$, and the two "augmented" vectors $\widetilde{x}$ and $\widetilde{y}$, we generate samples of parameters $\lambda, \gamma, \widetilde{\beta}, \eta, \alpha$ as well as $\widetilde{x}$ and $\widetilde{y}$ from their respective complete conditional distributions, as described in the next section.

19.5 Conditional Distributions

We derive forms for the complete conditional distributions of the model parameters as being proportional to (19.14); in each case, the conditional distribution is obtained by retaining only those quantities relevant to that parameter in the posterior density. Markov chain Monte Carlo algorithms then enable sampling from these distributions.

The complete conditional distribution of λ, given all other parameters and the data, is proportional to

$$f_1(\lambda) = \lambda^{\sum_{i=1}^{n}\sum_{j=1}^{m}\delta_{ij} + \rho - 1} \exp\left\{ -\lambda \sum_{i=1}^{n}\sum_{j=1}^{m} t_{ij}^{\gamma} \exp\left(\widetilde{\beta}' \widetilde{z}_{ij}\right) x_i - \lambda\rho \right\}. \quad (19.15)$$

The conditional distribution of λ is then

$$\text{Gamma}\left\{ \sum_{i=1}^{n}\sum_{j=1}^{m}\delta_{ij} + \rho,\ \sum_{i=1}^{n}\sum_{j=1}^{m} t_{ij}^{\gamma} \exp\left(\widetilde{\beta}'\widetilde{z}_{ij}\right) x_i + \rho \right\}.$$

The generation of samples of λ is straightforward. The complete conditional distribution of γ is proportional to

$$\begin{aligned} f_2(\gamma) &= \gamma^{\sum_{i=1}^{n}\sum_{j=1}^{m}\delta_{ij} + \kappa - 1} \prod_{i=1}^{n}\prod_{j=1}^{m}\left(t_{ij}^{\gamma-1}\right)^{\delta_{ij}} \\ &\quad \times \exp\left\{ -\lambda \sum_{i=1}^{n}\sum_{j=1}^{m} t_{ij}^{\gamma} \exp\left(\widetilde{\beta}'\widetilde{z}_{ij}\right) x_i - \gamma\kappa \right\}. \end{aligned} \quad (19.16)$$

The ratio of uniforms method [Wakefield, Gelfand and Smith (1991)] is used to generate samples from $f_2(\gamma)$. The complete conditional distribution of β_s is proportional to

$$\begin{aligned} f_3(\beta_s) &= \exp\left\{ \sum_{i=1}^{n}\sum_{j=1}^{m} \widetilde{\beta}'\widetilde{z}_{ij}\delta_{ij} - \lambda \sum_{i=1}^{n}\sum_{j=1}^{m} t_{ij}^{\gamma} \right. \\ &\quad \left. \times \exp\left(\widetilde{\beta}'\widetilde{z}_{ij}\right) x_i - \frac{(\beta_s - e_s)^2}{2d_s} \right\}, \end{aligned} \quad (19.17)$$

where e_s is the prior mean of β_s and d_s is the prior variance of β_s for $s = 1, \ldots, p$. The complete conditional distribution of x_i is proportional to

$$\begin{aligned} f_4(x_i) &= |x_i|^{\sum_{j=1}^{m}\delta_{ij} + \frac{1}{\alpha-1}} \exp\left\{ -\lambda \sum_{j=1}^{m} t_{ij}^{\gamma} \exp\left(\widetilde{\beta}'\widetilde{z}_{ij}\right) x_i \right. \\ &\quad \left. - \left|\frac{x_i}{\tau_\alpha(y_i)}\right|^{\alpha/(\alpha-1)} \alpha^{\frac{1}{\alpha-1}}\eta - \eta x_i \right\}. \end{aligned} \quad (19.18)$$

The ratio of uniforms method is again used to generate samples from $f_3(\beta_s)$ and $f_4(x_i)$. The complete conditional distribution of η is proportional to

$$
\begin{aligned}
f_5(\eta) &= \eta^{n+c-1}\exp\left\{-\sum_{i=1}^{n}\left|\frac{x_i}{\tau_\alpha(y_i)}\right|^{\alpha/(\alpha-1)}\alpha^{\frac{1}{\alpha-1}}\eta \right. \\
&\quad \left. -\eta\sum_{i=1}^{n}x_i + \frac{n\eta}{\alpha} - \eta c\right\}. \qquad (19.19)
\end{aligned}
$$

Although this complete conditional *appears* to have the form of a Gamma distribution, this is not the case, since the values that would constitute the shape and scale parameters are not always positive. In particular, the term $\left[\sum_{i=1}^{n}\left|\frac{x_i}{\tau_\alpha(y_i)}\right|^{\alpha/(\alpha-1)}\alpha^{\frac{1}{\alpha-1}} + \sum_{i=1}^{n}x_i + c - \frac{n}{\alpha}\right]$ is not always positive. Straightforward draws from a standard density are hence not possible, and we have used the multiple-try-Metropolis-Hastings algorithm with a lognormal proposal in order to generate posterior samples for η. This algorithm incorporates a local optimization step into an MCMC sampler in continuous state space [Chen, Shao and Ibrahim (2000)].

The conditional distribution of α is proportional to

$$
\begin{aligned}
f_6(\alpha) &= \frac{\alpha^{n\alpha/(\alpha-1)}}{|\alpha-1|^n}\prod_{i=1}^{n}\left\{|x_i|^{1/(\alpha-1)}\left|\frac{1}{\tau_\alpha(y_i)}\right|^{\alpha/(\alpha-1)}\right\} \\
&\quad \times\exp\left\{-\sum_{i=1}^{n}\left|\frac{x_i}{\tau_\alpha(y_i)}\right|^{\alpha/(\alpha-1)}\alpha^{\frac{1}{\alpha-1}}\eta + \frac{n\eta}{\alpha}\right\}. \qquad (19.20)
\end{aligned}
$$

Again, this does not correspond to any standard form. We use the Metropolis-Hastings algorithm with a beta proposal [see Buckle (1995)] to generate posterior samples for α. The complete conditional distribution of y_i is proportional to

$$
f_7(y_i) = \left|\frac{1}{\tau_\alpha(y_i)}\right|^{\alpha/(\alpha-1)}\exp\left\{-\left|\frac{x_i}{\tau_\alpha(y_i)}\right|^{\alpha/(\alpha-1)}\alpha^{\frac{1}{\alpha-1}}\eta\right\}. \qquad (19.21)
$$

A rejection algorithm [Devroye (1986)] is employed. Some details of the sampling algorithms are given in the appendix.

19.6 Illustration

We illustrate our approach using data on times to first and second occurrence of infection in 38 patients on portable dialysis machines [McGilchrist and Aisbett (1991)]. Here δ_{ij} are binary variables representing respectively the censoring

Table 19.1: Posterior summary of model parameters

Parameter	Posterior Mean	Posterior Median	95% Credible Interval
PVF Frailty			
γ	1.253	1.240	$(0.945, 1.598)$
β	-1.455	-1.437	$(-2.681, -0.319)$
λ	0.014	0.013	$(0.003, 0.036)$
α	0.297	0.262	$(0.065, 0.735)$
η	0.784	0.733	$(0.181, 1.820)$
Positive Stable Frailty			
γ	1.357	1.353	(1.021,1.720)
β	-2.106	-2.114	(-3.689, -0.434)
λ	0.008	0.005	(0.001,0.026)
α	0.569	0.557	(0.402,0.773)
Gamma Frailty			
γ	1.066	1.046	$(0.844, 1.374)$
β	-1.105	-1.056	$(-2.242, -0.206)$
λ	0.019	0.017	$(0.003, 0.047)$
$1/v$	0.201	0.205	(0.107,1.99)

indicators for the first and second recurrences. Occurrence of infection is indicated by 1, and censoring by 0. The gender of the patients (0 indicating male, and 1 indicating female), is a covariate. Other covariates, such as age and disease type of each patient, are also available with this data, but are omitted in this analysis as their effect on infection times was shown to be insignificant in previous work of McGilchrist and Aisbett (1991).

For the PVF$(\alpha, \eta^{1-\alpha}, \eta)$ frailty specification, we choose a noninformative prior for each parameter in the model. The prior for α is $p(\alpha) = 1$, $0 < \alpha < 1$, while the prior for η is Gamma$(0.1, 0.1)$. We choose the value 0.001 for each of ρ and κ in the prior specification. The prior on β is Normal$(0, 10^3)$, where β is the coefficient corresponding to gender. For the lognormal proposal for the generation of η, we assume that the standard deviation of the normal distribution is 0.4.

Using Gibbs sampling, we generate samples from the complete conditional distributions [see (19.15)–(19.21) and the Appendix]. Considering 10,000 iterates for burn in, we use 20,000 additional iterates for making inference. The CPU time was approximately 4 hours and 45 minutes on the IBM SP model 9076-2A4. In Table 19.1, we present the posterior means and quantiles of each parameter.

The negative estimate of β implies that the female patients have a slightly lower risk of infection. The estimated posterior distribution of all parameters except β are very slightly skewed to the right. In terms of the original parametrization, the posterior parameter estimates of δ and θ are, respectively, 0.843 and 0.784. The posterior estimates of α, δ and θ give us an idea about which frailty (gamma, positive stable or PVF) will give a better fit. For this data, none of the estimated values of α, δ and θ fall in the restricted parameter regions corresponding to the gamma or positive stable distributions; therefore, it appears that PVF frailty gives a better fit than either of these two (gamma or positive stable) frailties.

The median concordance for the PVF frailty model is 0.2639 [see (19.4)], showing that the degree of dependence between infection times for each patient is not large. However, it does not indicate near independence either. The posterior estimate of Kendall's τ [see (19.3)] is 0.2561 and has the same interpretation as median concordance. We illustrate the behavior of $\widehat{\chi}(t_1, t_2)$, the Monte Carlo estimate of the cross-ratio function [see (19.5)] for bivariate survival times evaluated at the posterior means of $\gamma, \beta, \lambda, \alpha$ and η. Figure 19.1 shows the 3-dimensional plot of $\widehat{\chi}(.)$ over a grid of (t_1, t_2) values from 0 to 100, and supports the intermediate dependence property of the PVF frailty. The plot provides strong evidence that $\widehat{\chi}(t_1, t_2)$ (cross-ratio) decreases as t_1 and t_2 increase. However, the decrease in cross-ratio is not as large as in the positive stable frailty case (see Figure 19.2).

For comparison, estimates of the posterior means and quantiles of the parameters of a positive stable frailty model with Weibull baseline hazard are also shown in Table 19.1. The 95 % credible interval for β is wider than that corresponding to the PVF frailty model, while the median concordance and Kendall's τ are respectively 0.4308 and 0.4313 (indicating lack of strong dependence between infection times for each patient). Table 19.1 also presents posterior estimates under the Gamma(v, v) frailty model with Weibull baseline hazard; here, the median concordance and Kendall's τ are 0.0886 and 0.0913, respectively. Conditional Predictive Ordinate (CPO) plots indicate that the PVF frailty model is supported over the gamma frailty model by 62% and over the positive stable frailty model by 57% of the observations.

19.7 Concluding Remarks

This article develops inference for the PVF frailty model for multivariate survival data. Although we have considered a Weibull baseline hazard for simplicity, the development of the likelihood function via a tilted positive stable density approach is completely general and should be useful for modeling a wide variety of lifetimes data. Further, for the case when $\alpha < 0$, it is possible to use the compound Poisson representation of the PVF frailty density function to enable

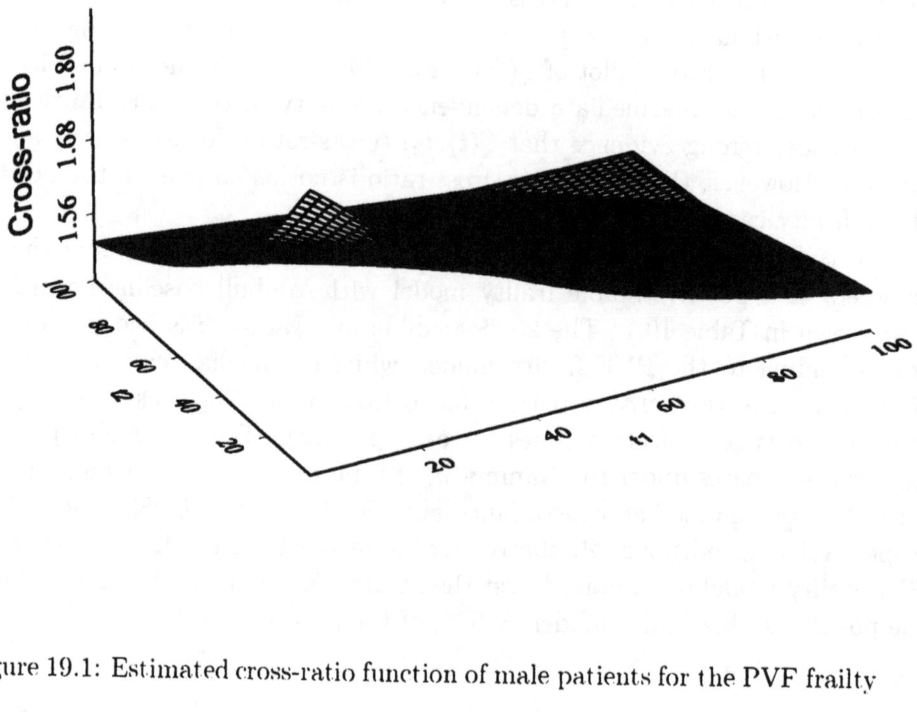

Figure 19.1: Estimated cross-ratio function of male patients for the PVF frailty

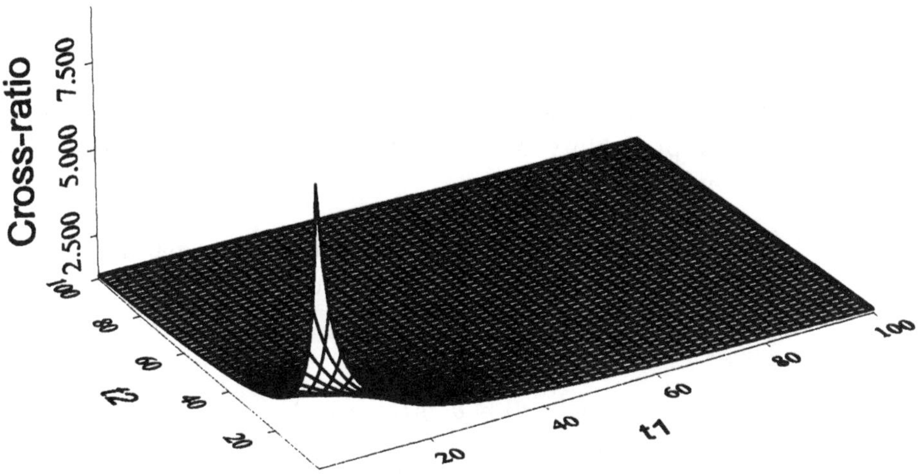

Figure 19.2: Estimated cross-ratio function of male patients for the positive stable frailty

Bayesian inference. This will be useful in competing risks models and in studies of death incidence where occurrence of the event under study is uncertain, and will be addressed in future work.

Appendix

We present details of some sampling algorithms employed in Section 19.5. For details of other algorithms, the reader is referred to Qiou, Ravishanker and Dey (1999).

The complete conditional distribution for η in (19.19) is not in standard form. We use the multiple-try-Metropolis-Hastings algorithm [Chen, Shao and Ibrahim (2000)], with a lognormal proposal. Let η_{k-1} denote the η value generated at the $(k-1)$th iteration. At the kth iteration, the corresponding normal mean is set at $\log(\eta_{k-1})$ and its variance is set at a reasonable value, so that the acceptance probability is not too high. At the kth iteration, a new η is generated as follows:

Step (i) Generate i samples $v_1, v_2, \ldots, v_i$ from a lognormal distribution whose mean and standard deviation are respectively determined from the normal mean $\log(\eta_{k-1})$ and normal standard deviation s. Compute $w(v_j,\ \eta_{k-1})$ for $j = 1, \ldots, i$ where

$$w(\eta, v) = p\,(\eta|D)\, f(\eta, v) \lambda\,(\eta, v),$$

$f(.,.)$ is the lognormal proposal, and $p\,(\eta|D)$ is the complete conditional distribution of η. Here, $\lambda\,(\eta, v)$ is a nonnegative symmetric function of η and v; $\lambda\,(\eta, v) > 0$ whenever $f(\eta, v) > 0$. We have considered $\lambda\,(\eta, v) = 1$.

Step (ii) Select v_l from among the v_js with probability proportional to $w(v_j, \eta_{k-1})$, $j = 1, \ldots, i$. Then draw $v_1^*, v_2^*, \ldots, v_{i-1}^*$ from the lognormal distribution with mean $\log(v_l)$ and standard deviation s, and let $v_i^* = \eta_{k-1}$.

Step (iii) Generate u from Uniform$(0, 1)$. Set $\eta^k = \eta^*$ if $u \le a$, and $\eta^k = \eta^{k-1}$ otherwise, where the acceptance probability is given by

$$a = \min\left[1, \frac{w(v_1,\eta_{k-1}) + w(v_2,\eta_{k-1}) + \cdots + w(v_i,\eta_{k-1})}{w\left(v_1^*,v_l\right) + w\left(v_2^*,v_l\right) + \cdots + w\left(v_i^*,v_l\right)}\right].$$

For the generation of y, we have used a rejection algorithm [Devroye (1986)]. Clearly $\tau_\alpha\left(\pm\frac{1}{2}\right) = \pm\infty$, and $\tau_\alpha\,(l_\alpha) = 0$, where $l_\alpha = -Z_\alpha/\pi\alpha$ and $Z_\alpha = \min(\alpha, 2-\alpha)\pi/2$ [Buckle (1995)]. Here, $\alpha \in (0, 1]$ and $x_i > 0$, so that in (19.21), $f_7\,(y_i)$ lies in the range $(-1/2, 1/2)$, and attains a maximum value of $\frac{\exp(-1)}{|x_i|^{\frac{\alpha}{\alpha-1}}\alpha^{\frac{1}{\alpha-1}}\eta}$ at $\tau_\alpha\,(y_i) = x_i \alpha^{\frac{1}{\alpha}} \eta^{\frac{\alpha-1}{\alpha}}$. Rewriting $f_7\,(y_i)$ from (19.21), we have

$$f_{77}(y_i) = |\tfrac{x_i}{\tau_\alpha(y_i)}|^{\alpha/(\alpha-1)} \alpha^{\frac{1}{\alpha-1}} \eta \ \exp\{1 - |\tfrac{x_i}{\tau_\alpha(y_i)}|^{\alpha/(\alpha-1)} \alpha^{\frac{1}{\alpha-1}} \eta\}.$$

We see that this expression attains its maximum value of 1. Using the rejection algorithm (Devroye, 1986), we implement the following steps:

Step (i) Generate y_i from Uniform$(-1/2, 1/2)$.
Step (ii) Generate u from Uniform$(0, 1)$.
Step (iii) Reject y_i until $u \leq f_{77}(y_i)$.

References

1. Buckle, D.J. (1995). Bayesian inference for stable distributions, *Journal of the American Statistical Association* **90**, 605–613.

2. Chen, M. H., Shao, Q.-M. and Ibrahim, J. G. (2000). *Monte Carlo Methods in Bayesian Computing*, Springer-Verlag, New York.

3. Clayton, D. and Cuzick, J. (1985). Multivariate generalizations of the proportional hazards model, *Journal of the Royal Statistical Society, Series A*, **148**, 82–117.

4. Devroye, L. (1986). *Non-uniform Random Variate Generation*, Springer-Verlag, New York.

5. Gustafson, P. (1997). Large hierarchical Bayesian analysis of multivariate survival data, *Biometrics*, **53,** 230–243.

6. Hougaard, P. (1986a). Survival models for heterogeneous populations derived from stable distributions, *Biometrika*, **73,** 387–396; Correction, **75**, pp. 395.

7. Hougaard, P. (1986b). A class of multivariate failure time distributions, *Biometrika*, **73,** 671–678.

8. Hougaard, P. (1995). Frailty models for survival data, *Lifetime Data Analysis*, **1**, 255–273.

9. Hougaard, P. (2000). *Analysis of Multivariate Survival Data*, Springer-Verlag, New York.

10. Hougaard, P., Harvald, B. and Holm, N. V. (1992). Measuring the similarities between the lifetime of adult Danish twins born between 1881–1930, *Journal of the American Statistical Association*, **87**, 17–24.

11. Jorgensen, B. (1987). Exponential dispersion model, *Journal of the Royal Statistical Society, Series B*, **49**, 137–162.

12. Lam, K. F. and Kuk, A. Y. C. (1997). A marginal likelihood approach to estimation in frailty models, *Journal of the American Statistical Association*, **92**, 985–990.

13. Manatunga, A. K. and Oakes, D. (1999). Parametric analysis for matched pair survival data, *Lifetime Data Analysis*, **5**, 371–387.

14. McGilchrist, C. A. and Aisbett, C. W. (1991). Regression with frailty in survival analysis, *Biometrics*, **47**, 461–466.

15. Oakes, D. (1994). Use of frailty models for multivariate survival data, *Proceedings of the XVIIth International Biometrics Conference*, pp. 275–286, Hamilton, Ontario, Canada.

16. Qiou, Z., Ravishanker, N. and Dey, D.K. (1999). Multivariate survival analysis with positive stable frailties, *Biometrics*, **55**, 637–644.

17. Wakefield, J. C., Gelfand, A. E. and Smith, A. F. M. (1991). Efficient generation of random variates via the ratio-of-uniforms method, *Statistics and Computing*, **1**, 139–133.

20

A Two-stage Design for Choosing Among Experimental Treatments in Clinical Trials

Linda Rollin[1] **and Pinyuen Chen**[2]

[1]*Bristol-Myers Squibb Company, Wallingford, CT U.S.A.*
[2]*Department of Mathematics, Syracuse University, Syracuse, NY U.S.A.*

Abstract: A two-stage selection and testing design is proposed for choosing among k (≥ 2) experimental treatments, provided it is better than a specific standard. In the first (selection) stage, ranking and selection formulation is adopted to select the one most promising treatment. In the second (testing) stage, hypothesis testing formulation is used to determine if the treatment selected at the first stage is better than the standard. The design allows for early termination at stage one if none of the treatments seem promising. The treatments are assumed to follow normal distributions with unknown means and unknown variances. "Better than the standard" means the population mean of an experimental treatment is larger than the standard. Appropriate definitions of size and power are given. Sample size requirements are compared with an analogous pure selection procedure of Taneja and Dudewicz (1992).

Keywords and phrases: Known standard, least favorable configuration, ranking and selection, unknown control

20.1 Introduction

There are many situations in which we may wish to select the one best among several experimental treatments. For example, a medical researcher may wish to choose the best among four cancer drugs. In this case, the best drug might be defined as the one producing the longest mean survival time in cancerous mice. However, it is often the case that even the best experimental treatment is not good enough. For example, potential serious effects and the cost of developing the new drug may prohibit its selection unless it is much better than the current standard of care for testing cancer. In this case a more appropriate goal is to select the best experimental treatment, but only if it is sufficiently better than a standard preset by the experimenter. This paper considers the setting in which k populations are normally distributed with unknown means and unknown variances. The goal is to select one of the populations, provided that it is better than a specific standard. If none of the k populations is better than the standard, then no population is to be chosen. We assume a large population mean is desirable and hence, "better than the standard" means that the population mean is larger than the standard. We propose a two-stage selection and testing design as a method of achieving this.

For selecting one of several experimental treatments, provided that it is better than a specific standard, Bechhofer and Turnbull (1978) proposed a ranking and selection formulation and two procedures: a single-stage procedure for the case of normal populations with common known variance and a two-stage procedure for the case of normal populations with a common unknown variance. Dunnett (1984) proposed a single-stage procedure for the same goal, with a different formulation for normal populations with a common known variance. Dunnett's approach is similar to that of Bechhofer and Turnbull, the main difference being that the known standard is replaced by an unknown control population.

The two-stage selection and testing approach was first introduced by Thall, Simon, and Ellenberg (1988, 1989) for the goal of comparing several binomial populations with a control population, the same goal studied by Dunnett (1984). Thall, Simon, and Ellenberg's two-stage design consists of a first (selection) stage, in which ranking and selection techniques are used to screen out the one most promising population, and a second (testing) stage, in which hypothesis testing is used to determine if the chosen population is better than the control population. The design allows for early termination of the procedure at stage one if none of the populations seem promising. Thall, Simon, and Ellenberg (1988) showed that their two-stage design for binomial populations required smaller sample sizes than the one-stage procedure of Dunnett (1984) for the same size and power requirements.

This paper adopts the same selection and testing design as in Thall, Simon, and Ellenberg (1988) for the goal of comparing several normal populations with a specific standard. The k (≥ 2) experimental normal populations are assumed to have unknown means and unequal and unknown variances. A two-stage Dudewicz-Dalal-type approach is utilized in each of the selection phase and testing phase of our proposed procedure. This approach was chosen over a Rinott-type approach because of the stochastically increasing property of the Dudewicz-Dalal test statistic. We shall see that this property plays a crucial role in the proof of the least favorable configuration given in Section 20.2.

This chapter is organized as follows. In Section 20.2 we state the assumptions, review the definitions and goals, and outline the details of the proposed procedure. In Section 20.3 the formulas for size and power are derived. An expression for the probability of early termination is also derived in this section. In Section 20.4 we give an explanation of related tables that appear at the end of the paper, and describe how the tables were constructed. The tables provide the parameter values necessary to implement the procedure. An illustrative example is given in Section 20.5. Section 20.6 provides a comparison of sample sizes for our procedure and for the analogous Dudewicz-Dalal-type procedure of Taneja and Dudewicz (1992).

20.2 Assumptions, Definitions, Goals, and Proposed Procedure

We assume that we have k normal populations, $\pi_1, \pi_2, \ldots, \pi_k$, having unknown means $\mu_1, \mu_2, \ldots, \mu_k$, respectively, and unknown variances $\sigma_1^2, \sigma_2^2, \ldots, \sigma_k^2$. We also assume we have a fixed known standard, μ_0, to which the μ_i are to be compared. The ranked values of the population means will be denoted by $\mu_{[1]} \leq \mu_{[2]} \leq \cdots \leq \mu_{[k]}$.

The experimenter will specify constants δ_1^* and δ_2^* $(0 \leq \delta_1^* < \delta_2^*)$ such that $\mu_0 + \delta_1^*$ represents only a marginal improvement over the standard, while $\mu_0 + \delta_2^*$ represents a practically significant improvement over the standard. A population, π_γ is considered 'acceptable' if its mean, μ_γ, satisfies $\mu_\gamma \geq \mu_0 + \delta_2^*$.

The goal is to select from the k populations an acceptable population if one exists, and to select no population if all population means are at most a marginal improvement over the standard. We express the goal in the form of a null and alternative hypothesis as follows.

$$
\begin{aligned}
H_0: &\quad \mu_{[1]} = \mu_{[2]} = \cdots = \mu_{[k]} = \mu_0. \\
H_a: &\quad \text{there is at least one acceptable population,} \\
&\quad \text{and no } \mu_i \text{ lies in the interval } (\mu_0 + \delta_1^*, \mu_0 + \delta_2^*). \qquad (20.1)
\end{aligned}
$$

We use the term 'preference zone' (PZ) to refer to the set of all parameter vectors $\boldsymbol{\mu} = (\mu_1, \mu_2, \ldots, \mu_k)$ satisfying the alternative hypothesis. A decision resulting in the selection of some population coincides with rejecting H_0, while a decision to select no population coincides with failing to reject H_0. Definitions of size (α), the power function $(1 - \beta(\boldsymbol{\mu}))$, and the least favorable configuration (LFC) are given below.

$\alpha = P(\text{selecting } \pi_i \text{ for some } i = 1, \ldots, k \text{ when } H_0 \text{ is true})$

$1 - \beta(\boldsymbol{\mu}) = P(\text{selecting an acceptable population } |\boldsymbol{\mu})$

LFC = the parameter vector $\boldsymbol{\mu}$ that minimizes the power function over all $\boldsymbol{\mu} \in PZ$

The selection and testing procedure which we propose when the σ_i^2 are unequal and unknown will be denoted by $\mathcal{P}_{UU}$. UU refers to the variance assumptions, with the first U standing for unequal, and the second U standing for unknown. For pre-specified integers $n_0 \geq 2$ and $m_0 \geq 2$ and positive constants y_1, y_2, h_1, and h_2, procedure $\mathcal{P}_{UU}$ is defined as follows.

Stage 1 (The Selection Stage):

1. Take n_0 observations $X_{i,1}, X_{i,2}, \ldots, X_{i,n_0}$ $(i = 1, \ldots, k)$ from each of the k populations $\pi_1, \pi_2, \ldots, \pi_k$.

2. For $i = 1, \ldots, k$ compute

$$\bar{X}_i(n_0) = \frac{\sum_{j=1}^{n_0} X_{ij}}{n_0},$$
$$S_i^2 = \frac{\sum_{j=1}^{n_0}(X_{ij} - \bar{X}_i(n_0))^2}{n_0 = 1},$$

and

$$n_i = \max\left\{n_0 + 1, \left[\frac{h_1^2 S_i^2}{y_1^2}\right] + 1\right\},$$

where $[x]$ denotes the greatest integer less than x.

3. Take $n_i - n_0$ additional observations $X_{i,n_0+1}, \ldots, X_{i,n_i}$ from each π_i, $(i = 1, \ldots, k)$, and define

$$\tilde{X}_i = \sum_{j=1}^{n_i} a_{ij} X_{ij} \quad (x = 1, \ldots, k),$$

where the a_{ij} $(j = 1, 2, \ldots, n_i;\ i = 1, 2, \ldots, k)$ are chosen so that

$$\sum_{j=1}^{n_i} a_{ij} = 1, \quad a_{i1} = a_{i2} = \cdots = a_{in_0}, \text{ and}$$
$$S_i^2 \sum_{j=1}^{n_i} a_{ij}^2 = \frac{y_1^2}{h_1^2}. \tag{20.2}$$

Denote the ordered values of the $\tilde{X}_i$ by $\tilde{X}_{[1]} \leq \tilde{X}_{[2]} \leq \cdots \leq \tilde{X}_{[k]}$.

4. If $\tilde{X}_{[k]} \geq \mu_0 + y_1$, then select the population π_γ which produced the largest test statistic, $\tilde{X}_{[k]}$, and proceed to stage 2.

If $\tilde{X}_{[k]} < \mu_0 + y_1$, then terminate the procedure and select no population. That is, do not reject H_0.

Stage 2 (The Testing Stage):

1. Take m_0 additional observations $W_1, W_2, \ldots, W_{m_0}$ from π_γ.

2. Compute

$$\bar{W}(m_0) = \frac{\sum_{j=1}^{m_0} W_j}{m_0},$$
$$S_W^2 = \frac{sum_{j=1}^{m_0}(W_j - \bar{W}(m_0))^2}{m_0 - 1}, \text{ and}$$
$$m = \max\left\{m_0 + 1, \left[\frac{h_2^2 W_W^2}{y_2^2}\right] + 1\right\},$$

where $[x]$ denotes the greatest integer less than x.

3. Take $m - m_0$ additional observations $W_{m_0+1}, \ldots, W_m$ from π_γ, and define

$$\tilde{W} = \sum_{j=1}^{m\cdot} b_j W_j,$$

where the b_j $(j = 1, 2, \ldots, m)$ are chosen so that

$$\sum_{j=1}^{m} b_j = 1, \qquad b_1 = b_2 = \cdots = b_{m_0},$$
$$S_W^2 \sum_{j=1}^{m} b_j^2 = \frac{y_2^2}{h_2^2}. \tag{20.3}$$

4. If $\frac{\tilde{X}_\gamma + \tilde{W}}{2} > \mu_0 + y_2$ then reject H_0 and select the population π_γ. If $\frac{\tilde{X}_\gamma + \tilde{W}}{2} \leq \mu_0 + y_2$ then select no population and do not reject H_0.

In the next section we will show how to choose the design parameters $(n_0, m_0, y_1, y_2, h_1, h_2)$ in order to achieve the desired levels of size and power. In order to implement procedure $\mathcal{P}_{UU}$, one must choose the constants a_{ij} and b_j to satisfy conditions (20.2) and (20.3), respectively. The fact that such constants can be chosen is proven in the next lemma.

Lemma 20.2.1 *Let z be any positive constant, and suppose an integer n is chosen so that $n = \max\left\{n_0 + 1, \left[\frac{S^2}{z}\right] + 1\right\}$, where $[x]$ denotes the greatest integer less than x. Then constants $a_1, a_2, \ldots, a_n$ can be chosen so that $\sum_{j=1}^{n} a_j = 1$, $a_1 = a_2 = \cdots = a_{n_0}$, and $S^2 \sum_{j=1}^{n} a_j^2 = z$.*

PROOF. Set $c = \frac{n_0}{n}\left(1+\sqrt{1-\frac{n}{n_0}\left(1-\frac{(n-n_0)z}{S^2}\right)}\right)$ and set $a_1 = a_2 = \cdots = a_{n_0} = \frac{c}{n_0}$ and $a_{n_0+1} = \cdots = a_n = \frac{1-c}{n-n_0}$. c is a real number as long as the quantity under the square root is non-negative. This is guaranteed because n has been chosen so that $n \geq \frac{S^2}{z}$ and $n - n_0 > 0$. Thus,

$$1 - \frac{n}{n_0}\left(1 - \frac{(n-n_0)z}{S^2}\right) \geq 1 - \frac{n}{n_0}\left(1 - \frac{(n-n_0)}{n}\right) = 0.$$

Now, clearly, $a_1 = a_2 = \cdots = a_{n_0}$ and

$$\sum_{j=1}^{n} a_j = \sum_{j=1}^{n_0} a_j + \sum_{j=n_0+1}^{n} a_j = n_0 \cdot \frac{c}{n_0} + (n-n_0)\cdot\frac{1-c}{n-n_0} = 1.$$

Finally,

$$\begin{aligned}
S^2\sum_{j=1}^{n} a_j^2 &= S^2\left(n_0\cdot\left(\frac{c}{n_0}\right)^2 + (n-n_0)\cdot\left(\frac{1-c}{n-n_0}\right)^2\right) \\
&= S^2\left(\frac{c^2}{n_0} + \frac{(1-c)^2}{n-n_0}\right) = S^2\left(\frac{(n-n_0)c^2 + n_0(1-c)^2}{(n-n_0)n_0}\right) \\
&= S^2\left(\frac{nc^2 + n_0 - 2n_0c}{(n-n_0)n_0}\right) = \frac{S^2}{n-n_0}\left(\frac{n}{n_0}c^2 + 1 - 2c\right) \\
&= \frac{S^2}{n-n_0}\left[\frac{n_0}{n}\left(1 + 2\sqrt{1-\frac{n}{n_0}\left(1-\frac{(n-n_0)z}{S^2}\right)}\right.\right. \\
&\quad \left. + 1 - \frac{n}{n_0}\left(1 - \frac{(n-n_0)z}{S^2}\right)\right) \\
&\quad \left. + 1 - 2\frac{n_0}{n}\left(1+\sqrt{1-\frac{n}{n_0}\left(1-\frac{(n-n_0)z}{S^2}\right)}\right)\right] \\
&= \frac{S^2}{n-n_0}\left[1\frac{n_0}{n} - 1 + \frac{(n-n_0)z}{S^2} + 1 - 2\frac{n_0}{n}\right] = z.
\end{aligned}$$

20.3 Size, Power, and the Least Favorable Configuration

In order to derive the least favorable configuration for procedure $\mathcal{P}_{UU}$, we shall use Mahamunulu (1967)'s result for stochastically increasing families of distribution functions. We will first need two lemmas. In the first lemma, we determine the distribution of the test statistics $\tilde{X}_i$ used in procedure $\mathcal{P}_{UU}$. In the second lemma, we show that the $\tilde{X}_i$ do indeed form a stochastically increasing family.

Lemma 20.3.1 *Let $X_1, X_2, \ldots, X_{n_0}, \ldots, X_n$ be iid $N(\mu, \sigma^2)$ random variables, $S^2 = \frac{\sum_{j=1}^{n_0}\left(X_j - \sum_{j=1}^{n_0} \frac{X_j}{n_0}\right)^2}{n_0 = 1}$, z a positive constant, and $a_1, a_2, \ldots, a_{n_0}, a_{n_0+1}, \ldots, a_n$ constants that satisfy the conditions*

$$\sum_{j=1}^{n} a_j = 1, \qquad a_1 = a_2 = \cdots = a_{n_0}, \qquad \textit{and}$$
$$S^2 \sum_{j=1}^{n} a_j^2 = a.$$

Let $\tilde{X}$ be defined by $\tilde{X} = \sum_{j=1}^{n} a_j X_j$, and U be defined by $U = \frac{\tilde{x}-\mu}{\sqrt{z}}$. Then U has a student-t distribution with $n_0 - 1$ degrees of freedom.

PROOF. Since $\sum_{j=1}^{n} a_j = 1$, we have

$$U = \frac{\tilde{X} - \mu}{\sqrt{z}} = \frac{\left(\sum_{j=1}^{n} a_j X_j\right) - \mu}{\sqrt{z}} = \frac{\sum_{j=1}^{n} a_j (X_j - \mu)}{\sqrt{z}}.$$

Now, $\frac{(n_0-1)S^2}{\sigma^2}$ has a chi-square distribution with $n_0 - 1$ degrees of freedom, and conditional on S^2, the a_js and $z = S^2 \sum_{j=1}^{n} a_j^2$ are constant. Thus, the conditional distribution of U given $S^2 = s_0^2$ is normal with a mean and variance given by

$$E(U \mid S^2 = s_0^2 0 = E\left(\frac{\sum_{j=1}^{n} a_j (X_j - \mu)}{\sqrt{z}} \middle| S^2 = s_0^2\right) = \frac{1}{\sqrt{z}} \sum_{j=1}^{n} a_j [E(X_j) - \mu] = 0,$$

$$\begin{aligned}
\mathrm{Var}(U|S^2 = s_0^2) &= \mathrm{Var}\left(\frac{\left(\sum_{j=1}^{n} a_j X_j\right) - \mu}{\sqrt{z}} \middle| S^2 = s_0^2\right) = \frac{1}{z} \sum_{j=1}^{n} \mathrm{Var}(a_j X_j) \\
&= \frac{1}{z} \sum_{j=1}^{n} a_j^2 \mathrm{Var}(X_j) = \frac{\sigma^2}{z} a_j^2 = \frac{\sigma^2}{s_0^2 \sum_{j=1}^{n} a_j^2} \sum_{j=1}^{n} a_j^2 = \frac{\sigma^2}{s_0^2}.
\end{aligned}$$

On the other hand, a student-t random variable, T, with $n_0 - 1$ degrees of freedom can be written as $T = \frac{Y}{S} = \frac{Y/\sigma}{S/\sigma}$ where Y is a $N(0, \sigma^2)$ random variable, $\frac{(n_0-1)S^2}{\sigma^2}$ has a chi-square distribution with $n_0 - 1$ degrees of freedom, and Y and S are independent. Thus, the conditional distribution of T given $S^2 = s_0^2$ is $N\left(0, \frac{\sigma^2}{s_0^2}\right)$. Since the conditional distribution of U given S^2 is the same as the conditional distribution of T given S^2, U and T must have the same distribution.

Lemma 20.3.2 *Let $F(\tilde{X}_i|\mu_i)$ denote the distribution function of $\tilde{X}_i$, and $F(\tilde{W}|\mu_\gamma)$ denote the distribution function of $\tilde{W}$. Then the collections $\{F(\tilde{X}_i|\mu_i) : i = 1, 2, \ldots, k\}$ and $\{G(\tilde{W}|\mu_\gamma) : \gamma = 1, 2, \ldots, k\}$ form two separate stochastically increasing families.*

PROOF. We need to show that for fixed x, $F(\tilde{X}_i|\mu_i) = P(\tilde{X}_i \leq x)$ is a non-increasing function of μ_i, and $G(\tilde{W}|\mu_\gamma) = P(\tilde{W} \leq x)$ is a non-increasing function of μ_γ. Let $\mu_1 < \mu_2$. Then

$$\begin{aligned} F(\tilde{X}_1|\mu_1) &= P(\tilde{X}_1 \leq x) = P\left\{\frac{\tilde{X}_1 - \mu_1}{\sqrt{z}} \leq \frac{x - \mu_1}{\sqrt{z}}\right\} = P\left\{T \leq \frac{x-\mu_1}{\sqrt{z}}\right\} \\ &= P\left\{\frac{\tilde{X}_2 - \mu_2}{\sqrt{z}} \leq \frac{x - \mu_1}{\sqrt{z}}\right\} = P\left\{\tilde{X}_2 \leq x + \mu_2 - \mu_1\right\} \\ &\geq P\{\tilde{X}_2 \leq x\} = F(\tilde{X}_2|\mu_2) \end{aligned}$$

where T is a Student-t random variable with $n_0 - 1$ degrees of freedom. The inequality in the last line follows from the fact that $\mu_2 - \mu_1 > 0$ by hypothesis. This shows that $F(\tilde{X}_i|\mu_i)$ is a non-increasing function of μ_i. By replacing n_0 with m_0, the same argument can be used to show that $G(\tilde{W}|\mu_\gamma)$ is a non-increasing function of μ_γ. ■

Theorem 20.3.1 *The least favorable configuration for procedure $\mathcal{P}_{UU}$ is given in the form:*

$$\mu_{[1]} = \mu_{[2]} = \cdots = \mu_{[k-1]} = \mu_0 + \delta_1^* < \mu_0 + \delta_2^* = \mu_{[k]}. \tag{20.4}$$

PROOF. We need to show that configuration (20.4) minimizes the power function over all possible configurations $(\mu_1, \mu_2, \ldots, \mu_k)$ within PZ. For convenience of notation we shall drop the square bracket notation and assume that $\mu_1 \leq \mu_2 \leq \cdots \leq \mu_k$. A typical configuration in PZ has the form:

$$\mu_1 \leq \mu_2 \leq \cdots \leq \mu_{k-m} \leq \mu_0 + \delta_1^* < \mu_0 + \delta_2^* \leq \mu_{k-m+1} \leq \cdots \leq \mu_k \tag{20.5}$$

where $1 \leq m \leq k$, the lower bound on m following from the fact that at least one $\mu_i \geq \mu_0 + \delta_2^*$. Let $A = \{k - m + 1, \ldots, k\}$ denote the set of indices of acceptable populations, and $A^c = \{1, \ldots, k - m\}$. Then

$$\begin{aligned} &P(\text{selecting an acceptable population}) \\ &= P\left(\max_{i\in A} \tilde{X}_i > \max_{j\in A^c} \tilde{X}_j; \max_{i\in A} \tilde{X}_i \geq y_1 + \mu_0; \frac{\tilde{X}_\gamma + \tilde{W}}{2} > y_2 + \mu_0\right) \end{aligned} \tag{20.6}$$

where γ is the index satisfying $\tilde{X}_\gamma = \max\{\tilde{X}_1, \tilde{X}_2, \ldots, \tilde{X}_k\}$. Let $F(\tilde{x}_h|\mu_j)$ denote the distribution function of $\tilde{X}_h$ for $h = 1, 2, \ldots, k$. That is, $F(\tilde{x}_h|\mu_h)$ is the distribution function of $\frac{y_1}{h_1}T + \mu_h$, where T is a Student-t random variable with $n_0 - 1$ degrees of freedom. Similarly, let $G(\tilde{w}|\mu_\gamma)$ denote the distribution function of $\tilde{W}, \gamma = 1, 2, \ldots, k$. That is, $G(\tilde{w}|\mu_\gamma)$ is the distribution function of $\frac{y_2}{h_2}T + \mu_\gamma$, where T is a Student-t random variable with $m_0 - 1$ degrees of freedom. Note that $\tilde{X}_1, \tilde{X}_2, \ldots, \tilde{X}_k$ and $\tilde{W}$ are independent random variables. Furthermore, by Lemma 20.3.2 $\{F(\tilde{X}_i|\mu_i) : i = 1, 2, \ldots, k\}$ and $\{G(\tilde{W}|\mu_\gamma) : \gamma = 1, 2, \ldots, k\}$ form two separate stochastically increasing families.

Define a function $\psi = \psi(\tilde{x}_1, \tilde{x}_2, \ldots, \tilde{x}_k, \tilde{w})$ as follows:

$$\psi = \left\{ \begin{array}{ll} 1 & \text{if } \max_{i \in A} \tilde{x}_i > \max_{j \in A^c} \tilde{x}_j \text{ and} \\ & \max_{i \in A} \tilde{x}_i \geq y_1 + \mu_0 \text{ and } \frac{\tilde{x}_\gamma + \tilde{w}}{2} > y_2 + \mu_0 \\ 0 & \text{otherwise} \end{array} \right\}. \tag{20.7}$$

Then by (20.7) we have

$$\begin{aligned} & P(\text{selecting an acceptable population}) \\ &= E\{\psi(\tilde{X}_1, \tilde{X}_2, \ldots, \tilde{X}_k, \tilde{W})\} \\ &= E\left[E\{\psi(\tilde{X}_1, \tilde{X}_2, \ldots, \tilde{X}_k, \tilde{W})|\tilde{x}_1, \tilde{x}_2, \ldots, \tilde{x}_k\}\right]. \end{aligned}$$

We will first show that ψ is a non-decreasing function of $\tilde{w}$ when all $\tilde{x}_h$ for $h = 1, 2, \ldots, k$ are held fixed. This will allow us to apply Mahamunulu's result (Lemma 2.1 in his article) to the innermost expectation. For this purpose, assume $\tilde{w} < \tilde{w}^*$ and $\tilde{x}_1, \tilde{x}_2, \ldots, \tilde{x}_k$ are fixed. Note that this implies that $\tilde{x}_\gamma = \max\{\tilde{x}_1, \tilde{x}_2, \ldots, \tilde{x}_k\}$ is also fixed. We wish to show that $\psi(\tilde{x}_1, \tilde{x}_2, \ldots, \tilde{x}_k, \tilde{w}) \leq \psi(\tilde{x}_1, \tilde{x}_2, \ldots, \tilde{x}_k, \tilde{w}^*)$. If $\psi(\tilde{x}_1, \tilde{x}_2, \ldots, \tilde{x}_k, \tilde{w}) = 0$ there is nothing to prove. If $\psi(\tilde{x}_1, \tilde{x}_2, \ldots, \tilde{x}_k, \tilde{w}) = 1$, then

$$\max\{\tilde{x}_{k-m+1}, \ldots, \tilde{x}_k\} > \max\{\tilde{x}_1, \ldots, \tilde{x}_{k-m}\}, \tag{20.8}$$

$$\max\{\tilde{x}_{k-m+1}, \ldots, \tilde{x}_k\} \geq y_1 + \mu_0 \tag{20.9}$$

and

$$\frac{\tilde{x}_\gamma + \tilde{w}^*}{2} > \frac{\tilde{x}_\gamma + \tilde{w}}{2} > y_2 + \mu_0. \tag{20.10}$$

Equations (20.8), (20.9) and the second inequality in (20.10) follow from the fact that $\psi(\tilde{x}_1, \tilde{x}_2, \ldots, \tilde{x}_k, \tilde{w}) = 1$. The first inequality in (20.10) follows from the assumption that $\tilde{w} < \tilde{w}^*$. By (20.7), (20.8) and (20.9), $\psi(\tilde{x}_1, \tilde{x}_2, \ldots, \tilde{x}_k, \tilde{w}^*) = 1$. Thus ψ is a non-decreasing function of $\tilde{w}$ when all $\tilde{x}_h$ for $h = 1, 2, \ldots, k$ are held fixed. Applying Mahamunulu's Lemma 2.1, we conclude that $E\{\psi|\tilde{x}_1, \tilde{x}_2, \ldots, \tilde{x}_k\}$ is a non-decreasing function of μ_γ. By the order-preserving property of expectation, it now follows that $E[E\{\psi|\tilde{x}_1, \tilde{x}_2, \ldots, \tilde{x}_k\}] = P(\text{selecting an acceptable population})$ is also a non-decreasing function of μ_γ.

Next, fix $i \in A = \{k-m+1, \ldots, k\}$ and write

$$\begin{aligned} & P(\text{selecting an acceptable population}) \\ = \; & E\{\psi(\tilde{X}_1, \tilde{X}_2, \ldots, \tilde{X}_k, \tilde{W})\} \\ = \; & E\left[E\{\psi|\tilde{x}_1, \ldots, \tilde{x}_{i-1}, \tilde{x}_{i+1}, \ldots, \tilde{x}_k, \tilde{w}\}\right]. \end{aligned}$$

We wish to show that ψ is a non-decreasing function of $\tilde{x}_i$ when $\tilde{w}$ and all $\tilde{x}_h$ for $h \neq i$ are held fixed. For this purpose, assume $\tilde{x}_i < \tilde{x}_i^*$ and fix $\tilde{w}$ and all $\tilde{x}_h$ for all $h \neq i$. For convenience of notation, denote $\psi(\tilde{x}_1, \tilde{x}_2, \ldots, \tilde{x}_k, \tilde{w})$ by $\psi(., \tilde{x}_i)$. We need to show that $\psi(., \tilde{x}_i) \leq \psi(., \tilde{x}_i^*)$. If $\psi(., \tilde{x}_i) = 0$ there is nothing to prove. If $\psi(., \tilde{x}_i) = 1$, then

$$\begin{aligned} \max\{\tilde{x}_{k-m+1}, \ldots, \tilde{x}_i^*, \ldots, \tilde{x}_k\} & \geq \max\{\tilde{x}_{k-m+1}, \ldots, \tilde{x}_i, \ldots, \tilde{x}_k\} \\ & > \max\{\tilde{x}_1, \ldots, \tilde{x}_{k-m}\}, \\ \max\{\tilde{x}_{k-m+1}, \ldots, \tilde{x}_i^*, \ldots, \tilde{x}_k\} & \geq \max\{\tilde{x}_{k-m+1}, \ldots, \tilde{x}_i, \ldots, \tilde{x}_k\} \geq y_1 + \mu_0 \end{aligned}$$

and

$$\frac{\max\{\tilde{x}_1, \ldots, \tilde{x}_i^*, \ldots, \tilde{x}_k\} + \tilde{w}}{2} \geq \frac{\max\{\tilde{x}_1, \ldots, \tilde{x}_i, \ldots, \tilde{x}_k\} + \tilde{w}}{2} > y_2 + \mu_0.$$

Therefore, $\psi(., \tilde{x}_i^*) = 1$. Thus we have shown that ψ is a non-decreasing function of $\tilde{x}_i$ when $\tilde{w}$ and all $\tilde{x}_h$ for $h \neq i$ are held fixed. Therefore, by Mahamunulus Lemma 2.1 $E\{\psi|\tilde{x}_1, \ldots, \tilde{x}_{i-1}, \tilde{x}_{i+1}, \ldots, \tilde{x}_k, \tilde{w}\}$ is a non-decreasing function of μ_i. By the order-preserving property of expectation it follows that $E[E\{\psi|\tilde{x}_1, \ldots, \tilde{x}_{i-1}, \tilde{x}_{i+1}, \ldots, \tilde{x}_k, \tilde{w}\}] = P(\text{selecting an acceptable population})$ is also a non-decreasing function of μ_i.

Finally, fix $j \in A^c = \{1, 2, \ldots, k-m\}$ and write

$$\begin{aligned} & P(\text{selecting an acceptable population}) \\ = \; & E\{\psi(\tilde{X}_1, \tilde{X}_2, \ldots, \tilde{X}_k, \tilde{W})\} \\ = \; & E[E\{\psi(\tilde{X}_1, \tilde{X}_2, \ldots, \tilde{X}_k, \tilde{W})|\tilde{x}_1, \ldots, \tilde{x}_{j-1}, \tilde{x}_{j+1}, \ldots, \tilde{x}_k, \tilde{w}\}]. \end{aligned}$$

We wish to show that ψ is a non-increasing function of $\tilde{x}_j$ when $\tilde{w}$ and all $\tilde{x}_h$ for $h \neq j$ are held fixed. For this purpose, assume $\tilde{x}_j < \tilde{x}_j^*$ and fix $\tilde{w}$ and $\tilde{x}_h$ for $h \neq j$. We need to show that $\psi(., \tilde{x}_j) \geq \psi(., \tilde{x}_j^*)$. If $\psi(., \tilde{x}_j^*) = 0$ there is nothing to prove. If $\psi(., \tilde{x}_j^*) = 1$, then

$$\begin{aligned} \max\{\tilde{x}_{k-m+1}, \ldots, \tilde{x}_k\} & > \max\{\tilde{x}_1, \ldots, \tilde{x}_j^*, \ldots, \tilde{x}_{k-m}\} \\ & \geq \max\{\tilde{x}_1, \ldots, \tilde{x}_j, \ldots, \tilde{x}_{k-m}\}, \quad (20.11) \\ \max\{\tilde{x}_{k-m+1}, \ldots, \tilde{x}_k\} & \geq y_1 + \mu_0 \quad (20.12) \end{aligned}$$

and

$$\frac{\tilde{x}_\gamma + \tilde{w}}{2} > y_2 + \mu_0. \quad (20.13)$$

Inequality (20.12) and the first inequality in (20.11) follow from the fact that $\psi(., \tilde{x}_j^*) = 1$. The second inequality in (20.11) follows from the assumption that $\tilde{x}_j < \tilde{x}_j^*$. Inequality (20.13) follows from the fact that $\psi(., \tilde{x}_j^*) = 1$ and from the fact that (20.11) implies $\gamma \in A$, and hence $\frac{\tilde{x}_\gamma + \tilde{w}}{2}$ does not depend upon the value taken by $\tilde{X}_j$. Thus by (20.11), (20.12), and (20.13) $\psi(., \tilde{x}_j) = 1$, and we conclude that ψ is a non-increasing function of $\tilde{x}_j$ when $\tilde{w}$ and all $\tilde{x}_h$ for $h \neq j$ are held fixed. By Lemma 20.2.1, $E\{\psi | \tilde{x}_1, \ldots, \tilde{x}_{j-1}, \tilde{x}_{j+1}, \ldots, \tilde{x}_k, \tilde{w}\}$ is a non-increasing function of μ_j. By the order-preserving property of expectation it follows that $E[E\{\psi | \tilde{x}_1, \ldots, \tilde{x}_{j-1}, \tilde{x}_{j+1}, \ldots, \tilde{x}_k, \tilde{w}\}] = P$(selecting an acceptable population) is also a non-increasing function of μ_j.

In summary, we have shown that P(selecting an acceptable population) is:

1. a non-decreasing function of μ_i for $i \in A$ and
2. a non-increasing function of μ_j for $j \in A^c$.

It follows that P(selecting an acceptable population) is minimized when μ_i ($i \in A$) is smallest and μ_j ($j \in A^c$) is largest. Subject to (20.5), this occurs when $\mu_i = \mu_0 + \delta_2^*$, $i \in A$, and $\mu_j = \mu_0 + \delta_1^*$, $j \in A^c$. Thus, for each $m = 1, 2, \ldots, k$, the power is minimized at the configuration:

$$\mu_1 = \mu_2 = \cdots = \mu_{k-m} = \mu_0 + \delta_1^* < \mu_0 + \delta_2^* = \mu_{k-m+1} = \cdots = \mu_k. \quad (20.14)$$

It remains to show that over all possible configurations (20.14), the minimum power is achieved when $m = 1$. For this purpose, consider the subset of all mean vectors $\boldsymbol{\mu} = (\mu_1, \mu_2, \ldots, \mu_k)$ that can be expressed in the form (20.14). For each $m = 1, 2, \ldots, k$ let

$$A_m = \{k - m + 1, \ldots, k\} \text{ and } A_m^c = \{1, 2, \ldots, k - m\}.$$

If $m_1 < m_2$, then

$$A_{m_1} \subset A_{m_2} \text{ and } A_{m_1}^c \supset A_{m_2}^c.$$

Then for any fixed values $(\tilde{x}_1, \tilde{x}_2, \ldots, \tilde{x}_k)$ we have

$$\begin{aligned} \max_{i \in A_{m_1}} \tilde{x}_i &\leq \max_{i \in A_{m_2}} \tilde{x}_i, \\ \max_{j \in A_{m_1}^c} \tilde{x}_j &\geq \max_{j \in A_{m_2}^c} \tilde{x}_j. \end{aligned}$$

Thus ,

$$\left\{ \max_{i \in A_{m_1}} \tilde{x}_i > \max_{j \in A_{m_1}^c} \tilde{x}_j \right\} \subset \left\{ \max_{i \in A_{m_2}} \tilde{x}_i > \max_{j \in A_{m_2}^c} \tilde{x}_j \right\}$$

and similarly,

$$\left\{ \max_{i \in A_{m_1}} \tilde{x}_i > y_1 + \mu_0 \right\} \subset \left\{ \max_{i \in A_{m_2}} \tilde{x}_i > y_1 + \mu_0 \right\}.$$

On the other hand, for any $(\tilde{x}_1, \tilde{x}_2, \ldots, \tilde{x}_k)$ satisfying $\max_{i\in A_m} \tilde{x}_i > \max_{j\in A_m^c} \tilde{x}_j$, the distribution of $\frac{\tilde{X}_\gamma+\tilde{W}}{2}$ is independent of the value of m since, in this case, $\gamma \in A_m$ and hence $\mu_\gamma = \mu_0 + \delta_2^*$. Thus for every value of m, $\tilde{X}_\gamma$ is distributed as $\frac{y_1}{h_1}T_1 + \mu_0 + \delta_2^*$, and $\tilde{W}$ is distributed as $\frac{y_2}{h_2}T_2 + \mu_0 + \delta_2^*$, where T_1 and T_2 are Student-t random variables with $n_0 - 1$ and $m_0 - 1$ degrees of freedom, respectively.

It follows that $m_1 < m_2 \Rightarrow$

$$P\left(\max_{i\in A_{m_1}} \tilde{X}_i > \max_{j\in A_{m_1}^c} \tilde{X}_j;\ \max_{i\in A_{m_1}} \tilde{X}_i \geq y_1 + \mu_0;\ \frac{\tilde{X}_\gamma + \tilde{W}}{2} > y_2 + \mu_0\right)$$
$$\leq P\left(\max_{i\in A_{m_2}} \tilde{X}_i > \max_{j\in A_{m_2}^c} \tilde{X}_j;\ \max_{i\in A_{m_2}} \tilde{X}_i \geq y_1 + \mu_0;\ \frac{\tilde{X}_\gamma + \tilde{W}}{2} > y_2 + \mu_0\right).$$

That is, (20.6) is an increasing function of m. Thus, the minimum occurs when $m = 1$.

Theorem 20.3.2 *Let δ_1^*, δ_2^* $(0 \leq \delta_1^* < \delta_2^*)$, α, and $1-\beta$ be given. Then a size of α and, for any parameter vector $\boldsymbol{\mu} \in PZ$, a power of $1-\beta$ can be achieved by choosing design parameters $(n_0, m_0, y_1, y_2, h_1, h_2)$ which simultaneously satisfy the equations:*

$$\int_{h_1 - \frac{h_1\delta_2^*}{y_1}}^{\infty} F_{n_0}^{k-1}\left(t + \frac{h_1}{y_1}(\delta_2^* - \delta_1^*)\right) F_{m_0}\left(\frac{h_2 y_1}{h_1 y_2}t - 2h_2 + 2\frac{h_2\delta_2^*}{y_2}\right) f_{n_0}(t)dt$$
$$= 1 - \beta \tag{20.15}$$

and

$$k\int_{h_1}^{\infty} F_{n_0}^{k-1}(t) F_{m_0}\left(\frac{h_2 y_1}{h_1 y_2}t - 2h_2\right) f_{n_0}(t)dt = \alpha \tag{20.16}$$

where $F_{n_0}(\cdot)$ and $f_{n_0}(\cdot)$ denote the distribution function and density function, respectively, of a Student-t random variable with $n_0 - 1$ degrees of freedom.

PROOF. Let $\boldsymbol{\mu}$ be an arbitrary vector in PZ. Let $\pi_{(i)}$ denote the population having population mean, $\mu[i]$ and let $\bar{X}_{(i)}(n_0)$, $S_{(i)}^2$, $n_{(i)}$, $a_{(i)j}$, and $\tilde{X}_{(i)}$ denote the corresponding statistics produced by $\pi_{(i)}$ $(i = 1, \ldots, k)$. Note that under the LFC, there is only one acceptable population, namely $\pi_{(k)}$.

For (20.15),

$$
\begin{aligned}
&1-\beta(\boldsymbol{\mu}) \geq 1-\beta(LFC)\\
&= P(\text{selecting acceptable population}|LFC) = P(\text{selecting } \pi_{(k)}|LFC)\\
&= P\Big(\tilde{X}_{(k)} > \tilde{X}_{(i)},\ i=1,\ldots,k-1; \tilde{X}_{(k)} \geq y_1+\mu_0;\\
&\qquad \frac{\tilde{X}_{(k)}+\tilde{W}}{2} > y_2+\mu_0|LFC\Big)\\
&= P\Big(\frac{\tilde{X}_{(k)}-\mu_{[k]}}{y_1/h_1} > \frac{\tilde{X}_{(i)}-\mu_{[i]}}{y_1/h_1} + \frac{\mu_{[i]}-\mu_{[k]}}{y_1/h_1},\\
&\qquad i=1,\ldots,k-1;\ \frac{\tilde{X}_{(k)}-\mu_{[k]}}{y_1/h_1} \geq \frac{y_1+\mu_0-\mu_{[k]}}{y_1/h_1}\\
&\qquad \frac{y_1}{h_1}\frac{\tilde{X}_{(k)}-\mu_{[k]}}{2y_1/h_1} + \frac{y_2}{h_2}\frac{\tilde{W}-\mu_{[k]}}{2y_2/h_2} > y_2+\mu_0-\mu_{[k]}|LFC\Big). \qquad (20.17)
\end{aligned}
$$

Let $T_i = \frac{\tilde{X}_{(i)}-\mu_{[i]}}{y_1/h_1}$, $i=1,2,\ldots,k$ and $V = \frac{\tilde{W}-\mu_{[k]}}{y_2/h_2}$. Then $T_1, T_2, \ldots, T_k, V$ are independent random variables, and by Lemma 20.3.2 the T_is have Student-t distributions with n_0-1 degrees of freedom, and V has a Student-t distribution with m_0-1 degrees of freedom. Therefore, (20.17) can be written as

$$
\begin{aligned}
1-\beta(\boldsymbol{\mu}) \ &\geq\ P\Big(T_k > T_i - \frac{h_1}{y_1}(\mu_{[k]}-\mu_{[i]},\ i=1,\ldots,k-1;\\
&\qquad T_k \geq \frac{h_1}{y_1}(y_1+\mu_0-\mu_{[k]});\\
&\qquad \frac{y_1}{2h_1}T_k + \frac{y_2}{2h_2}V > y_2+\mu_0-\mu_{[k]}|LFC\Big)\\
&=\ P\Big(T_k > T_i - \frac{h_1}{y_1}(\delta_2^*-\delta_1^*),\ i=1,\ldots,k-1;\ T_k \geq \frac{h_1}{y_1}(y_1-\delta_2^*);\\
&\qquad \frac{y_1}{2h_1}T_k + \frac{y_2}{2h_2}V > y_2-\delta_2^*\Big). \qquad (20.18)
\end{aligned}
$$

Conditioning on T_k, (20.18) becomes

$$
\begin{aligned}
&\int_{\frac{h_1}{y_1}(y_1-\delta_2^*)}^{\infty} P\Big(T_i < t + \frac{h_1}{y_1}(\delta_2^*-\delta_1^*),\ i=1,\ldots,k-1;\\
&\qquad V > \frac{2h_2}{y_2}\left(y_2-\delta_2^*-\frac{y_1}{2h_1}t\right)\Big) f_{n_0}(t)dt\\
&= \int_{\frac{h_1}{y_1}(y_1-\delta_2^*)}^{\infty} F_{n_0}^{k-1}\left(t+\frac{h_1}{y_1}(\delta_2^*-\delta_1^*)\right)\\
&\qquad \times F_{m_0}\left(-2h_2+\frac{2h_2\delta_2^*}{y_2}+\frac{h_2y_1}{h_1y_2}t\right) f_{n_0}(t)dt.
\end{aligned}
$$

By hypothesis, the above expression has been set equal to the desired power.

For (20.16),

$$\alpha = P(\pi_i \text{ is selected for some } i = 1, \ldots, k | H_0).$$

Thus, the size α can be obtained by evaluating (20.15) at $\delta^* + 1 = \delta_2^* = 0$ and multiplying by k. This gives

$$\alpha = k \int_{h_1}^{\infty} F_{n_0}^{k-1}(t) F_{m_0}\left(\frac{h_2 y_1}{h_1 y_2} t - 2h_2\right) f_{n_0}(t) dt,$$

as desired.

Of special interest is the case $\delta_1^* = 0$. Letting $d_1 = \frac{\delta_2^*}{y_1}$ and $d_2 = \frac{\delta_2^*}{y_2}$, we then have that (d_1, d_2, h_1, h_2) are the parameters that simultaneously satisfy the equations

$$\int_{h_1(1-d_1)}^{\infty} F_{n_0}^{k-1}(t + f_1 d_1) F_{m_0}\left(\frac{h_2 d_2}{h_1 d_1} t - 2h_2(1 - d_2)\right) f_{n_0}(t) dt = 1 - \beta \tag{20.19}$$

and

$$k \int_{h_1}^{\infty} F_{n_0}^{k-1}(t) F_{m_0}\left(\frac{h_2 d_2}{h_1 d_1} t - 2h_2\right) f_{n_0}(t) dt = \alpha \tag{20.20}$$

where $F_{n_0}(\cdot)$ and $f_{n_0}(\cdot)$ denote the distribution function and density function, respectively, of a Student-t random variable with $n_0 - 1$ degrees of freedom.

A useful feature of procedure $\mathcal{P}_{UU}$ is its capacity to terminate early if no population shows evidence of being superior to the standard. The probability of such early termination under the null hypothesis is given in the following corollary.

Corollary 20.3.1 *Let τ_0 denote the probability of terminating at stage 1 when H_0 is true. Then $\tau_0 = F_{n_0}^k(h_1)$, where $F_{n_0}(\cdot)$ denotes the distribution function of a Student-t random variable with $n_0 - 1$ degrees of freedom.*

PROOF.

$$\begin{aligned}
P(\text{terminating early}) &= P\left(\tilde{X}_{[k]} < \mu_0 + y_1\right) \\
&= P\left(\tilde{X}_{(i)} < \mu_0 + y_1,\ i = 1, \ldots, k\right) \\
&= P\left(\frac{\tilde{X}_{(i)} - \mu_{[i]}}{y_1/h_1} < \frac{\mu_0 - \mu_{[i]} + y_1}{y_1/h_1},\ i = 1, \ldots, k\right) \\
&= \prod_{i=1}^{k} P\left(T_i < \frac{h_1}{y_1}(\mu_0 - \mu_{[i]} + y_1)\right) \qquad (20.21)
\end{aligned}$$

where T_i are Student-t random variables with $n_0 - 1$ degrees of freedom as defined in the proof of Theorem 20.3.1. Evaluating the above expression at the null hypothesis, $H_0 : \mu_1 = \mu_2 = \cdots = \mu_k = \mu_0$, we obtain $P(\text{terminating early} \,|H_0) = F_{n_0}^k(h_1)$. This completes the proof of the corollary. ∎

20.4 Table

For the special case, Table 20.1 gives design parameters $(n_0, m_0, d_1, d_2, h_1, h_2)$ which simultaneously satisfy equations (20.19) and (20.20) for $\alpha = .05$, $k = 2$ and 3, and $1 - \beta = .90$. For each design, $\tau_0 = P(\text{terminating at stage } 1|H_0)$ is provided, using the result from Corollary 20.3.1. Extra tables for other n_0 values between 3 and 30 and for $1 - \beta = .95$ were also computed and they are available from the authors upon request.

In order to obtain the table values, a FORTRAN code was written in which we first fixed the values of n_0, m_0, d_1, and d_2, and then used the DNEQNJ subroutine from IMSL to find the simultaneous solutions (h_1, h_2) to the integral equations (20.19) and (20.20). The DNEQNJ subroutine in IMSL requires a user-supplied Jacobian matrix. The required Jacobian matrix is given by:

$$\begin{bmatrix} \frac{\partial f_1}{\partial h_1} & \frac{\partial f_1}{\partial h_2} \\ \frac{\partial f_2}{\partial h_1} & \frac{\partial f_2}{\partial h_2} \end{bmatrix}$$

where

$$\begin{aligned} f_1(h_1, h_2) &= \int_{h_1(1-d_1)}^{\infty} F_{n_0}^{k-1}(t + h_1 d_1) F_{m_0}\left(\frac{h_2 d_2}{h_1 d_1} t - 2h_2(1 - d_2)\right) \\ &\qquad \times f_{n_0}(t)dt - (1 - \beta), \\ f_2(h_1, h_2) &= k \int_{h_1}^{\infty} F_{n_0}^{k-1}(t) F_{m_0}\left(\frac{h_2 d_2}{h_1 d_1} t - 2h_2\right) f_{n_0}(t)dt - \alpha. \end{aligned}$$

The partial derivatives, computed by using a combination of Leibniz's Rule, The Fundamental Theorem of Calculus, and the chain rule, are given below.

$$\begin{aligned} &\frac{\partial f_1}{\partial h_1} \\ &= (d_1 - 1) F_{n_0}^{k-1}(h_1) F_{m_0}\left[h_2 d_2 \left(\frac{1 - d_1}{d_1}\right) - 2h_2(1 - d_2)\right] f_{n_0}(h_1(1 - d_1)) \\ &\quad + d_1 \int_{h_1(1-d_1)}^{\infty} f_{n_0}(t) \left\{ \frac{-h_2 d_2}{h_1^2 d_1^2} t F_{n_0}^{k-1}(t + h_1 d_1) f_{m_0}\left[\frac{h_2 d_2}{h_1 d_1} t - 2h_2(1 - d_2)\right] \right. \\ &\quad \left. + (k - 1) F_{m_0}\left[\frac{h_2 d_2}{h_1 d_1} t - 2h_2(1 - d_2)\right] F_{n_0}^{k-2}(t + h_1 d_1) f_{n_0}(t + h_1 d_1) \right\} dt, \end{aligned}$$

$$\frac{\partial f_1}{\partial h_2} = \int_{h_1(1-d_1)}^{\infty} F_{n_0}^{k-1}(t + h_1 d_1) f_{n_0}(t) \times \left\{ f_{m_0}\left[\frac{h_2 d_2}{h_1 d_1} t - 2h_2(1-d_2)\right]\left[\frac{d_2}{h_1 d_1} t - 2(1-d_2)\right]\right\} dt,$$

$$\frac{\partial f_2}{\partial h_1} = -kF_{n_0}^{k-1}(h_1) F_{m_0}\left(\frac{h_2 d_2}{d_1} - 2h_2\right) f_{n_0}(h_1) - k\frac{h_2 d_2}{h_1^2 d_1}\int_{h_1}^{\infty} F_{n_0}^{k-1}(t) f_{n_0}(t) f_{m_0}\left(\frac{h_2 d_2}{h_1 d_1} t - 2h_2\right) t\,dt,$$

$$\frac{\partial f_2}{\partial h_2} = k\int_{h_1}^{\infty} F_{n_0}^{k-1}(t) f_{n_0}(t)\left\{ f_{m_0}\left(\frac{h_2 d_2}{h_1 d_1} t - 2h_2\right)\left(\frac{d_2 t}{h_1 d_1} - 2\right)\right\} dt.$$

The single integrals in expressions (20.19) and (20.20) were evaluated using the IMSL subroutine DQDAGI. This subroutine is specifically designed to handle infinite limits of integration. DQDAGI allows the user to specify the values of two variables, ERRABS and ERRREL, such that the resulting integral approximation (RESULT) satisfies, We specified ERRABS = ERRREL =.0001. The Student-t distribution function, required for the computation of (20.19), (20.20), and in Corollary 20.3.1, was evaluated using the IMSL subroutine DTDF.

20.5 An Example

Suppose that a consumer is to decide whether or not to purchase one lot of bolts from among three that are being offered for his consideration. Assume that the tensile strength of a bolt from lot i ($i = 1, 2, 3$) is normally distributed with unknown mean μ_i and unknown variance σ_i^2. A lot is deemed acceptable only if the bolts in the lot have a mean tensile strength of at least 60,000 psi. Thus, $k = 3$ and $\mu_0 = 60,000$ psi. Suppose that the consumer requests $\alpha = .05$, $1 - \beta = .90$, $\delta_i^* = 0$, and $\delta_2^* = 250$ psi. Suppose that $n_0 = 15$ bolts are randomly sampled from each of the three lots, the tensile strength of each bolt is measured, and the sample standard deviations are $S_1 = 350$, $S_2 = 450$, and $S_3 = 470$. Consulting Table 20.1 with $k = 3$, power = .90, and $n_0 = 15$, we find the procedure parameters

$$m_0 = 10,\ d_1 = 2.0,\ d_2 = 2.5,\ h_1 = 1.8540,\ h_2 = 0.6194.$$

Thus, we tabulate $y_1 = \frac{\delta_2^*}{d_1} = \frac{250}{2.0} = 125$, and compute

$$\begin{aligned} n_1 &= \max\left\{16, \left[\frac{(1.8540)^2(350)^2}{(125)^2}\right] + 1\right\} = 27, \\ n_2 &= \max\left\{16, \left[\frac{(1.8540)^2(450)^2}{(125)^2}\right] + 1\right\} = 45, \\ n_3 &= \max\left\{16, \left[\frac{(1.8540)^2(470)^2}{(125)^2}\right] + 1\right\} = 49. \end{aligned}$$

Accordingly, we take an additional $27 - 15 = 12$ observations from lot 1, $45 - 15 = 30$ observations from lot 2, and $49 - 15 = 34$ observations from lot 3. Following the approach suggested in the proof of Lemma 20.2.1, we set

$$\begin{aligned} c_1 &= \frac{n_0}{n_1}\left(1 + \sqrt{1 - \frac{n_1}{n_0}\left(1 - \frac{(n_1 - n_0)\frac{y_1^2}{h_1^2}}{S_1^2}\right)}\right) \\ &= \frac{15}{27}\left(1 + \sqrt{1 - \frac{27}{15}\left(1 - \frac{12(\frac{125}{1.8540})^2}{(350)^2}\right)}\right) = .5773 \end{aligned}$$

and set

$$\begin{aligned} a_{1,1} &= a_{1,2} = \cdots = a_{1,15} = \frac{c_1}{n_0} = \frac{.5773}{15} = .0385 \text{ and} \\ a_{1,16} &= \cdots = a_{1,27} = \frac{1 - c_1}{n - n_0} = \frac{.4227}{12} = .0352. \end{aligned}$$

Similar computations show that $c_2 = .3808$, $a_{2,1} = a_{2,2} = \cdots = a_{2,15} = .0254$, $a_{2,16} = \cdots = a_{2,45} = .0206$, $c_3 = .3482$, $a_{3,1} = \cdots = a_{3,15} = .0232$, $a_{3,16} = \cdots = a_{3,49} = .0192$. We now compute $\tilde{X}_i = \sum_{j=1}^{n_i} a_{ij}x_{ij}$ $(i = 1, 2, 3)$, and rank the resulting statistics $\tilde{X}_{[1]} \leq \tilde{X}_{[2]} \leq \tilde{X}_{[3]}$. If $\tilde{X}_{[3]} < 60,125$ we terminate the procedure and do not buy any of the lots. Otherwise, we proceed to stage 2 and take an additional $m_0 = 10$ bolts from the lot which produced $\tilde{X}_{[3]}$. Suppose the sample standard deviation based on these 10 additional bolts is $S_w = 420$. Then we compute $y_2 = \frac{\delta_2^*}{d_2} = \frac{250}{2.5} = 100$, and tabulate

$$m = \max\left\{11, \left[\frac{(0.6194)^2(420)^2}{(100)^2}\right] + 1\right\} = 11.$$

Thus we take 1 additional bolt from the 'best' lot. Again, following Lemma 20.2.1 we set

$$
\begin{aligned}
c &= \frac{m_0}{m}\left(1+\sqrt{1-\frac{m}{m_0}\left(1-\frac{(m-m_0)\frac{y_2^2}{h_2^2}}{S_w^2}\right)}\right) \\
&= \frac{10}{11}\left(1+\sqrt{1-\frac{11}{10}\left(1-\frac{1(\frac{100}{0.6194})^2}{(420)^2}\right)}\right) = 1.1364
\end{aligned}
$$

and set

$$
\begin{aligned}
b_1 &= b_2 = \cdots b_{10} = \frac{c}{m_0} = \frac{1.1364}{10} = .11364 \text{ and} \\
b_{11} &= \frac{1-c}{m-m_0} = \frac{-.1364}{1} = -.1364.
\end{aligned}
$$

We now compute $\tilde{W} = \sum_{j=1}^{m} b_j W_j$. If $\frac{\tilde{X}_{[3]}+\tilde{W}}{2} > 60,100$, then buy the selected lot. If $\frac{\tilde{X}_{[3]}+\tilde{W}}{2} \leq 60,100$, then buy none of the lots.

20.6 Sample Size Comparison

We shall now perform a sample size comparison between our procedure $\mathcal{P}_{UU}$ and the Dudewicz-Dalal-type procedure of Taneja and Dudewicz (1992) for the case of unequal, unknown variances. The reason that we chose to compare with the Dudewicz-Dalal-type procedure of Taneja and Dudewicz (1992), rather than the Rinott-type procedures of Wilcox (1984) or Taneja and Dudewicz (1992), is that an expression for expected sample size was already provided by the authors for that procedure. On the other hand, no such expressions were provided by the authors for the other two procedures. Therefore, to avoid taking time away from the study of our own procedure in order to derive expected sample sizes for other procedures, we chose to utilize the expressions that were readily available. As before, the formulation of the problem for Taneja and Dudewicz's procedure is slightly different than for our procedure. The goal of Taneja and Dudewicz's procedure is to select the population with the largest mean, while the goal of our procedure is to select any acceptable population. Furthermore, Taneja and Dudewicz's preference zone consists of all mean vectors satisfying $\mu_{[k]} \geq \mu_0 + \delta_1^*$ and $\mu_{[k]} \geq \mu_{[k-1]} + \delta_2^*$, while our preference zone consists of all mean vectors such that at least one $\mu_i \geq \mu_0 + \delta_2^*$ and no μ_i lies in the interval $(\mu_0+\delta_1^*, \mu_0+\delta_2^*)$. We obtain a special case under which the two formulations are comparable by setting Taneja and Dudewicz's $\delta_0^* = 0$, our $\alpha =$ their$(1 - P_0^*)$, our $(1 - \beta) =$ theirP_1^*, our $\delta_2^* =$ theirδ_1^*, and our $\delta_1^* =$ their$(\delta_1^* - \delta_2^*)$.

Under these conditions we perform comparisons for $k = 2$ and 3; $\sigma_i = 0.5 \times i$, $i = 1, \ldots, k$; $\alpha = .05$; our $\delta_1^* = 0$; our $\delta_2^* = .20$; and $1-\beta = .90$ and .95. Since the required total sample size of each procedure is a random variable, we examine the expected total sample sizes of the two procedures, comparing designs with the same initial sample sizes,n_0. Procedure $\mathcal{P}_{UU}$ offers several designs for the same probability requirements and the same initial sample size n_0 (see Table 20.1). For the purpose of sample size comparison, we chose the design with the smallest upper bound on expected total sample size, given that τ_0 (the probability of terminating early under H_0) is at least .5.

In order to compute the expected sample sizes of Taneja and Dudewicz's procedure, our original intention was to use the extensive tables of procedure parameter values, $(\hat{h}, \hat{g})$, that were provided by the authors in their 1992 paper. However, during the course of the sample size comparison, we discovered an error in the tabled values, rendering the existing tables unusable. In light of this, we chose to perform sample size comparisons for just a few cases, computing the correct values of $(\hat{h}, \hat{g})$ just for those cases. Accordingly, we calculated upper and lower bounds for the expected total sample size of Taneja and Dudewicz's procedure as follows. We first obtained the simultaneous solution (h, g) to equations (7) and (8) given on p. 68 of their 1992 paper. We accomplished this by writing a FORTRAN code similar to the one used to solve our equations (20.19) and (20.20), as described in Section 20.4. For reference, the values of (h, g) that we obtained are given in Table 20.2.

We then used the above values of (h, g) in the expression given for $E(N_T)$ on p. 67 of Taneja and Dudewicz (1992), with all $\theta_i = 0$ and all $\theta_i = 1$, respectively, to obtain lower and upper bounds. The comparison results are given in Table 20.3. In order to verify the accuracy of the table entries, we also provide for each of the procedures simulation results for size, power, and expected sample size under both LFC and H_0. The simulation was based on 20,000 runs and used appropriately scaled pseudorandom numbers from a standard normal distribution generated by the IMSL random number generator RNNOR.

The figures in the tables indicate that neither procedure is uniformly better than the other with respect to expected sample size. In general, for small values of n_0 the upper bound on $E(N)$ for procedure $\mathcal{P}_{UU}$ is smaller than the lower bound for Taneja and Dudewicz's procedure. However for larger values of n_0, the simulation results suggest that Taneja and Dudewicz's procedure performs better under the LFC, while procedure $\mathcal{P}_{UU}$ performs better under H_0.

Table 20.1: Design parameters for procedure $\mathcal{P}_{UU}$

k = 2	Power = .90	$\alpha = 0.05$	$\delta_1^* = 0.00$			
n_0	m_0	d_1	d_2	h_1	h_2	τ_0
2	2	3.0	2.0	4.7990	3.3170	.8735
	3	3.5	1.5	2.8346	5.1861	.7957
	4	3.0	1.5	3.2805	4.0354	.8205
	5	4.0	1.5	2.4833	3.3062	.7711
	6	4.0	1.5	2.4807	3.0676	.7709
	10	3.5	1.5	2.8240	2.7507	.7951
6	2	3.0	2.5	1.5380	1.3005	.8239
	3	3.0	2.0	1.2483	1.3722	.7507
	4	2.0	2.5	1.8740	.9760	.8838
	5	2.0	2.5	1.8866	.8556	.8856
	6	2.5	2.0	1.4109	1.1389	.7944
	8	2.5	2.0	1.4381	1.0401	.8011
	10	2.5	2.0	1.4515	.9942	.8043
10	2	3.0	2.5	1.4050	1.2299	.8158
	3	3.0	2.0	1.1371	1.3270	.7354
	4	2.0	2.5	1.6688	1.0330	.8747
	5	2.0	2.5	1.6843	.8911	.8776
	6	2.0	2.5	1.6919	.8237	.8790
	8	2.0	2.5	1.6994	.7586	.8804
	10	2.0	2.5	1.7031	.7268	.8810
15	2	3.0	2.5	1.3552	1.2006	.8129
	3	3.0	2.0	1.0982	1.3006	.7305
	4	3.0	2.0	1.1672	1.0247	.7546
	5	2.0	2.5	1.6065	.9108	.8738
	6	2.0	2.5	1.6148	.8384	.8755
	8	2.0	2.5	1.6230	.7695	.8771
	10	2.0	2.5	1.6270	.7361	.8779
20	2	3.0	2.5	1.3333	1.1874	.8116
	3	3.0	2.0	1.0821	1.2867	.7287
	4	3.0	2.0	1.1503	1.0160	.7532
	5	2.0	2.5	1.5719	.9213	.8719
	6	2.0	2.5	1.5807	.8461	.8738
	8	2.0	2.5	1.5892	.7750	.8756
	10	2.0	2.5	1.5934	.7408	.8765
30	2	3.0	2.5	1.3133	1.1752	.8105
	3	3.0	2.0	1.0677	1.2725	.7272
	4	3.0	2.0	1.1349	1.0075	.7519
	5	2.0	2.5	1.5399	.9323	.8701
	6	2.0	2.5	1.5491	.8539	.8722
	8	2.0	2.5	1.5580	.7806	.8742
	10	2.0	2.5	1.5623	.7455	.8751

Table 20.1 continued

k = 3 Power = .90 $\alpha = 0.05$ $\delta_1^* = 0.00$						
n_0	m_0	d_1	d_2	h_1	h_2	τ_0
2	2	2.0	2.0	9.9721	3.9283	.9076
	3	1.5	2.0	12.9335	2.2652	.9281
	4	2.0	2.0	10.5941	1.2599	.9128
	5	2.0	2.0	10.6725	1.1303	.9134
	6	2.0	2.0	10.7166	1.0631	.9138
	8	2.0	2.0	10.7640	.9946	.9141
	10	3.5	1.5	4.3077	2.3706	.7976
6	2	2.5	2.5	1.9237	1.3969	.8407
	3	2.0	2.5	2.1587	.9995	.8801
	4	2.0	2.5	2.1835	.7942	.8837
	5	2.0	2.5	2.1933	.7173	.8851
	6	2.0	2.0	1.7964	1.6071	.8143
	8	2.0	2.0	1.7950	1.4945	.8140
	10	2.0	2.0	1.7937	1.4413	.8137
10	2	2.5	2.5	1.7175	1.3262	.8306
	3	2.0	2.5	1.8987	1.0192	.8709
	4	2.0	2.5	1.9254	.8019	.8761
	5	2.0	2.5	1.9358	.7222	.8780
	6	2.0	2.5	1.9413	.6811	.8790
	8	2.0	2.5	1.9469	.6392	.8801
	10	2.0	2.5	1.9497	.6180	.8806
15	2	2.5	2.5	1.6415	1.2958	.8267
	3	2.0	2.5	1.8014	1.0308	.8666
	4	2.0	2.5	1.8292	.8062	.8727
	5	2.0	2.5	1.8398	.7250	.8750
	6	2.0	2.5	1.8454	.6833	.8761
	8	2.0	2.5	1.8511	.6409	.8773
	10	2.0	2.5	1.8540	.6194	.8779
20	2	2.5	2.5	1.6084	1.2819	.8250
	3	2.0	2.5	1.7587	1.0370	.8645
	4	2.0	2.5	1.7870	.8084	.8711
	5	2.0	2.5	1.7978	.7265	.8736
	6	2.0	2.5	1.8034	.6844	.8748
	8	2.0	2.5	1.8092	.6418	.8761
	10	2.0	2.5	1.8121	.6202	.8767
30	2	2.5	2.5	1.5782	1.2687	.8235
	3	2.0	2.5	1.7194	1.0435	.8625
	4	2.0	2.5	1.7484	.8108	.8696
	5	2.0	2.5	1.7592	.7280	.8722
	6	2.0	2.5	1.7649	.6856	.8736
	8	2.0	2.5	1.7707	.6427	.8749
	10	2.0	2.5	1.7737	.6210	.8756

Table 20.2: (h, g) value for Taneja and Dudewicz's procedure

	$k=2$		$P_0^* = .95$		$P_1^* = .90$		
n_0	2	3	4	5	6	7	8
H	12.5447	4.2731	3.1659	2.7640	2.5601	2.4375	2.3560
G	16.3100	6.3145	4.8847	4.3525	4.0790	3.9133	3.8024
n_0	9	10	15	20	25	30	
H	2.2978	2.2543	2.1380	2.0866	2.0577	2.0392	
G	3.7232	3.6637	3.5038	3.4330	3.3930	3.3673	
	$k=2$		$P_0^* = .95$		$P_1^* = .95$		
n_0	2	3	4	5	6	7	8
H	12.5447	4.2731	3.1659	2.7640	2.5601	2.4375	2.3560
G	21.4040	7.5306	5.6627	4.9837	4.6390	4.4318	4.2939
n_0	9	10	15	20	25	30	
H	2.2978	2.2544	2.1379	2.0866	2.0577	2.0391	
G	4.1957	4.1222	3.9254	3.8387	3.7898	3.7585	
	$k=3$		$P_0^* = .95$		$P_1^* = .90$		
n_0	2	3	4	5	6	7	8
H	18.7589	5.2918	3.7158	3.1686	2.8971	2.7362	2.6302
G	22.9176	7.4133	5.4729	4.7813	4.4337	4.2260	4.0884
n_0	9	10	15	20	25	30	
H	2.5551	2.4992	2.3509	2.2861	2.2498	2.2266	
G	3.9907	3.9177	3.7233	3.6381	3.5902	3.5595	
	$k=3$		$P_0^* = .95$		$P_1^* = .95$		
n_0	2	3	4	5	6	7	8
H	18.7589	5.2918	3.7159	3.1686	2.8971	2.7362	2.6302
G	29.5656	8.7472	6.2863	5.4291	5.0035	4.7511	4.5847
n_0	9	10	15	20	25	30	
H	2.5551	2.4992	2.3509	2.2861	2.2498	2.2266	
G	4.4669	4.3793	4.1466	4.0449	3.9879	3.9515	

Table 20.3: Comparison of sample sizes for Taneja and Dudewicz's two-stage selection procedure and procedure $\mathcal{P}_{UU}$ at $\alpha = .05$, $\delta_1^* = 0.00$, $\delta_2^* = 0.20$, and $\sigma_i = 0.5 \times i$, $i = 1, \ldots, k$

			P_{UU}					Taneja & Dudewicz						
	n_0	m_0	Simulated Power	Simulated Size	Theoretical E(n) UB	Simulated E(n|LFC)	Simulated E(n|Ho)	Simulated Power P_1^*	Simulated Size $(1-P_0^*)$	Theoretical E(n) LB	Theoretical E(n) UB	Simulated E(n|LFC)	Simulated E(n|H0)
k = 2	2	10	0.8989	0.0515	3482	3433.42	3113.63	0.8972	0.0497	8314	8316	8385.28	8286.91
Power = .90	3	6	0.9021	0.05	1098	1085.96	990.85	0.9015	0.0501	1247	1249	1242.35	1256.61
	4	8	0.8984	0.0524	694	687.42	599.03	0.9044	0.0477	746	748	741.68	747.26
	5	10	0.9022	0.0496	572	566.53	480.85	0.9025	0.0496	593	595	589.73	595.46
	6	10	0.9006	0.0505	515	508.21	423.91	0.9036	0.0511	520	522	520.09	521.13
	7	10	0.8985	0.0499	500	488.39	422.46	0.8995	0.0498	479	481	481.77	480.49
	8	10	0.9031	0.0512	478	470.65	398.39	0.8992	0.0511	452	454	451.65	451.96
	9	10	0.9005	0.0488	462	455.18	382.72	0.8983	0.0495	434	436	433.92	436.24
	10	10	0.9058	0.0493	450	440.61	370.13	0.901	0.0495	420	422	420.11	420.18
	15	10	0.9045	0.0499	421	412.27	339.08	0.902	0.0509	384	386	384.25	384.04
	20	10	0.9006	0.0513	408	398.36	323.27	0.9	0.051	369	371	368.5	369.01
	25	10	0.9016	0.0483	401	392.1	318.17	0.8997	0.0476	360	362	361.48	361.94
	30	10	0.899	0.0498	397	388.15	313.02	0.9011	0.0481	355	357	355.59	354.85
k = 2	2	10	0.9528	0.0485	9476	9465.42	8827.4	0.9494	.0493.	14317	14319	14155.3	14321.36
Power = .95	3	7	0.9495	0.0478	1510	1498.49	1274.84	0.9481	0.0507	1773	1775	1771.07	1767.03
	4	8	0.9487	0.0507	942	927.52	777.32	0.9476	0.051	1003	1005	1014.85	1009.13
	5	7	0.9503	0.0471	782	771.42	618.61	0.9521	0.05	777	779	779.32	774.24
	6	10	0.9497	0.0504	683	677.56	560.75	0.9483	0.0494	673	675	671.97	673.12
	7	10	0.9491	0.049	638	631.04	521.35	0.9488	0.0513	614	616	617.85	615.91
	8	10	0.9499	0.0497	608	601.08	497.62	0.9487	0.0491	577	579	577.06	578.44
	9	10	0.9513	0.0463	588	581.86	476.86	0.9502	0.0486	551	553	550.8	552.21
	10	10	0.949	0.0504	573	568.55	466.86	0.9512	0.051	532	534	531.87	532.94
	15	10	0.9486	0.0492	533	528.56	430.94	0.9513	0.0479	482	484	484.65	482.7
	20	9	0.9499	0.0501	519	511.12	414.24	0.9494	0.0498	461	463	460.46	462.13
	25	10	0.9507	0.0508	509	504.84	405.79	0.9509	0.0521	449	451	449.39	450.5
	30	10	0.9539	0.0481	501	495.79	402.95	0.9477	0.0511	442	444	442.81	441.58

Table 20.3 continued

	n_0	m_0	P_{UU} Simulated Power	P_{UU} Simulated Size	P_{UU} Theoretical E(n) UB	P_{UU} Simulated E(n|LFC)	P_{UU} Simulated E(n|Ho)	Taneja & Dudewicz Simulated Power P_1^*	Taneja & Dudewicz Simulated Size $(1 - P_0^*)$	Taneja & Dudewicz Theoretical E(n) LB	Taneja & Dudewicz Theoretical E(n) UB	Taneja & Dudewicz Simulated E(n|LFC)	Taneja & Dudewicz Simulated E(n|H0)
k = 3 Power = .90	2	10	0.8973	0.0484	20606	20576.17	20087.9	0.9002	0.0509	45957	45960	45994.52	46114.37
	3	6	0.8996	0.0507	3885	3849.07	3562.62	0.9038	0.0471	4809	4812	4784.4	4820.46
	4	4	0.8972	0.0512	2702	2608.33	1668.96	0.9002	0.0475	2621	2624	2612.49	2619.97
	5	8	0.8995	0.05	1815	1771.74	1316.55	0.8974	0.0483	2001	2004	1997.72	1999.84
	6	10	0.9	0.0507	1598	1555.12	1168.74	0.9013	0.0482	1721	1724	1723.32	1724.62
	7	8	0.8949	0.045	1532	1488.51	1095.52	0.9005	0.0508	1563	1566	1562.25	1556.77
	8	10	0.8968	0.045	1430	1391.62	1040	0.8992	0.0488	1463	1466	1461.82	1463.14
	9	9	0.8948	0.0444	1398	1359.74	1001.71	0.8966	0.0503	1394	1397	1395.28	1396.85
	10	9	0.8954	0.0474	1364	1326.48	974.2	0.8975	0.0517	1343	1346	1341.86	1336.21
	15	10	0.9011	0.0498	1346	1330.22	1215.09	0.8977	0.0501	1214	1217	1212.61	1213.41
	20	10	0.8977	0.0489	1293	1281.09	1157.74	0.8999	0.0502	1159	1162	1162.89	1158.33
	25	10	0.9024	0.0491	1263	1252.8	1129.69	0.9045	0.0542	1128	1131	1129.37	1128.79
	30	10	0.8986	0.0517	1245	1232.61	1109.57	0.8991	0.0476	1109	1112	1109.95	1108.84
k = 3 Power = .95	2	10	0.951	0.0474	55643	55982.73	54513.43	0.9505	0.0505	76486	76489	76937.79	77159.74
	3	6	0.9507	0.0502	6098	6055.65	5338.98	0.9508	0.0515	6695	6698	6673.21	6697.18
	4	4	0.9481	0.0501	3469	3470.14	3092.35	0.9492	0.0493	3458	3461	3449.38	3482.24
	5	8	0.9471	0.0489	2275	2266.46	1955.57	0.9506	0.0513	2580	2583	2587.75	2592.22
	6	10	0.9485	0.0492	2027	2017.61	1732.64	0.9504	0.051	2191	2194	2198.64	2192.05
	7	8	0.948	0.0476	1847	1829.46	1541.95	0.9479	0.0507	1976	1979	1967.52	1976.19
	8	10	0.9483	0.0501	1927	1916.84	1686.12	0.9521	0.0502	1840	1843	1845.29	1836.01
	9	9	0.9518	0.0513	1897	1897.93	1663.47	0.9506	0.0505	1746	1749	1750.32	1741.76
	10	9	0.9516	0.0509	1731	1717.52	1430.58	0.9493	0.0474	1679	1682	1680.94	1681.98
	15	10	0.9469	0.0491	1687	1674.89	1351.02	0.9495	0.0508	1505	1508	1502.58	1507.97
	20	10	0.9485	0.0511	1591	1584.97	1290.92	0.9478	0.0503	1432	1435	1429.54	1432.01
	25	10	0.952	0.0501	1469	1457.26	1140.29	0.95	0.0506	1392	1395	1390.44	1394.38
	30	10	0.9487	0.0488	1581	1527.73	1299.06	0.9487	0.0519	1367	1370	1363.97	1370.14

References

1. Bechhofer, R. E. and Turnbull, B. W. (1978). Two $(k+1)$-decision selection procedures for comparing k normal means with a specified standard, *Journal of the American Statistical Association*, **73**, 385–392.

2. Dunnett, C. W. (1984). Selection of the best treatment in comparison to a control with an application to a medical trial, In *Design of Experiments: Ranking and Selection* (Ed., T. J. Santner and A. C. Tamhane), pp. 47–66, Marcel Dekker, New York.

3. Mahamunulu, D. M. (1967). Some fixed-sample ranking and selection problems, *Ann. Math. Statist.*, **38**, 1079–1091.

4. Taneja, B. K. and Dudewicz, E. J. (1992). Selection of the best experimental category provided it is better than a standard: The heterscedastic method solution, In *The Frontiers of Modern Statistical Inference Procedures*, Vol. II, pp. 44–94. American Sciences Press, Inc.

5. Thall, P. F., Simon, R. and Ellenberg, S. S. (1988). Two-stage selection and testing designs for comparative clinical trials, *Biometrika*, **75**, 303 310.

6. Thall, P. F., Simon, R. and Ellenberg, S. S. (1989). A two-stage design for choosing among several experimental treatments and a control in clinical trials, *Biometrics*, **45**, 537–547.

7. Wilcox, R. R. (1984). Selecting the best population, provided it is better than a standard: The unequal variance case, *Journal of the American Statistical Association*, **79**, 887–891.

Index

www.ingramcontent.com/pod-product-compliance
Lightning Source LLC
LaVergne TN
LVHW062108200926
848484LV00007B/1621
9780817632328